# MODERN MATHEMATICS FOR THE ENGINEER

Second Series

# MODERN MATHEMATICS FOR THE ENGINEER

Second Series

*Edited by*

EDWIN F. BECKENBACH

*With an Introduction by*

MAGNUS R. HESTENES

Dover Publications, Inc., Mineola, New York

*Bibliographical Note*

*Modern Mathematics for the Engineer: Second Series,* first published by Dover Publications, Inc., in 2013, is an unabridged republication of the work originally published by McGraw-Hill Book Company, New York, in 1961.

*International Standard Book Number*

*ISBN-13: 978-0-486-49747-1*
*ISBN-10: 0-486-49747-X*

Manufactured in the United States by Courier Corporation
49747X01
www.doverpublications.com

# The Authors

Arthur Erdélyi, Dr. Ver. Nat., D.Sc., Professor of Mathematics, California Institute of Technology

Bernard Friedman, Ph.D., Professor of Mathematics, University of California, Berkeley

John W. Miles, Ph.D., Professor of Engineering and Geophysics, University of California, Los Angeles

Ralph S. Phillips, Ph.D., Professor of Mathematics, Stanford University

J. Barkley Rosser, Ph.D., Professor of Mathematics, Cornell University

William Feller, Ph.D., Eugene Higgins Professor of Mathematics, Princeton University

David Blackwell, Ph.D., Professor of Statistics, University of California, Berkeley

Richard Bellman, Ph.D., Research Mathematician, The RAND Corporation

George B. Dantzig, Ph.D., Research Mathematician, The RAND Corporation

Samuel Karlin, Ph.D., Professor of Mathematics and Statistics, Stanford University

Stanislaw M. Ulam, Ph.D., Research Advisor, Los Alamos Scientific Laboratory

Raymond Redheffer, Ph.D., Professor of Mathematics, University of California, Los Angeles

Subrahmanyan Chandrasekhar, Ph.D., Sc.D., Morton D. Hull Distinguished Service Professor in the Departments of Physics and Astronomy, University of Chicago

Paul R. Garabedian, Ph.D., Professor of Mathematics, New York University

David Young, Ph.D., Professor of Mathematics and Director of the Computation Center, University of Texas

George Pólya, Ph.D., Emeritus Professor of Mathematics, Stanford University

# Foreword to the 1961 Edition

The physical and economic world in which the modern engineer operates continues to grow more complex, putting ever greater demands on the mathematical models representing that world. During the five years since University Extension last offered a lecture series in "Modern Mathematics for the Engineer," reliance on a variety of these models has grown in an amazing fashion, due in no small measure to the adaptation of advanced mathematical techniques for use in connection with high-speed computing machines. Hilbert-space methods, always meaningful to the mathematicians who developed them, have now become useful also to the engineer and applied scientist, and the acceptance of probabilistic as well as deterministic analyses has become commonplace.

The present series was conceived with the objective of presenting some exciting aspects of modern mathematics. The course was designed for nonspecialists with training in engineering or science, high-school and college teachers of mathematics, and others desiring to remain *au courant* concerning mathematical developments. The material is intended to be quite understandable in the large, although not necessarily in complete detail, on the basis of the single lectures, and all of it should be applicable either now or in the reasonably near future to science and engineering.

We are pleased to share the stimulating experience of the second "Modern Mathematics for the Engineer" lecture series with you, the reader, through the pages of this book.


PAUL H. SHEATS
Professor of Education
Dean, University Extension
University of California

L. M. K. BOELTER
Professor of Engineering
Dean, College of Engineering
University of California
Los Angeles

MORROUGH P. O'BRIEN
Professor of Engineering
Dean, College of Engineering
University of California
Berkeley

# Preface to the 1961 Edition

For several years the University of California, through University Extension, has been conducting a highly successful series of lecture courses devoted to Modern Science for the Engineer. The lectures are available in book form in the *University of California Engineering Extension Series,* of which six volumes (see page ii) have preceded the present contribution.

The timeliness of the series—or perhaps the scientific alertness of the Eastern European countries—is attested to by the fact that, during its first three years of existence, the initial volume of "Modern Mathematics for the Engineer" was translated into Hungarian, Polish, and Russian.

Turning a second time to mathematics, University Extension appointed an Advisory Committee composed of individuals representing various universities and industrial organizations in California to plan the 1958–59 lecture series.

The Committee adopted as its objective the presentation of some exciting aspects of modern mathematics that either are presently applicable or promise soon to be applicable to science and engineering.

To achieve this objective, the Committee sought to obtain the services of a group of outstanding speakers who were experts on advanced applicable mathematics. The topics chosen were, for the most part, those that have had recent spectacular applications in mathematics, that have been applied or are likely soon to be applied in physical, sociological, and biological sciences, and that involve a degree of mathematical subtlety.

The high success of the course was ensured by the enthusiastic response of the distinguished group of lecturers who participated in the series. It was presented at five locations throughout the state: at Berkeley, Corona, Los Angeles, Palo Alto, and San Diego.

The volume that has resulted is intended for engineers, scientists, mathematicians, students, high-school and college teachers, and others who desire to become or remain informed concerning current applicable mathematical developments.

The first volume of "Modern Mathematics for the Engineer" was somewhat arbitrarily divided into three parts: Mathematical Models, Probabilistic Problems, and Computational Considerations. The present volume is similarly divided into three parts: Mathematical Methods, Statistical and Scheduling Studies, and Physical Phenomena.

Certainly the foregoing partitioning is not a sharp one; rather, it is one of emphasis. Thus the broad and deep mathematical methods developed in Part 1 are clearly motivated by physical problems; the solutions of the physical problems treated in Part 3 call for the use of the most modern mathematical methods; and the powerful probabilistic processes described in Part 2 are strongly evident in Chaps. 11 and 12 of Part 3.

Chapter 12, for example, might equally well have been combined with Chaps. 1 to 4 to constitute a comprehensive part entitled Operational Observations, or it might have been put in Part 2 because of its probabilistic content; but its greatest emphasis seemed to be on physical propagation phenomena, and it could well furnish the basis of a one-semester course in this branch of mathematical physics. Again, Professor Pólya's delightful concluding Chap. 16 is concerned with physical observations, but these observations lead to general conjectures—and the conjectures demand proof, which in turn involves ingenious and penetrating mathematical methods.

There are numerous examples and exercises throughout the book. Some of the exercises are at the ends of chapters, others are interspersed with the text, and still others are incorporated in the text. It is hoped that they will aid the reader in his over-all assimilation of the material and that they will add to the usefulness of the book.

The editor is most grateful to the authors for their excellent and prompt contributions to this volume; to the other Advisory Committee members, John L. Barnes, Clifford Bell, L. M. K. Boelter, George W. Brown, John C. Dillon, Gerald Estrin, James C. Fletcher, Bernard Friedman, Magnus R. Hestenes, John W. Miles, Russell R. O'Neill, Louis A. Pipes, C. T. Singleton, Ivan S. Sokolnikoff, Thomas H. Southard, Angus E. Taylor, Charles B. Tompkins, and John D. Williams for their efforts and excellent ideas; to the Course Coordinators, John C. Bowman, Bernard Friedman, Stanley B. Schock, and Victor Twersky for their smooth handling of lecture arrangements; once again to Clifford Bell, the Statewide Coordinator of the course, for his unobtrusive but highly valued leadership; and especially the editor thanks his secretaries, Mildred Webb and Patti Hansen, for their careful and efficient work.

EDWIN F. BECKENBACH
University of California, Los Angeles

# Contents

## PART 3. PHYSICAL PHENOMENA

# Introduction to the 1961 Edition

MAGNUS R. HESTENES
PROFESSOR OF MATHEMATICS
UNIVERSITY OF CALIFORNIA, LOS ANGELES

During the last decade, there has been a remarkable expansion in the demand for advanced mathematics by the engineer. This demand has arisen in part because of the increasing complexities created by technological progress.

In advanced design it is frequently necessary to study a carefully constructed mathematical model before creating the physical model. The modern rockets and satellites, for example, could not have been built and successfully launched without careful mathematical analysis of the physical problem at hand.

Not only does mathematics enter into initial planning and design; it enters into testing programs as well. Data are collected and interpreted in accordance with a statistical theory. Once a product has been designed and tested, a mathematical theory of quality control is frequently used in the manufacture of the product.

More recently, mathematics has been found to be a useful tool in the field of production planning.

Thus mathematics enters into all phases of engineering and production.

The modern high-speed computing machine is playing an ever-increasing role in physical, biological, and social sciences, in engineering, and in business. The effective use of computers requires the aid of persons with a high degree of mathematical training and proficiency. Almost every branch of mathematics has been used on problems that have been successfully attacked with the help of computing machines.

These machines can be used for experimentation as well as for solving intricate mathematical problems. For example, a traffic problem has been simulated on a computing machine, and experiments have suggested means of traffic control that have significantly increased the flow of traffic in a congested area.

It is clear that persons responsible for the operations of computing machines must be proficient in mathematics and must have on hand source material necessary for the solution of their problems.

The role of mathematics in engineering has been most aptly presented by Dr. Royal Weller in the Introduction to "Modern Mathematics for the Engineer," First Series.

The topics selected for the present volume have been chosen to complement those found in the first volume. As is customary in mathematics, the mathematical theory is presented largely without regard to applications, except as illustrations of the theory. This is done in order to bring out the mathematical structure and in order to facilitate applications to problems that actually are of similar basic mathematical structure but that bear little superficial similarity to the applications being described. An attempt is made also to call attention to various mathematical fields that undoubtedly will play an important role in the science and engineering of the future.

Though it is not possible to cover all the phases of mathematics that are important for applications, it is hoped that this volume and its predecessor have exposed many of the most important and promising ones, have laid a firm foundation in modern mathematical thinking, and will stimulate the reader to further study of this most interesting and useful field of human endeavor.

# PART I
# Mathematical Methods

# 1

# From Delta Functions to Distributions

ARTHUR ERDÉLYI
PROFESSOR OF MATHEMATICS
CALIFORNIA INSTITUTE OF TECHNOLOGY

## 1.1 Introduction

In mathematical physics, one often encounters "impulsive" forces acting for a short time only. A unit impulse would be described by a function $p(t)$ that vanishes outside a short interval and is such that

$$\int_{-\infty}^{\infty} p(t)\,dt = 1$$

It is convenient to idealize such forces as "instantaneous" and to attempt to describe them by a function $\delta(t)$ that vanishes except for a single value of $t$ which we take to be $t = 0$, is undefined for $t = 0$, and for which

$$\int_{-\infty}^{\infty} \delta(t)\,dt = 1$$

Such a function, one convinces oneself, should possess the "sifting property"

$$\int_{-\infty}^{\infty} \delta(u)\phi(u)\,du = \phi(0) \tag{1.1}$$

for every continuous function $\phi$, and the corresponding property (obtained by integration by parts)

$$\int_{-\infty}^{\infty} \delta^{(k)}(u)\phi(u)\,du = (-1)^k\phi^{(k)}(0) \tag{1.2}$$

for every $k$ times continuously differentiable function $\phi$.

Unfortunately, it can be proved that no function, in the sense of the mathematical definition of this term, possesses the sifting property. Nevertheless, "impulse functions" postulated to have these or other similar properties are being used with great success in applied mathematics and mathematical physics.

The use of such improper functions can be defended as a kind of shorthand, or else as a heuristic means; it can also be justified by an appropriate mathematical theory. In Sec. 1.2 we shall indicate briefly some theories that can be employed to justify the use of the delta function. In order to provide a theoretical framework accommodating the great variety of improper functions occurring in contemporary investigations of partial differential equations, it seems necessary to widen the traditional concept of a mathematical function. The new concept, that of a "generalized function," is abstract and cannot reproduce all aspects of the older concept of a function. In particular, it is not possible to ascribe a definite value to a generalized function at a point. Nevertheless, we shall see that in some sense such generalized functions can be described. In particular, it makes perfectly good sense to say that $\delta(t)$, which is a generalized function, vanishes on any open interval not containing $t = 0$.

In this chapter we shall outline two theories of generalized functions. One, essentially algebraic in nature, is restricted to generalized functions on a half line; the other, more closely related to functional analysis, places less restrictions on the independent variable. We shall also mention briefly other theories of generalized functions.

## DELTA FUNCTIONS AND OTHER GENERALIZED FUNCTIONS

### 1.2 The Delta Function

Since the delta function is the idealization of functions that vanish outside a short interval, it is plausible to try to approximate the delta function by such functions. Let $s(t)$ be a function on $(-\infty, \infty)$ satisfying the following conditions:

(a) $$s(t) \geq 0$$
(b) $$s(t) = 0 \qquad \text{if } t \geq 1 \quad \text{or} \quad t \leq -1$$
(c) $$\int_{-\infty}^{\infty} s(t)\,dt = 1$$

Then the function

$$s_n(t) = ns(nt)$$

satisfies the conditions $a$ and $c$ and vanishes outside $(-1/n, 1/n)$, and it may be regarded as "approaching" the delta function as $n \to \infty$. Indeed, it can easily be proved that

$$\int_{-\infty}^{\infty} s_n(u)\phi(u)\,du \to \phi(0) \qquad \text{as } n \to \infty$$

for any continuous function $\phi$, or even for any function $\phi$ that is integrable over some interval containing 0 and is continuous at 0. Furthermore,

$$\int_{-\infty}^{\infty} s_n(u)\phi(t-u)\,du \to \phi(t) \qquad \text{as } n \to \infty \tag{1.3}$$

uniformly in $t$ over any finite interval, provided $\phi(t)$ is continuous over some larger interval. If $s(t)$ is $k$ times continuously differentiable, we also have

$$\int_{-\infty}^{\infty} s_n^{(k)}(u)\phi(t-u)\,du \to \phi^{(k)}(t) \qquad \text{as } n \to \infty$$

As a matter of fact, it is not necessary that $s$ have the property $b$. If $s$ has the properties $a$ and $c$, then the condition (1.3) will hold for all functions $\phi$ that are bounded and continuous for $-\infty < t < \infty$. Some examples of such approximations to the delta function that have been used by the great analysts of the last century are the following:

$$s(t) = \begin{cases} \tfrac{1}{2} & \text{for } -1 < t < 1 \\ 0 & \text{for } t \geq 1 \quad \text{or} \quad t \leq -1 \end{cases}$$

$$s(t) = \frac{1}{\pi(t^2+1)}$$

$$s(t) = \pi^{-\frac{1}{2}}e^{-t^2}$$

For a history of the delta function, see Ref. 16, Chap. V.

It may be remarked that we clearly have

$$\int_{-\infty}^{t} s_n(u)\,du \to \begin{cases} 0 & \text{if } t < 0 \\ 1 & \text{if } t > 0 \end{cases}$$

showing that in some sense $\delta(t)$ is the derivative of the unit function

$$U(t) = \begin{cases} 0 & \text{for } t < 0 \\ \tfrac{1}{2} & \text{for } t = 0 \\ 1 & \text{for } t > 0 \end{cases}$$

and indicating some connection between the theory of the delta function and that of generalized differentiation of discontinuous functions.

An entirely different kind of theory of the delta function was adumbrated by Heaviside (see Chaps. 2 and 3 and also Ref. 16, page 65) and more clearly pinpointed by Dirac (Ref. 2, pages 71 to 77); it was not carried out in detail, however, until more recently. According to this theory, the delta function is defined by its *action* on continuous functions, this action being given by the sifting property (1.1) or (1.3); any analytical operation that, acting on a continuous function $\phi$, produces $\phi(0)$ is then a representation of the delta function.

We have seen that it is impossible to construct such an analytical operation in the form of a Riemann (or Lebesgue) integral; but it is possible to express it as a *Stieltjes integral.* Indeed,

$$\int_{-\infty}^{\infty} \phi(t - u)\, dU(u) = \phi(t)$$

for all continuous functions $\phi$. If $U$ were differentiable, we should have

$$\int_{-\infty}^{\infty} \phi(t - u)\, dU(u) = \int_{-\infty}^{\infty} \phi(t - u) U'(u)\, du$$

so that here too the delta function appears as a generalized derivative of the unit function.

The two theories are not as far from each other as they might at first appear to be. Although the delta function cannot be expressed as an integral operation, it can be approximated by such operations, namely, the integral operations defined by means of the $s_n$. Indeed, this is exactly the burden of Eq. (1.3).

### 1.3 Other Generalized Functions

We have indicated theories of the delta function, the basic impulse function appropriate to functions on the line $-\infty < t < \infty$. Clearly there are corresponding basic impulse functions on an arbitrary finite or infinite interval; functions of two variables in a plane, where an impulse function may be concentrated at a point, along a curve, or on a more general set of points; functions of several variables; functions of a point on a curved surface or, more generally, on a manifold; and so on. While it should be possible to devise an appropriate theory for each of these impulse functions, it is clearly preferable to seek a general theory embracing all of them.

Other generalized functions occur in connection with Fourier analysis, the modern theory of partial differential equations, etc., and one should like these subjects to be included in any useful theory of generalized functions. We shall give a simple example to indicate the application of generalized functions to partial differential equations.

This example concerns the hyperbolic partial differential wave equation

$$u_{xx} - u_{yy} = 0$$

Clearly $f(x - y) + g(x + y)$ is a solution of this equation if $f$ and $g$ are twice continuously differentiable functions. Now, in many problems—e.g., problems of discontinuous wave motion—one should like to regard $f(x - y) + g(x + y)$ as a solution of the wave equation even if $f$ or $g$ fails to be twice continuously differentiable. There are several ways of considering such "weak" or "generalized" solutions. From the point of view adopted here, the most natural approach is that of a generalized theory of differentiation, according to which every function has generalized derivatives that are generalized functions and satisfy the partial differential equation.

We shall outline in this chapter two theories of generalized functions. The first of these is algebraic in nature; indeed, it closely imitates the widening of the concept of number from integers to rational numbers. It is most successful with functions of a single nonnegative variable, although it has been extended to functions of several such variables and to functions of a single variable on a finite interval. It has the further advantage of providing a very natural approach to operational calculus as well as to generalized functions and generalized differentiation. Its greatest drawback at present seems to be its inability to cope with functions of unrestricted real variables or with functions of several variables ranging over an arbitrary region.

The second theory belongs more to the domain of functional analysis. In a sense, it might be compared with the extension of the concept of number from rational to real numbers, but the comparison is somewhat farfetched. The principal advantage of this theory is its ability to cope with all generalized functions needed at present. There are several approaches to this theory; we shall outline one of them in the simplest case of functions of a single real variable and briefly mention some others. The considerable number of different approaches to this theory is partly due to an endeavor to remove a basic difficulty remaining in it—the difficulty in defining the product of two generalized functions—and largely due to a desire to make this concept of generalized functions more easily accessible to applied mathematicians and engineers.

An entirely different attempt to cope with the problem of the delta function may be mentioned here. Schmieden and Laugwitz[20] have enlarged the concept of real numbers. Their system of numbers contains infinitesimally small and infinitely large numbers, and the analysis based on this system leads to functions, in the mathematical sense of this word, that behave like the delta function. Moreover, in this system the multiplication of two functions presents no difficulties.

## MIKUSIŃSKI'S THEORY OF OPERATIONAL CALCULUS AND GENERALIZED FUNCTIONS

### 1.4 The Definition of Operators

In Secs. 1.4 to 1.10, which are based largely on Ref. 12, $t$ is a nonnegative variable, $f = \{f(t)\}$ denotes a function of this variable, $f(t)$ is the value of $f$ at $t$ (and hence a number), $\mathfrak{N}$ is the set of all (real or complex) numbers, $\mathcal{C}$ the set of all continuous functions of $t$, small Greek letters will denote scalars (numbers), $a \in \mathcal{C}$ will indicate that $a$ is an element of $\mathcal{C}$ with similar notation for other sets, $\Theta$ will tentatively denote the function vanishing identically (later we shall see that we may replace this notation by 0), and $l$ the function having a value equal to unity for every $t \geq 0$ [except for the value 1 at $t = 0$, this is the restriction of $U(t)$ to $t \geq 0$]. The sifting property of the delta function appropriate to the interval $0 \leq t < \infty$ may be expressed as

$$\int_0^t \delta(x)a(t - x)\,dx = a(t) \qquad \text{for } a \in \mathcal{C}$$

In $\mathcal{C}$ addition and multiplication by a scalar are defined in the obvious way, namely, for $a$ and $b \in \mathcal{C}$ and $\alpha$ and $\beta \in \mathfrak{N}$, $\alpha a + \beta b$ is the function whose value at $t$ is $\alpha a(t) + \beta b(t)$. These operations have the familiar properties. The *convolution* $a * b$, or simply $ab$, of two functions is defined by *Duhamel's integral*

$$ab(t) = \int_0^t a(x)b(t - x)\,dx$$

This operation has all the properties of multiplication, and it commutes with multiplication by a scalar; that is, $ab = ba$, $a(bc) = (ab)c$ and hence may be written as $abc$, $(a + b)c = ac + bc$, $(\alpha a)b = \alpha(ab)$, etc.

The set $\mathcal{C}$ with addition and multiplication by scalars forms a *vector space.* The same set with addition and convolution forms a *commutative ring,* which will be called the *convolution ring.*

Integral operations can be expressed in terms of convolutions. In fact, $la$ is the function having value at $t$ equal to

$$\int_0^t a(x)\,dx$$

We also have

$$ll = l^2 = \{t\}$$

and by induction,

$$l^n = ll^{n-1} = \left\{\frac{t^{n-1}}{(n-1)!}\right\}$$

for $n = 1, 2, 3, \ldots$, so that convolution with the latter function expresses the effect of $n$ successive integrations with fixed limit 0. More generally, for any $\alpha \in \mathfrak{N}$ with $\Re\alpha > 0$, where $\Re\alpha$ denotes the real part of $\alpha$, we may set

$$l^\alpha = \left\{\frac{t^{\alpha-1}}{\Gamma(\alpha)}\right\}$$

and call convolution with $l^\alpha$ *fractional integration of order* $\alpha$.

The convolution ring has *no unit element;* i.e., there is no $u \in \mathcal{C}$ such that $au = a$ for all $a \in C$. To see this, it is sufficient to note that $lu = l$ means

$$\int_0^t u(x)\,dx = 1$$

for all $t \geq 0$, which is clearly impossible. This means that the delta function appropriate to this case is certainly not a continuous function; actually, it is not any function.

A very important property of the convolution ring is contained in *Titchmarsh's theorem: For* $a,b \in \mathcal{C}$ *we have* $ab = \Theta$ *if and only if* $a = \Theta$ *or* $b = \Theta$ (or both these equations hold). An elementary proof of this theorem was given by Mikusiński in Ref. 12, Chap. 2, and is reproduced in Ref. 3, Sec. 2.1. Because of this property of $\mathcal{C}$, *division* is a meaningful, although not always a feasible, operation in $\mathcal{C}$. The equations $bu = a$, $bv = a$ imply $b(u - v) = \Theta$, and if $b \neq \Theta$, then it follows that $u = v$; that is, the convolution equation $bu = a$ has, with $b \neq \Theta$, at most one solution $u$. This solution may then be regarded as $a/b$. But, of course, $bu = a$ may have no solution in $\mathcal{C}$. Clearly, $bu(0) = 0$, and hence $bu = a$ will certainly have no solution if $a(0) \neq 0$; the equation may fail to have solutions in other cases as well.

The situation encountered here is very similar to that met upon the introduction of division of integers. There the feasibility of division (with the exception of division by zero) is ensured by the extension of the number concept from integers to rational numbers, and similarly here we shall ensure the existence of a unique solution of $bu = a$ with $b \neq \Theta$ by enlarging the convolution ring to a field of *convolution quotients.* Almost any construction of rational numbers from integers can be imitated; we shall follow the construction in terms of classes of equivalent ordered pairs of integers.

We shall consider ordered pairs $(a,b)$ of elements of $\mathcal{C}$, always assuming that the second element $\neq \Theta$. We call $(a,b)$ and $(c,d)$ *equivalent* if and only if $ad = bc$, denote by $\frac{a}{b}$ or $a/b$ the class of all ordered pairs equivalent to $(a,b)$, call $a/b$ a *convolution quotient,* and denote by $\mathfrak{F}$ the set of all convolution quotients. Clearly the cancellation law $(ac)/(bc) = a/b$ holds in $\mathfrak{F}$.

The elements of $\mathfrak{F}$ are abstract entities of which it is difficult to form a definite picture. It is, however, possible to point out that in a sense $\mathfrak{F}$ contains numbers, functions, and also the delta function and its derivatives. Thus, convolution quotients may be regarded as generalized functions, and they include the more common impulse functions on the half line $t \geq 0$. We *embed* $\mathcal{C}$ in $\mathfrak{F}$ by identifying $a \in \mathcal{C}$ with the convolution quotient $(ab)/b$ for any $b \neq \Theta$. Since

$$(ab)c = (ac)b \qquad \text{implies} \qquad \frac{(ab)}{b} = \frac{(ac)}{c}$$

this embedding is independent of $b$. We shall call a function $f$ *integrable* if it is absolutely integrable over every finite interval $0 \leq t \leq t_0$. For such a function, and $b \in \mathcal{C}$, the convolution $fb$ is defined and is a continuous function. It is thus natural to identify $f$ with the convolution quotient $(fb)/b$ for any $b \neq \Theta$. As in the previous case, the embedding is independent of $b$. Lastly, we embed $\mathfrak{N}$ in $\mathfrak{F}$ by an identification of $\alpha \in \mathfrak{N}$ with the convolution quotient $(\alpha b)/b$, an embedding that is independent of $b$. Now $b/b$, the image of the number 1, is the unit in $\mathfrak{F}$, and we shall see later that it acts as the delta function. Two integrable functions that differ only at a finite number of points (or, more generally, on a set of measure zero) will give the same convolution integral and hence will correspond to the same convolution quotient, thus being indistinguishable in $\mathfrak{F}$. This already shows that it is impossible to ascribe definite "values" to convolution quotients at a point $t$.

We now define the operations of addition, multiplication by a scalar, and (convolution) multiplication in $\mathfrak{F}$ by the equations

$$\frac{a}{b} + \frac{c}{d} = \frac{ad + bc}{bd} \qquad \alpha\frac{a}{b} = \frac{\alpha a}{b} \qquad \frac{a}{b}\frac{c}{d} = \frac{ac}{bd}$$

It is necessary to verify that these definitions are independent of the ordered pair used in the representation of the convolution quotients involved. Now, if

$$\frac{a_1}{b_1} = \frac{a_2}{b_2} \quad \text{and} \quad \frac{c_1}{d_1} = \frac{c_2}{d_2}$$

so that

$$a_1b_2 = b_1a_2 \quad \text{and} \quad c_1d_2 = d_1c_2$$

then

$$\begin{aligned}(a_1d_1 + b_1c_1)b_2d_2 &= a_1b_2d_1d_2 + b_1b_2c_1d_2 \\ &= a_2b_1d_1d_2 + b_1b_2c_2d_1 \\ &= (a_2d_2 + b_2c_2)b_1d_1\end{aligned}$$

and therefore

$$\frac{a_1d_1 + b_1c_1}{b_1d_1} = \frac{a_2d_2 + b_2c_2}{b_2d_2}$$

Hence the definition of addition is meaningful as an operation on convolution quotients. Similarly, the other two definitions can be proved meaningful, and the operations can be shown to have all the usual properties. Moreover, it is easy to verify that the embedding of $\mathcal{C}$ and $\mathfrak{N}$ in $\mathfrak{F}$ preserves all these operations. For this reason, we may write $f$ in place of $(fb)/b$ and $\alpha$ in place of $(\alpha b)/b$; in particular, we may write 1 in place of $b/b$. We also see that multiplication by 1 (which may be interpreted either as multiplication by the scalar 1 or as convolution multiplication by the convolution quotient corresponding to this number) reproduces $f$; hence we have the identification of the convolution quotient 1 with the delta function. Furthermore, the scalar 0 and the function $\Theta$ are mapped into the same convolution quotient, so that from now on we may write 0 indiscriminately for any of these three entities, which are conceptually entirely different yet operationally indistinguishable in $\mathfrak{F}$.

The set $\mathfrak{F}$ of convolution quotients is an *algebra;* i.e., it is a *vector space* under addition and multiplication by scalars and a *field* under addition and convolution multiplication. The set $\mathfrak{F}$ is closed under all these operations, and it is also closed under convolution division with the single exception of division by 0, which is prohibited. The rules of ordinary algebra hold in $\mathfrak{F}$.

From now on, elements of $\mathfrak{F}$ will be denoted by single letters, and in case of doubt it will be indicated whether an entity belongs to $\mathfrak{F}$ or to $\mathfrak{C}$.

The elements of $\mathfrak{F}$ will primarily be considered here as generalized functions, but we shall see in the next section that they may act as *operators*, thus providing a convenient approach to Heaviside's operational calculus.

## 1.5 Differential and Integral Operators

We have seen that convolution multiplication of a function by $l$ means integration of that function. It is a plausible conjecture that multiplication by the inverse element in $\mathfrak{F}$, that is to say by

$$s = l^{-1} = \frac{b}{(bl)}$$

with $b \in \mathfrak{C}$, $b \neq 0$, means differentiation. We shall see that this is not quite the case.

Let

$$a = \{a(t)\}$$

be a differentiable function and

$$\dot{a} = \left\{\frac{da}{dt}(t)\right\}$$

an integrable function. Then

$$a(t) = a(0) + \int_0^t \dot{a}(x)\,dx$$

or $a = a(0)l + l\dot{a}$. On multiplying by $s$, we obtain

$$sa = \dot{a} + a(0) \tag{1.4}$$

and thus see that, even in the case of a differentiable function $a$, the product $sa$ represents the derivative function only if $a$ vanishes at $t = 0$ (Sec. 2.1). This may be explained to some extent by interpreting all our functions as vanishing for $t < 0$, so that there is a jump of $a(0)$ at $t = 0$ that contributes $a(0)\delta(t)$ to the derivative. This also explains why, even for a differentiable function $a$, the product $sa$ is in general not a function; further, it shows some of the difficulties encountered in the early applications of Heaviside's operational calculus. By applying Eq. (1.4) several times, we obtain by induction

$$s^n a = a^{(n)} + a^{(n-1)}(0) + a^{(n-2)}(0)s + \cdots + a(0)s^{n-1}$$

for a function $a$ that is $n$ times differentiable and has an integrable $n$th derivative $a^{(n)}$. For such a function, $s^n a$ is in general a generalized function. On the other hand, $s^n a$ exists as a convolution quotient for any (not necessarily differentiable) function $a$, or any convolution quotient. The generalized function $s^n a$ may be called the extended or generalized $n$th derivative of $a$.

Since 1 corresponds to $\delta(t)$, $s^n$ is the extended $n$th derivative of the delta function, and a *polynomial* in $s$ with constant, i.e., scalar, coefficients is an impulse function.

Next we investigate simple rational functions of $s$. From Eq. (1.4) we have

$$s\{e^{\alpha t}\} = \{\alpha e^{\alpha t}\} + 1$$

or

$$(s - \alpha)\{e^{\alpha t}\} = 1$$

so that

$$(s - \alpha)^{-1} = \{e^{\alpha t}\}$$

From this, it can be proved by induction that

$$(s - \alpha)^{-n} = \left\{\frac{t^{n-1}}{(n-1)!} e^{\alpha t}\right\} \qquad n = 1, 2, \ldots \tag{1.5}$$

We are now ready to interpret any rational function of $s$. Such a function can be decomposed into a sum of a polynomial and partial fractions of the form (1.5), and every term of this decomposition can then be interpreted.

EXAMPLE 1.1. *To decompose the rational function $s^3/(s^2 + 1)$.*
Since

$$\frac{s^3}{s^2+1} = s - \frac{s}{s^2+1} = s - \frac{1}{2}\left(\frac{1}{s+j} + \frac{1}{s-j}\right)$$
$$= s - \tfrac{1}{2}(\{e^{-jt}\} + \{e^{jt}\})$$

where $j^2 = -1$, we have

$$\frac{s^3}{s^2+1} = s - \{\cos t\}$$

The operational calculus so developed may be used to solve ordinary linear differential equations with constant coefficients, and also to solve systems of such equations. It will be sufficient to illustrate the process by a simple example.

EXAMPLE 1.2. *To solve the differential equation*

$$\frac{d^2u}{dt^2} + \alpha^2 u(t) = f(t)$$

By applying Eq. (1.4) twice, we have

$$\ddot{u} = s^2 u - \dot{u}(0) - u(0)s$$

and hence

$$(s^2 + \alpha^2)u = f + u(0)s + \dot{u}(0) \qquad \text{and} \qquad u = \frac{1}{s^2+\alpha^2} f + \frac{u(0)s + \dot{u}(0)}{s^2+\alpha^2}$$

Now,

$$\frac{1}{s^2+\alpha^2} = \left\{\frac{1}{\alpha}\sin \alpha t\right\} \qquad \frac{s}{s^2+\alpha^2} = \{\cos \alpha t\}$$

and

$$\frac{1}{s^2+\alpha^2} f = \left\{\frac{1}{\alpha}\int_0^t f(x)\sin \alpha(t-x)\,dx\right\}$$

Hence we have the solution

$$u(t) = u(0)\cos \alpha t + \dot{u}(0)\frac{\sin \alpha t}{\alpha} + \int_0^t f(x)\frac{\sin \alpha(t-x)}{\alpha}\,dx$$

Such equations can be solved even if the right-hand sides are generalized functions, for instance, delta functions, as with the differential equation satisfied by the Green's function. For instance, the solution of

$$\frac{d^2u}{dt^2} + \alpha^2 u(t) = \delta(t)$$

or

$$(s^2 + \alpha^2)u = 1 + \dot{u}(0) + u(0)s$$

obtained in a manner similar to that for the above Example 1.2, is

$$u(t) = u(0)\cos \alpha t + \frac{1 + \dot{u}(0)}{\alpha}\sin \alpha t$$

We note that Eq. (1.5) is in agreement with the Laplace transform of $e^{\alpha t}t^{n-1}/(n-1)!$ in case $s$ denotes a complex variable. We shall see in Example 1.8 that this is not a coincidence. Thus, tables of Laplace transforms may be used in interpreting rational functions of $s$.

## 1.6 Limits of Convolution Quotients

It is fairly clear that in $\mathcal{C}$ we should use a notion of convergence of continuous functions under which the limit of a convergent sequence of such functions is again continuous. Uniform convergence on every finite interval offers itself as the simplest notion of convergence that preserves continuity. It is much less clear how convergence of a sequence of convolution quotients should be defined, and no simple notion of convergence in $\mathfrak{F}$ that has all the desirable properties is known. We shall follow Mikusiński in introducing a notion of convergence that is at any rate simple, has many of the desirable features of convergence, and appears to be adequate for the applications of convolution quotients to operational calculus and to partial differential equations. According to this concept of convergence, a sequence of convolution quotients is regarded as convergent if it has a common denominator and if the numerators, which are continuous functions, are convergent in the sense outlined above. Thus, we shall say that a sequence of convolution quotients $a_n$ converges to $a$, in symbols

$$a_n \to a \qquad \text{or} \qquad \lim a_n = a$$

if there is a $q \neq 0$ such that, for each $n$, $qa_n \in \mathfrak{C}$ and if furthermore the sequence of continuous functions $qa_n$ converges to $qa$ uniformly over every finite interval $0 \leq t \leq t_0$. Clearly $a$ itself is then a convolution quotient.

It is fairly easy to prove that the limit, if it exists, is unique and has most of the usual properties. In particular, the sequence $a, a, a, \ldots$ converges to $a$; the sum (product) of convergent sequences is convergent and tends to the sum (product) of the limits; and a sequence of scalars is convergent in the ordinary sense if and only if the corresponding sequence of convolution quotients converges in the sense outlined here. However,

$$\lim \frac{a_n}{b_n} = \frac{\lim a_n}{\lim b_n}$$

does not necessarily hold even if it is assumed that $b_n \neq 0$ and $\lim b_n \neq 0$.

We shall now give some examples and comment on them.

EXAMPLE 1.3. *To prove that* $\{\sin nt\} \to 0$.

We have

$$l\{\sin nt\} = \left\{\frac{1 - \cos nt}{n}\right\}$$

and the latter function converges to 0 uniformly (in this case over the entire nonnegative axis). This example shows that convergence in $\mathfrak{F}$, even for ordinary functions, demands much less than ordinary convergence. It thus allows us to ascribe limits to sequences of functions that would ordinarily be regarded as divergent, and it also opens the way to a representation of some convolution quotients as limits, in this sense, of ordinary functions. (See also Example 1.5.)

EXAMPLE 1.4. *To prove that for* $c \in \mathfrak{C}$, $c^n \to 0$ *as* $n \to \infty$.

For a fixed $t_0$, there exists an $M > 0$ such that $|c(t)| \leq M$ for $0 \leq t \leq t_0$.

We shall prove by induction that, for $n = 1, 2, \ldots,$

$$|c^n(t)| \leq \frac{M^n t^{n-1}}{(n-1)!} \qquad 0 \leq t \leq t_0 \tag{1.6}$$

This relationship clearly holds when $n = 1$. If it holds for $n - 1$, then

$$|c^n(t)| = \left| \int_0^t c(t-x)c^{n-1}(x)\,dx \right| \leq \int_0^t M \frac{M^{n-1}x^{n-2}}{(n-2)!}\,dx = \frac{M^n t^{n-1}}{(n-1)!}$$

and this completes the proof by induction of the inequality (1.6). Since the right-hand side of (1.6) converges to 0 uniformly for $0 \leq t \leq t_0$, we have $c^n \to 0$ in the sense of convergence in $\mathfrak{C}$.

EXAMPLE 1.5. *To prove that if $f(t)$ is absolutely integrable over $0 \leq t < \infty$ and*

$$\int_0^\infty f(x)\,dx = 1$$

*then*

$$\{nf(nt)\} \to 1$$

We set

$$\int_0^\infty |f(x)|\,dx = A$$

Now,

$$l\{nf(nt)\} = \left\{n \int_0^t f(nx)\,dx\right\} = \left\{\int_0^{nt} f(y)\,dy\right\}$$
$$l^2\{nf(nt)\} = l\left\{\int_0^{nt} f(y)\,dy\right\} = \left\{\int_0^t dx \int_0^{nx} f(y)\,dy\right\}$$

and we shall prove that this function approaches $\{t\} = l^2$ uniformly over every finite interval. Set

$$\begin{aligned} g_n(t) &= t - \int_0^t dx \int_0^{nx} f(y)\,dy \\ &= \int_0^t \left[1 - \int_0^{nx} f(y)\,dy\right] dx = \int_0^t dx \int_{nx}^\infty f(y)\,dy \end{aligned}$$

First assume that $0 \leq t \leq \delta$. Then

$$|g_n(t)| \leq \int_0^\delta dx \int_0^\infty |f(y)|\,dy \leq \delta A \qquad 0 \leq t \leq \delta$$

For $\delta \leq t \leq t_0$,

$$\begin{aligned} |g_n(t)| &\leq \int_0^\delta dx \int_0^\infty |f(y)|\,dy + \int_\delta^t dx \int_{nx}^\infty |f(y)|\,dy \\ &\leq \delta A + t_0 \int_{n\delta}^\infty |f(y)|\,dy \qquad \delta \leq t \leq t_0 \end{aligned}$$

In either case, for any $\delta$ between 0 and $t_0$,

$$|g_n(t)| \leq \delta A + t_0 \int_{n\delta}^\infty |f(y)|\,dy \qquad 0 \leq t \leq t_0$$

Given $t_0 > 0$ and $\epsilon > 0$, we first choose $\delta$ so that $0 < \delta < \epsilon/(2A)$ and then choose $N$ so that

$$\int_{N\delta}^{\infty} |f(y)|\, dy < \frac{\epsilon}{2t_0}$$

We then have $|g_n(t)| < \epsilon$ for $0 \leq t \leq t_0$ and $n \geq N$, showing that $g_n(t)$ converges to 0 uniformly for $0 \leq t \leq t_0$. Thus, $l^2\{nf(nt)\}$ converges to $l^2$ uniformly on every finite interval.

We have accordingly found a family of approximations to the delta function. This should be compared with the approximations discussed in Sec. 1.2. The result suggests that many other convolution quotients might be represented as limits of functions.

## 1.7 Operator Functions

We shall now consider convolution quotients that depend on *parameters*. For the sake of simplicity, we shall consider a single parameter $x$ varying over a closed and bounded interval $I: \alpha \leq x \leq \beta$, and we shall denote the domain $\alpha \leq x \leq \beta$, $t \geq 0$ of the $xt$ plane by $D$. An *operator function* $a(x)$ assigns to each $x \in I$ a convolution quotient $a(x)$. Mikusiński calls such a function a *parametric function* if each $a(x) \in \mathfrak{C}$, so that $a(x) = \{a(x,t)\}$, and considers $a(x)$ as a *continuous operator function* if there exists a $q \neq 0$ in $\mathfrak{F}$ such that $b(x) = qa(x)$ is a parametric function and $b(x,t)$ is continuous in $D$; he says $a(x)$ is $k$ times continuously differentiable with respect to $x$ if there exists a $q \neq 0$ in $\mathfrak{F}$ such that $b(x) = qa(x)$ is a parametric function that is $k$ times continuously differentiable with respect to $x$; and he sets

$$a^{(k)}(x) = q^{-1} \left\{ \frac{\partial^k b(x,t)}{\partial x^k} \right\}$$

Continuous and differentiable operator functions have many of the usual properties, and differentiation obeys the familiar rules. It is unnecessary for us to go into further details here. Instead of this, let us consider some examples.

EXAMPLE 1.6. *To discuss the function* $a(x) = \{\cos(x - t)\}$.

This is a continuous parametric function for any interval $I$. By virtue of the results in Example 1.2, this operator function can be expressed as

$$\frac{s \cos x + \sin x}{s^2 + 1}$$

The function is indefinitely differentiable with respect to $x$, and the reader may easily verify that its derivatives can be computed by differentiating the explicit form. The function $a(x)$ satisfies the operator differential equation

$$a''(x) + a(x) = 0$$

and the initial conditions

$$a(0) = \{\cos t\} \qquad a'(0) = \{\sin t\}$$

EXAMPLE 1.7. *To discuss the function $h_\alpha(x)$, defined as follows: For $x \geq 0$ let*

$$h(x,t) = 0 \quad \text{if } 0 \leq t < x \qquad h(x,t) = 1 \quad \text{if } x \leq t$$

*and set*

$$h(x) = \{h(x,t)\} \qquad h_\alpha(x) = l^\alpha h(x)$$

If $\Re\alpha > -1$, then $h_\alpha(x)$ is a parametric function, and

$$h_\alpha(x,t) = 0 \quad \text{for } 0 \leq t < x \qquad h_\alpha(x,t) = \frac{(t-x)^\alpha}{\Gamma(\alpha+1)} \quad \text{for } x \leq t$$

Thus $h_\alpha(x)$ is a continuous parametric function if $\Re\alpha > 0$, and it is $k$ times continuously differentiable if $\Re\alpha > k$; further,

$$h'_\alpha(x) = h_{\alpha-1}(x) = -sh_\alpha(x)$$

if $\Re\alpha > 1$. Since

$$l^\beta h_\alpha(x) = h_{\alpha+\beta}(x)$$

it follows that $h_\alpha(x)$ is infinitely differentiable, in the sense of differentiation of operator functions, with respect to $x$; and

$$h'_\alpha(x) = -sh_\alpha(x)$$

for all $\alpha$. This differential equation, together with

$$h_\alpha(0) = l^\alpha h(0) = l^{\alpha+1}$$

suggests writing

$$h_\alpha(x) = l^{\alpha+1}e^{-sx}$$

We shall justify this later.

Of particular importance is

$$h_{-1}(x) = e^{-sx}$$

For $f \in \mathfrak{C}$, we set

$$h(x)f = \{g(x,t)\}$$

and find

$$g(x,t) = 0 \quad \text{for } 0 \leq t < x \qquad g(x,t) = \int_0^{t-x} f(u)\,du \quad \text{for } x \leq t$$

Consequently, for

$$h_{-1}(x)f = sg(x) = g_1(x)$$

we have

$$g_1(x,t) = 0 \quad \text{for } 0 \leq t < x \qquad g_1(x,t) = f(t-x) \quad \text{for } x \leq t$$

Thus, $g_1(x,t)$ is simply the function $f(t)$ shifted by $x$, and $e^{-sx}$ is the *shift operator*.

By a direct computation, it can be verified that

$$h(x)h(y) = lh(x + y)$$

for $x \geq 0$, $y \geq 0$, and this relationship holds for all real $x$, $y$, provided that we define $h(x)$ for negative values of $x$ by the equation

$$h(x)h(-x) = l^2$$

We have thus defined $h_\alpha(x)$ for all complex $\alpha$ and for all real $x$. In particular,

$$h_{-1}(x)h_{-1}(-x) = 1$$

We now turn to *integration* of operator functions with respect to $x$. Let $\phi(x)$ be absolutely integrable over $I = [\alpha,\beta]$, and let $a(x)$ be a continuous operator function and $q \neq 0$ such that $qa(x) = b(x)$ is a continuous parametric function. We then set

$$\int_\alpha^\beta a(\lambda)\phi(\lambda)\,d\lambda = q^{-1}\left\{\int_\alpha^\beta b(x,t)\phi(x)\,dx\right\}$$

It can be proved that this definition is independent of $q$ and that the integral has all the usual properties. Infinite integrals may then be defined as limits of finite integrals.

EXAMPLE 1.8. *To prove that for any integrable function $f$, the integral $\int_0^\infty e^{-sx}f(x)\,dx$ exists and is equal to $f$.*

This result is the background for the coincidence noted at the end of Sec. 1.5. It should be remarked, though, that here $s$ is an operator, so that the integral is *not* a Laplace integral; also, it might be noted that the result to be proved holds without any restriction on the growth of $f(x)$. Since

$$le^{-sx} = h(x)$$

we have

$$l\int_0^\beta e^{-sx}f(x)\,dx = \left\{\int_0^\beta h(x,t)f(x)\,dx\right\}$$
$$= \left\{\int_0^{\min(\beta,t)} f(x)\,dx\right\}$$

As $\beta \to \infty$, the last function approaches $lf$ uniformly over every finite interval $0 \leq t \leq t_0$. Thus

$$l\int_0^\infty e^{-sx}f(x)\,dx$$

exists and is equal to $lf$.

## 1.8 Exponential Functions

Suppose that, for a certain $w \in \mathfrak{F}$, there exists an interval $I$ containing $x = 0$ and a differentiable operator function $e(x)$ on $I$ that satisfies the differential equation

$$e'(x) = we(x)$$

and the initial condition

$$e(0) = 1$$

We then say that $w$ is a *logarithm* and set

$$e(x) = e^{xw}$$

It is fairly easy to prove that in this case $e(x)$ exists for all real $x$, is infinitely differentiable, is uniquely defined by the differential equation and the initial conditions, and has the properties

$$e^{xw} \neq 0 \quad \text{for all } x \qquad (e^{xw})^{-1} = e^{-xw} \qquad e^{xw}e^{yw} = e^{(x+y)w}$$

### EXERCISE

Verify that $s$ is a logarithm and that (see Example 1.7)

$$e^{-xs} = h_{-1}(x)$$

Some but not all elements of $\mathfrak{F}$ are logarithms. The element $s$ is a logarithm, and so are real multiples of $s$, but it can be proved that $js$ is not a logarithm. All elements of $\mathfrak{C}$ are logarithms, and

$$e^{xw} = \sum_{n=0}^{\infty} \frac{1}{n!} x^n w^n \qquad w \in \mathfrak{C}$$

The series representation holds also for other $w$; thus it holds for integrable (rather than continuous) functions, or for $w = 1$. But it does not hold for all logarithms; for instance, the series fails to converge for $w = s$, which is a logarithm. If $u$ and $v$ are logarithms, then $\alpha u + \beta v$ is a logarithm for real, but not necessarily for complex, $\alpha$ and $\beta$.

Exponential functions arise in the solution of partial differential equations, in which they often correspond to fundamental solutions.

EXAMPLE 1.9. *To prove that $s^{1/2}$ is a logarithm.*

Let us set

$$Q(x) = \{Q(x,t)\} \qquad R(x) = \{R(x,t)\}$$

where

$$Q(x,t) = \frac{x}{2\pi^{1/2}t^{3/2}} \exp\left(-\frac{x^2}{4t}\right) \qquad x \geq 0 \quad t > 0$$

$$R(x,t) = \frac{1}{(\pi t)^{1/2}} \exp\left(-\frac{x^2}{4t}\right) \qquad x \geq 0 \quad t > 0$$

$$Q(x,0) = R(x,0) = 0 \qquad x \geq 0$$

Clearly, $Q(x) = -R'(x)$. Now the function

$$lQ(x) = \left\{\int_0^t Q(x,u)\,du\right\} = \left\{\frac{2}{\pi^{1/2}}\int_{x/(2t^{1/2})}^{\infty} e^{-v^2}\,dv\right\}$$

although a parametric function, fails to be continuous; but the function

$$l^2Q(x) = \left\{\frac{2}{\pi^{1/2}}\int_0^t du \int_{x/(2u^{1/2})}^{\infty} e^{-v^2}\,dv\right\}$$

is continuously differentiable with respect to $x$ when $x > 0$, and

$$[lQ(x)]' = -R(x)$$

Also,

$$lQ'(x) = -R(x)$$

On the other hand, we have

$$l^{1/2}Q(x) = \left\{\frac{1}{\pi^{1/2}}\int_0^t \frac{1}{(t-u)^{1/2}} Q(x,u)\,du\right\}$$

and upon introducing a new variable of integration $v$ by

$$\frac{x^2}{4u} = \frac{x^2}{4t} + v$$

we obtain

$$l^{1/2}Q(x) = \left\{\frac{2}{\pi t^{1/2}}\int_0^{\infty} \exp\left(-\frac{x^2}{4t} - v^2\right) dv\right\} = R(x)$$

so that

$$Q'(x) = -s^{1/2}Q(x)$$

Moreover, $l^2Q(x)$ approaches $\{t\} = l^2$ uniformly in every interval $0 \leq t \leq t_0$, and hence

$$e^{-xs^{1/2}} = Q(x) \qquad x > 0$$

In the course of this work we have also seen that

$$s^{-1/2}e^{-xs^{1/2}} = R(x)$$

and

$$s^{-1}e^{-xs^{1/2}} = \left\{\frac{2}{\pi^{1/2}}\int_{x/(2t^{1/2})}^{\infty} e^{-v^2}\,dv\right\} = \left\{\operatorname{erf}\frac{x}{2t^{1/2}}\right\} \qquad x \geq 0$$

where erf denotes the error function. For fixed $x > 0$, $Q(x,t)$ as a function of $t$ increases for $0 < t < x^2/6$ and decreases thereafter. Thus,

$$0 \leq Q(x,t) \leq Q\left(x, \frac{x^2}{6}\right) = \frac{3}{x^2}\left(\frac{6}{\pi e^3}\right)^{1/2}$$

Since this expression approaches zero, uniformly in $t$, as $x \to \infty$, we see that $\exp(-xs^{1/2}) \to 0$ as $x \to \infty$. It follows from this that, if for some $a \in \mathfrak{F}$, $a \exp(xs^{1/2})$ is a bounded continuous parametric function for $x \geq 0$, then $a = 0$; and if for $a \in \mathfrak{F}$ and $b \in \mathfrak{F}$, the function

$$a \exp(xs^{1/2}) + b \exp(-xs^{1/2})$$

is a bounded continuous parametric function for all real $x$, then $a = b = 0$.

## 1.9 The Diffusion Equation

Let us briefly indicate the application of the technique developed here to the *diffusion equation*

$$u_{xx}(x,t) = u_t(x,t)$$

in the half plane $-\infty < x < \infty$, $0 \leq t < \infty$ (subscripts indicate partial derivatives). If

$$u(x) = \{u(x,t)\}$$

is a parametric function possessing continuous partial derivatives with respect to $x$ and $t$, and a continuous second partial derivative with respect to $x$, our partial differential equation may be replaced by the operator differential equation

$$u''(x) = su(x) - \phi(x) \tag{1.7}$$

where $\phi(x) = u(x,0)$.

If $\phi(x)$ is an integrable function, this differential equation may be solved by the method of variation of parameters. Two solutions differ by an operator function of the form

$$a \exp(xs^{1/2}) + b \exp(-xs^{1/2})$$

where $a \in \mathfrak{F}$ and $b \in \mathfrak{F}$; and, by the remark at the end of the preceding section, a solution that is a bounded continuous parametric function is unique within the class of such functions. It may be verified that

$$u(x) = \frac{1}{2s^{1/2}} \int_{-\infty}^{\infty} e^{-|x-y|s^{1/2}} \phi(y)\, dy$$

is a bounded continuous parametric solution of Eq. (1.7) if $\phi$ is a bounded measurable function. In this sense, the function

$$u(x,t) = \frac{1}{2} \int_{-\infty}^{\infty} R(|x - y|, t)\phi(y)\, dy = \frac{1}{\pi^{1/2}} \int_{-\infty}^{\infty} e^{-u^2} \phi(x + 2ut^{1/2})\, du$$

may be regarded as a (unique) generalized solution of our boundary-value problem if $\phi(x) = u(x,0)$ is a bounded measurable function. Actually, this solution is differentiable, indeed analytic, for $t > 0$ and for all $x$; and although it is not a continuous function for $t \geq 0$, it satisfies the "initial condition" in the generalized sense that

$$u(x,t) \to \phi(x) \qquad \text{as } t \to 0 \quad t > 0$$

at least for all those $x$ at which $\phi$ is continuous. By a more refined analysis, this result can be extended to measurable functions that, instead of being bounded, are assumed to satisfy an inequality

$$|\phi(x)| \leq A \exp (B|x|^\alpha)$$

where $A$, $B$, and $\alpha$ are constants and $0 \leq \alpha < 2$.

Other problems involving parabolic equations, and problems involving the wave equation and other hyperbolic equations, can be solved by means of this operational calculus, but so far no significant and successful applications to elliptic partial differential equations, such as Laplace's equation, are known.

## 1.10 Extensions and Other Theories

Mikusiński[14] has extended this theory to functions of several variables, $t_1, \ldots, t_n$, ranging over the "cone"

$$t_1 \geq 0, \ldots, t_n \geq 0$$

He[13] has also developed the corresponding theory for convolution quotients of functions on a finite interval, $\alpha \leq t \leq \beta$.

An alternative theory has been proposed by J. D. Weston,[26,27] whose generalized functions are operators acting on certain "perfect functions" rather than convolution quotients of functions.

# DISTRIBUTIONS

## 1.11 Testing Functions

We now turn to generalized functions of an entirely different type, to *distributions.*[21] There are several different approaches to this theory, most of them resembling the theories of the delta function indicated in Sec. 1.2 in that distributions appear either as generalized limits of functions or else as characterized by their action on certain classes of functions. We shall present here the second point of view for generalized functions of a single real variable $t$ ranging over the entire real line $(-\infty, \infty)$. Alternative approaches and extensions will be mentioned in Sec. 1.18.

Since distributions will be defined in terms of their action on certain classes of functions, the resulting notion of generalized functions will depend on the class of *testing functions* on which distributions act. We shall use two spaces of testing functions: One has proved useful in applications to Fourier analysis, and the other has been employed in connection with partial differential equations. Other classes of testing functions have also been used.[5]

Let $\mathcal{S}$ be the set of infinitely differentiable functions decreasing rapidly as $t \to \pm\infty$. More precisely, $\phi$ is in $\mathcal{S}$ if all derivatives $\phi^{(k)}$ exist and if for any integer $k$ and any polynomial $P(t)$, $P(t)\phi^{(k)}(t) \to 0$ as $t \to \pm\infty$.

The set $\mathfrak{S}$ is a *vector space* in the sense that, for any two elements $\phi_1$ and $\phi_2$ of $\mathfrak{S}$ and any two real or complex numbers $c_1$ and $c_2$, the function $c_1\phi_1 + c_2\phi_2$ is defined (in the obvious way) and is again in $\mathfrak{S}$. We shall use 0 indiscriminately to denote the number 0 and the function identically equal to zero for all values of $t$. We now introduce a notion of convergence in $S$. A sequence of functions $\phi_n$ in $S$ is said to converge to 0 if, for any fixed $k$ and fixed polynomial $P(t)$, $P(t)\phi_n^{(k)}(t) \to 0$ uniformly for all real $t$ as $n \to \infty$. A sequence of functions $\phi_n$ is said to converge to $\phi$ if $\phi_n - \phi$ converges to 0. We shall indicate this by writing $\phi_n \to \phi$ as $n \to \infty$.

Let $c_n \to c$ (in the sense of convergence of numbers) and let $\phi_n \to \phi$ and $\theta_n \to \theta$, in the sense of convergence in $\mathfrak{S}$, as $n \to \infty$; then it is easy to see that also $c_n\phi_n \to c\phi$ and $\phi_n + \theta_n \to \phi + \theta$, in the sense of convergence in $\mathfrak{S}$, as $n \to \infty$. Thus multiplication by a number and addition of functions are continuous operations. From now on, we shall usually omit the qualifying phrases appearing in the parentheses above, since the nature of the entities involved will indicate which space we are in and which notion of convergence should be used.

Secondly, let $\mathfrak{D}$ be the set of infinitely differentiable functions $\phi$ vanishing outside some finite interval, the interval depending on $\phi$ and varying from element to element of $\mathfrak{D}$. There are such functions. As an example, let us define

$$\begin{aligned} \phi(t) &= 0 && \text{for } t \leq a \quad \text{or} \quad t \geq b \\ \phi(t) &= \exp\{-[c(t-a)^{-\alpha} + d(b-t)^{-\beta}]\} && \text{for } a < t < b \end{aligned}$$

where $a$ and $b$ are real numbers, $a < b$, and $c$, $d$, $\alpha$, and $\beta$ are positive numbers. Clearly, $\phi$ vanishes outside a finite interval and is infinitely differentiable except possibly at $a$ and $b$, and it is easy enough to show that $\phi$ is also infinitely differentiable at these two points. We now introduce a notion of convergence in $\mathfrak{D}$. A sequence of functions $\phi_n$ in $\mathfrak{D}$ is said to *converge* to 0 if there is a finite interval $I$ such that each $\phi_n$ vanishes outside $I$, and if, for each fixed nonnegative integer $k$, $\phi_n^{(k)}(t) \to 0$ uniformly for all real $t$ (or all $t$ in $I$) as $n \to \infty$. A sequence of functions $\phi_n$ is said to converge to $\phi$ if $\phi_n - \phi$ converges to 0, and we shall write $\phi_n - \phi \to 0$ and $\phi_n \to \phi$ as $n \to \infty$. Clearly, $\mathfrak{D}$ is a vector space, and the operations of multiplication of a function by a number and addition of two functions are continuous. Equally clearly, every element of $\mathfrak{D}$ belongs also to $\mathfrak{S}$, and convergence in $\mathfrak{D}$ implies convergence in $\mathfrak{S}$.

Elements of $\mathfrak{S}$ or $\mathfrak{D}$ will be called *testing functions.*

In studying the *action* of other functions on testing functions, we shall start with continuous functions of $t$ and proceed to functions that are merely *locally integrable* in the sense that they are integrable over every finite interval. In this context a function of $t$ will be said to be of *slow growth* if its growth is dominated by that of some polynomial; in other words, a function $f$ is of slow growth if $(1 + t^2)^{-N}f(t)$ is bounded for some $N$.

## 1.12 The Definition of Distributions

For continuous functions $f$ of slow growth,

$$f\langle\phi\rangle = \int_{-\infty}^{\infty} f(t)\phi(t)\,dt \tag{1.8}$$

converges for each $\phi$ in $\mathcal{S}$ and defines an "evaluation" of $f$ on all elements of $\mathcal{S}$. In classical analysis, we think of a function $f$ as characterized by its values $f(t)$ for all real $t$. We now claim that, alternatively, we can characterize such a function by its evaluations $f\langle\phi\rangle$ on all elements of $\mathcal{S}$. In order to substantiate this claim, we have to show that two continuous functions of slow growth possessing the same evaluations on all elements of $\mathcal{S}$ also possess the same values for all $t$ and hence are identical. It will be sufficient to show that $f\langle\phi\rangle = 0$ for all $\phi$ in $\mathcal{S}$ entails $f(t) = 0$ for all $t$. Indeed, suppose $f(t_0) \neq 0$ for some $t_0$, say $f(t_0) > 0$. Since $f$ is a continuous function, there is some interval $I$ around $t_0$ on which $f$ is positive. Take any interval $(a,b)$ in the interior of $I$ and define

$$\begin{aligned} \phi_{ab}(t) &= 0 && \text{for } t \leq a \quad \text{or} \quad t \geq b \\ \phi_{ab}(t) &= \exp\left(-\frac{1}{t-a} - \frac{1}{b-t}\right) && \text{for } a < t < b \end{aligned}$$

Then clearly $f\langle\phi_{ab}\rangle > 0$, and accordingly the assumption $f(t_0) \neq 0$ for some $t_0$ is inconsistent with $f\langle\phi\rangle = 0$ for all $\phi$ in $S$. Incidentally, we see that a continuous function of slow growth is completely characterized by its evaluations on the $\phi_{ab}$ for rational $a$ and $b$, but we shall continue to think of it in terms of its evaluations on all $\phi$ in $\mathcal{S}$.

Many discontinuous functions—namely, all locally integrable functions—of slow growth also possess evaluations on $\mathcal{S}$, and the proof given above shows that at points of continuity the values of such a function are completely determined by the evaluations of the function. On the other hand, the values at points of discontinuity are not at all determined. For instance, the two functions $f$ and $g$ defined by

$$\begin{aligned} f(t) &= 0 \quad \text{for } t \leq 0 \qquad \text{and} \qquad f(t) = 1 \quad \text{for } t > 0 \\ g(t) &= 0 \quad \text{for } t < 0 \qquad \text{and} \qquad g(t) = 1 \quad \text{for } t \geq 0 \end{aligned} \tag{1.9}$$

clearly have the same evaluations on all $\phi$ in $\mathcal{S}$, but their values at $t = 0$ differ. More generally, if $N$ is a *null function*—that is, $N$ is locally integrable and

$$\int_a^b |N(t)|\,dt = 0 \qquad \text{for all } a < b$$

(or, $N$ vanishes almost everywhere)—then $f$ and $f + N$ will have the same evaluations on all elements of $\mathcal{S}$. The situation is not unlike the one encountered in connection, say, with Fourier series or Laplace transforms, where $f$ and $f + N$ have the same Fourier series or Laplace transforms, as the case may be, and are thus indistinguishable. It so happens that in many situations the distinction between two functions differing by a null function is unimportant, and in such situations the evaluations of a function on $\mathcal{S}$ characterize it as far as a characterization is meaningful —i.e., up to a null function. In many problems in applied mathematics, the functions that most naturally arise in the data or in the final results either are continuous or else have simple types of discontinuities, and in the latter case the right-hand and left-hand limits of a function at a discontinuity do matter, while the value of the function itself at the discontinuity is usually artificial and has no physical significance. In such problems, then, $f\langle\phi\rangle$ is as satisfactory as $f(t)$ for a characterization of $f$.

The function $f(t)$ indicates a mapping of the real line into the space of real and complex numbers; similarly, $f\langle\phi\rangle$ indicates a mapping of $\mathcal{S}$ into the space of numbers in that it assigns a number, defined by Eq. (1.8), to each element of $\mathcal{S}$. Now, a mapping of a vector space into a space of numbers is called a *functional*, and in this sense we may say that the function $f$ is characterized by the functional on $\mathcal{S}$ that it generates. We shall henceforth use $f$ indiscriminately for either the function or the functional.

The functional $f\langle\phi\rangle$ defined by Eq. (1.8) is clearly *linear* in the sense that, for any two numbers $c_1$ and $c_2$ and any two testing functions $\phi_1$ and $\phi_2$, we have

$$f\langle c_1\phi_1 + c_2\phi_2\rangle = c_1 f\langle\phi_1\rangle + c_2 f\langle\phi_2\rangle$$

We claim that the function is also *continuous* in the sense that $\phi_n \to \phi$ in $\mathcal{S}$ entails $f\langle\phi_n\rangle \to f\langle\phi\rangle$, in the sense of convergence of numbers, as $n \to \infty$. On account of the linearity of the functional, it will be sufficient to show this in the special case $\phi = 0$. Now, $f$ being of slow growth, there exist numbers $A$ and $N$ such that

$$|f(t)| \leq A(1 + t^2)^N$$

Moreover, by the definition of convergence in $\mathcal{S}$,

$$(1 + t^2)^{N+1}\phi_n(t) \to 0$$

uniformly for all real $t$ as $n \to \infty$, and consequently for any positive $\epsilon$ we have

$$(1 + t^2)^{N+1}|\phi_n(t)| < \epsilon$$

for all $t$ and all sufficiently large $n$. But then

$$|f\langle\phi_n\rangle| \leq \int_{-\infty}^{\infty} \frac{|f(t)|}{(1 + t^2)^N} (1 + t^2)^{N+1}|\phi_n(t)| \frac{dt}{1 + t^2} < A\epsilon \int_{-\infty}^{\infty} \frac{dt}{1 + t^2}$$

for all sufficiently large $n$, showing that $f\langle\phi_n\rangle \to 0$ as $n \to \infty$.

Thus we see that every locally integrable function of slow growth determines uniquely a continuous linear functional on $\mathcal{S}$, and conversely such a function is determined up to a null function by its continuous linear functional.

There are many continuous linear functionals on $\mathcal{S}$ that are not generated by functions. For instance, for all nonnegative integers $k$, the equations

$$\delta\langle\phi\rangle = \phi(0) \qquad \delta^{(k)}\langle\phi\rangle = (-1)^k\phi^{(k)}(0) \qquad k = 0, 1, 2, \ldots \tag{1.10}$$

assign numbers to each testing function and thus determine functionals on $\mathcal{S}$ that are easily seen to be linear and continuous. Now, a process similar to the one carried out above for continuous functions shows that if this functional were generated by a function, then that function would have to vanish at all of its points of continuity other than the origin. Thus, it would vanish identically if it were continuous, and a familiar process of approximation of integrable functions by continuous ones shows that, in any event, it would have to be a null function. But a null function fails to generate the functional $\delta^{(k)}$, showing that this functional cannot be generated by a locally integrable function. Indeed, comparison with Eqs. (1.1) and (1.2) shows that it is the functional corresponding to the delta function for $k = 0$ and to its derivatives for $k \geq 1$.

These considerations suggest that we regard every continuous linear functional on $\mathcal{S}$ as defining a generalized function. The space of these continuous linear functionals will be denoted by $\mathcal{S}^*$; it clearly contains all locally integrable functions of slow growth, and it also contains the delta function and its derivatives.

In all the foregoing considerations, we could restrict the testing functions to $\mathfrak{D}$ and thus obtain the space $\mathfrak{D}^*$ of continuous linear functionals on $\mathfrak{D}$. (Note in this connection that $\phi_{ab}$ is in $\mathfrak{D}$.) Every continuous linear functional on $\mathcal{S}$ is also a continuous linear functional on $\mathfrak{D}$ and hence in $\mathfrak{D}^*$, and there are functionals in $\mathfrak{D}^*$ that are not in $\mathcal{S}^*$. The integral in Eq. (1.8) converges for any continuous or locally integrable function, not necessarily of slow growth, when $\phi$ is in $\mathfrak{D}$, thus showing that $\mathfrak{D}^*$ generalizes functions of arbitrary growth in the same sense in which $\mathcal{S}^*$ generalizes functions of slow growth.

We shall call the elements both of $\mathcal{S}^*$ and of $\mathfrak{D}^*$ *distributions;* in particular, we shall call $\delta$ the *delta distribution.* If it is necessary to distinguish between them, we shall call the elements of $\mathcal{S}^*$ *distributions of slow growth,* and those of $\mathfrak{D}^*$ we shall call *distributions of arbitrary growth* or, shortly, *distributions.* The name "distribution" was introduced by L. Schwartz. It is suggested by a consideration of an integrable function $f(t)$ as the density of a continuous distribution of mass, while the delta distribution corresponds to the unit mass concentrated at $t = 0$. Although it fails to offer a description of other generalized functions such as $\delta'$ and is not in accordance with the use of the same term in probability theory, the name "distribution" will be retained here as a convenient means for distinguishing generalized functions introduced here from those that appear, say, in Mikusiński's theory of operators.

## 1.13 Operations with Distributions

We shall generally denote distributions by capital letters, such as $R$, $S$, and $T$. Two distributions $S$ and $T$ are equal if $S\langle\phi\rangle = T\langle\phi\rangle$ for all testing functions $\phi$. We define multiplication of a distribution $T$ by a number $c$ and addition of two distributions $S$ and $T$ by

$$(cT)\langle\phi\rangle = cT\langle\phi\rangle \qquad (S + T)\langle\phi\rangle = S\langle\phi\rangle + T\langle\phi\rangle$$

These algebraic operations have the usual properties, and the set of distributions equipped with these operations is a vector space.

The product of two distributions, to correspond to multiplication of the values of two functions, cannot be defined in general. In particular cases, however, such a definition is possible. For instance, let $\theta(t)$ be infinitely differentiable for all real $t$. Then the function $\theta$ corresponds to a distribution, which we shall also denote by $\theta$. Given an arbitrary distribution $T$, we can define the product $\theta T$ by

$$(\theta T)\langle\phi\rangle = T\langle\theta\phi\rangle$$

noting that $\theta\phi \in \mathfrak{D}$ if $\phi \in \mathfrak{D}$. Moreover, if both the function $\theta$ and the distribution $T$ are of slow growth, then $\theta T$ will also be of slow growth.

The convolution of two distributions can also be defined, but its definition requires a consideration of distributions in two variables and will be considered only briefly, in Sec. 1.15.

If we think of distributions as generalized functions, we sometimes write $T(t)$, and in this spirit occasionally write symbolically

$$T\langle\phi\rangle = \int_{-\infty}^{\infty} T(t)\phi(t)\,dt$$

For instance, in this sense we may write

$$\phi(0) = \delta\langle\phi\rangle = \int_{-\infty}^{\infty} \delta(t)\phi(t)\,dt$$

Distributions do not in general have definite "values at points $t$." Nevertheless, a distribution may have "values" on an interval in the following sense. We shall say that the distribution $T$ vanishes on the open interval $(a,b)$ if $T\langle\phi\rangle = 0$ for all testing functions vanishing outside $(a,b)$. In such a case we also write

$$T(t) = 0 \qquad a < t < b$$

If the distribution $T$ is generated by a locally integrable function $f$, then $T$ will vanish on $(a,b)$ in the sense of our definition if and only if $f$ vanishes almost everywhere on $(a,b)$.

This definition can be extended to a local comparison of a distribution $T$ with a function $f$. Let $f$ be integrable on $(a,b)$, and extend $f$ to all values of $t$, for instance by setting it equal to zero outside $(a,b)$. The function $f$ so extended defines a distribution, which we shall again denote by $f$. We then say

$$T(t) = f(t) \qquad a < t < b \tag{1.11}$$

or $T = f$ on $(a,b)$ if the distribution $T - f$ vanishes on $(a,b)$ in the sense described above. Note that the vanishing or otherwise of $T - f$ on $(a,b)$ is independent of the manner of extending $f$.

In this sense, the delta distribution clearly vanishes on any open interval not containing the origin, and this circumstance may be expressed by stating $\delta(t) = 0$ for $t \neq 0$, without implying the existence of "values" of $\delta(t)$ for individual values of $t$. For the *Heaviside distribution*, defined by

$$H\langle\phi\rangle = \int_0^\infty \phi(t)\,dt$$

we clearly have $H(t) = 0$ for $t < 0$ and $H(t) = 1$ for $t > 0$, so that in this sense we can attribute values to $H$ everywhere except at the origin. As a matter of fact, $H$ can be generated by either of the two functions defined in Eqs. (1.9).

The open set consisting of all open intervals on which a distribution $T$ vanishes (i.e., the union of all such intervals) is called the *null set* of $T$. The collection of values of $t$ not in the null set (i.e., in the complement of the null set) is called the *support* of $T$. The support is the set of points on which $T(t)$ is essentially different from zero; it is a closed set, and it can be characterized either as the smallest closed set outside of which $T(t)$ vanishes or as the collection of points $t$ not contained in any open interval on which $T$ vanishes.

The support of the delta distribution is clearly the single point $t = 0$, and the support of the Heaviside distribution $H$ is the nonnegative half $t \geq 0$ of the real line.

We now come to the differentiation of distributions. If $f$ is continuously differentiable, then $f$ and $f'$ generate distributions, and we have, by integration by parts,

$$f'\langle\phi\rangle = \int_{-\infty}^{\infty} f'(t)\phi(t)\,dt = -\int_{-\infty}^{\infty} f(t)\phi'(t)\,dt = f\langle-\phi'\rangle \tag{1.12}$$

and similarly for higher derivatives. This suggests the following definition for the *derivative* $T'$ of an arbitrary distribution $T$:

$$T'\langle\phi\rangle = T\langle-\phi'\rangle$$

and for the derivatives of higher order,

$$T^{(k)}\langle\phi\rangle = T\langle(-1)^k\phi^{(k)}\rangle \tag{1.13}$$

In this connection, it should be noted that the derivative of a testing function (rapidly decreasing testing function) is again a testing function (rapidly decreasing testing function).

EXAMPLE 1.10. *To determine the derivative $H'$ of Heaviside's distribution $H$.*

According to the definition, we have

$$H'\langle\phi\rangle = H\langle-\phi'\rangle = -\int_0^\infty \phi'(t)\,dt = \phi(0) = \delta\langle\phi\rangle$$

for all testing functions, and hence $H' = \delta$. It can be verified similarly that the derivatives of the delta distribution are the distributions defined in Eqs. (1.10).

According to our definition, every distribution is infinitely differentiable, and if the distribution is of slow growth, its derivatives will also be of slow growth. Since every locally integrable function generates a distribution, every such function possesses "distribution derivatives" of arbitrary order, but these derivatives need not be functions. Thus, Heaviside's function and its derivative, the delta distribution, exemplify this. If, on an interval $(a,b)$, $f$ possesses a locally integrable derivative, however, then the distribution derivative and the derivative in the usual sense are equal on $(a,b)$. To show this, denote for the moment the distribution derivative of $f$ by $T'$, so that

$$T'\langle\phi\rangle = f\langle-\phi'\rangle$$

For testing functions vanishing outside $(a,b)$, the computation (1.12) remains true and shows that

$$T'(t) = f'(t)$$

on $(a,b)$. For this reason, we may write $f'$ for the distribution derivative without ambiguity. In particular,

$$H'(t) = \delta(t)$$

is both meaningful and correct in this sense.

EXAMPLE 1.11. *To investigate the derivatives of $t^\alpha$, $\alpha > -1$.*

Since the values of this function fail to be real for negative $t$ when $\alpha$ is fractional, we shall first discuss the function

$$f(t) = t^\alpha H(t)$$

and the distribution $T$ generated by it. If $k$ is a positive integer and $\alpha - k > -1$, then $T^{(k)}$ is everywhere equal to the locally integrable function

$$f_k(t) = \alpha(\alpha - 1) \cdots (\alpha - k + 1)t^{\alpha-k}H(t)$$

and this equality holds for all positive integers $k$ if $\alpha$ is a nonnegative integer. If $\alpha$ is not an integer and $\alpha - k < -1$, then $T^{(k)}$ will no longer be a function; nevertheless, the distribution $T^{(k)}$ will be equal to the function $f_k$ when $t \neq 0$.

The following definition is then suggested. If $[t^\alpha H(t)]$ represents the distribution generated by $t^\alpha H(t)$ when $\alpha > -1$, then the formula

$$[t^\alpha H(t)] = \frac{1}{(\alpha + 1)(\alpha + 2) \cdots (\alpha + k)} [t^{\alpha+k}H(t)]^{(k)}$$

in which $[\cdots]^{(k)}$ denotes the $k$th distribution derivative, can be proved for $\alpha > -1$, $k = 1, 2, \ldots$, and it may be taken as the *definition* of the left-hand side when $\alpha < -1$, $k$ is a positive integer, and $\alpha + k > -1$. It is easy to see that this definition is independent of $k$ and that, with $[t^\beta H(t)]$ so defined, the formula holds for all nonintegral $\alpha$ and all positive integers $k$.

The definition of $[t^\alpha]$ for a negative integer $\alpha$ involves logarithms and will not be given here.

In order to extend fractional powers also to negative $t$, we introduce

$$\operatorname{sgn} t = H(t) - H(-t) = \begin{cases} 1 & \text{if } t > 0 \\ -1 & \text{if } t < 0 \end{cases}$$

and obtain the further formulas

$$[|t|^\alpha] = \frac{1}{(\alpha + 1) \cdots (\alpha + k)} [|t|^{\alpha+k}(\operatorname{sgn} t)^k]^{(k)}$$

$$[|t|^\alpha \operatorname{sgn} t] = \frac{1}{(\alpha + 1) \cdots (\alpha + k)} [|t|^{\alpha+k}(\operatorname{sgn} t)^{k+1}]^{(k)}$$

the discussion of which is left to the reader.

The differentiation of distributions obeys the usual rules. If $c$ is a constant, if $\theta(t)$ is an infinitely differentiable function and $\theta$ the distribution that it generates, if $S$ and $T$ are distributions, and if $\theta T$ is the product defined at the beginning of this section, then

$$(cT)' = cT' \qquad (S + T)' = S' + T' \qquad (\theta T)' = \theta' T + \theta T'$$

It will be sufficient to prove the last of these statements:

$$\begin{aligned}(\theta T)'\langle\phi\rangle &= (\theta T)\langle-\phi'\rangle = T\langle-\theta\phi'\rangle = T\langle-(\theta\phi)'\rangle + T\langle\theta'\phi\rangle \\ &= T'\langle\theta\phi\rangle + T\langle\theta'\phi\rangle = (\theta T')\langle\phi\rangle + (\theta' T)\langle\phi\rangle\end{aligned}$$

Finally, we can define *primitives* or *antiderivatives* of distributions, corresponding to indefinite integrals of functions. The distribution $T_1$ is called a primitive of $T$ if it satisfies $T_1' = T$. It is clear from the corresponding situation with regard to functions that a distribution may possess several primitives; we shall presently pursue this question further. Meanwhile, let us prove that every distribution possesses at least one primitive. In order to do this, let us fix a $\phi_0$ in $\mathfrak{D}$ for which

$$\int_{-\infty}^{\infty} \phi_0(u)\, du = 1$$

For this, we shall arbitrarily set $T_1\langle\phi_0\rangle = 0$. Corresponding to every testing function $\phi$, we define a function $\phi_1$ by the equation

$$\phi_1(t) = \int_{-\infty}^{t} \phi(u)\, du - \int_{-\infty}^{\infty} \phi(u)\, du \int_{-\infty}^{t} \phi_0(u)\, du$$

It is easy to verify that $\phi_1$ is again a testing function and that every testing function can be represented in the form $\phi_1$. We further define a new distribution $T_1$ by

$$T_1\langle\phi\rangle = T\langle-\phi_1\rangle$$

Clearly, $T_1\langle\phi_0\rangle = 0$, and since

$$\phi_1'(t) = \phi(t) - \phi_0(t) \int_{-\infty}^{\infty} \phi(u)\, du \tag{1.14}$$

we have $\quad T_1'\langle\phi_1\rangle = T_1\langle-\phi_1'\rangle = T_1\langle-\phi\rangle = T\langle\phi_1\rangle$

so that $T_1$ is a primitive of $T$.

It may be noted that in this discussion $\phi$ may be either in $\mathfrak{D}$ or in $\mathfrak{S}$; $\phi_1$ will be in the same class of testing functions, so that for distributions of slow growth we have demonstrated the existence of a primitive of slow growth.

A constant $c$ is an infinitely differentiable function and hence generates a distribution, which we shall again denote by $c$; thus,

$$c\langle\phi\rangle = c\int_{-\infty}^{\infty} \phi(t)\, dt$$

Such a distribution will be called a *constant distribution*. Clearly, the constant distribution defined by $c$ is equal to $c$ everywhere in the sense of the definition (1.11), and $cT$ may be used unambiguously for either the constant multiple of the distribution $T$ or the product of the two distributions.

The derivative of a constant distribution is zero, since

$$c'\langle\phi\rangle = c\langle -\phi'\rangle = -c\int_{-\infty}^{\infty} \phi'(t)\,dt = 0$$

We shall now prove that, conversely, the condition $T' = 0$ or, equivalently, $T\langle\phi'\rangle = 0$ for all testing functions entails that $T$ is a constant distribution. To prove this, fix $\phi_0$, and for every testing function $\phi$ define the testing function $\phi_1$, as above. Since $T\langle\phi_1'\rangle = 0$, it follows from Eq. (1.14) that

$$T\langle\phi\rangle = T\langle\phi_0\rangle \int_{-\infty}^{\infty} \phi(t)\,dt$$

Hence $T$ is the constant distribution that is equal to $T\langle\phi_0\rangle$ everywhere.

If $T_1$ is a primitive of $T$ and $c$ is any constant distribution, then $T_1 + c$ is also a primitive of $T$, and it follows from the result of the preceding paragraph that any two primitives of a given distribution differ by a constant distribution.

These considerations can be extended to repeated primitives. A distribution generated by a polynomial of degree $n$ being called a polynomial distribution of degree $n$, it is clear that derivatives and primitives of polynomial distributions are again polynomial distributions and have the appropriate degrees. In particular, the $k$th derivative of a polynomial distribution of degree $k - 1$ or less is zero; if $T_k$ is a $k$th primitive of $T$ and if $p$ is any polynomial distribution of degree $k - 1$ or less, then $T_k + p$ is also a $k$th primitive of $T$; conversely, any two $k$th primitives of a given distribution differ at most by a polynomial distribution of degree $k - 1$ or less.

## 1.14 Convergence of Distributions

We have seen that distributions form a vector space. In this vector space we introduce a notion of convergence, saying that the sequence of distributions $T_n$ converges provided that, for every testing function $\phi$, the sequence of numbers $T_n\langle\phi\rangle$ converges; further, we say that $T_n$ converges to $T$, in symbols $T_n \to T$ as $n \to \infty$, provided that, for every testing function $\phi$, we have $T_n\langle\phi\rangle \to T\langle\phi\rangle$. Let $T_n$ be a convergent sequence of distributions; then, for each $\phi$, $\lim T_n\langle\phi\rangle$ exists and defines a functional $T\langle\phi\rangle$. This functional is clearly linear, and it can also be proved to be continuous; thus, a convergent sequence always has a limit.

The notion of convergence defined above is consistent with the vector-space structure in that it makes the addition of two distributions and the multiplication of a distribution by a number continuous operations; that is, if $c_n \to c$, $S_n \to S$, and $T_n \to T$ as $n \to \infty$, then also $c_nT_n \to cT$ and $S_n + T_n \to S + T$ as $n \to \infty$. Moreover, a sequence of constant distributions $T_n = c_n$ converges in the sense of distributions if and only if the sequence of numbers $c_n$ converges in the sense of convergence of numbers, and the two limits correspond.

Since every locally integrable function determines a distribution, convergence of distributions may be used to define generalized limits of sequences of functions, it being understood that the generalized limit, if it exists, is in general a distribution rather than a function. The connection between ordinary pointwise limits and generalized limits of sequences of functions is by no means simple. Either of these limits may exist when the other does not, and even if both exist, they need not be equal. We shall exemplify these points.

EXAMPLE 1.12. *To show that if* $f_n(t) = \sin nt$, *then* $f_n \to 0$ *in the sense of convergence of distributions.*

Since every testing function $\phi$ is absolutely integrable, we have

$$f_n\langle\phi\rangle = \int_{-\infty}^{\infty} \phi(t) \sin nt \, dt \to 0 \qquad \text{as } n \to \infty$$

by the Riemann-Lebesgue lemma. Note that every derivative of a testing function is again absolutely integrable and that therefore it can be proved through integrations by parts that the sequence of functions $n^a \sin nt$, for any value of the constant $a$, also tends to 0 in the sense of convergence of distributions. Note also that for a fixed $t$ that is not an even multiple of $2\pi$, $\lim f_n(t)$ fails to exist as $n \to \infty$.

EXAMPLE 1.13. *To prove that the function* $f_n(t)$, *defined by*

$$\begin{aligned} f_n(t) &= 0 \qquad \text{if } t \le -n^{-1} \quad \text{or} \quad t = 0 \quad \text{or} \quad t \ge n^{-1} \\ f_n(t) &= n^2 \qquad \text{if } -n^{-1} < t < 0 \quad \text{or} \quad 0 < t < n^{-1} \end{aligned}$$

*does not converge in the sense of distributions.*

Here, clearly, $f_n(t) \to 0$, for every fixed $t$, as $n \to \infty$, but

$$f_n\langle\phi\rangle = n^2 \int_{-n^{-1}}^{n^{-1}} \phi(t) \, dt$$

fails to have a limit as $n \to \infty$ if $\phi(0) \neq 0$; thus, we have pointwise limits everywhere and yet no distribution limit.

EXAMPLE 1.14. *To prove that for the function* $f_n(t)$, *defined by*

$$\begin{aligned} f_n(t) &= 0 \qquad \text{if } t \le -n^{-1} \quad \text{or} \quad t = 0 \quad \text{or} \quad t \ge n^{-1} \\ f_n(t) &= \frac{n}{2} \qquad \text{if } -n^{-1} < t < 0 \quad \text{or} \quad 0 < t < n^{-1} \end{aligned}$$

*we have* $f_n \to \delta$ *as* $n \to \infty$.

Since

$$f_n\langle\phi\rangle = \frac{n}{2}\int_{-n^{-1}}^{n^{-1}} \phi(t)\,dt \to \phi(0)$$

it follows that $f_n \to \delta$, in the sense of convergence of distributions, as $n \to \infty$. On the other hand, for each fixed $t$, $f_n(t) \to 0$ as $n \to \infty$. Thus both limits exist, though they do not agree.

Note that in the foregoing example the discrepancy was caused by the nonuniformity of the pointwise limit at $t = 0$; indeed the very existence of the limit at $t = 0$ was enforced only by an artificial definition of the function $f_n(t)$ at that point. Under more stringent conditions, for instance under the condition that $f_n(t) \to f(t)$ uniformly for all $t$ as $n \to 0$ or that

$$\int_{-\infty}^{\infty} |f_n(t) - f(t)|\,dt \to 0 \qquad \text{as } n \to \infty$$

it can be proved that the existence of the pointwise limit entails that of the distribution limit, and that the two limits are equal.

## EXERCISE

Show that if $f(t)$ is absolutely integrable over $(-\infty, \infty)$, $\int_{-\infty}^{\infty} f(t)\,dt = 1$ and $f_n(t) = nf(nt)$, then $f_n \to \delta$ as $n \to \infty$. (Use Example 1.2 as a hint.)

One of the more remarkable properties of the convergence of distributions is its insensitivity to differentiation. If $T_n \to T$ as $n \to \infty$, then

$$T'_n\langle\phi\rangle = T_n\langle-\phi'\rangle \to T\langle-\phi'\rangle = T'\langle\phi\rangle \qquad \text{as } n \to \infty$$

and hence also $T'_n \to T'$ as $n \to \infty$.

As an application of this property, convergent series of distributions (and likewise infinite integrals of distributions) may always be differentiated term by term. The differentiated series are then always convergent in the sense of distributions. This causes a very wide class of trigonometric series, in particular all Fourier series, to converge in the sense of distributions even though they may fail to converge in the ordinary sense. A trigonometric series such as

$$\sum_{n=-\infty}^{\infty} c_n e^{jnt}$$

always converges in the sense of distributions provided that $c_n = O(n^k)$

for some fixed integer $k$ as $n \to \infty$; for in this case the series

$$\sum_{n=-\infty}^{\infty} n^{-k-2} c_n e^{jnt}$$

converges uniformly for all $t$ and its differentiation $k + 2$ times produces the given series. In particular, for a Fourier series the $c_n$ are bounded, and this process may be used with $k = 0$, so that the distribution sum of a Fourier series is at worst the second distribution derivative of a continuous periodic function.

In the definition of the relation $T_n \to T$, the integer-valued parameter $n$ may be replaced by a continuously varying parameter tending to some limit, and we shall take this extension for granted.

The notion of the distribution limit may be used in order to interpret divergent sums and integrals.

Example 1.15. *To show that*

$$\lim_{a \to \infty} \frac{1}{\pi} \int_0^a \cos tx \, dx$$

*exists and is equal to the delta distribution.*

Set

$$f_a(t) = \int_0^a \cos tx \, dx = \frac{\sin ta}{t}$$

Then $$f_a\langle\phi\rangle = \int_{-\infty}^{\infty} \frac{\sin at}{t} \phi(t) \, dt \to \pi\phi(0) \qquad \text{as } a \to \infty$$

by Fourier's single-integral theorem, and hence $f_a \to \pi\delta$ as $a \to \infty$. This proves the result and shows that the formula

$$\int_0^{\infty} \cos tx \, dx = \pi\delta(t)$$

frequently used in applied mathematics is correct in the sense of distribution convergence.

The notion of the distribution limit may also be used to establish

$$T'(t) = \lim_{h \to 0} \frac{T(t + h) - T(t)}{h}$$

as a meaningful formula. Here $T(t)$ is the distribution $T$, and $T'(t)$ is the distribution $T'$ defined by Eq. (1.13). The interpretation of $T(t + h)$ as a distribution $T_h$ is suggested by the formula

$$\int_{-\infty}^{\infty} T(t + h)\phi(t) \, dt = \int_{-\infty}^{\infty} T(t)\phi(t - h) \, dt$$

which is certainly valid when $T(t)$ is a function. For any testing function $\phi$ let us define a new testing function $\phi_h$ by

$$\phi_h(t) = \phi(t - h)$$

and then define, for any distribution $T$, a new distribution $T_h$ by

$$T_h\langle\phi\rangle = T\langle\phi_h\rangle$$

We then wish to prove that

$$\frac{T_h - T}{h} \to T' \qquad \text{as } h \to 0$$

Now,

$$\frac{1}{h}(T_h\langle\phi\rangle - T\langle\phi\rangle) - T'\langle\phi\rangle = T\left\langle\frac{\phi_h - \phi}{h} + \phi'\right\rangle$$

and thus we have to show that

$$\theta_h = \frac{\phi_h - \phi}{h} + \phi' \to 0 \qquad \text{as } h \to 0$$

in the sense of convergence in $\mathfrak{D}$. Now, if $\phi$ is in $\mathfrak{D}$, then, for all values of $h$ such that $|h| \leq h_0$, $\theta_h$ vanishes outside a finite interval that is independent of $h$, and we need to show that for each nonnegative integer $k$ we have $\theta_h^{(k)}(t) \to 0$, uniformly for all $t$, as $h \to 0$. We note that

$$\begin{aligned}\theta_h^{(k)}(t) &= \frac{1}{h}\{\phi^{(k)}(t - h) - \phi^{(k)}(t)\} + \phi^{(k+1)}(t) \\ &= \frac{1}{h}\int_{t-h}^{t}\{\phi^{(k+1)}(t) - \phi^{(k+1)}(u)\}\,du\end{aligned}$$

and by the mean-value theorem of differential calculus we thus obtain

$$\theta_h^{(k)}(t) = \frac{1}{h}\int_{t-h}^{t}(t - u)\phi^{(k+2)}(\bar{u})\,du$$

where $\bar{u}$ is between $u$ and $t$. The function $\phi^{(k+2)}(t)$ is bounded, say

$$|\phi^{(k+2)}(t)| \leq A_{k+2}$$

for all $t$. It then follows that

$$|\theta_h^{(k)}(t)| \leq \frac{1}{h}A_{k+2}\left|\int_{t-h}^{t}(t - u)\,du\right| = \tfrac{1}{2}A_{k+2}h$$

so that, for each $k$, $\theta_h^{(k)}(t) \to 0$ uniformly for all $t$ as $h \to 0$, and the result is established. This consideration shows that the definition of $T'$ is consistent with the definition more closely resembling that of the derivative of a function.

## 1.15 Further Properties of Distributions

Let us consider a distribution depending on a parameter $u$, and let us denote such a distribution by $T_u$. Here $u$ may be a real or complex parameter, or a parameter of a more general kind. The theory of convergence of distributions makes it possible to speak of continuous dependence of $T_u$ on $u$, of the partial derivative of $T_u$ with respect to $u$, of integrating $T_u$ with respect to $u$, and so on. We shall take these developments for granted without enlarging upon them here.

The concept of distributions can easily be extended to functions of several variables, together with the notions of equality, convergence of distributions in several variables, partial derivatives of such distributions with respect to these variables, and so on. In particular, mixed partial derivatives of distributions are always independent of the order of differentiation, since mixed partial derivatives of testing functions have this property.

It will be sufficient to make a few remarks on distributions in two variables, $s$ and $t$, based on testing functions $\phi(s,t)$. In this case the notation of generalized functions, $R(s,t)$, $S(s)$, $T(t)$, is especially useful in that it indicates whether we are considering distributions in one or the other variable or in both variables. Similarly, for the testing functions we shall write $\rho(s,t)$, $\sigma(s)$, $\tau(t)$. Partial derivatives may then be defined by

$$\frac{\partial R}{\partial s}\langle\rho\rangle = R\langle-\frac{\partial\rho}{\partial s}\rangle \qquad \frac{\partial R}{\partial t}\langle\rho\rangle = R\langle-\frac{\partial\rho}{\partial t}\rangle$$

We shall say that the distribution $R(s,t)$ is independent of $s$, or depends only on $t$, if there exists a fixed distribution $T(t)$ such that $R = T$ on all open sets of the $st$ plane, or, alternatively, if

$$R\langle\sigma\tau\rangle = T\langle\tau\rangle \int_{-\infty}^{\infty} \sigma(s)\, ds$$

whenever $\sigma$ is a testing function of $s$ and $\tau$ is a testing function of $t$ (and hence $\sigma\tau$ is a testing function of $s$ and $t$).

It can easily be proved that a distribution $R$ is independent of $s$ if and only if $\partial R/\partial s = 0$. In a similar manner we may speak of distributions depending only on $s + t$, say, and prove that a distribution depends only on $s + t$ if and only if $\partial R/\partial s - \partial R/\partial t = 0$. These concepts have applications in the theory of partial differential equations. For example, if $f$ and $g$ are two locally integrable functions of one variable, then $f(s + t)$ and $g(s - t)$ generate distributions in the two variables $s$ and $t$. The first of these distributions depends only on $s + t$, the second only on $s - t$, and both satisfy the differential equation

$$\frac{\partial^2 R}{\partial s^2} - \frac{\partial^2 R}{\partial t^2} = 0$$

so that their sum is also a solution of this differential equation in the sense of distributions. Conversely, it can be shown that any distribution that satisfies this equation is a sum of a distribution depending only on $s + t$ and one depending only on $s - t$. Thus, the general distribution solution of the one-dimensional wave equation has been obtained, and we are in a position to handle discontinuous wave motions (see Sec. 1.3).

Now $S$ and $T$ determine a distribution

$$R(s,t) = S(s)T(t)$$

in two variables, which is known as the *direct product* of $S$ and $T$. It may be defined by

$$R\langle\sigma\tau\rangle = S\langle\sigma\rangle T\langle\tau\rangle$$

where $\sigma$ is a testing function of $s$ and $\tau$ a testing function of $t$. It may also be shown that, for any testing function $\phi = \phi(s,t)$ and any fixed $s$ (which makes $\phi$ a testing function of $t$), $T\langle\phi\rangle$ exists; as $s$ varies, $T\langle\phi\rangle$ may be shown to be a testing function of $s$, so that $S\langle T\langle\phi\rangle\rangle$ exists and may be used as a definition of $R\langle\phi\rangle$.

These concepts may be used to define the *convolution* of two distributions. To motivate this definition, consider two functions $f$ and $g$ of a single variable and define their convolution $h = f * g$ by

$$h(u) = \int_{-\infty}^{\infty} f(u - t)g(t)\, dt$$

Now let $\phi$ be a testing function in one variable. At least formally,

$$h\langle\phi\rangle = \int_{-\infty}^{\infty} h(u)\phi(u)\, du = \int_{-\infty}^{\infty} \int_{-\infty}^{\infty} f(u - t)g(t)\phi(u)\, du\, dt$$

and if we set $u = s + t$, we get

$$f * g\langle\phi\rangle = \int_{-\infty}^{\infty} \int_{-\infty}^{\infty} f(s)g(t)\phi(s + t)\, ds\, dt \tag{1.15}$$

Now let $S$ and $T$ be two distributions in a single variable and let $S(s)T(t)$ be their direct product in the sense explained above. We then define the convolution $S * T$ as that distribution of a single variable for which

$$S * T\langle\phi\rangle = S(s)T(t)\langle\phi(s + t)\rangle$$

Already in the case of locally integrable functions $f$ and $g$ a difficulty arises in that $h$ need not be locally integrable (indeed the integral defining $h$ need not exist), so that Eq. (1.15) does not really define a distribution. The corresponding difficulty in the definition of $S * T$ shows up in the form that $\phi(s + t)$ is constant along the lines $s + t = \text{const}$; it certainly does not vanish at infinity (unless it vanishes identically), and hence it is not a testing function. The situation can be saved if suitable restrictions are placed on $S$ and $T$, for instance if the support of one of these distributions, say that of $T$, is contained in a finite interval. In this case it can be shown that $T(t)\langle\phi(s+t)\rangle$ is a testing function of $s$, so that $S(s)\langle T(t)\langle\phi(s+t)\rangle\rangle$ is a valid definition of $S * T\langle\phi\rangle$.

EXAMPLE 1.16. *To show that for any distribution $T(t)$, $T * \delta$ exists and is equal to $T$.*

Here

$$\delta(t)\langle\phi(s+t)\rangle = \phi(s)$$

is clearly a testing function, and

$$T(s)\langle\delta(t)\langle\phi(s+t)\rangle\rangle = T\langle\phi\rangle$$

Similarly, it can be proved that

$$T * \delta^{(k)} = T^{(k)} \qquad k = 1, 2, \ldots$$

The convolution $S * T$ is a continuous function of each of its two factors.

We have seen in Example 1.14 that the delta distribution can be represented as a distribution limit of functions. It can be shown that this is true of every distribution and, moreover, that the approximating functions may be chosen as infinitely differentiable. If in the exercise following Example 1.14 we take $f$ as a testing function $\alpha$, it is seen that there are sequences of testing functions $\alpha_n$ such that $\alpha_n \to \delta$ as $n \to \infty$. We then consider the distributions generated by the $\alpha_n$, denoting them again by $\alpha_n$, and with these distributions we consider the convolutions $T * \alpha_n$. By the continuity of the convolution, $T * \alpha_n \to T * \delta = T$. Moreover, the distribution $T * \alpha_n$ is equal to the function $T(u)\langle\alpha_n(t-u)\rangle$, and it is not difficult to prove that this function of $t$ is infinitely differentiable. Thus we see that every distribution is the limit, in the sense of distributions, of infinitely differentiable functions.

Every locally integrable function has distribution derivatives of all orders, and some distributions can thus be represented as distribution derivatives of locally integrable functions. Such distributions are known as *distributions of finite order*, and the least integer $r$ for which $T = f^{(r)}$ for some locally integrable $f$ is called the *order* of $T$. In this sense, locally integrable functions are distributions of order zero, the delta distribution is of order one, and so on. Not all distributions are of finite order; for instance, the distribution $T$ defined by

$$T\langle\phi\rangle = \sum_{n=1}^{\infty} \phi^{(n)}(n)$$

is not of finite order. Nevertheless, it can be proved that, given a distribution $T$ and any finite interval $I$, there exists an integrable function $f$ and an integer $r$ such that $T = f^{(r)}$ on $I$. Thus, locally—i.e., on every finite interval—distributions are of finite order, but the order may increase indefinitely as the interval $I$ is made to expand.

The results briefly described in the last two paragraphs give us an added understanding of the nature and structure of distributions, and they suggest alternative approaches to the theory of distributions. These will be taken up in Sec. 1.18.

## APPLICATIONS AND EXTENSIONS

### 1.16 Application to Fourier Transforms

In this section, "testing function" will mean a rapidly decreasing infinitely differentiable function, i.e., an element of $\mathcal{S}$, and "distribution" will mean a distribution of slow growth, i.e., an element of $\mathcal{S}^*$.

For a function $f$ that is absolutely integrable over $(-\infty, \infty)$ the *Fourier transform* $\tilde{f}$ is defined by

$$\tilde{f}(t) = \mathfrak{F}[f(t)] = \int_{-\infty}^{\infty} e^{ist} f(s)\, ds \tag{1.16}$$

As a result of the uniform convergence of the infinite integral, $\tilde{f}(t)$ is a continuous function of $t$, and by the appraisal

$$|\tilde{f}(t)| \leq \int_{-\infty}^{\infty} |f(s)|\, ds$$

$\tilde{f}(t)$ is bounded. For two absolutely integrable functions $f$ and $g$ and their Fourier transforms $\tilde{f}$ and $\tilde{g}$ we have *Parseval's relationship*

$$\int_{-\infty}^{\infty} \tilde{f}(t) g(t)\, dt = \int_{-\infty}^{\infty} f(t) \tilde{g}(t)\, dt$$

Since $f$ and $g$ are absolutely integrable and $\tilde{f}$ and $\tilde{g}$ are bounded, the infinite integrals on both sides of this equation converge.

For functions that fail to be integrable on $(-\infty, \infty)$, this definition of Fourier transform does not apply. Nevertheless, the infinite integral evaluated in Example 1.15 suggests that the Fourier transform of the function $f$, for which $f(t) = 1$ for all $t$, exists in the distribution sense and is equal to $2\pi\delta(t)$. We shall show that Fourier transforms can be defined for all distributions of slow growth, and that they are in general again distributions of slow growth. It may be remarked here that Fourier transforms of functions of slow growth were investigated (Ref. 1, Chap. VI) before the development of a general theory of distributions, and it might be noted further that a very elegant and elementary presentation[9] of

Fourier transforms of distributions of slow growth has recently been given.

Let us start with a testing function $\phi$ and its Fourier transform

$$\tilde{\phi}(t) = \mathfrak{F}[\phi(t)] = \int_{-\infty}^{\infty} e^{jst}\phi(s)\, ds$$

Formal differentiation of this relationship leads to

$$\tilde{\phi}^{(k)}(t) = \mathfrak{F}[(jt)^k\phi(t)] = \int_{-\infty}^{\infty} e^{jst}(js)^k\phi(s)\, ds \tag{1.17}$$

and formal repeated integrations by parts to

$$(-jt)^k\tilde{\phi}(t) = \mathfrak{F}[\phi^{(k)}(t)] = \int_{-\infty}^{\infty} e^{jst}\phi^{(k)}(s)\, ds \tag{1.18}$$

Since $\phi$ is rapidly decreasing, repeated differentiation of $\tilde{\phi}$ may be justified by the uniform convergence of all integrals involved, and it shows that the Fourier transform $\tilde{\phi}$ is again infinitely differentiable. The rapid decrease of all derivatives of $\phi$ shows that repeated integrations by parts are legitimate, that the integrated parts vanish, and that Eq. (1.18) holds. By combining Eqs. (1.17) and (1.18), we have

$$(-jt)^m\tilde{\phi}^{(k)}(t) = \mathfrak{F}\left(\frac{d^m}{dt^m}[(jt)^k\phi(t)]\right)$$

If $P(t)$ is any polynomial, we can now express $P(t)\tilde{\phi}^{(k)}(t)$ as a Fourier integral involving a sum of polynomials multiplied by derivatives of $\phi$, and by estimating this Fourier integral we can show that $\tilde{\phi}$ is rapidly decreasing. Thus, the Fourier transform of a testing function is again a testing function.

Since $\tilde{\phi}$ is again absolutely integrable, Fourier's inversion formula

$$\phi(t) = \frac{1}{2\pi}\int_{-\infty}^{\infty} e^{-jst}\tilde{\phi}(s)\, ds$$

applies in this case and shows that

$$\tilde{\tilde{\phi}}(t) = 2\pi\phi(-t)$$

Thus, every testing function is the Fourier transform of some testing function, so that the Fourier transformation is a one-to-one mapping of the space $\mathcal{S}$ of testing functions onto itself.

This mapping is clearly linear, and it is continuous, that is, $\tilde{\phi}_n \to 0$ if and only if $\phi_n \to 0$ as $n \to \infty$. Because of the essential symmetry of the relationship between $\phi$ and $\tilde{\phi}$, it will be sufficient to indicate the proof going one way. Since $\phi_n \to 0$ in the sense of convergence in $\mathcal{S}$, given any $\epsilon > 0$ we have

$$(1 + t^2)|\phi_n(t)| < \epsilon$$

for all sufficiently large $n$, and hence

$$|\bar{\phi}_n(t)| \leq \epsilon \int_{-\infty}^{\infty} \frac{dt}{1+t^2} = \frac{\pi}{2}\epsilon$$

for all sufficiently large $n$. This shows that $\bar{\phi}_n(t) \to 0$ uniformly for all $t$ as $n \to \infty$. To show that $\bar{\phi}_n \to 0$ in the sense of convergence in $\mathcal{S}$, we have to show that for any integer $k$ and any polynomial $P(t)$

$$P(t)\bar{\phi}_n^{(k)}(t) \to 0$$

uniformly for all $t$ as $n \to \infty$. Now, it has been pointed out above that $P(t)\bar{\phi}_n^{(k)}(t)$ is the Fourier transform of a finite number of expressions of the form $Q(t)\phi_n^{(m)}(t)$, where $Q(t)$ again denotes a polynomial. Each of these expressions can be made less than $\epsilon/(1 + t^2)$ by making $n$ sufficiently large, thus proving as above that

$$P(t)\bar{\phi}_n^{(k)}(t) \to 0$$

uniformly for all $t$ as $n \to \infty$.

We now return to Fourier transforms of locally integrable functions. If $f$ is absolutely integrable, then its Fourier transform is continuous and bounded; hence it is locally integrable and of slow growth. Thus, both $f$ and $\bar{f}$ can be evaluated on testing functions, and Parseval's relationship shows that

$$\bar{f}\langle\phi\rangle = f\langle\bar{\phi}\rangle$$

This suggests that we define the Fourier transform $\bar{T}$ of a distribution of slow growth $T$ by

$$\bar{T}\langle\phi\rangle = T\langle\bar{\phi}\rangle$$

Clearly, $\bar{T}$ as thus defined is a linear functional on testing functions. This linear functional is continuous, since $\phi_n \to \phi$ entails $\bar{\phi}_n \to \bar{\phi}$ and consequently

$$\bar{T}\langle\phi_n\rangle = T\langle\bar{\phi}_n\rangle \to T\langle\bar{\phi}\rangle = \bar{T}\langle\phi\rangle \qquad \text{as } n \to \infty$$

According to this definition, every distribution of slow growth possesses a Fourier transform that is again a distribution of slow growth. The relationship

$$\bar{\bar{T}}(t) = T(-t)$$

holds and shows that, conversely, every distribution of slow growth is the Fourier transform of some such distribution. Using the same symbol $\mathfrak{F}$ for the Fourier transformation of distributions and for the Fourier transformation of functions, we may write the definition given above also as

$$\mathfrak{F}T\langle\phi\rangle = T\langle\mathfrak{F}\phi\rangle$$

or more briefly as

$$\mathfrak{F}T = T\mathfrak{F}$$

The relationships

$$(\mathfrak{F}T)^{(k)} = \mathfrak{F}((jt)^k T) \qquad (-jt)^n \mathfrak{F}T = \mathfrak{F}T^{(k)}$$

are analogous to and follow from Eqs. (1.17) and (1.18); their proof is left as an exercise for the reader.

EXAMPLE 1.17. *To determine the Fourier transforms of the slowly increasing functions $t^n$, $n = 0, 1, \ldots$.*

The expression

$$\mathfrak{F}[t^n]\langle\phi\rangle = [t^n]\langle\tilde{\phi}\rangle = \int_{-\infty}^{\infty} t^n \tilde{\phi}(t)\, dt$$

gives the value at $t = 0$ of the Fourier transform, in the sense of Eq. (1.16), of the integrable function $t^n\tilde{\phi}(t)$. By Eq. (1.18),

$$t^n\tilde{\phi}(t) = \mathfrak{F}[j^m \phi^{(n)}(t)]$$

and by Fourier's inversion formula,

$$\mathfrak{F}^2[j^n\phi^{(n)}(t)] = 2\pi j^n \phi^{(n)}(-t)$$

so that $\quad \mathfrak{F}[t^n]\langle\phi\rangle = 2\pi j^n \phi^{(n)}(0) \quad$ and $\quad \mathfrak{F}[t^n] = 2\pi j^{-n}\delta^{(n)}$

### EXERCISE

Given that $\mathfrak{F}[|t|^{-\nu}] = 2\Gamma(1-\nu)\sin\frac{\nu\pi}{2}[|t|^{\nu-1}]$; $0 < \nu < 1$. $\mathfrak{F}[|t|^{-\nu}\operatorname{sgn} t] = 2j\Gamma(1-\nu)\cos\frac{\nu\pi}{2}[|t|^{\nu-1}\operatorname{sgn} t]$; $0 < \nu < 2$ the reader should verify that these formulas hold for all nonintegral values of $\nu$. (See Example 1.11 for the definitions involved here.)

The interested reader will find further material on Fourier transforms in Refs. 9 and 5; the latter reference also gives applications to partial differential equations.

## 1.17 Application to Differential Equations

In this section, "testing functions" will mean elements of $\mathfrak{D}$ and "distributions" will mean elements of $\mathfrak{D}^*$.

Consider a linear ordinary differential equation

$$x^{(n)} + a_1x^{(n-1)} + \cdots + a_nx = f \tag{1.19}$$

in which we assume the $a_i$, $i = 1, \ldots, n$, to be infinitely differentiable functions of $t$ and $f$ to be a locally integrable function of $t$. It is known from classical analysis that this differential equation possesses an infinity of solutions all of which are $n - 1$ times continuously differentiable, with $x^{(n-1)}$ absolutely continuous, so that $x^{(n)}$ exists and is locally integrable. It is also known that suitable initial conditions, for instance the values of $x, x', \ldots, x^{(n-1)}$ at a fixed point $t_0$, determine a unique solution.

We may now regard Eq. (1.19) as a differential equation in distributions. If $x$ is any distribution, $x^{(k)}$ is again a distribution; since $a_{n-k}$ is infinitely differentiable, $a_{n-k}x^{(k)}$ is defined so that $x$ may be substituted in the left-hand side of Eq. (1.19). In this sense, every function that satisfies the differential equation in the classical sense is also a distribution solution of that equation. In this context it is natural to ask whether Eq. (1.19) has distribution solutions that are not included among the classical solutions either because $x$, although a function, lacks the appropriate differentiability properties or because $x$ itself is a distribution. The answer to this question is an emphatic no: As long as $f$ is locally integrable, the classical solutions are the only distribution solutions of Eq. (1.19).

To prove this, consider Eq. (1.19) on a fixed closed finite interval. On this interval, the distribution $x^{(n)}$ is of a fixed order $r$; that is, it can be represented as the $r$th distribution derivative of some locally integrable function. If $r \geq 1$, then $x^{(n-1)}, x^{(n-2)}, \ldots$ are of order $r - 1$ at most, and so is

$$f - a_1x^{(n-1)} - \cdots - a_nx$$

But this last expression, being equal to $x^{(n)}$, is of order $r$; this contradiction shows that $r$ cannot be $\geq 1$ and hence must be 0. Thus, $x^{(n)}$ is of order zero on every finite interval and hence is a locally integrable function.

The situation is different if we now consider the same differential equation with a right-hand side $f$ that is itself a distribution. In this case, every solution is necessarily a distribution solution. If $f$ is of order $r$ on an interval, then the argument of the preceding paragraph shows that $x^{(n)}$ is also of order $r$; if $r \geq n$, then $x$ itself is of order $r - n$; and if $r < n$, then $x$ is $n - r - 1$ times continuously differentiable and $x^{(n-r-1)}$ is absolutely continuous so that $x^{(n-r)}$ is locally integrable. The difference of two solutions satisfies the homogeneous equation

$$x^{(n)} + a_1x^{(n-1)} + \cdots + a_nx = 0$$

and is a solution in the classical sense, so that the general distribution solution of Eq. (1.19) is the sum of a particular distribution solution and of the classical general solution of the homogeneous equation.

Such considerations are relevant with regard to Green's functions (see, for instance, Ref. 4, Chap. 3) that satisfy differential equations of the form

$$x^{(n)} + a_1x^{(n-1)} + \cdots + a_nx = \delta(t - \tau)$$

and also satisfy appropriate boundary conditions. Since $\delta(t - \tau)$ is of order zero and infinitely differentiable on any interval not including $\tau$, and $\delta(t - \tau)$ is of order 1 on any interval including that point, it follows that, under our assumptions on $a_1, \ldots, a_n$, the Green's function is infinitely differentiable except at $t = \tau$; at this point, it possesses $n - 2$ continuous derivatives while $x^{(n-1)}$ has a unit jump.

Similar statements hold for the differential equation

$$a_0x^{(n)} + a_1x^{(n-1)} + \cdots + a_nx = f$$

in which $a_0, \ldots, a_n$ are infinitely differentiable functions, as long as $a_0(t) \neq 0$. Zeros of $a_0(t)$ are singularities of the differential equation, and at such singularities a different behavior may arise.

As an example, we shall consider the differential equation of the first order (see Ref. 6, Sec. 8)

$$tx' + x = 0$$

The classical solution $x = ct^{-1}$, where $c$ is an arbitrary constant, exists on $(0,\infty)$ and on $(-\infty,0)$ but fails to be integrable in any neighborhood of $t = 0$. The distribution suggested by the classical solution is $c[\log t]'$, where the prime indicates a distribution derivative. To prove that this is indeed a solution, we evaluate

$$\begin{aligned}(t[\log t]'')\langle\phi\rangle + [\log t]'\langle\phi\rangle &= [\log t]''\langle t\phi\rangle + [\log t]'\langle\phi\rangle \\ &= [\log t]\langle(t\phi)'' - \phi'\rangle = [\log t]\langle t\phi'' + \phi'\rangle \\ &= [\log t]\langle(t\phi')'\rangle\end{aligned}$$

and obtain, through integration by parts,

$$\int_{-\infty}^{\infty} \log t(t\phi')'\,dt = -\int_{-\infty}^{\infty} \phi'\,dt = 0$$

So far, the only complication caused by the singularity was the necessity of replacing $t^{-1}$ by a distribution that, although equal to $t^{-1}$ for all $t \neq 0$, is not generated by it. But more is to come. The integrated form of the differential equation is $(tx)' = 0$, or $tx = c$, and it can be verified that $c[\log t]'$ satisfies this equation. However, the solution of the equation is not unique, for

$$t\delta\langle\phi\rangle = \delta\langle t\phi\rangle = 0$$

for all testing functions $\phi$, showing that $t\delta = 0$. Thus, $c[\log t]' + a\delta(t)$ satisfies $tx' + x = 0$ for arbitrary values of $a$ and $c$, so that this differential equation of the first order possesses a two-parameter family of distribution solutions.

Distributions are even more important in the theory of partial differential equations, but this topic is not within the scope of the present consideration.

## 1.18 Extensions and Alternative Theories

Different classes of generalized functions may be obtained by using different classes of testing functions. For instance, by using infinitely differentiable testing functions of a fixed period, one obtains generalizations of periodic integrable functions; by using $k$ times differentiable testing functions, one obtains $k$ times (rather than infinitely) differentiable distribution functions, etc. Distributions have also been defined on finite intervals, on arbitrary regions of $n$-dimensional space, on surfaces, and more generally on manifolds. For all these variants and extensions, the reader may consult Refs. 21 and 5. Vector-valued distributions and distributions in more abstract situations have also been studied, but these lie outside the scope of the present introduction. We mention only Ref. 15, which envisages applications to quantum mechanics.

In this chapter, we have chosen to follow Schwartz in presenting distributions as functionals on classes of testing functions. There are several alternative theories more or less equivalent to the one outlined here. The largest single group of these is inspired by the theorem that states that every distribution is the distribution limit of some sequence of functions (see Sec. 1.15) and indicates that it might be possible to construct distributions as generalized limits of functions.

Temple[23] attributes a very general method of defining generalized functions as "weak limits" of ordinary functions to Mikusiński.[11] This method was taken up by Ravetz;[17] from the point of view of applied mathematics, it was treated by Temple[24,25] and Saltzer;[18] and from the point of view of Fourier analysis, it was discussed by Lighthill.[9] A somewhat more direct approach is inspired by a combination of approximation to distributions through functions and the identification of distributions on finite intervals with generalized derivatives of functions; this was the viewpoint adopted by Sikorski[22] and Korevaar.[8] The latter author defines distributions by sequences of integrable functions that are convergent in the generalized sense that on any given finite interval the sequence may be made uniformly convergent through a suitable number of integrations, the number of integrations depending on the interval.

A rather different approach is based on the identification of distributions, on finite intervals, with generalized derivatives of locally integrable functions. Here distributions are represented by formal series

$$\sum_{n=0}^{\infty} D^n f_n(t)$$

in which $D^n$ is the symbol of generalized differentiation of order $n$ and the $f_n$ are locally integrable functions of which all but a finite number vanish identically on any given finite interval. This approach was briefly indicated by Halperin[6] and elaborated by König.[7] It was adopted by Sauer,[19] who used it in the solution of boundary-value problems.

Some of the major difficulties encountered in studying the mathematical theory of distributions are due to the circumstance that no norm exists for distributions, since none exists for the vector space of testing functions. There are several attempts at overcoming this difficulty, if necessary by restricting attention to a smaller class of generalized functions, and by basing the theory of distributions on the comparatively well-known and accessible theory of Banach spaces. We mention in this connection the work of E. R. Love[10] and an unpublished theory of M. Riesz.

## REFERENCES


1. Bochner, S., "Vorlesungen über Fouriersche Integrale," Akademie-Verlag G.m.b.H., Berlin, 1932.
2. Dirac, P. A. M., "The Principles of Quantum Mechanics," 2d ed., Oxford University Press, New York, 1935.
3. Erdélyi, A., "Operational Calculus," Mathematics Department, California Institute of Technology, Pasadena, Calif., 1955.
4. Friedman, B., "Principles and Techniques of Applied Mathematics," John Wiley & Sons, Inc., New York, 1956. "Operational Calculus and Generalized Functions," California Institute of Technology, Pasadena, Calif., 1959.
5. Gelfand, I. M., and G. E. Šilov, Fourier Transforms of Rapidly Increasing Functions and Questions of Uniqueness of the Solution of Cauchy's Problem, *Uspehi Mat. Nauk*, n.s., vol. 8, no. 6 (58), pp. 3–54, 1953; *Amer. Math. Soc. Transl.*, ser. 2, vol. 5, pp. 221–274, 1957.
6. Halperin, I., "Introduction to the Theory of Distributions," University of Toronto Press, Toronto, Canada, 1925. Based on lectures by Laurent Schwartz.
7. König, H., Neue Begründung der Theorie der "Distributionen" von L. Schwartz, *Math. Nachr.*, vol. 9, pp. 129–148, 1953.
8. Korevaar, J., Distributions Defined from the Point of View of Applied Mathematics, *Nederl. Akad. Wetensch. Proc.*, ser. A, vol. 58, pp. 368–389, 483–503, 663–674, 1955.
9. Lighthill, M. J., "Introduction to Fourier Analysis and Generalized Functions," Cambridge University Press, New York, 1958.
10. Love, E. R., A Banach Space of Distributions, *J. London Math. Soc.*, vol. 32, pp. 483–498, 1957; vol. 33, pp. 288–306, 1958.
11. Mikusiński, J. G., Sur la méthode de généralisation de M. Laurent Schwartz et sur la convergence faible, *Fund. Math.*, vol. 35, pp. 235–239, 1948.
12. ———, "Rachunek Operatorów" [The Calculus of Operators], Monografie Matematyczne, Tom XXX, Polskie Towarzystwo Matematycne, Warszawa, 1953. "Operational Calculus," Pergamon Press, Inc., New York, 1959.
13. ———, Le Calcul opérationnel d'intervalle fini, *Studia Math.*, vol. 15, pp. 225–251, 1956.
14. ——— and C. Ryll-Nardzewski, Un théorème sur le produit de composition des fonctions de plusieurs variables, *Studia Math.*, vol. 13, pp. 62–68, 1953.

15. Nikodým, O. M., Summation of Quasi-vectors on Boolean Tribes and Its Applications to Quantum Theories. I. Mathematically Precise Theory of the Genuine P. A. M. Dirac's Delta Function, *Rend. Sem. Mat., Univ. Padova* (in preparation).
16. Pol, Balth. van der, and H. Bremmer, "Operational Calculus, Based on the Two-sided Laplace Integral," Cambridge University Press, New York, 1950.
17. Ravetz, J. R., Distributions Defined as Limits, *Proc. Cambridge Phil. Soc.*, vol. 53, pp. 76–92, 1957.
18. Saltzer, C., The Theory of Distributions, *Advances in Appl. Mech.*, vol. 5, pp. 91–110, 1958.
19. Sauer, R., "Anfangswertprobleme bei partiellen Differentialgleichungen," 2d ed., Springer-Verlag OHG, Berlin, 1958.
20. Schmieden, C., and D. Laugwitz, Eine Erweiterung des Infinitesimalkalküls, *Math. Z.*, vol. 69, pp. 1–39, 1958.
21. Schwartz, L., "Théorie des distributious," 2 vols., Hermann & Cie, Paris, 1950, 1951.
22. Sikorski, R., A Definition of the Notion of Distribution, *Bull. Acad. Polon. Sci., Cl. III*, vol. 2, pp. 209–211, 1954.
23. Temple, G., Theories and Applications of Generalized Functions, *J. London Math. Soc.*, vol. 28, pp. 134–148, 1953.
24. ———, La Théorie de la convergence généralisée et les fonctions généralisées et leur application á la physique mathématique, *Rend. Mat. e Appl.*, ser. 5, vol. 11, pp. 113–122, 1953.
25. ———, The Theory of Generalized Functions, *Proc. Roy. Soc. London*, ser. A, vol. 228, pp. 175–190, 1955.
26. Weston, J. D., An Extension of the Laplace-transform Calculus, *Rend. Circ. Mat. Palermo*, ser. 2, vol. 6, pp. 1–9, 1957.
27. ———, Operational Calculus and Generalized Functions, *Proc. Roy. Soc. London*, ser. A, vol. 250, pp. 460–471, 1959.

# 2

# Operational Methods for Separable Differential Equations

**BERNARD FRIEDMAN**
PROFESSOR OF MATHEMATICS
UNIVERSITY OF CALIFORNIA, BERKELEY

## 2.1 Introduction

From the earliest development of calculus, mathematicians have attempted to apply algebraic manipulations to differential operators. They have tried to add, multiply, and divide operators; to expand them in infinite series; and, in short, to treat them as if they were algebraic quantities. Leibnitz[7] himself, when he discovered differential calculus in 1695, was inspired enough to introduce the suggestive notation $Dy$ for the first derivative, $D^2y$ for the second derivative, and so on. Because of the algebraic properties of the operator $D$, it was easy to derive the formula for the $n$th derivative of a product of two functions $u$ and $v$. The procedure might have been as follows:

The well-known formula for the derivative of $uv$ is

$$D(uv) = uDv + vDu$$

We introduce two operators $D_1$ and $D_2$, where $D_1$ operates only on $u$ and $D_2$ operates only on $v$; then this formula may be written as

$$D(uv) = D_1uv + D_2uv = (D_1 + D_2)(uv)$$

Consequently, we have

$$D = D_1 + D_2$$

and naturally

$$D^n = (D_1 + D_2)^n = D_1{}^n + \binom{n}{1} D_1{}^{n-1}D_2 + \cdots + D_2{}^n$$

By applying this result to the product $uv$, we get

$$D^n uv = D_1^n uv + \binom{n}{1} D_1^{n-1} D_2 uv + \cdots + D_2^n uv$$
$$= v(D^n u) + n(D^{n-1}u)(Dv) + \cdots + (D^n v)$$

In 1859, George Boole[1] published a book on differential equations containing a calculus of operators. Some parts of this calculus are still taught in elementary courses in differential equations. With the methods he developed, it is possible to find particular solutions of ordinary differential equations. For example, consider the equation

$$u'' + k^2 u = x^2 \tag{2.1}$$

We write $D$ for the differentiation operator, and then this equation can be represented as follows:

$$(D^2 + k^2)u = x^2$$

If we may treat $D$ as an algebraic quantity, the solution of the foregoing equation is

$$u = \frac{1}{D^2 + k^2} x^2 \tag{2.2}$$

To interpret this result, we use ordinary algebraic division to obtain

$$\frac{1}{D^2 + k^2} = \frac{1}{k^2} - \frac{D^2}{k^4} + \frac{D^4}{k^4(k^2 + D^2)}$$

By substituting in Eq. (2.2), we get

$$u = \frac{x^2}{k^2} - \frac{D^2 x^2}{k^4} + \frac{1}{k^4(k^2 + D^2)} D^4 x^2 = \frac{x^2}{k^2} - \frac{2}{k^4} + 0$$

since $D^4 x^2 = 0$. It is easy to verify that the polynomial $(x/k)^2 - 2k^{-4}$ is indeed a solution of Eq. (2.1).

Such remarkable successes encouraged further attempts at developing a calculus of operators. If $D$ represents differentiation, then it was natural to let $D^{-1}$ represent integration, because with that convention we have

$$DD^{-1}u(x) = D \int^x u(\xi)\, d\xi = u(x) \tag{2.3}$$

However, when we consider that

$$D^{-1}Du(x) = D^{-1}u'(x) = \int^x u'(\xi)\, d\xi = u(x) + c \tag{2.4}$$

where $c$ is an arbitrary constant of integration, we run into difficulties (see Sec. 1.5). One difficulty is that from Eq. (2.3) we should write $DD^{-1} = 1$, but from Eq. (2.4) we see that $D^{-1}D \neq 1$. Another difficulty is that we obtain particular solutions and not the complete solution of the equation. Boole of course recognized these difficulties and introduced notation such as $D^{-1}0$ to represent the constant of integration. This notation is unsatisfactory because it contradicts the well-known algebraic fact that, when any quantity is multiplied by zero, the product is zero. These difficulties, among others, caused Boole's treatment to be neglected.

## 2.2 Heaviside Theory

The theory of operators was completely revitalized by O. Heaviside[6] in a series of papers in the 1890s. His first contribution was to interpret $D^{-1}$ not as an indefinite integral, but as a definite integral, so that for him $D^{-1}u$ meant

$$\int_0^t u(\tau)\, d\tau$$

With this interpretation, we still have

$$DD^{-1}u = D\int_0^t u(\tau)\, d\tau = u$$

but now

$$D^{-1}Du = D^{-1}u' = \int_0^t u'(\tau)\, d\tau = u(t) - u(0) = u(t)$$

if $u(0) = 0$. Therefore, if $u(0) = 0$, we may write

$$DD^{-1} = D^{-1}D = 1$$

By using this definition of $D^{-1}$, Heaviside showed that algebraic methods could be used to find solutions of differential equations satisfying given initial conditions.

We shall illustrate Heaviside's algebraic approach. Let us follow Heaviside's notation by using $p$ to represent differentiation and $p^{-1}$ to represent the definite integral from 0 to $t$. From the usual rule for differentiating a product, we see that

$$pe^{at}f(t) = \frac{d}{dt}[e^{at}f(t)] = e^{at}\frac{d}{dt}f(t) + ae^{at}f(t)$$
$$= e^{at}(p + a)f(t)$$

or

$$e^{-at}pe^{at}f(t) = (p + a)f(t)$$

If we omit the function $f(t)$, this result may be written in the following operational form:

$$e^{-at}pe^{at} = p + a \tag{2.5}$$

Notice that, by using Eq. (2.5) repeatedly, we get

$$(p + a)^2 = (e^{-at}pe^{at})^2 = e^{-at}pe^{at}e^{-at}pe^{at} = e^{-at}p^2e^{at}$$

and

$$(p + a)^n = e^{-at}p^ne^{at}$$

It is easy to generalize this result to obtain

$$\phi(p + a) = e^{-at}\phi(p)e^{at}$$

if $\phi(p)$ is an analytic function of $p$. In particular,

$$\frac{1}{p + a} = e^{-at}\frac{1}{p}e^{at} \tag{2.6}$$

We shall use Eq. (2.6) to find the solution of the differential equation

$$u' + au = e^{-t} \tag{2.7}$$

that vanishes when $t = 0$. We write the equation as

$$(p + a)u = e^{-t}$$

and then, by Eq. (2.6),

$$\begin{aligned} u &= \frac{1}{p + a}e^{-t} = e^{-at}\frac{1}{p}e^{at}e^{-t} = e^{-at}\int_0^t e^{(a-1)\tau}\,d\tau \\ &= e^{-at}\frac{e^{(a-1)t} - 1}{a - 1} = \frac{e^{-t} - e^{-at}}{a - 1} \end{aligned}$$

The verification that this is the correct solution of the differential equation will be left to the reader.

Heaviside's successes in the use of his techniques were remarkable. Many objections, however, were advanced against his methods, especially by professional mathematicians. One objection was that he did not have a systematic way of evaluating functions of the operator $p$, such as $e^{-\sqrt{p}}/\sqrt{p}$, but instead treated each case individually by means of some trick adapted to it. Another and more serious objection was that his methods were frankly experimental. If the methods achieved the desired results, he continued to use them; if they did not work, he modified them until they did. This meant that he could not really explain the principles underlying his methods. Heaviside's famous answer to this argument was that he didn't stop eating just because he didn't understand the process of digestion (see Chap. 3). The rebuttal, however, is that even though we do not stop eating, we try to learn how to avoid eating anything that may upset our digestion, and some of Heaviside's methods are still indigestible.

The first objection to Heaviside's methods, namely, that there is no systematic method for evaluating a function of an operator, was overcome when Bromwich,[2] in 1919, introduced the contour integral that inverts the integral now known as the Laplace transform. By this integral, to any given function $F(p)$ of the operator $p$ there corresponds a function $f(t)$ defined by the contour integral

$$f(t) = \frac{1}{2\pi j}\int F(p)e^{pt}\,dp \tag{2.8}$$

For example, the operational solution of

$$u'' + k^2 u = 1 \tag{2.9}$$

is

$$u = \frac{1}{p^2 + k^2}$$

and by Eq. (2.8) this corresponds to

$$u = \frac{1}{2\pi j} \int \frac{e^{pt}}{p^2 + k^2}\, dp = \frac{1 - \cos kt}{k^2}$$

Thus we obtain the solution $u$ of Eq. (2.9) that satisfies the conditions $u(0) = u'(0) = 0$.

One final step was necessary before a rigorous theory of some of Heaviside's work could be obtained. The need for such a step may be seen by referring to Eq. (2.7). The operational solution of Eq. (2.7) is

$$u = \frac{1}{p + a}\, e^{-t} \tag{2.10}$$

Bromwich's integral cannot be used here because the right-hand side of Eq. (2.10) is not solely a function of $p$ but is a function of $p$ acting on a function of $t$. Wagner,[10] in 1916, and Carson,[3] in 1922, however, introduced the Laplace transform that established a correspondence between a function $f(t)$ and a function $F(p)$ as follows:

$$F(p) = \int_0^\infty e^{-pt} f(t)\, dt$$

For Eq. (2.7), let $f(t) = e^{-t}$, so that $F(p) = 1/(p + 1)$; the solution of Eq. (2.7) can then be obtained from the operational form $(p + a)^{-1}(p + 1)^{-1}$ by means of Bromwich's integral. We find

$$u = \frac{1}{2\pi j} \int \frac{e^{pt}\, dp}{(p + a)(p + 1)} = \frac{e^{-t} - e^{-at}}{a - 1}$$

In practical applications, the operational technique is not used. Instead, the Laplace transform of Eq. (2.7) is taken. If we let $\bar{u}$ denote the Laplace transform of $u$, we get immediately

$$\bar{u}(p + a) = (p + 1)^{-1}$$

and therefore

$$\bar{u} = \frac{1}{(p + a)(p + 1)}$$

By using Bromwich's integral or a table of Laplace transforms, we again obtain the result

$$u = \frac{e^{-t} - e^{-at}}{a - 1}$$

The work of Bromwich, Carson, and Wagner thus provided a justification for a large part of Heaviside's work, but there still are parts that cannot be explained. It should be noticed that the Laplace transform and the Heaviside calculus can be applied principally to differential equations containing time derivatives and not to differential equations containing space derivatives. The basic reason for this is that Heaviside's method was designed for differential equations involving initial conditions and not for those involving boundary conditions.

## 2.3 Domain of an Operator

It is the purpose of this chapter to develop an operational calculus of the Heaviside type for problems involving space derivatives and boundary conditions.[5] For this purpose, it is necessary to look very carefully at the concept of an operator. Notice, first, that just as a pianist must have a piano on which to play, an operator must have something on which to operate. The set of functions on which the operator acts will be called the *domain* of the operator.

For illustrative purposes, let us consider the operator $L = -d^2/dx^2$. What kind of functions $u(x)$ could be in its domain? Clearly, the functions $u(x)$ must have at least a second derivative. For mathematical convenience, it is useful to restrict the domain of $L$ further by requiring that

$$\int [u(x)]^2 \, dx$$

be finite, that $u''(x)$ exist, and that

$$\int [u''(x)]^2 \, dx$$

be finite.

Consider the differential equation

$$u'' + k^2 u = f(x) \tag{2.11}$$

We may write it as

$$(L + k^2)u = f(x)$$

and then we are tempted to divide by $L + k^2$ to get

$$u = \frac{1}{L + k^2} f$$

But we know that the solution of Eq. (2.11) is not unique. To make it unique, we may specify that $u(x)$ satisfy two conditions—e.g., initial conditions such as $u(0) = 1$ and $u'(0) = 2$, or boundary conditions such as $u(0) = 1$ and $u(1) = 2$. Because our purpose is to discuss operators with boundary conditions, we restrict the domain of $L$ still further by requiring that $u(x)$ in the domain of $L$ must satisfy a sufficient number of boundary conditions to make the solution of Eq. (2.11) unique. There is another restriction that arises from the further requirement that both $L$ and $L^{-1}$ acting on the function zero should give zero. To satisfy this requirement, we must confine our discussion to homogeneous boundary conditions only. Later, we shall show how nonhomogeneous boundary conditions may be included.

Let us consider an arbitrary differential operator $L$ defined over the interval $(a,b)$. For mathematical convenience, we shall assume that the operator acts only on a collection $\mathfrak{F}$ of real-valued functions $u(x)$, defined for $a \leq x \leq b$, that are such that $Lu$ exists and the integrals

$$\int_a^b u^2\,dx \qquad \int_a^b (Lu)^2\,dx$$

are finite. The domain of the operator will be those functions of $\mathfrak{F}$ that satisfy a pair of homogeneous boundary conditions. The following are examples of such boundary conditions:

$$(a) \qquad u(0) = u(1) = 0$$
$$(b) \qquad u'(0) = u'(1) = 0$$
$$(c) \qquad u(0) + \alpha_1 u'(0) = u(1) + \alpha_2 u'(1) = 0$$

where $\alpha_1$ and $\alpha_2$ are arbitrary scalars.

## 2.4 Linear Operators

The domains referred to above have the very important property of being *linear vector spaces*.[8] A linear vector space of functions is defined as a collection of functions such that, if $u_1(x)$ and $u_2(x)$ are any two functions in the collection, then any linear combination $\alpha u_1(x) + \beta u_2(x)$, where $\alpha$ and $\beta$ are scalars, is also in the collection. If we had used nonhomogeneous boundary conditions, the domains would not have been linear vector spaces. We shall see later that the use of this concept of linear vector space enables us to give a more intuitive and geometric description of the properties of the operator.

We shall also restrict our attention to the consideration of linear differential operators,[4] i.e., operators $L$ that act on functions $u(x)$ in such a way that

$$L[\alpha u(x)] = \alpha L u(x) \qquad \text{and} \qquad L[u_1(x) + u_2(x)] = Lu_1(x) + Lu_2(x)$$

Examples of linear differential operators are $d/dx$, $-d^2/dx^2$, and

$$p(x)\frac{d^2}{dx^2} + q(x)\frac{d}{dx} + r(x)$$

When this last operator acts on a function $u(x)$, the result is

$$p(x)u''(x) + q(x)u'(x) + r(x)u(x)$$

Note that an operator such as $Au$, defined by

$$Au = u\frac{du}{dx}$$

is *not* a linear operator, because

$$A(u_1 + u_2) = (u_1 + u_2)\frac{d(u_1 + u_2)}{dx} \neq Au_1 + Au_2$$

Later, we shall also consider partial differential operators, e.g.,

$$\Delta w(x,y) = \frac{\partial^2 w}{\partial x^2} + \frac{\partial^2 w}{\partial y^2}$$

To summarize, hereafter when using the term "operator" we shall mean a linear differential operator, usually of the second order, together with a linear vector space as its domain. Both the differential operator and the domain are required in the definition of the operator. For us, the same differential operators but with different domains will be considered as different operators. For example, the second derivative on a domain of functions $u(x)$ such that $u(0) = u(1) = 0$ is not the same as the second derivative on a domain of functions $u(x)$ such that $u'(0) = u'(1) = 0$. The failure to distinguish between two such operators has caused a great deal of confusion and even error in the applications of the calculus of operators.

## 2.5 Functions of Operators

Now that we have a suitable definition of operator, our next aim will be to define functions of an operator. If $L$ represents an operator as defined above, what will be meant by operators such as $L^2$ or $e^{-\alpha L}$ or $\sin\sqrt{k^2 - L}$? We proceed first to define powers of $L$. Suppose that $u$ is in the domain of $L$; then $v = Lu$ exists. If $v$ is also in the domain of $L$, then $Lv = w$ exists. We put

$$w = Lv = L^2u$$

Thus we see that, if $u$ is such that $u$ and $Lu$ are in the domain of $L$, then $L^2u$ is defined as the result of the repeated application of the operator $L$ on $u$. For example, if $L = -d^2/dx^2$ with the boundary condition $u(0) = u(1) = 0$, then we can define $L^2u$ for the function $u_1 = \sin \pi x$ but not for the function $u_2 = x(1 - x)$. Note that both $u_1$ and $u_2$, and also $Lu_1 = \pi^2 \sin \pi x$, are in the domain of $L$, but $Lu_2 = 2$ is not; consequently $L^2u_1 = \pi^4 \sin \pi x$, whereas $L^2u_2$ is not defined.

It is clear that this process may be generalized to higher powers of $L$. If $u, Lu, L^2u, \ldots, L^{n-1}$ are all in the domain of $L$, then $L^nu$ is defined as follows:

$$L^nu = L(L^{n-1}u)$$

Naturally, then, if $p(L)$ is a polynomial in $L$ of the $n$th degree, say

$$p(L) = a_0 + a_1L + \cdots + a_nL^n$$

then $$p(L)u = a_0u + a_1Lu + \cdots + a_nL^nu$$

We may go even further and define analytic functions of $L$. Suppose $\phi(\lambda)$ is an analytic function of $\lambda$ having a convergent power-series expansion

$$\phi(\lambda) = a_0 + a_1\lambda + a_2\lambda^2 + \cdots$$

Then if $L^nu$ is in the domain of $u$ for all values of $n$, we put

$$\phi(L)u = a_0u + a_1Lu + a_2L^2u + \cdots$$

provided that the resulting series of functions converges.

## 2.6 Eigenfunctions and Self-adjoint Operators

The foregoing method of defining functions of the operator, although natural, is nevertheless not the most convenient. We shall see that a better way to define the functions of the operator is to begin with the concept of an *eigenfunction* of the operator. An eigenfunction of an operator $L$ is a function $\phi(x)$, not identically zero, in the domain of $L$ such that $L$ acting on $\phi(x)$ leaves the function unchanged except for multiplication by a scalar $\lambda$,

$$L\phi = \lambda\phi$$

The scalar $\lambda$ is called an *eigenvalue* of $L$; cf. Ref. 9 for the analogous concept in matrix theory. For example, if $L_1$ is $-d^2/dx^2$ on the domain of functions $u(x)$ such that $u(0) = u(1) = 0$, then the functions

$$\phi_n(x) = \alpha_n \sin n\pi x$$

where $n = 1, 2, 3, \ldots$ and the $\alpha_n$ are constants $\neq 0$, are eigenfunctions of $L_1$ with corresponding eigenvalues $n^2\pi^2$, because

$$\frac{-d^2}{dx^2}\phi_n(x) = n^2\pi^2\phi_n(x)$$

In many cases, the eigenfunctions are *mutually orthogonal;* i.e., if $\phi_n(x)$ and $\phi_m(x)$ are eigenfunctions of $L$, then

$$\int \phi_n(x)\phi_m(x)\,dx = 0$$

if $n \neq m$. For example,

$$\int_0^1 \sin n\pi x \sin m\pi x\,dx = \int_0^1 \frac{\cos(n-m)\pi x - \cos(n+m)\pi x}{2}\,dx$$

$$= \frac{\sin(n-m)\pi x}{2(n-m)\pi}\bigg|_0^1 - \frac{\sin(n+m)\pi x}{2(n+m)\pi}\bigg|_0^1 = 0$$

We shall give a characterization of the class of operators that have mutually orthogonal eigenfunctions. An operator $L$ defined over the interval $(a,b)$ is called *self-adjoint* if, for all functions $u(x)$ and $v(x)$ in the domain of $L$, we have

$$\int_a^b u(x)Lv(x)\,dx = \int_a^b v(x)Lu(x)\,dx \tag{2.12}$$

Notice that this is a sort of generalized integration by parts. For example, the operator $L_1$ we have considered previously is self-adjoint because, by using integration by parts twice, we get

$$\int_0^1 u(-v'')\,dx = -(uv' - vu')\Big|_0^1 + \int_0^1 v(-u'')\,dx$$

If $u(x)$ and $v(x)$ are in the domain of $L_1$, then

$$u(0) = u(1) = v(0) = v(1) = 0$$

and therefore
$$\int_0^1 u(-v'')\,dx = \int_0^1 v(-u'')\,dx$$

which shows that Eq. (2.12) is satisfied when $L = L_1$.

We now prove that, for a self-adjoint operator, eigenfunctions corresponding to different eigenvalues are mutually orthogonal:

THEOREM. *Suppose that $\phi_n(x)$ is an eigenfunction of a self-adjoint operator $L$ with eigenvalue $\lambda_n$ and that $\phi_m(x)$ is also an eigenfunction with eigenvalue $\lambda_m$; then*

$$\int_a^b \phi_n(x)\phi_m(x)\,dx = 0$$

*provided that $\lambda_n \neq \lambda_m$.*

*Proof.* From the definition of eigenfunctions we have

$$L\phi_n = \lambda_n\phi_n \qquad L\phi_m = \lambda_m\phi_m$$

If we multiply the first equation by $\phi_m$ and the second equation by $\phi_n$, subtract, and then integrate, we get

$$\int_a^b (\phi_m L\phi_n - \phi_n L\phi_m)\,dx = (\lambda_n - \lambda_m)\int_a^b \phi_n\phi_m\,dx$$

Since $L$ is self-adjoint, Eq. (2.12) shows that the left-hand side of the above equation is zero; therefore, since $\lambda_n \neq \lambda_m$, we have

$$\int_a^b \phi_n(x)\phi_m(x)\,dx = 0 \tag{2.13}$$

as desired.

Since $\alpha_n\phi_n(x)$, where $\alpha_n$ is a constant $\neq 0$, is an eigenfunction if $\phi_n(x)$ is, we can always *normalize* the eigenfunctions, i.e., so choose the constant multiplying factor $\alpha_n$ that

$$\int_a^b (\alpha_n\phi_n)^2\,dx = 1$$

For example, for the eigenfunctions of $L_1$ we have

$$\int_0^1 (\alpha_n \sin n\pi x)^2\, dx = \alpha_n^2 \int_0^1 \sin^2 n\pi x\, dx = \tfrac{1}{2}\alpha_n^2 = 1$$

and therefore $\alpha_n = \sqrt{2}$; consequently, if we take the eigenfunctions of $L_1$ as $\phi_n = \sqrt{2} \sin n\pi x$, they will be normalized.

Hereafter, we shall assume that the eigenfunctions are both orthogonal and normalized, so that both Eq. (2.13) and the following equation hold:

$$\int_a^b \phi_n^2\, dx = 1 \qquad n = 1, 2, 3, \ldots \tag{2.14}$$

The set of eigenfunctions $\phi_1(x)$, $\phi_2(x)$, . . . will thus correspond exactly to a set of mutually perpendicular unit vectors.

## 2.7 Spectral Representation

Self-adjoint operators are important because in many cases the eigenfunctions of such an operator form a *basis* for the space of functions under consideration; i.e., any function of the space can be written as a sum, perhaps with an infinite number of terms, of linear multiples of the eigenfunctions. For example, we have seen that for the operator $L_1$ the normalized eigenfunctions are $\sqrt{2} \sin n\pi x$, where $n = 1, 2, 3, \ldots$. Let $u(x)$ be any function such that

$$\int_0^1 u^2\, dx < \infty$$

Then there exists an expansion

$$u(x) = \sqrt{2}\,(c_1 \sin \pi x + c_2 \sin 2\pi x + \cdots) \tag{2.15}$$

This equation is to be understood in the following sense: The partial sums of the series on the right approximate the functions on the left in the mean-square sense, so that

$$\lim_{n\to\infty} \int_0^1 \left[u(x) - \sqrt{2} \sum_{k=1}^{n} c_k \sin \pi k x\right]^2 dx = 0$$

Because the eigenfunctions are orthogonal and normalized, it is easy to find an explicit formula for $c_n$ in Eq. (2.15). Multiply this equation by the normalized eigenfunction $\sqrt{2} \sin n\pi x$ and integrate. We find

$$\int_0^1 \sqrt{2}\, u(x) \sin n\pi x\, dx = 2 \sum_{k=1}^{\infty} c_k \int_0^1 \sin k\pi x \sin n\pi x\, dx = c_n \tag{2.16}$$

by the use of Eqs. (2.13) and (2.14) in this special case. Note that Eq. (2.15) is just the Fourier sine series for the function $u(x)$ and Eq. (2.16) is the formula for the $n$th coefficient.

For a general self-adjoint operator $L$, the results will be very much the same as they are for $L_1$. Suppose that $\phi_1(x)$, $\phi_2(x)$, . . . are the normalized eigenfunctions for $L$ corresponding to the eigenvalues $\lambda_1, \lambda_2, \ldots$, respectively. If $u(x)$ is a function in the domain of $L$, then $u(x)$ can be expanded, in the mean-square sense, as follows:

$$u(x) = c_1\phi_1(x) + c_2\phi_2(x) + c_3\phi_3(x) + \cdots$$

where, because of Eqs. (2.13) and (2.14),

$$c_n = \int_a^b u(x)\phi_n(x)\,dx$$

Furthermore, by the definition of an eigenfunction, we have

$$Lu(x) = c_1\lambda_1\phi_1(x) + c_2\lambda_2\phi_2(x) + c_3\lambda_3\phi_3(x) + \cdots \tag{2.17}$$

A formula such as (2.17), which uses an eigenfunction expansion to represent the action of an operator $L$ on an arbitrary function in its domain, will be called the *spectral representation* of the operator $L$.

Once we have the spectral representation (2.17), it is easy to define functions of $L$. For example,

$$\begin{aligned} L^2u = L(Lu) &= L(c_1\lambda_1\phi_1 + c_2\lambda_2\phi_2 + \cdots) \\ &= c_1\lambda_1^2\phi_1 + c_2\lambda_2^2\phi_2 + \cdots \end{aligned}$$

Similarly,

$$L^nu = c_1\lambda_1^n\phi_1 + c_2\lambda_2^n\phi_2 + \cdots$$

and if

$$p(\lambda) = a_0 + a_1\lambda + \cdots + a_n\lambda^n$$

then

$$p(L)u = c_1p(\lambda_1)\phi_1 + c_2p(\lambda_2)\phi_2 + \cdots$$

We make a natural extension of this formula. If $q(\lambda)$ is a function of $\lambda$ that is defined for $\lambda = \lambda_1, \lambda_2, \ldots$, then we define $q(L)$ as follows:

$$q(L)u = c_1q(\lambda_1)\phi_1 + c_2q(\lambda_2)\phi_2 + \cdots \tag{2.18}$$

Of course, it is understood that this definition applies only when the right-hand side converges in the mean-square sense.

As an illustration, consider again the operator $L_1$. What is to be meant by the operator

$$\frac{e^{-\alpha L_1}}{L_1}$$

where $\alpha$ is a positive scalar? Since the eigenfunctions of $L_1$ are $\sqrt{2}\sin \pi nx$ and correspond to the eigenvalues $n^2\pi^2$, $n = 1, 2, \ldots$, we find from Eq. (2.18) that

$$\frac{e^{-\alpha L_1}}{L_1}u = \sqrt{2}\sum_{k=1}^{\infty} \frac{e^{-\alpha n^2\pi^2}}{n^2\pi^2} c_n \sin n\pi x$$

where

$$c_n = \sqrt{2}\int_0^1 u(x)\sin n\pi x\,dx$$

## 2.8 A Partial Differential Equation

We shall now use these ideas concerning spectral representations and functions of operators to solve the following partial differential equation, which is the mathematical formulation of the problem of determining the equilibrium distribution of temperature $T(x,y)$ in a rectangular plate of length 1 and width $h$ if three sides are kept at temperature zero but the fourth side has a prescribed temperature distribution $f(x)$:

Find a function $T(x,y)$ such that

$$\frac{\partial^2 T}{\partial x^2} + \frac{\partial^2 T}{\partial y^2} = 0 \qquad 0 \leq x \leq 1 \quad 0 \leq y \leq h \tag{2.19}$$

and $T(0,y) = T(1,y) = T(x,h) = 0$ but $T(x,0) = f(x)$.

Equation (2.19) may be represented in the following operational form:

$$(L_1 + L_2)T = 0$$

where $L_1 = -\partial^2/\partial x^2$ on the domain of functions $u(x,y)$ that as functions of $x$ satisfy the boundary conditions of vanishing for both $x = 0$ and $x = 1$, and where $L_2 = -\partial^2/\partial y^2$ on the domain of functions $u(x,y)$ that as functions of $y$ satisfy a zero boundary condition for $y = h$ and a given nonzero boundary condition at $y = 0$. Notice that $L_2$ is not a linear operator, because the boundary condition for $y = 0$ is not homogeneous. This suggests writing Eq. (2.19) as follows:

$$\left(\frac{\partial^2}{\partial y^2} - L_1\right) T = 0 \tag{2.20}$$

where $T = f(x)$ for $y = 0$ and $T = 0$ for $y = h$.

Since in Eq. (2.20) the operator $L_1$ has no effect on the $y$ dependence of $T$ but acts on $T$ only as a function of $x$ and has its boundary conditions dependent only on the values of $x$, we may treat $L_1$ in Eq. (2.20) as if it were a constant independent of $y$. If $L_1$ were a constant, however, the solution of Eq. (2.20) would be

$$T = Ae^{ky} + Be^{-ky}$$

where $k = L_1^{1/2}$ and where $A$ and $B$ are quantities independent of $y$. The boundary conditions at $y = 0$ and $y = h$ imply that

$$f(x) = A + B$$
$$0 = Ae^{kh} + Be^{-kh}$$

When these equations are solved for $A$ and $B$ and the expression for $T$ is rearranged slightly, we find that

$$T = \frac{\sinh L_1^{1/2}(h - y)}{\sinh L_1^{1/2}h} f(x) \tag{2.21}$$

is the solution of Eq. (2.20) satisfying the prescribed boundary conditions.

In Eq. (2.21), the function $T(x,y)$ is represented as a function of a differential operator in $x$ acting on a function of $x$. It would be desirable to interpret this function of the operator and so obtain $T$ as an explicit function of $x$ and $y$. This interpretation can be achieved by using the spectral representation of the operator $L_1$. We have seen that the eigenfunctions of $L_1$ are the functions $\sqrt{2} \sin n\pi x$, for $n = 1, 2, \ldots$, and that they correspond to the eigenvalues $n^2\pi^2$. By using Eqs. (2.15) and (2.16), we see that

$$f(x) = 2 \sum_{n=1}^{\infty} \sin n\pi x \int_0^1 f(\xi) \sin n\pi\xi \, d\xi$$

and then by using Eq. (2.18) we find that

$$T(x,y) = 2 \sum_{n=1}^{\infty} \frac{\sinh n\pi(h - y)}{\sinh n\pi h} \int_0^1 f(\xi) \sin n\pi\xi \, d\xi$$

It is easy to verify that this is the desired solution of Eq. (2.19).

## 2.9 Types of Spectral Representation

So far, we have considered the spectral representation of only those operators having a set of eigenfunctions that can form a basis for the space. For many operators, no eigenfunctions exist, and even if they do exist, there may not be enough of them to form a basis for the space. For example, consider the operator $-d^2/dx^2$ on the domain of functions $u(x)$ that are defined and have second derivatives for $-\infty < x < \infty$ and that are such that the integrals

$$\int_{-\infty}^{\infty} [u(x)]^2 \, dx \qquad \int_{-\infty}^{\infty} [u''(x)]^2 \, dx \tag{2.22}$$

are finite. This operator has no eigenfunctions; i.e., there do not exist any functions $\phi(x) \not\equiv 0$ in the domain such that

$$\frac{-d^2\phi}{dx^2} = \lambda\phi \tag{2.23}$$

because any nonnull solution of Eq. (2.23) will cause the integrals (2.22) to become infinite.

It is also easy to give examples of differential operators that have only a finite number of eigenfunctions. It is clear that such a finite number of functions cannot be a basis for a function space containing an infinite number of linearly independent functions.

In order to use the techniques of the previous sections in interpreting functions of operators, we must extend the concept of spectral representation. We shall say that the operator $L$ has a spectral representation if, for any function $u(x)$ in the domain of $L$, there exists a representation of $u(x)$ as a sum or as an integral or, perhaps, as a combination of both sum and integral,

$$u(x) = \Sigma \alpha_k \phi_k(x) + \int \beta(\lambda)\phi(x,\lambda)\, d\lambda$$

and if the expansion is such that

$$Lu(x) = \Sigma \alpha_k \lambda_k \phi_k(x) + \int \lambda\beta(\lambda)\phi(x,\lambda)\, d\lambda$$

For example, if $L = -d^2/dx^2$ on the domain defined by Eqs. (2.22), then, for any $u(x)$ in the domain, we have

$$u(x) = \frac{1}{2\pi}\int_0^\infty \beta(\lambda)\cos\sqrt{\lambda}\,x\,\frac{d\lambda}{\sqrt{\lambda}}$$

and

$$Lu(x) = \frac{1}{2\pi}\int_0^\infty \lambda\beta(\lambda)\cos\sqrt{\lambda}\,x\,\frac{d\lambda}{\sqrt{\lambda}}$$

These formulas reduce to the well-known ones for the Fourier integral if we put $\lambda = k^2$. They become

$$u(x) = \frac{1}{2\pi}\int_{-\infty}^\infty \beta(k^2)e^{ikx}\, dk$$

and

$$Lu = \frac{1}{2\pi}\int_{-\infty}^\infty k^2\beta(k^2)e^{ikx}\, dk$$

It is clear that these formulas can be used to define functions of the operator $L$. We find

$$f(L)u = \frac{1}{2\pi}\int_{-\infty}^\infty f(k^2)\beta(k^2)e^{ikx}\, dk$$

We shall use these formulas to find the acoustic field produced in the region $-\infty < x < \infty$, $0 \le y < \infty$ by sources of given strength $f(x,y)$ and frequency $\omega$ if the line $y = 0$ represents a rigid surface. Mathematically, the problem is to find that solution of

$$\Delta u + \omega^2 c^2 u = f(x,y) \tag{2.24}$$

satisfying the condition

$$u_y(x,0) = 0 \tag{2.25}$$

for $-\infty < x < \infty$. Here $c$ is the velocity of sound in the medium. Proceeding as before, we put

$$L = \frac{-\partial^2}{\partial x^2}$$

over the interval $(-\infty,\infty)$ and then Eq. (2.24) becomes

$$\frac{\partial^2 u}{\partial y^2} + (\omega^2 c^2 - L)u = f(x,y)$$

with the condition (2.25). By the method of variation of constants, we find that the solution of this ordinary differential equation is

$$u = \frac{\exp\,[j(\omega^2c^2 - L)^{1/2}y]}{j(\omega^2c^2 - L)^{1/2}} \int_0^y \cos\,(\omega^2c^2 - L)^{1/2}\eta f(x,\eta)\,d\eta + \cos\,(\omega^2c^2 - L)^{1/2}y \int_y^\infty \frac{\exp\,[j(\omega^2c^2 - L)^{1/2}\eta]}{j(\omega^2c^2 - L)^{1/2}} f(x,\eta)\,d\eta$$

This result can be interpreted by means of the spectral representation of $L$. The function $f(x,\eta)$ is represented in terms of the Fourier integral as follows:

$$f(x,\eta) = \frac{1}{2\pi} \int g(k,\eta)e^{jkx}\,dk$$

Now, remembering that

$$Le^{jkx} = k^2e^{jkx}$$

we find that

$$u(x,y) = \frac{1}{2\pi j} \int_{-\infty}^{\infty} \frac{dke^{jkx}}{(\omega^2c^2 - k^2)^{1/2}} \left\{\exp\,[j(\omega^2c^2 - k^2)^{1/2}y] \int_0^y \cos\,(\omega^2c^2 - k^2)\eta + \cos\,(\omega^2c^2 - k^2)^{1/2}y \int_y^\infty \exp\,[j(\omega^2c^2 - k^2)^{1/2}\eta]\right\} f(x,\eta)\,d\eta$$

## 2.10 Conclusion

The preceding discussion has shown how separable partial differential equations may be solved by operational means. The procedure is as follows: One of the differential operators is treated as a constant and the resulting simpler differential equation is solved. The solution of this simpler differential equation will contain functions of the operator that was treated as a constant. These functions of the operator are interpreted by the use of the spectral representation of the operator; thus, the solution to the partial differential equation is obtained.

The usual transform methods are applications of this theory. For example, the Laplace transform defines the spectral representation for the operator $d/dt$ on the domain of functions $u(t)$ satisfying the initial condition $u(0) = 0$; the Fourier transform defines the spectral representation for the operator $-d^2/dx^2$ on the domain of functions $u(x)$ defined over $(-\infty,\infty)$ such that the integrals

$$\int_{-\infty}^{\infty} u^2\,dx \qquad \int_{-\infty}^{\infty} u''^2\,dx$$

are finite, and the Hankel transform defines the spectral transformation for the operator

$$J = \frac{d^2}{dr^2} + \frac{1}{r}\frac{d}{dr} + k^2$$

on the domain of functions $u(r)$, defined over $(0, \infty)$, such that the integrals

$$\int_0^\infty u^2 r\,dr \qquad \int_0^\infty (Ju)^2 r\,dr$$

are finite. These transforms will be studied in detail in Chap. 3. It is clear, then, that transform methods and operational calculus depend on the study of the spectral representation of differential operators.

## REFERENCES

1. Boole, George, "Treatise on Differential Equations," Cambridge University Press, London, 1859.
2. Bromwich, T. J. I'A., Operational Methods in Mathematical Physics, *Phil. Mag.*, ser. 6, vol. 37, pp. 407–419, 1919.
3. Carson, J. P., Heaviside Operational Calculus, *Bell System Tech. J.*, vol. 1, pp. 1–13, 1922.
4. Dunford, Nelson, and Jacob T. Schwartz, "Linear Operators," Interscience Publishers, Inc., New York, 1958.
5. Friedman, Bernard, "Principles and Techniques of Applied Mathematics," John Wiley & Sons, Inc., New York, 1956.
6. Heaviside, O., "Electromagnetic Theory," vol. 1, The Electrician, London, 1893.
7. Leibnitz, G. W., "Mathematische Schriften," vol. 1, edited by C. J. Gerhardt, Berlin, 1849.
8. Morrey, Charles B., Jr., Nonlinear Methods, chap. 16 in "Modern Mathematics for the Engineer," First Series, edited by E. F. Beckenbach, McGraw-Hill Book Company, Inc., New York, 1956.
9. Pipes, Louis A., Matrices in Engineering, chap. 13 in "Modern Mathematics for the Engineer," First Series, edited by E. F. Beckenbach, McGraw-Hill Book Company, Inc., New York, 1956.
10. Wagner, K. W., Über eine Formel von Heaviside zur Berechnung von Einschaltvorgange, *Ark. Electrotechnik*, vol. 4, pp. 159–193, 1916.

# 3

# Integral Transforms

JOHN W. MILES
PROFESSOR OF ENGINEERING AND GEOPHYSICS
UNIVERSITY OF CALIFORNIA, LOS ANGELES

## 3.1 Introduction

We define

$$F(p) = \int_a^b K(p,x)f(x)\,dx \tag{3.1}$$

to be an *integral transform* of the function $f(x)$, with $K(p,x)$, a prescribed function of $p$ and $x$, as the *kernel* of the transform; see Chap. 2. The introduction of such a transform in a particular problem may be advantageous if the determination or manipulation of $F(p)$ is simpler than that of $f(x)$, much as the introduction of log $x$ in place of $x$ is advantageous in certain arithmetical operations. The representation of $f(x)$ by $F(p)$ is, in many cases, merely a way of organizing a solution more efficiently, as in the introduction of logarithms for multiplication, but in some instances it affords solutions to otherwise apparently intractable problems, just as in the introduction of logarithms for manually extracting the 137th root of a given number.

Integral transforms in applied mathematics find their antecedents in the classical methods of Fourier and in the operational methods of Heaviside—antecedents that had rather different receptions by contemporary mathematicians. The classical eleventh edition of the "Encyclopedia Britannica" devotes five pages to Fourier series but does not mention Heaviside's operational calculus; indeed, no direct entry appears for Heaviside in that edition, although his name is mentioned peripherally. The fourteenth edition does contain a brief biographical entry on Heaviside, but the only reference to his operational calculus is the rather oblique statement that "he made use of unusual methods of his own in solving his problems."

Fourier's theorem has constituted one of the cornerstones of mathematical physics from the publication (1822) of his "La Théorie analytique de la chaleur,"[18] and its importance was quickly appreciated by mathematicians and physicists alike. For example, as quoted by Campbell and Foster,[10] Thomson and Tait remarked:

> . . . Fourier's Theorem, which is not only one of the most beautiful results of modern analysis, but may be said to furnish an indispensable instrument in the treatment of nearly every recondite question in modern physics. To mention only sonorous vibrations, the propagation of electrical signals along a telegraph wire, and the conduction of heat by the earth's crust, as subjects in their generality intractable without it, is to give but a feeble idea of its importance.

The concept of an integral transform follows directly from Fourier's theorem (see Secs. 3.2 to 3.8), but the historical approach, at least to modern applications, was largely through operational methods.

Operational methods in mathematical analysis, having been introduced originally by Leibnitz, are nearly as old as the calculus, but their widespread use in modern technology stems almost entirely from the solitary genius of Oliver Heaviside (1850–1923). To be sure, the bases of Heaviside's method, as he recognized and stated, lay in the earlier work of Laplace (1779) and Cauchy (1823), but it was Heaviside who recognized and exposed its power not only in circuit analysis but also in partial differential equations. Unlike Fourier, however, Heaviside had no university training and was not a recognized mathematician; indeed, he scorned not only mathematical rigor ("Shall I refuse my dinner because I do not fully understand the process of digestion?") but, it sometimes appeared, mathematicians ("Even Cambridge mathematicians deserve justice."). This lack of rapport with mathematicians perhaps helped cause the full importance of his work to be appreciated only gradually, and even some modern mathematicians have been reluctant to give Heaviside his due; thus, Van der Pol and Bremmer[9] take Doetsch to task for his description[6] of Heaviside as merely "*ein englischer Electroingenieur,*" using methods that were "*mathematisch sehr unzulänglich*" and "*allerdings mathematisch unzureichend.*"

Today, the *Laplace transform*, which is the case $K = e^{-px}$, $a = 0$, and $b = \infty$ in Eq. (3.1), namely†

$$F(p) = \int_0^\infty e^{-px} f(x)\, dx \tag{3.2}$$

may be claimed as a working tool for the solution of ordinary differential equations by every well-trained engineer. We shall consider here its application to partial differential equations, along with the *Fourier transform* (also called the *exponential Fourier transform* or the *complex Fourier*

† In Heaviside's form of operational calculus, the right-hand side of Eq. (3.2) appears multiplied by $p$, but the Laplace form gradually has gained the ascendancy.

*transform*),

$$F(p) = \int_{-\infty}^{\infty} e^{-ipx} f(x)\, dx \tag{3.3}$$

the *Fourier cosine* and *Fourier sine transforms*,

$$F(p) = \int_0^{\infty} f(x) \cos px\, dx \tag{3.4}$$

and

$$F(p) = \int_0^{\infty} f(x) \sin px\, dx \tag{3.5}$$

the *Hankel transform* (also called the *Bessel* or *Fourier-Bessel transform*),

$$F(p) = \int_0^{\infty} f(x) J_n(px) x\, dx \tag{3.6}$$

and, rather briefly, the *Mellin transform*,

$$F(p) = \int_0^{\infty} x^{p-1} f(x)\, dx \tag{3.7}$$

Our definitions are those of Erdélyi, Magnus, Oberhettinger, and Tricomi[12] (hereinafter abbreviated as EMOT) except for Eq. (3.6); other definitions of the Fourier transforms, differing from those of Eqs. (3.3) to (3.5) by constant factors, are not uncommon. Notations for the transforms themselves vary widely, with $\bar{f}(p)$ a frequent variant for $F(p)$; it also may prove expedient to introduce subscripts such as $c$, $s$, and $n$ in the definitions of Eqs. (3.4) to (3.6), respectively; finally, symbols other than $p$ often are used in defining the transforms of Eqs. (3.3) to (3.6), and $s$ often appears in place of $p$ in those of Eqs. (3.2) and (3.7).

The transforms defined by Eqs. (3.2) to (3.7) are the only infinite ones in widespread use at this time, but many others have been studied and tabulated (see EMOT,[12] vol. 2), and still more may be introduced in the future. If one's goal is merely to produce formal solutions, usually accompanied by the phrase, so satisfying to mathematicians but so frustrating to physical workers, "The problem is now reduced to quadrature," it suffices to know the inversion formula that determines $f(x)$ from $F(p)$, but for wide usefulness in applied mathematics, extensive tabulations of $f(x)$ versus $F(p)$ are essential. Returning to our analogue of the logarithm, we note that formal analysis requires only the knowledge that the inverse of $y = \log_a x$ is $x = a^y$, but for aid in numerical computation a table of $x$ versus $y$ is indispensable.

In addition to the possibility of defining new transforms through new kernels, there is also the possibility of adopting finite limits in Eq. (3.2), thereby obtaining so-called *finite transforms*. If, for example, we replace the upper limits in Eqs. (3.4) and (3.5) by $\pi$, the inversion formulas are ordinary Fourier cosine and sine series summed over integral values of $p$. More generally, if the kernel $K(p,x)$ in Eq. (3.1) yields a set of functions orthogonal, with suitable weighting function, over the interval $a,b$ for an infinite discrete set of values $p$, then the inversion formula defines a Fourier-type series.

The result of introducing a finite Fourier transform in a given problem is merely to mechanize the classical technique of Fourier series, but it is generally true that the direct, though more tedious, solution of the problem by the classical technique is straightforward, albeit often calling for greater ingenuity. (Compare the use of Lagrange's equations in mechanics.) This is to be contrasted with the application of infinite transforms, which frequently offer entirely new insight and reduce transcendental to algebraic operations, thereby affording solutions to problems that either have resisted the previous forays of classical techniques or that could have been solved only with an ingenuity of Newtonian or Laplacian magnitude.

We conclude this introduction by contrasting the approaches of the pure mathematician and the pragmatist to transform theory. At one extreme we have Titchmarsh's statement, in the preface to his treatise,[8] that "I have retained, as having a certain picturesqueness, some references to 'heat,' 'radiation,' and so forth; but the interest is purely analytical, and the reader need not know whether such things exist." At the other, we have Heaviside's cavalier statement, "The mathematicians say this series diverges; therefore, it should be useful." In the following condensed presentation of integral-transform theory, we† shall follow the line set down by Lord Rayleigh:[19]

> In the mathematical investigation I have usually employed such methods as present themselves naturally to a physicist. The pure mathematician will complain, and (it must be confessed) sometimes with justice, of deficient rigour. But to this question there are two sides. For, however important it may be to maintain a uniformly high standard in pure mathematics, the physicist may occasionally do well to rest content with arguments which are fairly satisfactory and conclusive from his point of view. To his mind, exercised in a different order of ideas, the more severe procedure of the pure mathematician may appear not more but less demonstrative. And further, in many cases of difficulty to insist upon the highest standard would mean the exclusion of the subject altogether in view of the space that would be required.

## INVERSION FORMULAS AND TRANSFORM PAIRS

### 3.2 Fourier's Integral Formulas

We first give a formal derivation of Fourier's theorem in complex form, following in all essential respects the argument offered by Fourier himself.

† The reader should note that the writer of this chapter is an engineer, sometimes defined as "one who assumes everything but the responsibility."

Let $f(x)$ be represented by the complex Fourier series

$$f(x) = \sum_{n=-\infty}^{\infty} c_n e^{ik_n x} \qquad -\tfrac{1}{2}\lambda < x < \tfrac{1}{2}\lambda \tag{3.8}$$

where

$$c_n = \frac{1}{\lambda} \int_{-\lambda/2}^{\lambda/2} f(\xi) e^{-ik_n \xi}\, d\xi \tag{3.9}$$

and

$$k_n = \frac{2n\pi}{\lambda} \tag{3.10}$$

This representation is evidently periodic with a wavelength $\lambda$. Now we allow $\lambda$ to tend to infinity, noting that the consecutive $k_n$ are separated by the increment $\Delta k = 2\pi/\lambda$; then, by combining Eqs. (3.8) and (3.9), we have

$$f(x) = \lim_{\lambda \to \infty} \sum_{n=-\infty}^{\infty} e^{ik_n x} \frac{\Delta k}{2\pi} \int_{-\lambda/2}^{\lambda/2} f(\xi) e^{-ik_n \xi}\, d\xi$$

The sum may be replaced by an integral in the limit, and we obtain

$$f(x) = \frac{1}{2\pi} \int_{-\infty}^{\infty} dk \int_{-\infty}^{\infty} f(\xi) e^{ik(x-\xi)}\, d\xi \tag{3.11}$$

By expressing the exponential in terms of its trigonometric components and invoking the even and odd nature of $\cos k(x - \xi)$ and $\sin k(x - \xi)$, respectively, as functions of $k$, we obtain

$$f(x) = \frac{1}{\pi} \int_{0}^{\infty} dk \int_{-\infty}^{\infty} f(\xi) \cos k(x - \xi)\, d\xi \tag{3.12}$$

which is *Fourier's integral formula.* Fourier's derivation differed from the above only in starting from the trigonometric form of his series. We emphasize that the order of integration in Eqs. (3.11) and (3.12) must be preserved, because its reversal would lead to meaningless integrals; on the other hand, we assume that $f(x)$ vanishes with sufficient rapidity for large $|x|$ to ensure the existence of the double integrals as written. Actually, in typical, not-too-idealized, physical problems, $f$ vanishes exponentially.

If $f(x)$ is either an even or an odd function (and any function that is not even or odd can be split into a sum of two such functions) and the cosine in Eq. (3.12) is expanded, we obtain *Fourier's cosine formula*

$$f(x) = f(-x) = \frac{2}{\pi} \int_{0}^{\infty} \cos kx\, dk \int_{0}^{\infty} f(\xi) \cos k\xi\, d\xi \tag{3.13}$$

or *Fourier's sine formula*

$$f(x) = -f(-x) = \frac{2}{\pi} \int_{0}^{\infty} \sin kx\, dk \int_{0}^{\infty} f(\xi) \sin k\xi\, d\xi \tag{3.14}$$

### 3.3 Fourier Transform

Equation (3.11) may be resolved directly into the Fourier pair

$$F(k) = \int_{-\infty}^{\infty} f(x)e^{-jkx}\,dx \tag{3.15a}$$

and

$$f(x) = \frac{1}{2\pi}\int_{-\infty}^{\infty} F(k)e^{jkx}\,dk \tag{3.15b}$$

thereby providing the inversion of Eq. (3.3). The location of the factor $(2\pi)^{-1}$ is essentially arbitrary; from the viewpoint of establishing an analogy between Fourier series and Fourier integrals, it would appear preferable to place it in Eq. (3.15*a*) rather than (3.15*b*), while from an aesthetic viewpoint a symmetric disposal of the identical factors $(2\pi)^{-1/2}$ would be desirable. Each of these conventions has been adopted by various writers, but the form chosen in Eqs. (3.15*a*) and (3.15*b*) has two major advantages: First, it agrees with the notation adopted by Campbell and Foster[10] in their very extensive table (see also EMOT,[12] vol. 2), and, secondly, it affords a direct transition to the accepted definition of the Laplace-transform pair (see Sec. 3.5). A third advantage, of especial interest in electric-circuit or wave-motion problems, is that, if $x$, implicitly a space variable in the foregoing discussion, be replaced by the time variable $t$ and $k$ by $2\pi\nu$, where $\nu$ is a frequency, Eqs. (3.15*a*) and (3.15*b*) go over to the symmetric pair

$$F(\nu) = \mathfrak{F}\{f\} = \int_{-\infty}^{\infty} f(t)e^{-2\pi j\nu t}\,dt \tag{3.16a}$$

and

$$f(t) = \mathfrak{F}^{-1}\{F\} = \int_{-\infty}^{\infty} F(\nu)e^{2\pi j\nu t}\,d\nu \tag{3.16b}$$

in which $f(t)$ is represented as a spectral superposition of simple harmonic oscillations having the complex amplitude $F(\nu)$. We remark that a similar form for the space-variable pair of Eqs. (3.15*a*) and (3.15*b*) results from the substitution $k = 2\pi/\lambda$.

### 3.4 Fourier Cosine and Sine Transforms

Equations (3.13) and (3.14) yield the transform pairs

$$F(k) = \mathfrak{F}_c\{f\} = \int_0^{\infty} f(x)\cos kx\,dx \tag{3.17a}$$

$$f(x) = \mathfrak{F}_c^{-1}\{F\} = \frac{2}{\pi}\int_0^{\infty} F(k)\cos kx\,dk \tag{3.17b}$$

and

$$F(k) = \mathfrak{F}_s\{f\} = \int_0^{\infty} f(x)\sin kx\,dx \tag{3.18a}$$

$$f(x) = \mathfrak{F}_s^{-1}\{F\} = \frac{2}{\pi}\int_0^{\infty} F(k)\sin kx\,dk \tag{3.18b}$$

thereby providing the inversions of Eqs. (3.4) and (3.5), respectively. Again, the $2/\pi$ factors may be disposed differently—in particular, symmetrically; the notation adopted here is that of EMOT.[12]

## 3.5 Laplace Transform

The path of integration for Eq. (3.15*b*) may be deformed into a complex $k$ plane in any manner that ensures the convergence of the integrals for both $F(k)$ and $f(x)$. Suppose, to cite the most important particular case, that $\exp(-cx)f(x)$ vanishes appropriately at both limits; in particular, $f(x)$ may vanish identically at one limit, usually $-\infty$. Then the modified transform

$$F(k) = \int_{-\infty}^{\infty} f(x)e^{-jkx}\,dx \qquad \mathfrak{I}\{k\} = -c \tag{3.19a}$$

exists, where $\mathfrak{I}\{k\}$ denotes the imaginary part of $k$, and $f(x)$ is given by

$$f(x) = \frac{1}{2\pi}\int_{-\infty-jc}^{\infty-jc} F(k)e^{jkx}\,dk \tag{3.19b}$$

Thus, Fourier's integral formula is extended to functions for which Eq. (3.11) might not be valid. In the most important physical applications, $c$ is positive and the path of integration appears as in Fig. 3.1.

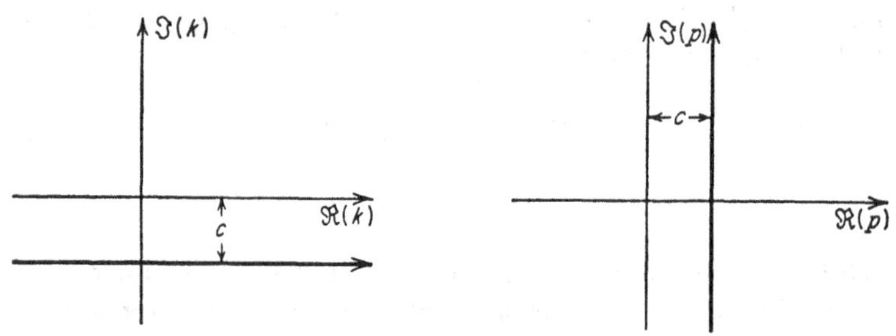


**Fig. 3.1** Path of integration for Fourier integral inversion,

$$f(x) = \frac{1}{2\pi}\int_{-\infty-jc}^{\infty-jc} F(k)e^{jkx}\,dk$$

**Fig. 3.2** Path of integration for Laplace transform inversion,

$$f(x) = \frac{1}{2\pi j}\int_{c-j\infty}^{c+j\infty} F(p)e^{px}\,dp$$

If we now rotate the path of integration through a right angle (see Fig. 3.2), introduce $p = jk$, and, at the same time, replace $F(k)$ by $F(p)$, Eqs. (3.19*a*) and (3.19*b*) go over to the *two-sided Laplace-transform pair*

$$F(p) = \int_{-\infty}^{\infty} f(x)e^{-px}\,dx \tag{3.20a}$$

and

$$f(x) = \frac{1}{2\pi j}\int_{c-j\infty}^{c+j\infty} F(p)e^{px}\,dp \tag{3.20b}$$

Finally, we suppose $f(x)$ to vanish for $f(x) < 0$, so that the lower limit in Eq. (3.20*a*) may be replaced by zero. The resulting integral converges if the real part of $p$ exceeds some minimum value, say $\gamma$, not necessarily nonnegative, such that all singularities of $F(p)$ lie in $\Re p < \gamma$; the inverse transform (3.20*b*) exists for all $c > \gamma$, and we have the Laplace-transform pair

$$F(p) = \mathfrak{L}\{f(x)\} = \int_0^\infty e^{-px} f(x)\,dx \qquad \Re\{p\} > \gamma \tag{3.21a}$$

and $$f(x) = \mathfrak{L}^{-1}\{F(p)\} = \frac{1}{2\pi j}\int_{c-j\infty}^{c+j\infty} F(p)e^{px}\,dp \qquad c > \gamma \tag{3.21b}$$

### 3.6 Mellin Transform

Put $x = -\log y$ in Eqs. (3.20*a*) and (3.20*b*) to obtain

$$F(p) = \int_0^\infty y^{p-1} f(-\log y)\,dy$$

and $$f(-\log y) = \frac{1}{2\pi j}\int_{c-j\infty}^{c+j\infty} F(p)y^{-p}\,dp$$

Then the replacement of $f(-\log y)$ by $f(x)$ and the dummy variable $y$ by $x$ yields the Mellin-transform pair

$$F(p) = \int_0^\infty x^{p-1} f(x)\,dx \qquad \Re p > \gamma \tag{3.22a}$$

and $$f(x) = \frac{1}{2\pi j}\int_{c-j\infty}^{c+j\infty} F(p)x^{-p}\,dp \qquad c > \gamma \tag{3.22b}$$

### 3.7 Multiple Fourier Transforms

The Fourier-transform pair may be extended to a function of two variables to obtain

$$F(k_1,k_2) = \int_{-\infty}^\infty \int_{-\infty}^\infty f(x,y)e^{-j(k_1x+k_2y)}\,dx\,dy \tag{3.23a}$$

and $$f(x,y) = \frac{1}{4\pi^2}\int_{-\infty}^\infty \int_{-\infty}^\infty F(k_1,k_2)e^{j(k_1x+k_2y)}\,dk_1\,dk_2 \tag{3.23b}$$

More generally, if $\mathbf{r}$ denotes a vector having the cartesian components $x_1, x_2, \ldots, x_n$ in an $n$-dimensional space, and $\mathbf{k}$ a similar vector in the wave-number space $k_1, k_2, \ldots, k_n$, we obtain

$$F(\mathbf{k}) = \int_{-\infty}^\infty \cdots \int_{-\infty}^\infty f(\mathbf{r})e^{-j\mathbf{k}\cdot\mathbf{r}}\,dx_1 \cdots dx_n \tag{3.24a}$$

and $$f(\mathbf{r}) = \left(\frac{1}{2\pi}\right)^n \int_{-\infty}^\infty \cdots \int_{-\infty}^\infty F(\mathbf{k})e^{j\mathbf{k}\cdot\mathbf{r}}\,dk_1 \cdots dk_n \tag{3.24b}$$

### 3.8 Hankel Transforms

Further transform pairs may be obtained from Eqs. (3.23*a*), (3.23*b*), (3.24*a*), and (3.24*b*) by coordinate transformation. In particular, if we introduce polar coordinates according to

$$x = r\cos\theta \qquad y = r\sin\theta \qquad k_1 = k\cos\alpha \qquad k_2 = k\sin\alpha$$

then Eqs. (3.23*a*) and (3.23*b*) go over, with appropriate changes in the functional notation, to

$$F(k,\alpha) = \int_0^\infty \int_0^{2\pi} f(r,\theta)e^{-jkr\cos(\theta-\alpha)}r\,dr\,d\theta \tag{3.25a}$$

and

$$f(r,\theta) = \frac{1}{4\pi^2}\int_0^\infty \int_0^{2\pi} F(k,\alpha)e^{jkr\cos(\theta-\alpha)}k\,dk\,d\alpha \tag{3.25b}$$

We remark that if $f(r,\theta)$ is multiplied by $\exp(-j\omega t)$, Eq. (3.25*b*) represents a packet of plane waves having the amplitude distribution $F(k,\alpha)$, the wave speeds $\omega/k$, and wavefront normals inclined at the angle $\alpha$ to the $x$ axis.

We now assume that

$$f(r,\theta) = f_n(r)e^{jn\theta} \tag{3.26a}$$

and make use of the integral representation

$$J_n(kr) = \frac{1}{2\pi}\int_0^{2\pi} e^{j(n\varphi - kr\sin\varphi)}\,d\varphi$$

for Bessel's function of order $n$ to obtain

$$F(k,\alpha) = 2\pi e^{jn(\alpha-\pi/2)}F_n(k) \tag{3.26b}$$

where

$$F_n(k) = \mathcal{H}_n\{f_n\} = \int_0^\infty f_n(r)J_n(kr)r\,dr \tag{3.27a}$$

and

$$f_n(r) = \mathcal{H}_n^{-1}\{F_n\} = \int_0^\infty F_n(k)J_n(kr)k\,dk \tag{3.27b}$$

the latter result following from the substitution of Eqs. (3.26*a*) and (3.26*b*) in Eq. (3.25*b*); Eqs. (3.27*a*) and (3.27*b*) constitute the Hankel-transform pair of order $n$.

Another form of the Hankel transform, especially important in that it is used in EMOT,[12] vol. 2, is given by

$$G_n(k) = \int_0^\infty g_n(r)J_n(kr)(kr)^{1/2}\,dr$$

and

$$g_n(r) = \int_0^\infty G_n(k)J_n(kr)(kr)^{1/2}\,dk$$

This evidently can be reconciled with Eqs. (3.27*a*) and (3.27*b*) by setting $g_n = r^{1/2}f_n$ and $G_n = k^{1/2}F_n$. As it stands, it reduces, except for the constant factor $\sqrt{2/\pi}$, to the Fourier sine or cosine transform for $n = \frac{1}{2}$ or $-\frac{1}{2}$, respectively.

The properties of the foregoing infinite transforms are summarized in Table 3.1.

## Table 3.1 Infinite Integral Transforms

| Transform (references†) | $F(p) = T\{f(x)\}$ | $f(x) = T^{-1}\{F(p)\}$ | Transforms of typical derivatives | Convolution theorem $T^{-1}\{F(p)G(p)\}$ |
|---|---|---|---|---|
| Laplace (1–14) | $\int_0^\infty f(x)e^{-px}\,dx$ | $\frac{1}{2\pi j}\int_{c-j\infty}^{c+j\infty} F(p)e^{px}\,dp$ | $T\{f^{(n)}\} = p^nF(p) - \sum_{m=0}^{n-1} p^m f^{(n-m-1)}(0)$ | $\int_0^x f(\xi)g(x-\xi)\,d\xi$ |
| Fourier (5, 7, 8, 10, 12, 13) | $\int_{-\infty}^\infty f(x)e^{-jpx}\,dx$ | $\frac{1}{2\pi}\int_{-\infty}^\infty F(p)e^{jpx}\,dp$ | $T\{f^{(n)}\} = (jp)^nF(p)$ | $\int_{-\infty}^\infty f(\xi)g(x-\xi)\,d\xi$ |
| Fourier cosine (5, 7, 8, 12) | $\int_0^\infty f(x)\cos px\,dx$ | $\frac{2}{\pi}\int_0^\infty F(p)\cos px\,dp$ | $T\{f^{(2n)}\} = (-1)^n p^{2n}F(p) - \sum_{m=0}^{n-1} (-1)^m p^{2m} f^{(2n-2m-1)}(0)$ | $\frac{1}{2}\int_0^\infty f(\xi)[g(\lvert x-\xi\rvert) + g(x+\xi)]\,d\xi$ |
| Fourier sine (5, 7, 8, 12) | $\int_0^\infty f(x)\sin px\,dx$ | $\frac{2}{\pi}\int_0^\infty F(p)\sin px\,dp$ | $T\{f^{(2n)}\} = (-1)^n p^{2n}F(p) - \sum_{m=1}^{n} (-1)^m p^{2m-1} f^{(2n-2m)}(0)$ | None available, but see Ref. 7, Sec. 3.6, for related theorems |
| Mellin (5, 7, 8, 12) | $\int_0^\infty f(x)x^{p-1}\,dx$ | $\frac{1}{2\pi j}\int_{c-j\infty}^{c+j\infty} F(p)x^{-p}\,dp$ | $T\{x^n f^{(n)}\} = (-1)^n p(p+1)\cdots(p+n-1)F(p)$ | $\int_0^\infty f(\xi)g\left(\frac{x}{\xi}\right)\xi^{-1}\,d\xi$ |
| Hankel (5, 7, 8, 12) | $\int_0^\infty f(x)J_n(px)x\,dx$ | $\int_0^\infty F(p)J_n(px)p\,dp$ | $T\left\{f'' + \frac{1}{x}f' - \frac{n^2}{x^2}f\right\} = -p^2F$ | No simple result |

† Numbers refer to the list of references at the end of the chapter.

## THE LAPLACE TRANSFORM

### 3.9 Introduction

We now proceed to consider some special properties and applications of the Laplace transform. There exist many textbooks and treatises dealing with both the theory and the applications of this transform; and, while there are many specific examples in which other transforms prove more expedient, the Laplace transform is generally the most powerful and flexible in dealing with initial-value problems and is the most extensively tabulated. We first consider a few of the properties that invest it with this power and flexibility and then apply it to three typical problems governed by partial differential equations. Many more examples are given in the textbooks listed at the end of the chapter, especially in Ref. 1.

### 3.10 Transforms of Derivatives

We require†

$$\mathcal{L}\{f^{(n)}(t)\} = \int_0^\infty e^{-pt} f^{(n)}(t)\, dt$$

where $f^{(n)}(t)$ denotes the $n$th derivative of $f(t)$ and $f(0), f'(0), \ldots, f^{(n-1)}(0)$ are prescribed as initial conditions as $t \to 0+$. The required result follows upon repeated integration by parts, the explicit results for the first and second derivatives being

$$\mathcal{L}\{f'(t)\} = pF(p) - f(0) \tag{3.28a}$$

and

$$\mathcal{L}\{f''(t)\} = p^2F(p) - pf(0) - f'(0) \tag{3.28b}$$

We emphasize that the Heaviside operational rule

$$\mathcal{L}^{-1}\{p^nF(p)\} = f^{(n)}(t)$$

is not valid except in the special case, always implied in Heaviside's work, for which $f(0) = f'(0) = \cdots = f^{(n-1)}(0) = 0$. A related result, which, however, is generally valid, is

$$\mathcal{L}^{-1}\{p^{-n}F(p)\} = \int_0^t \cdots \int_0^t f(t)(dt)^n \tag{3.29}$$

### 3.11 Heaviside's Shifting Theorem

Disturbances frequently arise at times other than zero, it being implicit that $t = 0$ is chosen to correspond to the first event of interest, the effects of all prior events being included in the prescribed initial conditions. It

† We now use $t$ as the independent variable in recognition of the fact that usually that variable is time; nevertheless, initial-value problems may be encountered in which $t$ is a space variable—e.g., those in linearized, supersonic flow.

follows directly from the definition of the Laplace transform that

$$\mathcal{L}\{f(t-a)\} = e^{-ap}F(p) \qquad a > 0 \tag{3.30a}$$

and

$$\mathcal{L}^{-1}\{e^{-ap}F(p)\} = f(t-a) \qquad t > a \tag{3.30b}$$

where $F(p)$ is the transform of $f(t)$ and $f$ vanishes for negative values of its argument. Many closely related theorems are to be found in EMOT,[12] vol. 1, and in other references listed at the end of this chapter.

We add that Eqs. (3.30a) and (3.30b) apply also to the exponential Fourier transform with $p = jk$.

## 3.12 Convolution Theorem

It frequently proves expedient to resolve a Laplace transform into a product of two transforms, either because the inversions of the latter transforms are known or because one of them represents an arbitrary function—typically an input to some physical system. For a pair of given inverse transforms $f_1(t)$ and $f_2(t)$, the convolution theorem states that

$$\mathcal{L}^{-1}\{F_1(p)F_2(p)\} = \int_0^t f_1(t-\tau)f_2(\tau)\,d\tau \tag{3.31}$$

To prove Eq. (3.31), we multiply the defining integrals for $F_1$ and $F_2$ to obtain

$$F_1(p)F_2(p) = \int_0^\infty \int_0^\infty e^{-p(\sigma+\tau)}f_1(\sigma)f_2(\tau)\,d\sigma\,d\tau$$

By introducing the change of variable $\sigma = t - \tau$ and invoking the requirement that $f_1$ must vanish for negative values of its argument, we get

$$F_1(p)F_2(p) = \int_0^\infty e^{-pt}\left[\int_0^t f_1(t-\tau)f_2(\tau)\,d\tau\right]dt$$

whence the transform of the quantity in brackets is $F_1F_2$, the inversion of which yields Eq. (3.31).

We remark that, in typical applications, the right-hand side of Eq. (3.31) represents a superposition of effects of magnitude $f_2(\tau)$, arising at $t = \tau$, for which $f_1(t - \tau)$ is the influence function, i.e., the response to a unit impulse. Indeed, it constitutes the extension to impulsive inputs of Duhamel's superposition theorem (1833) for step inputs. Such a unit impulse is known as the Dirac delta function and has the formal properties

$$\delta(x-a) = 0 \qquad x \neq a \tag{3.32a}$$

and

$$\int_{-\infty}^{\infty} f(x)\delta(x-a)\,dx = f(a) \tag{3.32b}$$

By letting

$$f_2(\tau) = \delta(\tau)$$

in Eq. (3.31), we see that

$$\mathcal{L}\{\delta(\tau)\} = 1 \tag{3.33}$$

As defined by Eqs. (3.32*a*) and (3.32*b*), the delta function is highly improper (see Chap. 1 for a more sophisticated treatment), but it can be defined as the limit of a proper function and was so introduced by both Cauchy and Poisson in their independent but almost simultaneous (1815) derivations of the Fourier integral theorem.† The function used by Cauchy was

$$\delta(x - a) = \lim_{y \to 0+} f(x,y) \qquad \pi f(x,y) = \frac{y}{(x - a)^2 + y^2}$$

where $f(x,y)$ may be identified as a solution to Laplace's equation for a doublet source located at $(x,y) = (a,0)$.

### 3.13 Inversion Procedures

The most direct manner of inversion is through a suitable table of transform pairs, but it frequently happens that the required entry cannot be found in any existing table. Aside from such extensions of the tables as may be achieved through the shifting, convolution, and other theorems, the most powerful methods of inversion are provided by complex-variable theory.

We consider first the case where $F(p)$ is a meromorphic function of the complex variable $p$—that is, its only singularities in the entire finite $p$ plane are poles—that exhibits the asymptotic behavior

$$F(p) = O(|p|^{-b}) \qquad |p| \to \infty \quad b > 0$$

Here, the relationship $w = O(z)$ in the neighborhood of $z = a$ means $\overline{\lim_{z \to a}} |w/z| < \infty$; the most frequently occurring values of $a$, usually clear from the context, are 0 and $\infty$. If $F(p)$ satisfies the foregoing condition, the line integral of Fig. 3.2 may be closed at infinity as in Fig. 3.3; it then follows from Cauchy's residue theorem that $f$ is given by $2\pi j$ times the sum of the residues at the poles of $F(p)e^{pt}$, all of which, by hypothesis, must lie in $\Re(p) < c$. In particular, if

$$f(p) = \frac{G(p)}{H(p)}$$

where $G$ has no singularities and $H$ has $N$ simple zeros at $p_1, p_2, \ldots, p_N$,

† *Adumbration* might be more accurate than *derivation*, in that the equivalent of Fourier's integral theorem appeared as one step in obtaining a general solution to the wave equation. Its use in the present context is, of course, due to Fourier himself.

$N$ finite or infinite, we have

$$\mathcal{L}^{-1}\left\{\frac{G(p)}{H(p)}\right\} = \sum_{n=1}^{N} \frac{G(p_n)}{H'(p_n)} e^{p_n t} \tag{3.34}$$

a result due essentially to Heaviside.

If the transform $F(p)$ contains branch-point singularities, the contour of Fig. 3.3 must be deformed around the appropriate branch cuts, as illustrated in Fig. 3.4 for the important special case of a branch point at

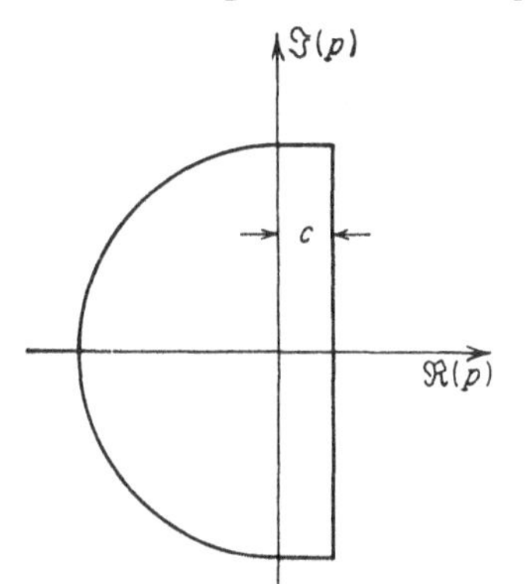


**Fig. 3.3** Contour of integration for meromorphic Laplace transform.

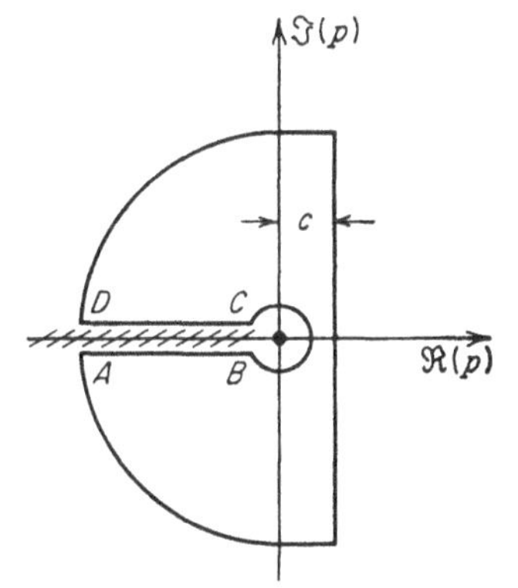


**Fig. 3.4** Contour of integration for Laplace transform having branch point at the origin.

$p = 0$. More generally, branch points of Laplace transforms are likely to be on the imaginary axis, but only occasionally elsewhere. The contributions of the poles, if any, may be evaluated as before, in particular from Eq. (3.34), but it also is necessary to include the integrals over both sides of the branch cut. A specific example will be considered in Sec. 3.15, but the ultimate form of the integrals over the cut of Fig. 3.4 will be

$$I = I_{AB} + I_{CD} = \int_0^\infty e^{-xt}\phi(x)\,dx \tag{3.35a}$$

where $\phi$ is derived from $F(p)$. It is possible that this integral may be evaluated from a table of Laplace transforms with $I$ appearing as the transform of $\phi$, but whether or not this is true, it often suffices to obtain an asymptotic approximation for large $t$. Actually, such approximations sometimes prove satisfactory for surprisingly small values of $t$.

The asymptotic evaluation (see Chap. 5) of $I$ is especially simple if $\phi(x)$ is an analytic function of $x$, regular in some circle ($|x| < R$), where it has the representation

$$\phi(x) = \sum_{n=0}^{\infty} a_n x^{\mu+n} \tag{3.35b}$$

If positive real numbers $C$ and $a$ exist such that $|\phi| < Ce^{ax}$, $x \to \infty$, then *Watson's lemma* states that $I$ has the asymptotic expansion

$$I \sim \sum_{n=0}^{\infty} a_n \Gamma(\mu + n + 1) t^{-(\mu+n+1)} \tag{3.35c}$$

If $\mu$ is a positive integer, the same result may be obtained by repeatedly integrating Eq. (3.35a) by parts. More generally, the expansion of $\phi(x)$ about $x = 0$ may contain logarithmic terms, and a formal asymptotic expansion for $I$ may be obtained by integrating term by term.

Other methods of evaluating branch-cut integrals, of which $I$ is only one—albeit the most important—form, are discussed in the monographs of Erdélyi[16] and Copson.[15] We also remark that approximations valid for small $t$ may be obtained by an inverse application of Watson's lemma to the Laplace transform $F(p)$; that is, if

$$F(p) \sim \sum_{n=0}^{\infty} b_n p^{-(\mu+n+1)} \tag{3.36a}$$

then

$$f(t) = \sum_{n=0}^{\infty} \frac{b_n t^{\mu+n}}{\Gamma(\mu + n + 1)} \tag{3.36b}$$

Finally, we note that the numerical evaluation of integrals such as that in Eq. (3.35a) may be entirely practical by virtue of the exponential convergence, although it may be necessary first to separate out the singularity at the origin.

### 3.14 A Problem in Wave Motion

A uniform bar of unit cross section is at rest and unstressed for $t < 0$, with one end fixed at $x = 0$ (see Fig. 3.5). At $t = 0$, a force of magnitude $P$ is applied to the free end, where $x = l$, in the direction of the positive $x$ axis. We require the subsequent motion.

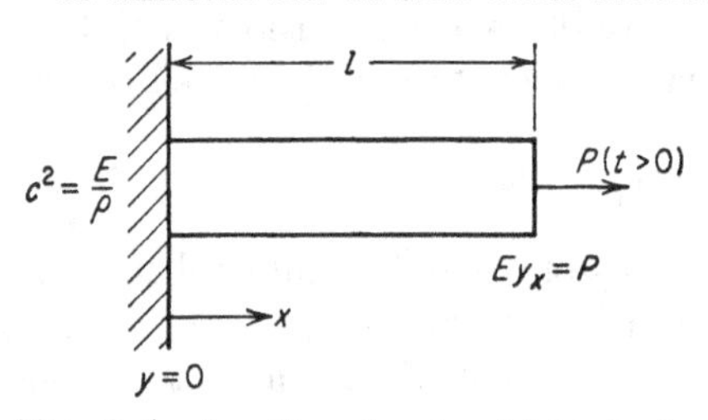

**Fig. 3.5** A uniform bar to which a load $P$ is applied abruptly at time $t = 0$.

Let $y(x,t)$ denote the displacement from equilibrium of any section initially at $x$; it is known that $y$ satisfies the wave equation

$$y_{xx} = c^2 y_{tt} \qquad c^2 = \frac{E}{\rho} \tag{3.37}$$

where $c$, $E$, and $\rho$ denote bar velocity, Young's modulus, and density, respectively. The initial conditions are

$$y = y_t = 0 \qquad \text{at } t = 0 \quad \text{and} \quad 0 < x < l$$

while the boundary conditions are

$$y = 0 \qquad \text{at } x = 0 \qquad \text{and} \qquad Ey_x = P \qquad \text{at } x = l \quad t > 0$$

Transforming Eq. (3.37) subject to the initial conditions yields

$$Y_{xx} - \left(\frac{p}{c}\right)^2 Y = 0$$

while transforming the boundary conditions yields

$$Y = 0 \quad \text{at } x = 0 \qquad \text{and} \qquad Y_x = \frac{P}{Ep} \quad \text{at } x = l$$

The solution that satisfies these boundary conditions is

$$Y = \frac{Pc \sinh (px/c)}{Ep^2 \cosh (pl/c)} \tag{3.38}$$

as may be verified by direct substitution.

We may invert Eq. (3.38) via Eq. (3.34), with

$$G = \frac{Pc}{Ep} \sinh \frac{px}{c} \qquad \text{and} \qquad H = p \cosh \frac{pl}{c}$$

We note that $H = 0$ at $p = 0$ and at $p = \pm(2n + 1)(\pi jc/2l)$, $n = 0$, 1, 2, . . . . Substitution in Eq. (3.34) yields

$$y(x,t) = \frac{Pl}{E}\left[\frac{x}{l} - \frac{8}{\pi^2}\sum_{n=0}^{\infty} \frac{(-1)^n}{(2n + 1)^2} \sin \frac{(2n + 1)\pi x}{2l} \cos \frac{(2n + 1)\pi ct}{2l}\right] \tag{3.39}$$

The first term in Eq. (3.39)—viz., $Px/E$—represents the ultimate static deflection of the bar, while the remaining terms represent standing waves that gradually would die out if friction were admitted. In the absence of friction, however, the displacement continues to oscillate about the static displacement, as shown for the loaded end ($x = l$) in Fig. 3.5.

An alternative solution to the problem may be obtained by expressing the hyperbolic functions in terms of exponentials and using the expansion

$$\frac{1}{\cosh (pl/c)} = \frac{2e^{-pl/c}}{1 + e^{-2pl/c}} = 2\sum_{n=0}^{\infty} (-1)^n e^{-(2n+1)(pl/c)}$$

to obtain

$$Y = \frac{Pc}{Ep^2} \sum_{n=0}^{\infty} (-1)^n \left\{ \exp\left[ -\frac{p}{c}((2n+1)l - x) \right] - \exp\left[ -\frac{p}{c}((2n+1)l + x) \right] \right\}$$

Now the inverse transform of $p^{-2}$ is $t$, whence the shifting theorem, Eq. (3.30b), yields

$$y(x,t) = \frac{Pc}{E} \sum_{n=0}^{\infty} (-1)^n \left\{ \left[ t - \frac{1}{c}((2n+1)l - x) \right] - \left[ t - \frac{1}{c}((2n+1)l + x) \right] \right\} \tag{3.40}$$

where, by definition, the square brackets vanish identically if their contents are negative.

Equation (3.40) exhibits the solution as a series of traveling waves, the first and second sets moving respectively toward and away from $x = 0$. Such a representation is valuable not only because it presents the solution in a finite number of terms (since only a finite number of the square

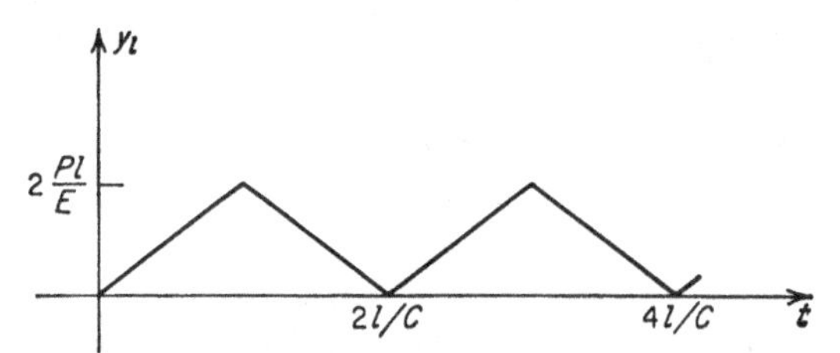


**Fig. 3.6** Displacement of the loaded end of the bar shown in Fig. 3.5.

brackets are positive at any finite time), thereby rendering numerical computation simpler for small $ct/l$, but also because it provides additional insight into the physical problem. It is, indeed, one of the great virtues of the Laplace-transform solution that it comprises both the standing- and traveling-wave representations.

At $x = l$, where the load acts, the displacement given by Eq. (3.40) reduces to

$$y(l,t) = \frac{Pc}{E} \left\{ [t] + 2 \sum_{n=1}^{\infty} (-1)^n \left[ t - \frac{2nl}{c} \right] \right\}$$

corresponding to the triangular wave shown in Fig. 3.6, with a mean value equal to the static displacement $Pl/E$.

## 3.15 A Problem in Heat Conduction

We consider now the classical problem of a semi-infinite solid, $x > 0$, that is initially at temperature $v = 0$ and for which the boundary, $x = 0$, is maintained at temperature $v = v_0$ for $t > 0$. We require a solution to the diffusion equation

$$v_t = \kappa v_{xx} \tag{3.41}$$

where $\kappa$ is the diffusivity, subject to the initial condition

$$v = 0 \qquad \text{at } t = 0 \quad \text{and} \quad x > 0$$

and the boundary conditions

$$v = v_0 \qquad \text{at } x = 0 \qquad \text{and} \qquad v < \infty \qquad \text{as } x \to \infty$$

By taking the Laplace transform of these equations, we obtain

$$V_{xx} - \frac{pV}{\kappa} = 0$$

$$V = \frac{v_0}{p} \qquad \text{at } x = 0 \qquad \text{and} \qquad V < \infty \qquad \text{as } x \to \infty$$

to which the solution is

$$V = \frac{v_0}{p} \exp\left[-\left(\frac{p}{\kappa}\right)^{1/2} x\right] \tag{3.42}$$

This may be inverted directly from standard tables, but we shall use it to illustrate the general procedure.

The right-hand side of Eq. (3.42) has a square-root branch point at $p = 0$; accordingly, we choose a branch cut and close the path of integration as in Figs. 3.4 and 3.7. The function $V$ has no singularities inside the closed contour, whence the contour integral of $V$ exp $(pt)$ around the closed path is zero; further, since the arcs at infinity contribute null values, the inversion integral for $V$ is equal to the integral over the branch cut plus that over the small circle around the origin taken in a *counterclockwise* direction, viz.,

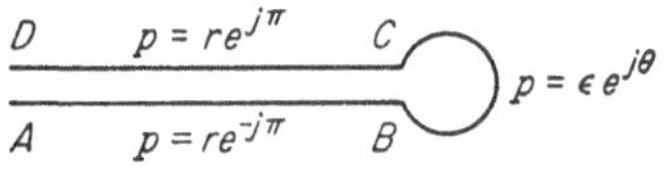

**Fig. 3.7** Contour $ABCD$ of integration for

$$v = \frac{v_0}{2\pi j}\int_{ABCD} p^{-1} \exp\left[pt - \left(\frac{p}{\kappa}\right)^{1/2} x\right] dp$$

$$v = \frac{v_0}{2\pi j}\int_{ABCD} p^{-1} \exp\left[pt - \left(\frac{p}{\kappa}\right)^{1/2} x\right] dp$$

In the neighborhood of the origin, $V$ tends to infinity like $p^{-1}v_0$, and the contribution of the path $BC$ as its radius tends to zero is simply $v_0$ [as may be proved by setting $p = \epsilon \exp(j\theta)$ in the integrand, letting $\epsilon \to 0$, and integrating between $-\pi$ and $\pi$]. We emphasize that $V$ does *not* have a Laurent-series representation in the neighborhood of $p = 0$, and the contribution of the path $BC$ must be treated more or less *ad hoc.* If $F(p)$ is of the form $p^{-1}F_1(p)$, where $F_1(p)$ has a branch point at $p = 0$ but is finite there, the contribution to $f(t)$ of the path $BC$ as its radius tends to zero is simply $F_1(0)$, but if the infinity is of higher order, the limiting forms of the integrals over $AB$, $BC$, and $CD$ all may be improper, in which case the limit as $\epsilon \to 0$ may be taken only after the contributions of $AB$ and $CD$ have been combined with that of $BC$.

To evaluate the contributions of $AB$ and $CD$, we set

$$p = re^{-j\pi} \qquad \text{and} \qquad p = re^{j\pi}$$

respectively, to obtain

$$\frac{1}{2\pi j}\lim_{\epsilon\to 0}\int_\epsilon^\infty e^{-rt}\left\{\exp\left[-j\left(\frac{r}{\kappa}\right)^{1/2}x\right] - \exp\left[j\left(\frac{r}{\kappa}\right)^{1/2}x\right]\right\} r^{-1}\,dr$$

$$= -\frac{1}{\pi}\int_0^\infty e^{-rt}\sin\left[\left(\frac{r}{\kappa}\right)^{1/2}x\right] r^{-1}\,dr \qquad (3.43a)$$

$$= -\frac{2}{\pi}\int_0^\infty e^{-k^2\kappa t}k^{-1}\sin kx\,dk \qquad (3.43b)$$

This last integral may be transformed to the error-function integral (see Ref. 1, page 94), and the end result is

$$\frac{v}{v_0} = 1 - \operatorname{erf}\frac{x}{2(\kappa t)^{1/2}}$$

Having this representation, we could obtain numerical results directly from tabulations of the error function, while approximations useful for small or large values of $(\kappa t)^{-1/2}x$ may be obtained from the appropriate series representations. It is only in the simplest problems, however, that the branch-cut integrals lead to tabulated functions, and it generally is necessary to revert directly to series approximations; more generally, asymptotic approximations to these integrals may be deduced from Watson's lemma. Thus, by comparing the integral on the right-hand side of Eq. (3.43*a*) to $I$ in Eq. (3.35), with $r$ substituted for $x$ therein, we have

$$\phi(r) = -(\pi r)^{-1}\sin\frac{r^{1/2}x}{\kappa^{1/2}}$$

$$= -\frac{1}{\pi}\sum_{n=0}^{\infty}\frac{(-1)^n\kappa^{-n-1/2}x^{2n+1}r^{n-1/2}}{(2n+1)!}$$

whence

$$\mu = -\tfrac{1}{2}$$

and

$$\frac{v}{v_0} \sim 1 - \frac{1}{\pi}\sum_{n=0}^{\infty}\frac{(-1)^n\Gamma(n+\frac{1}{2})}{(2n+1)!}\left(\frac{x}{\sqrt{\kappa t}}\right)^{2n+1}$$

We remark that this same result could have been obtained directly by expanding the exponential in Eq. (3.42) in ascending powers of $p$ and inverting term by term, but then we would have had to assume that the inverse transform of a positive integral power of $p$ vanishes identically for $t > 0$.

### 3.16 A Problem in Supersonic Flow

The unsteady flow past a two-dimensional supersonic airfoil is governed, in linear approximation, by the partial differential equation

$$\phi_{xx} + \phi_{yy} = c^{-2}\left(U\frac{\partial}{\partial x} + \frac{\partial}{\partial t}\right)^2 \phi \tag{3.44}$$

where $\phi$ denotes the perturbation velocity potential, $x$ and $y$ cartesian coordinates in a reference frame moving with the airfoil, $c$ the velocity of sound, and $U$ the flight speed directed along the negative $x$ axis, with $U > c$. We shall solve Eq. (3.44) subject to the boundary condition

$$\phi_y = w(x)e^{j\omega t} \qquad \text{at } y = 0 \qquad \text{and} \qquad x > 0 \qquad -\infty < t < \infty$$

corresponding to a prescribed transverse velocity of complex amplitude $w(x)$ for an oscillating airfoil having leading edge at $x = 0$. The airfoil may be assumed to terminate at the trailing edge $x = l$; but the flow in $x < l$ cannot be influenced by conditions in $x > l$, by virtue of the supersonic flight speed.

We first remark that $\phi$ may be assumed to exhibit the time dependence $\exp(j\omega t)$; taking the Laplace transform of Eq. (3.44) with respect to $x$ then yields

$$\Phi_{yy} - p^2\Phi = \left(\frac{Up + j\omega}{c}\right)^2 \Phi$$

while the transform of the boundary condition is

$$\Phi_y = W(p)e^{j\omega t}$$

The required solution, subject to the condition that $\phi$ must remain bounded as $|y| \to \infty$, is

$$\Phi = -\lambda^{-1}W(p)e^{j\omega t - \lambda y} \qquad y > 0$$

where

$$\lambda = \left[\left(\frac{Up + j\omega}{c}\right)^2 - p^2\right]^{1/2}$$
$$= (M^2 - 1)^{1/2}[(p + j\nu M)^2 + \nu^2]^{1/2}$$

with

$$M = \frac{U}{c} \qquad \text{and} \qquad \nu = \frac{\omega}{(M^2 - 1)c}$$

That branch of $\lambda$ for which $\Re\lambda > 0$ is implied, while if $y < 0$, the sign of $\lambda$ must be changed.

We may invert $\Phi$ with the aid of the convolution theorem. Considering first the factor $\lambda^{-1} \exp(-\lambda y)$, we apply the shifting theorem to obtain

$$\mathcal{L}^{-1}\{\lambda^{-1}e^{-\lambda y}\} = (M^2 - 1)^{-1/2}e^{-j\nu Mx}\mathcal{L}^{-1}\{(p^2 + \nu^2)^{-1/2} \exp[-(p^2 + \nu^2)^{1/2}(M^2 - 1)^{1/2}y]\}$$

which then may be inverted from tables (see, for example, Ref. 10, No. 866) to yield, by means of the convolution theorem, the end result

$$\phi = -(M^2 - 1)^{-1/2} \int_0^{x-(M^2-1)^{1/2}y} K(x - \xi, y)w(\xi)\, d\xi \qquad x > (M^2 - 1)^{1/2}y$$

where

$$K(x,y) = e^{j(\omega t - \nu Mx)}J_0\{\nu[x^2 - (M^2 - 1)y^2]\}$$

## FOURIER TRANSFORMS

### 3.17 Introduction

We have seen that the Laplace transformation is especially suited to initial-value problems,[17] in that the transform of the $n$th derivative incorporates the initial values of the first $n - 1$ derivatives. The Fourier transformation, on the other hand, appears to best advantage in boundary-value problems associated with semi-infinite or infinite domains, with the appropriate selection depending on the boundary conditions or symmetry considerations. We also recall that the convergence restrictions for Fourier transforms generally are more stringent than for Laplace transforms.

### 3.18 Transforms of Derivatives

With the three types of Fourier transforms defined as in Eqs. (3.15) to (3.18), integration by parts yields

$$\mathfrak{F}\{f^{(n)}(x)\} = (jk)^nF(k) \tag{3.45a}$$

$$\mathfrak{F}_c\{f''(x)\} = -k^2F_c(k) - f'(0) \tag{3.45b}$$

$$\mathfrak{F}_s\{f''(x)\} = -k^2F_s(k) + kf(0) \tag{3.45c}$$

Analogous results may be established for the cosine and sine transforms of higher derivatives of even order, but the cosine (sine) transform of an odd derivative involves the sine (cosine) transform of the original function. Thus, as implied directly by their trigonometric kernels, these transforms are intrinsically suited to differential equations having only even derivatives with respect to the variable in question. Moreover, the cosine (sine) transform of such a differential equation incorporates only the values of the odd (even) derivatives at $x = 0$; the values of other derivatives at $x = 0$ could be incorporated as constants to be determined, but the most satisfactory applications are those in which the unincorporated boundary conditions are null conditions at $x = \infty$.

The exponential transform, on the other hand, may be applied to all derivatives, but it incorporates no boundary values and therefore arises naturally only for infinite domains; to be sure, it may be applied to semi-infinite domains, but then it becomes essentially a Laplace transform.

The cosine or sine transform also may be advantageously applied to an infinite domain if $f(x)$ is an even or odd function of $x$, respectively, in which case $f'(0)$ or $f(0)$ vanishes in consequence of symmetry.

### 3.19 Application to a Semi-infinite Domain

It is evident from the foregoing remarks that the cosine and sine transforms are less flexible than the Laplace transform when applied to a semi-infinite domain; nevertheless, where applicable they may offer distinct advantages. We shall illustrate this last assertion for the heat-conduction problem of Eqs. (3.41) et seq., although it should be emphasized that we are comparing the application of the Laplace transform relative to $t$ with the Fourier transform relative to $x$; the Laplace transform is not well suited to $x$, nor is the Fourier transform to $t$.

By taking the sine transform of Eq. (3.41) and incorporating the boundary condition at $x = 0$ in accordance with Eq. (3.45$c$), we obtain

$$V_t + \kappa k^2 V = \kappa k v_0$$
$$V = 0 \qquad \text{at } t = 0$$

to which the solution is

$$V = \frac{v_0}{k}[1 - \exp(-k^2 \kappa t)] \tag{3.46}$$

This solution could be obtained via a Laplace transform with respect to $t$, but would be a rather trivial application thereof. Substituting in the inversion formula (3.18$b$) and using the known result

$$\int_0^\infty k^{-1} \sin(kx)\, dk = \frac{\pi}{2} \qquad x > 0$$

yields

$$\frac{v}{v_0} = 1 - \frac{2}{\pi}\int_0^\infty e^{-k^2\kappa t} \sin kx k^{-1}\, dk \tag{3.47}$$

in agreement with the results of Eq. (3.43$b$).

### 3.20 Initial-value Problem for One-dimensional Wave Equation

A classical problem in wave motion requires the solution to the wave equation

$$c^2 \phi_{xx} = \phi_{tt} \tag{3.48}$$

for the initial values

$$\phi = f(x) \qquad \text{and} \qquad \phi_t = g(x) \qquad \text{for } -\infty < x < \infty$$

We may imagine $f(x)$ and $g(x)$ to be the initial displacement and velocity of an infinitely long string.

By taking the Fourier transform with respect to $x$ and using the result (3.45a), we obtain

$$\Phi_{tt} + (kc)^2\Phi = 0$$

and

$$\Phi = F(k) \qquad \text{and} \qquad \Phi_t = G(k) \qquad \text{at } t = 0$$

the solution to which is

$$\Phi = F(k) \cos (kct) + (kc)^{-1}G(k) \sin (kct) \tag{3.49a}$$
$$= \tfrac{1}{2}[F(k) + (jkc)^{-1}G(k)]e^{jkct} + \tfrac{1}{2}[F(k) - (jkc)^{-1}G(k)]e^{-jkct} \tag{3.49b}$$

By applying the shifting theorem (3.30b) with $p = jk$ and $a = \pm jkc$ and noting that division by $jk$ transforms to integration with respect to $x$ [cf. Eq. (3.29)], we get

$$\phi = \tfrac{1}{2}[f(x + ct) + f(x - ct)] + \frac{1}{2c} \int_{x-ct}^{x+ct} g(\xi)\, d\xi \tag{3.50}$$

## HANKEL TRANSFORMS

### 3.21 Introduction

The Hankel transformation arises naturally in connection with the wave equation in cylindrical polar coordinates, viz.,

$$\phi_{xx} + \phi_{rr} + \frac{1}{r}\,\phi_r + \frac{1}{r^2}\,\phi_{\theta\theta} = \frac{1}{c^2}\,\phi_{tt} \tag{3.51a}$$

or, under the assumption that $\phi$ varies as either $\cos n\theta$ or $\sin n\theta$,

$$\phi_{xx} + \phi_{rr} + \frac{1}{r}\,\phi_r - \frac{n^2}{r^2}\,\phi = \frac{1}{c^2}\,\phi_{tt} \tag{3.51b}$$

Taking the $n$th-order Hankel transform of Eq. (3.51b), we require

$$\mathcal{H}_n\left\{\phi_{rr} + \frac{1}{r}\,\phi_r - \frac{n^2}{r^2}\,\phi\right\} = \int_0^\infty \left\{\frac{1}{r}\,(r\phi_r)_r - \frac{n^2}{r^2}\,\phi\right\} J_0(kr)r\, dr$$
$$= -k^2\mathcal{H}_n\{\phi\} \tag{3.52}$$

as follows after twice integrating by parts and invoking Bessel's differential equation

$$\left\{\frac{d^2}{dr^2} + \frac{1}{r}\frac{d}{dr} + \left(k^2 - \frac{n^2}{r^2}\right)\right\} J_0(kr) = 0$$

### 3.22 Problem of an Oscillating Piston

We shall illustrate the Hankel transformation by applying it to the problem of an oscillating piston of radius $a$ in an infinite baffle at $x = 0$, as shown in Fig. 3.8; this frequently is used as a simple model of a loudspeaker. On introducing the velocity potential $\phi$, we require a solution to the wave equation (3.51*a*) subject to the boundary condition

$$\phi_x = \begin{cases} j\omega A e^{j\omega t} & \text{at } x = 0 \text{ and } 0 \leq r < a \\ 0 & \text{at } x = 0 \text{ and } r > a \end{cases} \tag{3.53}$$

where $c$ denotes the velocity of sound and $\omega$ the angular frequency of oscillation. In accordance with the usual convention, $A$ is the complex displacement, assumed small, of the piston from its equilibrium position, and the actual displacement is given by the real part of $A$ exp $(j\omega t)$; similarly, only the real part of $\phi$ is to be retained in the end result.

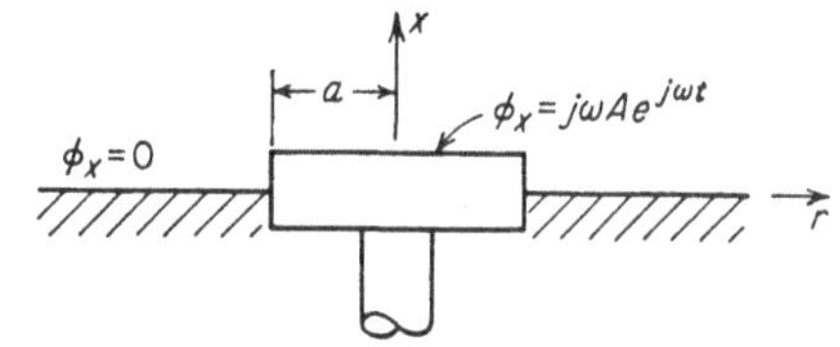


**Fig. 3.8** An oscillating circular piston in an infinite baffle.

We first observe that, by virtue of radial symmetry, $\phi$ must be independent of $\theta$ and therefore $\phi$ satisfies Eq. (3.51*b*) with $n = 0$; accordingly, we introduce a Hankel transform of order zero. We also remark that $\phi$ must be proportional to exp $(j\omega t)$ in order to satisfy Eq. (3.53), so that $\phi_{tt} = -\omega^2\phi$. Transforming Eqs. (3.51*b*) and (3.53) then yields

$$\Phi_{xx} - \left[k^2 - \left(\frac{\omega}{c}\right)^2\right]\Phi = 0$$

and

$$\Phi_x\Big|_{x=0} = j\omega A e^{j\omega t}\int_0^a J_0(kr)r\,dr$$
$$= j\omega A a k^{-1} J_1(ka)e^{j\omega t}$$

to which a solution is

$$\Phi = -j\omega\frac{Aa}{k\lambda}J_1(ka)e^{j\omega t - \lambda x} \qquad \lambda = \sqrt{k^2 - \left(\frac{\omega}{c}\right)^2} \tag{3.54}$$

where $\lambda$ must be a positive real number for $k > \omega/c$ and a positive imaginary number ($\lambda = i\mu$, $\mu > 0$) for $k < \omega/c$, the latter condition following from the fact that the disturbance must satisfy the radiation conditions of propagating outward from $x = 0$. Inverting $\Phi$ via Eq. (3.27*b*) then yields

$$\phi = -j\omega A a \int_0^\infty \lambda^{-1}J_1(ka)J_0(kr)e^{j\omega t - \lambda x}\,dk \tag{3.55}$$

The pressure on the piston is given by

$$p = -\rho\phi_t\Big|_{x=0} = -\rho\omega^2 A a e^{j\omega t}\int_0^\infty \lambda^{-1}J_1(ka)J_0(kr)\,dk$$

while the total force thereon is expressed as

$$2\pi \int_0^a pr\,dr = -2\pi\rho\omega^2 A a^2 e^{j\omega t} \int_0^\infty (k\lambda)^{-1} J_1^2(ka)\,dk$$

After some manipulation, this last integral may be reduced to the complex sum of a Bessel function and a Struve function, in agreement with the result obtained originally, in an entirely different fashion, by Lord Rayleigh.[19]

## FINITE FOURIER TRANSFORMS

### 3.23 Introduction

The transforms considered thus far have been applicable to semi-infinite or infinite domains and have had a common antecedent in Fourier's integral formula (3.12). It is natural to inquire whether transforms can be defined by Eq. (3.1) and their inverses derived from the theory of Fourier series. The essential result of this theory (Chap. 2) is that if $\psi_1(x)$, $\psi_2(x)$, . . . constitute a complete orthogonal set of functions for the interval $a < x < b$ and the weighting function $w(x)$, corresponding to a discrete set of eigenvalues $p_1$, $p_2$, . . . —that is, if

$$\int_a^b \psi_m(x)\psi_n(x)w(x)\,dx = \delta_{nm}N(p_n)$$

where $\delta_{nm}$ is the Kronecker delta, defined by

$$\delta_{nm} = 0,\ m \neq n;\ \delta_{nm} = 1,\ m = n$$

and $N(p_n)$ denotes the integral of $[\psi_n(x)]^2 w(x)$—then it may be shown that

$$F(p) = \int_a^b f(x)\psi(x,p)w(x)\,dx \tag{3.56a}$$

and

$$f(x) = \sum_{p_n} \frac{F(p_n)}{N(p_n)}\psi_n(x) \qquad \text{where } \psi_n(x) = \psi(x,p_n) \tag{3.56b}$$

corresponding to $K(p,x) = \psi(x,p)w(x)$ in Eq. (3.1). Equations (3.56*a*) and (3.56*b*) evidently constitute a generalized finite-Fourier-transform pair.

The choice of the orthogonal functions $\psi(x,p)$ depends both on the differential equation and on the boundary conditions to be satisfied by $f(x)$, just as with the infinite transforms. We shall consider here only finite sine and cosine transforms, appropriate to differential equations containing only even derivatives with respect to $x$ and implying a weighting function equal to unity. The only other finite transform that has been applied extensively is that of Hankel (see especially Ref. 7), which is appropriate to the radial derivatives in $\nabla^2 f$, viz., $f_{rr} + r^{-1}f_r$, and implies a weighting function $w = r$ corresponding to that in the element of area $r\,dr\,d\theta$ in polar coordinates. The technique is, however, clearly applicable to all functions and boundary conditions of the Sturm-Liouville type and serves to mechanize much of the time-consuming detail associated with the determination of the unknown coefficients in the classical procedure that begins with separation of variables.

### 3.24 Finite Cosine and Sine Transforms

The simplest finite cosine and sine transforms, introduced originally (1935) by Doetsch, are those for which $a = 0$, $b = \pi$, and $p = n$ in Eqs. (3.56*a*) and (3.56*b*), which then reduce to

$$F(n) = \int_0^\pi f(x) \begin{matrix}\cos\\ \sin\end{matrix}\, nx\, dx \tag{3.57a}$$

and

$$f(x) = \frac{1}{\pi} \sum_{n=0}^{\infty} (2 - \delta_{n0}) F(n) \begin{matrix}\cos\\ \sin\end{matrix}\, nx \tag{3.57b}$$

The corresponding transforms of the second derivatives are given by

$$\int_0^\pi f''(x) \cos nx\, dx = -n^2 F(n) + (-1)^n f'(\pi) - f'(0) \tag{3.58a}$$

and

$$\int_0^\pi f''(x) \sin nx\, dx = -n^2 F(n) + n[f(0) - (-1)^n f(\pi)] \tag{3.58b}$$

whence these transforms are expedient for problems where $f'(x)$ or $f(x)$, respectively, is prescribed at the end points of the interval. The generalization to an interval of length other than $\pi$ merely requires a scale transformation.

More general forms of these transforms, corresponding to Fourier's own generalization of his series, are given by

$$F(k) = \int_0^l f(x) \begin{matrix}\cos\\ \sin\end{matrix}\, kx\, dx \tag{3.59a}$$

$$f(x) = \sum_k \left[ \frac{(2 - \delta_{k0})(k^2 + h^2)}{h + (k^2 + h^2)l} \right] F(k) \begin{matrix}\cos\\ \sin\end{matrix}\, kx \tag{3.59b}$$

$$k \begin{matrix}\tan\\ \cot\end{matrix}\, kl = \pm h \tag{3.59c}$$

$$\int_0^l f''(x) \begin{matrix}\cos\\ \sin\end{matrix}\, kx\, dx = -k^2 F(k) + [f'(l) + hf(l)] \begin{matrix}\cos\\ \sin\end{matrix}\, kl \begin{matrix} -f'(0)\\ +kf(0)\end{matrix} \tag{3.60}$$

where either upper or lower alternatives must be taken together. These results are applicable to problems in heat conduction in which radiation takes place at $x = l$ and to problems involving lumped parameters at the boundaries of electrical or mechanical systems, all of which prescribe $f'(l) + hf(l)$.

We remark that $h$ usually is nonnegative in physical problems, by virtue of which Eq. (3.59*c*) has only real roots, which occur in pairs of equal magnitude and opposite sign, with only the positive values being included in the summation of Eq. (3.59*b*). But if $h$ is negative, Eq. (3.59*c*) has a pair of conjugate imaginary roots, one of which must appear in Eq. (3.59*b*) and accordingly must introduce a hyperbolic function therein; we note, however, that $f(x)$ depends only on $k^2$ and therefore remains real. If $h = 0$, then the value $k = 0$ appears as a nontrivial root of Eq. (3.59*c*) for the cosine transform.

These results all have their counterparts for Hankel transforms. See Ref. 5 or Ref. 7 for $h > 0$. Hankel transforms for $h < 0$ do not appear to have been considered in the literature, but the properties of the corresponding orthogonal expansions, the Dini series, are well known; see pages 596ff. of Ref. 20.

### 3.25 A Problem in Wave Motion

We now consider, by way of illustration, the application of the finite sine transform of Eq. (3.59) to the problem of Sec. 3.14. The prescribed boundary conditions for this problem are on $f(0)$ and $f'(1)$, whence Eq. (3.60) indicates a sine transform with $h = 0$ and $kl$ an odd multiple of $\pi/2$, viz.,

$$Y(k,t) = \int_0^l y(x,t) \sin kx \, dx \tag{3.61}$$

where $$kl = \frac{\pi}{2}(2n + 1) \qquad n = 0, 1, \ldots$$

Transforming the differential equation (3.37) then yields

$$Y_{tt} + (kc)^2 Y = \frac{Pc^2}{E} \sin kl$$

to which a solution satisfying the initial conditions $Y = Y_t = 0$ is given by

$$Y(k,t) = \frac{P}{k^2 E}(1 - \cos kct) \sin kl$$

Substituting this result in Eq. (3.59*b*) with $Y = F$, $h = 0$, and

$$kl = \frac{\pi}{2}(2n + 1)$$

gives

$$y(x,t) = \frac{8}{\pi^2}\frac{Pl}{E}\sum_{n=0}^{\infty}\frac{(-1)^n}{(2n+1)^2}\left[1 - \cos\frac{(2n+1)\pi ct}{2l}\right] \times \sin\frac{(2n+1)\pi x}{2l} \tag{3.62}$$

The terms in the series that are not time-dependent may be identified as the Fourier series representation of $Px/E$, whence Eq. (3.62) may be reduced to Eq. (3.39). A comparison of the detailed solutions, only outlines of which have been given here, indicates the superiority of the finite-transform method in obtaining the modal expansion.

## 3.26 Conclusion

The superiority of the finite-transform method, in obtaining solutions as expansions of natural modes, over either the classical procedure of separation of variables or the Laplace transformation tends to increase with the complexity of the problem and is even more evident for the Hankel transform. In particular, the finite-transform method always provides the modal expansion of the static solution, for example, $Px/E$ in the above problem, although this may not always be an advantage. We emphasize, nevertheless, that the Laplace transform is both more flexible and more powerful. Not only does it incorporate alternative interpretations, such as the traveling-wave expansion of Eq. (3.40), but it also places less stringent conditions on the boundary conditions that may be accommodated. To be sure, the expansions provided by Laplace transforms for non-Sturm-Liouville problems might be used to develop finite transforms of a more general type, but such an *ad hoc* procedure scarcely appears worthwhile.

The properties of several finite transforms are summarized in Table 3.2.

## EXERCISES

**1.** A uniform bar of unit cross section is at rest and unstressed for $t < 0$ and has free ends at both $x = 0$ and $x = l$. When $t = 0$, a force $P$ is applied at $x = l$. Use both the Laplace-transform and finite-Fourier-transform methods to show that the subsequent displacement of any section initially at $x$ is given by

$$y(x,t) = \frac{Pt^2}{2m} + \frac{2Pl}{\pi^2 E}\sum_{n=1}^{\infty}\frac{(-1)^n}{n^2}\left(1 - \cos\frac{n\pi ct}{l}\right)\cos\frac{n\pi x}{l}$$

where $c^2 = E/\rho$, $m = \rho l$, and $\rho$ and $E$ denote density and Young's modulus. (*Hint:* The center of mass of any body, not necessarily rigid, moves as if all external forces were applied there; this component of the motion may be separated out in the finite-

**Table 3.2 Finite Integral Transforms**

| Transform (references†) | $F(p) = T\{f(x)\}$ | $f(x) = T^{-1}\{F(p)\}$ | Eigenvalue equation for $p_n$ ($p_n \geq 0$) | Transforms of typical derivatives |
|---|---|---|---|---|
| Cosine (2, 5, 7) | $\int_0^l f(x) \cos px \, dx$ | $\sum_{p_n} \frac{(2 - \delta_{p0})(p^2 + h^2)F(p) \cos px}{h + l(p^2 + h^2)}$ | $p \tan pl = h$ | $T\{f''(x)\} = -p^2F(p) - f'(0) + [f'(l) + hf(l)] \cos pl$ |
| Sine (2, 5, 7) | $\int_0^l f(x) \sin px \, dx$ | $2 \sum_{p_n} \frac{(p^2 + h^2)F(p) \sin px}{h + l(p^2 + h^2)}$ | $p \cot pl = -h$ | $T\{f''(x)\} = -p^2F(p) + pf(0) + [f'(l) + hf(l)] \sin pl$ |
| Hankel (5, 7) | $\int_0^a f(x)J_n(px)x \, dx$ | $2 \sum_{p_n} \frac{p^2F(p)J_n(px)}{[(h^2 + p^2)a^2 - n^2]J_n^2(pa)}$ $(h > 0)$ | $pJ_n'(pa) + hJ_n(pa) = 0$ | $T\left\{f'' + \frac{1}{x}f' - \frac{n^2}{x^2}f\right\} = -p^2F(p) + [f'(a) + hf(a)]aJ_n(pa)$ |
| Annular Hankel (5, 7) | $\int_b^a f(x)B_n(px) \, dx$<br>$B_n(px) = Y_n(pa)J_n(px) = -J_n(pa)Y_n(px)$ | $\frac{\pi^2}{2} \sum_{p_n} \frac{p^2J_n^2(pb)F(p)B_n(px)}{J_n^2(pa) - J_n^2(pb)}$ | $B_n(pb) = 0$ | $T\left\{f'' + \frac{1}{x}f' - \frac{n^2}{x^2}f\right\} = -p^2F(p) + \frac{2}{\pi}\left[\frac{J_n(pa)}{J_n(pb)} f(b) - f(a)\right]$ |
| Legendre (5) | $\int_{-1}^1 f(x)P_n(x) \, dx$ | $\frac{1}{2} \sum_{n=0}^{\infty} (2n + 1)F(n)P_n(x)$ | $p = n$ | $T\{[(1 - x^2)f'(x)]'\} = -n(n + 1)F(n)$ |

† Numbers refer to the list of references at the end of the chapter.

transform approach.) Obtain also an expression for the motion in terms of traveling waves.

**2.** Let $v_0$ be a function of $t$ in the problem of Sec. 3.15. Use the convolution theorem to obtain the solution

$$v(x,t) = \frac{x}{2\sqrt{\pi\kappa}} \int_0^t v_0(\tau)(t-\tau)^{-3/2} \exp\left[-\frac{x^2}{4\kappa(t-\tau)}\right] d\tau$$

**3.** Obtain the asymptotic solution to the problem of Sec. 3.15 by expanding Eq. (3.42) in ascending powers of $p$ and inverting term by term.

**4.** Use the Fourier sine transform to obtain a solution to Laplace's equation

$$\phi_{xx} + \phi_{yy} = 0 \qquad \text{in } x > 0 \quad \text{and} \quad 0 < y < b$$

subject to the boundary conditions

$$\begin{aligned} \phi &= f(x) && \text{for } x > 0 \quad \text{and} \quad y = 0 \\ &= 0 && \text{for } x > 0 \quad \text{and} \quad y = b \\ &= 0 && \text{for } x = 0 \quad \text{and} \quad 0 < y < b \end{aligned}$$

*Answer:*

$$\phi(x,y) = \frac{2}{\pi} \int_0^\infty f(\xi)\, d\xi \int_0^\infty \frac{\sinh k(b-y)}{\sinh kb} \sin kx \sin k\xi \, dk$$

**5.** Steady, axially symmetric heat conduction is governed by the equation

$$\nabla^2 v = v_{xx} + v_{rr} + \frac{1}{r} v_r = 0$$

Heat is supplied over a circular area on the surface of a semi-infinite solid, $x > 0$, yielding the boundary condition

$$Kv_x = -Q \qquad \text{for } 0 \le r < a \quad \text{and} \quad x = 0$$

Show that the steady temperature distribution is given by

$$v(x,r,t) = \frac{Qa}{K} \int_0^\infty k^{-1} J_1(ka) J_0(kr) e^{-kx}\, dk$$

**6.** A slab of infinite area extends from $x = 0$ to $x = 1$ and is initially at zero temperature. The temperature of the face at $x = 0$ is raised to $v = v_0$ when $t = 0$, while radiation takes place at the other face in accordance with

$$v_x + hv = 0 \qquad \text{at } x = l$$

Show that the subsequent temperature is given by

$$v(x,t) = 2v_0 \sum_k \frac{k^2 + h^2}{k[h + (k^2 + h^2)l]} (1 - e^{-k^2\kappa t}) \sin kx$$

where the $k$ are determined by

$$k \cot kl = -h$$

## REFERENCES

### Annotated Bibliography on Integral Transforms

*a. Texts*

1. Carslaw, H. S., and J. C. Jaeger, "Operational Methods in Applied Mathematics" Oxford University Press, New York, 1953. Deals only with the Laplace transform but contains extensive collection of worked and unworked problems involving both ordinary and partial differential equations.
2. Churchill, R. V., "Operational Mathematics," 2d ed., McGraw-Hill Book Company, Inc., New York, 1958. Elementary text dealing with the Laplace transform, finite Fourier transforms, and, briefly, complex-variable theory. Reasonable mathematical rigor is maintained, but the problems are more elementary than in Ref. 1.
3. McLachlan, N. W., "Complex Variable Theory and Transform Calculus," Cambridge University Press, New York, 1953. Excellent, albeit specialized, presentation of complex-variable theory required for handling inversion integrals. Uses Heaviside notation for Laplace transforms.
4. Thomson, W. T., "Laplace Transformation," Prentice-Hall, Inc., Englewood Cliffs, N.J., 1950. Similar in scope to Ref. 2; less rigorous mathematics but more elaborate engineering applications.
5. Tranter, C. J., "Integral Transforms in Mathematical Physics," Methuen & Co., Ltd., London, 1956. A brief (133-page) but not elementary coverage of all the commonly used transforms with more sophisticated physical examples than Refs. 1–4, above; an excellent supplement to any of Refs. 1–4.

*b. Treatises*

The following references do not contain exercises for the student but are otherwise more extensive and more advanced than Refs. 1–4, above.

6. Doetsch, G., "Theorie und Anwendung der Laplace-transformation," Dover Publications, New York, 1943. Originally published by Springer-Verlag, Berlin, Vienna, 1937. Standard mathematical treatise on the Laplace transform; extensive bibliography of pre-1937 works.
7. Sneddon, I. N., "Fourier Transforms," McGraw-Hill Book Company, Inc., New York, 1951. Extensive applications of various transforms to physical problems at research-paper level.
8. Titchmarsh, E. C., "Introduction to the Theory of Fourier Integrals," Oxford University Press, New York, 1948. Standard treatise on Fourier, including Fourier-Bessel or Hankel, integrals and transforms; largely complementary to Ref. 6; extensive bibliography of pre-1948 works.
9. Van der Pol, B., and H. Bremmer, "Operational Calculus Based on the Two-sided Laplace Integral," Cambridge University Press, New York, 1950. A modern, rigorous presentation of Heaviside's operational calculus *as operational calculus*. Advanced and stimulating applications in such diverse fields as electric circuits and number theory.

*c. Tables*

10. Campbell, G. A., and R. M. Foster, "Fourier Integrals for Practical Application," John Wiley & Sons, Inc., New York, 1948. The most extensive table of exponential Fourier integrals; many entries are effectively Laplace-transform pairs and are presented as such.

11. Erdélyi, A., and J. Cossar, "Dictionary of Laplace Transforms," Department of Scientific Research and Experiment, Admiralty Computing Service, London, 1944. Most of the material from these tables has been included in Ref. 12.
12. [EMOT] Erdélyi, A., (ed.), with W. Magnus, F. Oberhettinger, and F. Tricomi, "Tables of Integral Transforms," 2 vols., McGraw-Hill Book Company, Inc., 1954. The most comprehensive tables of integral transforms presently available. Volume 1 contains Fourier-exponential, -cosine, and -sine, Laplace, and Mellin transforms; vol. 2 contains Hankel transforms, along with many transforms not introduced in the foregoing treatment.
13. Magnus, W., and F. Oberhettinger, "Formulas and Theorems for the Special Functions of Mathematical Physics," Chelsea Publishing Co., New York, 1949. This indispensable (for the applied mathematician) compendium contains short but well-selected tables of both Fourier- and Laplace-transform pairs. Ideal for graduate students.
14. McLachlan, N. W., and P. Humbert, "Formulaire pour le calcul symbolique," Gauthier-Villars, Paris, 1950. Extensive table of Heaviside ($p$-multiplied Laplace) transforms.

*d. Asymptotic Expansions*

15. Copson, E. T., "The Asymptotic Expansion of a Function Defined by a Definite Integral or Contour Integral," Department of Scientific Research and Experiment, Admiralty Computing Service, London, 1946.
16. Erdélyi, A., "Asymptotic Expansions," Dover Publications, New York, 1956. The treatments of integrals given in Ref. 15 and in Chap. 2 of Ref. 16 are approximately similar in scope and give a coordinated treatment of material that otherwise can be found only in widely scattered sources.

## Other References

17. Barnes, John L., Functional Transformations for Engineering Design, chap. 14 in "Modern Mathematics for the Engineer," First Series, edited by E. F. Beckenbach, McGraw-Hill Book Company, Inc., New York, 1956.
18. Fourier, Joseph, "La Théorie analytique de la chaleur," Paris, 1822, translated by A. Freeman, Cambridge, 1878. Reprinted by Dover Publications, New York, 1955.
19. Rayleigh, Lord, "The Theory of Sound," London, 1894. Reprinted by Dover Publications, New York, 1945.
20. Watson, G. N., "Bessel Functions," The Macmillan Company, New York, 1948.

# 4

# Semigroup Methods in the Theory of Partial Differential Equations

**RALPH S. PHILLIPS**
PROFESSOR OF MATHEMATICS
STANFORD UNIVERSITY

## 4.1 Introduction

The theory of semigroups of operators is intimately connected with the *initial-value problem* for systems of partial differential equations. Mathematical physics abounds in problems of this kind; for instance, the initial-value problem for the heat equation,

$$\begin{gathered} u_t = u_{xx} \qquad -\infty < x < \infty \quad t > 0 \\ u(x,0+) = f(x) \end{gathered} \tag{4.1}$$

and the initial-value problem for the wave equation,

$$\begin{gathered} u_{tt} = u_{xx} \qquad -\infty < x < \infty \quad t > 0 \\ u(x,0+) = f(x) \qquad u_t(x,0+) = g(x) \end{gathered} \tag{4.2}$$

are both of this type. J. Hadamard has called such a problem *well formulated* if there is a unique solution for the given initial data and if the solution varies continuously with the initial data; see pages 93 to 94 and 111 to 112 of Ref. 1. These two requirements are certainly reasonable on physical grounds, the existence and uniqueness being merely an affirmation of the principle of scientific determinism, whereas the continuous dependence is an expression of the stability of the solution of the problem. In order that the initial-value problem for a given system of partial differential equations may be well formulated, it is frequently necessary to limit the set of initial data, for instance by imposing certain boundary conditions. Thus, for the heat equation over a finite interval, a well-formulated problem would be

$$\begin{aligned} u_t &= u_{\chi\chi} \qquad 0 < \chi < 1 \quad t > 0 \\ u(\chi,0+) &= f(\chi) \\ u(0,t) &= 0 \qquad u_\chi(1,t) + u(1,t) = 0 \quad t > 0 \end{aligned}$$

The purpose of this chapter is to give a precise, but somewhat abstract, formulation to the initial-value problem, also called the *Cauchy problem.*

If we assume that the system of differential equations is independent of time—or, equivalently, that the corresponding physical mechanism is time-invariant—then the Cauchy problem leads directly to a semigroup of operators. In fact, suppose a class of suitable initial data is denoted by $\mathfrak{D}$ and let the solution to the problem at time $t > 0$ with initial data $y$ in $\mathfrak{D}$ be denoted by $S(t)y$. We must also assume that $\mathfrak{D}$ is invariant in the sense that $S(t)y$ lies in $\mathfrak{D}$ for each $y$ in $\mathfrak{D}$; this guarantees that the solution is continuable. We can then determine the solution at time $t_1 + t_2$, taking both $t_1$ and $t_2 > 0$, either directly as $S(t_1 + t_2)y$ or indirectly by using $S(t_1)y$ as initial data and determining the solution at time $t_2$ later, as $S(t_2)[S(t_1)y]$. The uniqueness of the solution implies that

$$S(t_1 + t_2)y = S(t_2)[S(t_1)y]$$

or briefly,

$$S(t_1 + t_2) = S(t_2)S(t_1) \qquad t_1 > 0 \quad t_2 > 0 \tag{4.3}$$

This is the *semigroup property* for the operators $[S(t); t > 0]$.

In some physical problems, the initial data determine the entire past as well as the future of the mechanism. For such problems, the restriction $t_1 > 0$, $t_2 > 0$ in Eq. (4.3) need not be imposed and the resulting family of operators $[S(t)]$ defined for all $t$, $-\infty < t < \infty$, forms a *group* of operators. In particular, $S(t)$ and $S(-t)$ will be inverses of each other, so that

$$S(t)S(-t) = I$$

the identity operator.

The semigroup property of solutions to the time-invariant Cauchy problem is reflected in certain addition theorems. Thus, as is well known, in the case of the previously mentioned heat equation (4.1) we have

$$S(t)f = \int_{-\infty}^{\infty} k(\chi - \sigma, t)f(\sigma)\, d\sigma \tag{4.4}$$

and the kernel

$$k(\chi,t) = (\pi t)^{-1/2} \exp\left(\frac{-\chi^2}{t}\right)$$

exhibits the semigroup property of the operators $[S(t); t > 0]$ in the form

$$k(\chi, t_1 + t_2) = \int_{-\infty}^{\infty} k(\chi - \sigma, t_2)k(\sigma,t_1)\, d\sigma$$

since

$$\int_{-\infty}^{\infty} k(\chi - \sigma, t_1 + t_2)f(\sigma)\, d\sigma = S(t_1 + t_2)f$$
$$= S(t_2)[S(t_1)f] = \int_{-\infty}^{\infty} k(\chi - \omega, t_2) \left[\int_{-\infty}^{\infty} k(\omega - \sigma, t_1)f(\sigma)\, d\sigma\right] d\omega$$

A similar result holds for the wave-equation initial-value problem (4.2). If we write $\eta^1 = u_\chi$, $\eta^2 = u_t$, then Eq. (4.2) takes the form

$$\begin{gathered} \eta^1{}_t = \eta^2{}_\chi \qquad \eta^2{}_t = \eta^1{}_\chi \qquad -\infty < \chi < \infty \qquad t > 0 \\ \eta^1(\chi,0+) = f'(\chi) \qquad \eta^2(\chi,0+) = g(\chi) \end{gathered} \tag{4.5}$$

and the solution can be written as

$$S(t)F = \int_{-\infty}^{\infty} k(\chi - \sigma, t)F(\sigma)\, d\sigma$$

Here $F(\chi)$ is the column vector $F(\chi) = [f'(\chi), g(\chi)]$ and

$$K(\chi,t) = \frac{1}{2}\begin{pmatrix} \delta(\chi + t) + \delta(\chi - t) & \delta(\chi + t) - \delta(\chi - t) \\ \delta(\chi + t) - \delta(\chi - t) & \delta(\chi + t) + \delta(\chi - t) \end{pmatrix}$$

where $\delta(\chi)$ denotes the Dirac delta function (see Chap. 1). Again the fact that the solution satisfies the semigroup property results in the kernel satisfying the relationship

$$K(\chi, t_1 + t_2) = \int_{-\infty}^{\infty} K(\chi - \sigma, t_2)K(\sigma,t_1)\, d\sigma$$

Stability requires that $S(t)y_n$ converges to $S(t)y$ whenever $y_n$ converges to $y$, whence we may conclude that the operator $S(t)$ is continuous on $\mathfrak{D}$. Moreover, the statement that $y$ in $\mathfrak{D}$ is the initial value for the solution of a Cauchy problem means that the solution at time $t$, namely $S(t)y$, approaches $y$ as $t$ goes to $0+$, in other words that

$$\lim_{t \to 0+} S(t)y = y$$

We shall in addition assume that the problem is linear; i.e., for $y$ and $w$ in $\mathfrak{D}$, and $a$ and $b$ complex numbers, we assume that $ay + bw$ lies in $\mathfrak{D}$ and that the solution satisfies the equation

$$S(t)(ay + bw) = aS(t)y + bS(t)w \qquad t > 0 \tag{4.6}$$

This will be the case when the system of partial differential equations is linear and the boundary conditions determining $\mathfrak{D}$ are homogeneous.

Uniqueness, stability, and linearity constitute our principal assumptions, and it is evident that a large class of problems in partial differential equations meet these conditions and consequently have solutions that can be described by semigroups of linear operators continuous on $\mathfrak{D}$ for each $t > 0$ and such that

$$\lim_{t\to 0+} S(t)y = y$$

for each $y$ in $\mathfrak{D}$.

## 4.2 Semigroups of Operators on Finite-dimensional Spaces

We now proceed to a discussion of one-parameter families of operators of the sort described in Sec. 4.1, and we begin by considering the simplest case, namely, that for which $\mathfrak{D}$ is one-dimensional. In effect, then, $\mathfrak{D}$ will be the complex-number field, and for each $t > 0$, $S(t)$ will be merely a multiplicative factor, also a complex number. Since $S(t + \tau) = S(t)S(\tau)$ and $S(\tau) \to 1$ as $\tau \to 0+$, we see that $S(t)$ is continuous on the right and hence integrable. Clearly we have

$$\lim_{t\to 0+} t^{-1} \int_0^t S(\tau)\, d\tau = 1$$

so that in particular, for $t$ sufficiently small, $S(t)$ must satisfy the condition

$$\int_0^t S(\tau)\, d\tau \neq 0$$

For $0 < \epsilon < t$, we have

$$\begin{aligned} \epsilon^{-1}[S(\epsilon) - 1] \int_0^t S(\tau)\, d\tau &= \epsilon^{-1}\left[\int_0^t S(\tau + \epsilon)\, d\tau - \int_0^t S(\tau)\, d\tau\right] \\ &= \epsilon^{-1} \int_t^{t+\epsilon} S(\tau)\, d\tau - \epsilon^{-1} \int_0^\epsilon S(\tau)\, d\tau \end{aligned}$$

The right-hand member has the limit $S(t) - 1$ as $\epsilon \to 0+$, and therefore

$$\lim_{\epsilon\to 0+} \epsilon^{-1}[S(\epsilon) - 1] \equiv A \tag{4.7}$$

exists. Accordingly, we obtain

$$S(t) = 1 + A \int_0^t S(\tau)\, d\tau$$

and successive substitutions into the right-hand member yield

$$S(t) = \sum_{n=0}^{\infty} \frac{(At)^n}{n!} \equiv \exp(At) \tag{4.8}$$

Although the relationship (4.8) has been established only for $t$ sufficiently small, the semigroup property shows directly that it holds for all $t > 0$. Thus in the one-dimensional case we can show that all semigroups of operators are exponentials.

The argument given above also applies when $\mathfrak{D}$ is finite-dimensional, with the difference that now $S(t)$ and $A$ are to be interpreted as matrices. The relationships (4.7) and (4.8) continue to hold, however, and we obtain in this way a general form for the solution to problems that arise in the theory of small vibrations and in electric-circuit theory. Thus, if $y$ is a given initial vector and we set $y(t) = S(t)y$, then it follows that

$$\begin{aligned} \frac{dy}{dt} &= Ay \qquad t > 0 \\ y(0) &= y \end{aligned} \tag{4.9}$$

which is the general form of the initial-value problem in the finite-dimensional case. As an example of the solution to Eq. (4.9) for

$$A = \begin{pmatrix} \alpha & 0 & 0 \\ 1 & \alpha & 0 \\ 0 & 1 & \alpha \end{pmatrix} \tag{4.10}$$

we have

$$S(t) = \begin{pmatrix} e^{\alpha t} & 0 & 0 \\ te^{\alpha t} & e^{\alpha t} & 0 \\ \frac{t^2}{2} e^{\alpha t} & te^{\alpha t} & e^{\alpha t} \end{pmatrix} \tag{4.11}$$

### 4.3 Hilbert Space

Applications of semigroup theory to problems such as the Cauchy problem for the heat and wave equations require that we extend the foregoing theory to the infinite-dimensional case. The resulting theory is much more refined and considerably richer in detail. It is now necessary, however, that we be precise in defining our concepts. For example, we have repeatedly used the notion of a limit in $\mathfrak{D}$ without saying what was meant by this concept. As a matter of fact, almost all definitions of limit in the finite-dimensional case are equivalent. This is not true of the infinite-dimensional case, and so we shall start by describing a suitable setting for our theory.

The simplest and perhaps the most useful of the infinite-dimensional spaces is the Hilbert space $H$ (see, for instance, the book by F. Riesz and B. Sz.-Nagy[9]); $H$ is a direct generalization of the familiar complex euclidean space and accordingly one has considerable insight into its geometry right from the start. Thus $H$ is a vector space with complex numbers as scalar multipliers; more precisely, $H$ consists of elements $y$, $z$, $w$, etc., called *vectors;* to every pair $y$ and $z$ of vectors there corresponds a vector $w$, called their *sum*, $w = y + z$; and to every vector $y$ and complex number $a$ there corresponds a vector $z$, called their *product*, $z = ay$. The sum and product satisfy the following properties:

*a.* $y + z = z + y$
*b.* $y + (z + w) = (y + z) + w$

*c*. There exists a unique *zero* vector, denoted by 0, such that for all $y$ in $H$,

$$y + 0 = y$$

*d*. To each $y$ in $H$ there corresponds a unique vector, denoted by $-y$, with the property

$$y + (-y) = 0$$

*e*. $a(y + z) = ay + az$
*f*. $(a + b)y = ay + by$
*g*. $a(by) = (ab)y$
*h*. $0y = 0$ and $1y = y$

What distinguishes the Hilbert space $H$ from other vector spaces is the existence of a complex-valued *inner product* $(y,z)$ defined for all ordered pairs $y$ and $z$ in $H$ and having the properties

*a*. $(ay,z) = a(y,z)$
*b*. $(y + z, w) = (y,w) + (z,w)$
*c*. $(y,z) = \overline{(z,y)}$ where $\overline{\alpha + j\beta} = \alpha - j\beta$ and $\alpha$, $\beta$ are real
*d*. $(y,y) \geq 0$ the equality sign holding if and only if $y = 0$

The *norm*

$$\|y\| = [(y,y)]^{1/2}$$

is a measure of the length of the vector $y$. We note that

$$\|y\| \geq 0$$

the equality sign holding if and only if $y = 0$. Further, we have

$$\|ay\| = |a| \, \|y\|$$

The fact that the cosine of the angle between two vectors is of absolute value less than or equal to 1 is expressed by the inequality of Schwarz (see pages 126, 134, 166, 175, and 400 of Ref. 1):

$$|(y,z)| \leq \|y\| \, \|z\|$$

In order to establish this inequality, we note that, if $y$ and $z$ are both equal to the zero element, then the result is obviously valid. If $z \neq 0$, then

$$0 \leq (y + az, y + az) = (y,y) + a(z,y) + \bar{a}(y,z) + |a|^2(z,z)$$

and by setting $a = -(y,z)/(z,z)$ we obtain the desired result. The so-called triangle inequality

$$\|y + z\| \leq \|y\| + \|z\|$$

now follows directly; in fact, we have

$$\|y + z\|^2 = \|y\|^2 + (y,z) + (z,y) + \|z\|^2 \leq \|y\|^2 + 2\|y\| \, \|z\| + \|z\|^2$$

The distance between two elements $y$ and $z$ is defined as $\|y - z\|$. It is clear that this function is symmetric and nonnegative, zero if and only if $y = z$, and it follows from the triangle inequality that

$$\|y - z\| \leq \|y - w\| + \|w - z\|$$

It also follows from the triangle inequality that

$$\big| \|y\| - \|z\| \big| \leq \|y - z\|$$

from which we see that $\|y\|$ is a continuous function. The inner product can also be shown to be a continuous function of both of its arguments.

It is assumed that $H$ is *complete* in terms of the above distance function; that is, given a *Cauchy sequence* $\{y_n\}$, which by definition is a sequence such that

$$\lim_{m,n\to\infty} \|y_m - y_n\| = 0$$

then there is a $y$ in $H$ for which

$$\lim_{n\to\infty} \|y_n - y\| = 0$$

Finally, it is assumed that $H$ is *not finite-dimensional.*

The simplest example of a Hilbert space is the space of complex-valued sequences

$$y = \{\eta^1, \eta^2, \ldots\} \qquad z = \{\zeta^1, \zeta^2, \ldots\}$$

for which the sum of the squares of the absolute values of the components is finite; here the inner product is defined as

$$(y,z) = \sum_{n=1}^{\infty} \eta^n \overline{\zeta^n}$$

Another example of a Hilbert space is given by $L_2(G)$, the Lebesgue-measurable complex-valued functions $y = y(\chi)$ defined on a domain $G$ in an $m$-dimensional real euclidean space $E_m$ and having the property that

$$\int_G |y(\chi)|^2 \, d\chi < \infty$$

In this case, the inner product is defined for $y$ and $z$ in $L_2(G)$ as

$$(y,z) = \int_G y(\chi)\overline{z(\chi)} \, d\chi$$

In order that the inequality $(y,y) > 0$ be satisfied when $y \neq 0$, it is necessary to identify all functions that differ only on sets of measure zero. The resulting classes of functions in $L_2(G)$ then define a Hilbert space. It should be noted that in such a space it is possible to have a sequence

$\{y_n\}$ for which $\int_G |y_n(\chi)|^2\, d\chi$ converges to zero even though $y_n(\chi)$ does not converge for any point $\chi$ in $G$.

A *subspace* $L$ contained in $H$ is simply a vector space, i.e., a subset with the property that for all $y$ and $z$ in $L$ and for all complex numbers $a$ and $b$, the vector $ay + bz$ lies in $L$. A *closed subspace* is a subspace that contains all its limit points. It is easy to see that a closed subspace of a Hilbert space is complete and therefore is itself a Hilbert space when it is not finite-dimensional. Two vectors $y$ and $z$ of $H$ are said to be *orthogonal* if $(y,z) = 0$; it follows that the zero vector is orthogonal to every vector of $H$. Two sets $S_1$ and $S_2$ are said to be orthogonal if every element of $S_1$ is orthogonal to every element of $S_2$. The *orthogonal complement* of a set $S$, in symbols $S^\perp$, is the set of all vectors orthogonal to $S$. It is readily verified that the orthogonal complement of a set is a closed subspace of $H$.

If $L$ is a closed subspace of $H$ and $y$ is an arbitrary element in $H$, then there is in $L$ a unique closest element $y_1$ to $y$. In fact, let

$$d = \inf\,(\|y - z\|;\, z \text{ in } L)$$

Then there is a sequence $\{z_n\}$ in $L$ such that $\|y - z_n\|$ converges to $d$. Now it can readily be verified that

$$\|z_m - z_n\|^2 = 2(\|y - z_n\|^2 + \|y - z_m\|^2) - 4\|y - \tfrac{1}{2}(z_n + z_m)\|^2$$

and since $\|y - z_n\|^2$ and $\|y - z_m\|^2$ converge to $d^2$, and

$$\|y - \tfrac{1}{2}(z_n + z_m)\|^2 \geq d^2$$

[because $(\frac{1}{2})(z_n + z_m)$ lies in $L$], it follows that $\{z_n\}$ forms a Cauchy sequence. Thus $\{z_n\}$ converges to some $y_1$ in $L$ with $\|y - y_1\| = d$. Now, if there were a second value $w_1$ in $L$ with $\|y - w_1\| = d$, then the above identity applied to $y_1$ and $w_1$ shows that $\|y_1 - w_1\| = 0$, which proves that $y_1$ is unique.

To pursue the foregoing matter a bit further, suppose $y$ and $y_1$ are defined as above and let $z$ be an arbitrary element in $L$. Then we have

$$\begin{aligned} d^2 = \|y - y_1\|^2 &\leq \|y - y_1 + az\|^2 \\ &= \|y - y_1\|^2 + a(z, y - y_1) + \bar{a}(y - y_1, z) + |a|^2\|z\|^2 \end{aligned}$$

This inequality can hold for all complex $a$ if and only if $(y - y_1, z) = 0$. Thus $y_2 = y - y_1$ lies in the orthogonal complement to $L$; accordingly, to each $y$ in $H$ there correspond a $y_1$ in $L$ and a $y_2$ in $L^\perp$ such that $y = y_1 + y_2$. Since 0 is the only vector common to $L$ and $L^\perp$, it is clear that this decomposition of $y$ is unique. The vector $y_1$ is usually called the *projection* of $y$ on $L$; in symbols, we write

$$P_L y = y_1$$

Now, because $y_1$ and $y_2$ are orthogonal, we have

$$\|y\|^2 = \|y_1\|^2 + \|y_2\|^2$$

so that

$$\|P_L y\| \leq \|y\|$$

Further, it is easily seen that

$$P_L(ay + bz) = aP_L y + bP_L z \quad P_L P_L y = P_L y \quad (P_L y,z) = (y,P_L z)$$

It is obvious from the definition of orthogonal complements that $(S^\perp)^\perp \supset S$ for an arbitrary subset $S$. In the case of a closed subspace $L$, however, if $y$ lies in $(L^\perp)^\perp$, then $y$ is orthogonal to $L^\perp$; by writing $y = y_1 + y_2$, $y_1$ in $L$ and $y_2$ in $L^\perp$, we see that

$$\|y_2\|^2 = (y,y_2) - (y_1,y_2) = 0$$

Hence in this case $y$ lies in $L$ and we have $L = (L^\perp)^\perp$.

It is convenient to introduce the notion of a product space $H \times H$ consisting of all ordered pairs $[y,z]$ with $y$ and $z$ in $H$. This becomes a vector space under the convention

$$a[y,z] + b[u,v] = [ay + bu,\ az + bv]$$

And if we define an inner product as

$$([y,z],[u,v]) = (y,u) + (z,v)$$

then $H \times H$ actually becomes a Hilbert space.

We come now to the subject of transformations or operators on $H$ to itself. An operator $T$ with domain $\mathfrak{D}(T)$ and range $\mathfrak{R}(T)$ is said to be linear if $\mathfrak{D}(T)$ is a linear subspace and if

$$T(ay + bz) = aTy + bTz \qquad \text{for } y \text{ and } z \text{ in } \mathfrak{D}(T)$$

The *graph* of $T$, in symbols $\mathcal{G}(T)$, is defined as

$$\mathcal{G}(T) = \{[y,Ty];\ y \text{ in } \mathfrak{D}(T)\}$$

If $T$ is a linear operator, it is clear that $\mathcal{G}(T)$ is a linear subspace of $H \times H$. For an operator $T$, if we have $y_n \to y$ and $Ty_n \to z$, we shall say that $[y_n,Ty_n]$ converges to $[y,z]$ in the *graph topology*. In particular if, for every convergent sequence of this type, $y$ lies in $\mathfrak{D}(T)$ and $Ty = z$, then $T$ is said to be a *closed operator;* in other words, $T$ is closed if and only if $\mathcal{G}(T)$ is a closed subspace of $H \times H$. Continuity is a much more restrictive concept, since an operator is *continuous* if $y_n \to y$ in $\mathfrak{D}(T)$ implies that $Ty_n \to Ty$. A continuous operator need not be closed; however, it always has a smallest closed extension that is continuous and has a closed domain. If $T$ is linear and continuous, then there is a $\delta > 0$

such that $\|Tx\| < 1$ if only $\|x\| < \delta$. For any $y$ in $H$, it is clear that

$$x = \frac{\delta y}{2\|y\|}$$

is of norm less than $\delta$ and hence

$$\frac{\delta\|Ty\|}{2\|y\|} = \|Tx\| < 1$$

Thus, for a continuous linear operator, the *norm* or *bound*

$$\|T\| = \sup \{\|Ty\|; y \text{ in } \mathfrak{D}(T) \text{ and } \|y\| \leq 1\}$$

is finite. It is clear that

$$\|Ty\| \leq \|T\| \, \|y\| \qquad \text{for all } y \text{ in } \mathfrak{D}(T) \tag{4.12}$$

and that $\|T\|$ is the smallest number for which this holds. Thus if $T$ is continuous and linear, and if $\{y_n\}$ is a Cauchy sequence of elements in $\mathfrak{D}(T)$, then the inequality

$$\|Ty_n - Ty_m\| = \|T(y_n - y_m)\| \leq \|T\| \, \|y_n - y_m\| \tag{4.13}$$

shows that $\{Ty_n\}$ also forms a Cauchy sequence. We can therefore define an extension $\bar{T}$ of $T$ as

$$\bar{T}\,(\lim y_n) = \lim Ty_n$$

for all such Cauchy sequences. This is the smallest closed extension of $T$, to which we alluded above; it is readily shown that $\bar{T}$ is again continuous and linear and that $\mathfrak{D}(\bar{T})$ is a closed subspace of $H$. The inequality (4.13) shows, incidentally, that a linear operator satisfying the relationship (4.12) is necessarily continuous.

In order to illustrate the foregoing discussion, we note that the operator $S(t)$ defined as in Eq. (4.4) on $H = L_2(-\infty, \infty)$ and representing the solution to the initial-value problem (4.1) for the heat equation is linear and continuous. In fact, we have $\|S(t)\| \leq 1$ since

$$\int k(\chi,t)\, d\chi = 1$$

and since, by the Schwarz inequality,

$$\begin{aligned}\|S(t)f\|^2 &= \int_{-\infty}^{\infty} \left| \int_{-\infty}^{\infty} [k(\chi - \sigma, t)^{1/2}][k(\chi - \sigma, t)^{1/2} f(\sigma)]\, d\sigma \right|^2 d\chi \\ &\leq \int_{-\infty}^{\infty} \left[\int_{-\infty}^{\infty} k(\chi - \sigma, t)\, d\sigma\right] \left[\int_{-\infty}^{\infty} k(\chi - \sigma, t)|f(\sigma)|^2\, d\sigma\right] d\chi \\ &= \|f\|^2\end{aligned}$$

On the other hand, the associated diffusion operator $A$ [also on the space $H = L_2(-\infty, \infty)$], defined by

$$Ay = y_{\chi\chi} \qquad -\infty < \chi < \infty \tag{4.14}$$

$$\mathfrak{D}(A) = \{y; y \text{ and } y_x \text{ absolutely continuous with } y \text{ and } y_{xx} \text{ in } L_2(-\infty,\infty)\}$$

can be shown to be closed but not continuous.

Suppose $T$ is a closed linear operator and let $\lambda$ be a complex number. If $\lambda I - T$ has a continuous inverse with domain all of $H$, then $\lambda$ is said to belong to the *resolvent set* of $T$; the inverse is called the *resolvent* of $T$ at the point $\lambda$ and is denoted by $R(\lambda;T)$. It is clear that

$$(\lambda I - T)R(\lambda;T) = I \qquad R(\lambda;T)(\lambda I - T)y = y \qquad y \in \mathfrak{D}(T)$$

As an example, we consider the operator $A$ defined as in Eq. (4.14) and take $\lambda > 0$. Then

$$(\lambda I - A)y = \lambda y - y_{xx}$$

has an inverse if and only if the equation $\lambda y - y_{xx} = 0$ has no nontrivial solution in $\mathfrak{D}(A)$. The only nontrivial solutions of this equation for which $y$ and $y_x$ are absolutely continuous, however, are linear combinations of $\exp(\lambda^{1/2}x)$ and $\exp(-\lambda^{1/2}x)$, none of which lie in $L_2(-\infty,\infty)$. Thus $\lambda I - A$ has an inverse; in fact, the equation

$$(\lambda I - A)y = f$$

has a solution given by

$$y = R_\lambda f = \frac{1}{2\lambda^{1/2}} \int_{-\infty}^{\infty} \exp(-\lambda^{1/2}|x - \sigma|)f(\sigma)\, d\sigma$$

One can verify directly that the solution $y$ given here belongs to $\mathfrak{D}(A)$ and also that $R_\lambda$ is a continuous linear operator with bound $1/\lambda$. Thus $R_\lambda$ is a right inverse of $\lambda I - A$. In order to prove that $R_\lambda$ is actually the resolvent of $A$ at $\lambda$, it remains only to verify that $\mathfrak{R}(R_\lambda) = \mathfrak{D}(A)$. To this end, let $z$ in $\mathfrak{D}(A)$ be given, and set

$$f = \lambda z - Az \qquad y = R_\lambda f$$

According to what has already been established, we have

$$(\lambda I - A)(y - z) = 0$$

Hence it follows from the above uniqueness argument that

$$z = y = R_\lambda f$$

This proves that

$$\mathfrak{R}(R_\lambda) = \mathfrak{D}(A)$$

We note parenthetically that $\mathfrak{D}(A)$ is dense in $H$; that is, every element in $H$ is the limit of a sequence of elements in $\mathfrak{D}(A)$.

## 4.4 Semigroups of Operators on a Hilbert Space

Returning now to semigroup theory, we suppose that we have a semigroup of linear operators $[S(t); t > 0]$ defined on a domain $\mathfrak{D}$ that we assume to be contained and dense in $H$. Stability implies that each $S(t)$ is continuous on $\mathfrak{D}$, and hence $S(t)$ can be uniquely extended to be linear and continuous on $H$. We shall denote the extended operator again by $S(t)$, and it is readily verified that the extended operators also have the semigroup property (4.3).

It is convenient at this point to make one further assumption, namely, that the operators $S(t)$ satisfy the inequality

$$\|S(t)\| \leq 1$$

Such operators are called *contraction* operators. In this case, continuity with respect to $t$ at $t = 0$ also carries over for the extended operators; in other words,

$$\lim_{t \to 0+} S(t)y = y \qquad y \text{ in } H \tag{4.15}$$

Actually, more is true. It can be shown on the basis of Eq. (4.15) that $S(t)y$ is a continuous function of $t$, $t > 0$, for each $y$ in $H$, and if we set $S(0) = I$, then the assertion holds for all $t \geq 0$.

We now define the *infinitesimal generator* $A$ of the semigroup $[S(t)]$ by setting

$$Ay \equiv \lim_{\epsilon \to 0+} \frac{S(\epsilon)y - y}{\epsilon}$$

The domain of $A$ consists of all $y$ in $H$ for which this limit exists. It is easy to verify that $A$ is a linear operator, and it can be shown that $A$ is actually closed with a dense domain $\mathfrak{D}(A)$. Now

$$\epsilon^{-1}[S(t + \epsilon)y - S(t)y] = \epsilon^{-1}[S(\epsilon) - I]S(t)y = S(t)\{\epsilon^{-1}[S(\epsilon)y - y]\}$$

For each $y$ in $\mathfrak{D}(A)$, the right-hand member converges to $S(t)Ay$ as $\epsilon \to 0+$. This shows that the middle member also converges and hence that $S(t)y$ lies in $\mathfrak{D}(A)$; we may therefore assert that the right-hand derivative satisfies

$$\frac{dS(t)y}{dt} = AS(t)y = S(t)Ay \qquad t > 0 \tag{4.16}$$

Upon writing

$$\frac{S(t)y - S(t - \epsilon)y}{\epsilon} = S(t - \epsilon)\,\frac{S(\epsilon)y - y}{\epsilon}$$

we see that the left-hand derivative also exists, and since it equals $S(t)Ay$, we conclude that Eq. (4.16) holds for the two-sided derivative.

We note that the derivative with respect to $t$ is taken in the sense of the metric in $H$. Thus suppose that $H = L_2(-\infty,\infty)$ and set $u(\chi,t)$ equal to a function in $H$ that corresponds to $S(t)f$ for a given $f$ in $\mathfrak{D}(A)$ and $t \geq 0$. We can then take the difference quotient of $u(\chi,t)$ with respect to $t$ and pass to the limit in the mean-square sense as the $t$ increment goes to zero. This gives a generalized partial derivative with respect to $t$, and according to Eq. (4.16) the resulting derivative is obtained by having $A$ act on $u(\chi,t)$. If $A$ were the operator (4.14), say, then the semigroup $S(t)$ restricted to $\mathfrak{D}(A)$ would provide a solution to the Cauchy problem for the heat equation (4.1), at least in this generalized sense. It will be the aim of this chapter to show how the theory of semigroups can be used to establish the existence of such solutions to a large class of Cauchy problems defined by means of certain differential operators.

The semigroup method is in essence an abstract analogue of the Laplace-transform approach to the initial-value problem for time-invariant partial differential equations and as such should be not entirely unfamiliar to the reader (see Chap. 3). The Laplace transform

$$R_\lambda y \equiv \int_{-\infty}^{\infty} \exp(-\lambda t) S(t) y \, dt$$

can be defined in a straightforward manner and converges for $\lambda > 0$. Since $S(t)$ is very close to $\exp(At)$, and at least heuristically

$$\int_{-\infty}^{\infty} \exp(-\lambda t) \exp(At) = (\lambda I - A)^{-1}$$

it is not surprising that $R_\lambda$ turns out to be the resolvent of $A$, namely, $R(\lambda;A)$. Here, as in the classical Laplace-transform approach, it is the inverse problem that is important. In other words, given an operator $A$, for which the resolvent exists when $\lambda$ is sufficiently large, the problem is to determine whether $A$ is the infinitesimal generator of a semigroup of operators. It is precisely at this point that the abstract semigroup theory provides a better answer than the classical theory, at least better in the sense that the abstract criterion is easier to verify than the classical criterion. The first result of this kind was obtained independently by E. Hille and K. Yosida (see pages 360 to 364 of Ref. 6) and can be expressed as follows:

THEOREM 4.1. *A necessary and sufficient condition that a closed linear operator L with dense domain generate a semigroup of contraction operators is that the resolvent $R(\lambda;L)$ satisfy the inequality*

$$\|R(\lambda;L)\| \leq \lambda^{-1} \qquad \lambda > 0$$

The Hille-Yosida result is valid in more general settings than a Hilbert space $H$, so that it is to be expected that a simpler criterion holds in the case of $H$. By way of motivating the terminology in this simpler criterion, we note that, in the model that we shall use, the inner product $(y,y)$ turns out to be the energy associated with the state $y$; thus, assuming $S(t)$ to be a contraction operator is the same as assuming that the process is dissipative in the sense that no energy enters the system. Since for $\tau > 0$ we have

$$\|S(t+\tau)y\| = \|S(\tau)S(t)y\| \leq \|S(t)y\|$$

we see that $\|S(t)y\|$ is nonincreasing and hence for $y \in \mathfrak{D}(A)$ we obtain

$$\frac{d}{dt}(S(t)y,\, S(t)y)\bigg|_{t=0} = (Ay,y) + (y,Ay) \leq 0 \tag{4.17}$$

We shall call an operator $L$ *dissipative* if

$$(Ly,y) + (y,Ly) \leq 0 \qquad y \in \mathfrak{D}(L)$$

and *maximal dissipative* if it is not the proper restriction of any other dissipative operator. We can prove the following theorems.[7]

THEOREM 4.2. *A necessary and sufficient condition for a dissipative operator $L$ with dense domain to be maximal dissipative is that the range of $I - L$ be all of $H$.*

THEOREM 4.3. *A necessary and sufficient condition that a linear operator $L$ generate a semigroup of contraction operators is that $L$ be maximal dissipative with dense domain.*

These results are readily verified in the case of the finite-dimensional analogue of $H$, namely, the $k$-dimensional complex euclidean space $Z_k$ with elements

$$y = (\eta^1, \eta^2, \ldots, \eta^k) \qquad z = (\zeta^1, \zeta^2, \ldots, \zeta^k)$$

and inner product

$$\langle y,z\rangle = \sum_{i=1}^{k} \eta^i \bar{\zeta}^i$$

As we have already seen, each semigroup of matrices $[S(t)]$ defined on $Z_k$ and continuous on the right can be represented as

$$S(t) = \exp(At) \tag{4.18}$$

where the matrix $A$ is the infinitesimal generator of the semigroup. If, in addition, the operators $S(t)$ are contraction operators, then, according to the relationship (4.17), $A$ will be dissipative, actually maximal dissipative since $A$ is defined on all $Z_k$. In the finite-dimensional case this simply means that the eigenvalues of the real part of $A$ are nonpositive, in symbols

$$B = \tfrac{1}{2}(A + A^*) \leq \Theta$$

where $A^*$ is the transpose of $A$ and $\Theta$ is the zero matrix. Conversely, if $B$ is negative, then for $S(t)$ defined as in Eq. (4.18) and for arbitrary $y$ in

$Z_k$ we have

$$\frac{d}{dt}\langle S(t)y, S(t)y\rangle = \langle AS(t)y, S(t)y\rangle + \langle S(t)y, AS(t)y\rangle$$
$$= 2\langle BS(t)y, S(t)y\rangle \leq 0$$

and since

$$\langle S(t)y, S(t)y\rangle = \langle y,y\rangle \qquad \text{for } t = 0$$

we see that

$$\|S(t)y\| \leq \|y\| \qquad \text{for all } t > 0$$

Thus $A$ generates a semigroup of contraction operators in accordance with Theorem 4.3. For example, the semigroup (4.11) generated by Eq. (4.10) will consist of contraction operators if and only if

$$\alpha + \bar{\alpha} \leq -\sqrt{2}$$

We further note for $A$ dissipative, $\lambda > 0$, and $f = \lambda y - Ay$, that

$$\langle f,y\rangle + \langle y,f\rangle = \langle \lambda y - Ay, y\rangle + \langle y, \lambda y - Ay\rangle$$
$$= 2[\lambda\langle y,y\rangle - \langle By,y\rangle] \geq 2\lambda\langle y,y\rangle$$

from which we conclude that

$$\lambda\|y\| \leq \|f\|$$

It follows from this that $\lambda I - A$ is nonsingular, so that

$$\mathcal{R}(\lambda I - A) = Z_k$$

in accordance with Theorem 4.2, and it further follows that

$$R(\lambda;A) = (\lambda I - A)^{-1}$$

is of norm less than or equal to $\lambda^{-1}$ in accordance with Theorem 4.1.

For additional material on the theory of semigroups of operators, we refer the reader to the treatise by E. Hille and R. S. Phillips,[6] entitled "Functional Analysis and Semi-groups," and to the excellent set of notes by K. Yosida,[11] entitled "Lectures on Semi-group Theory and Its Applications to Cauchy's Problem in Partial Differential Equations."

## 4.5 Hyperbolic Systems of Partial Differential Equations

We shall apply the previous discussion to symmetric hyperbolic systems of partial differential equations. Such a system can be described as follows: Let $G$ be a bounded domain in an $m$-dimensional euclidean space with points

$$\chi = (\chi^1, \chi^2, \ldots, \chi^m)$$

and again let $Z_k$ be a $k$-dimensional complex euclidean space with elements

$$y = (\eta^1, \eta^2, \ldots, \eta^k) \qquad z = (\zeta^1, \zeta^2, \ldots, \zeta^k)$$

and inner product

$$\langle y,z\rangle = \sum_{i=1}^{k} \eta^i \overline{\zeta^i} \tag{4.19}$$

For functions $y(\chi,t)$ defined on $G \times (0,\infty)$, we shall consider the Cauchy problem

$$\begin{aligned} y_t &= \sum_{i=1}^{m} (A^i y)_i + By \qquad \chi \in G \quad t > 0 \\ y(\chi,0+) &= f(\chi) \end{aligned} \tag{4.20}$$

Moreover, $y(\chi,t)$ is restricted so as to satisfy certain homogeneous lateral conditions that suffice to make the solution to the initial-value problem unique. Throughout the remainder of this chapter, the subscript $i$ will denote differentiation with respect to $\chi^i$. In Eq. (4.20), $A^i$ and $B$ are matrix-valued functions of $\chi$ alone, the $A^i$ are symmetric and continuously differentiable in $\bar{G}$, whereas $B$ is merely continuous in $\bar{G}$. The coefficients are subject to one further condition, namely,

$$D = B + B^* + \sum_{i=1}^{m} A^i{}_i \leq \Theta \qquad \chi \text{ in } G \tag{4.21}$$

Here $B^*$ is the transpose of $B$ relative to the inner product (4.19). As we shall see, the hypothesis (4.21) is a consequence of the fact that we deal only with energy-dissipative systems.

We note that the wave equation can readily be transformed into a symmetric hyperbolic system. We have already shown in Eqs. (4.5) that this is true of Eq. (4.2). More generally, we have

$$u_{tt} + u_t = u_{11} + u_{22} \tag{4.22}$$

and if we set

$$\eta^1 = u_1 \qquad \eta^2 = u_2 \qquad \text{and} \qquad \eta^3 = u_t$$

then Eq. (4.22) becomes

$$\begin{aligned} \eta^1{}_t &= \eta^3{}_1 \\ \eta^2{}_t &= \eta^3{}_2 \\ \eta^3{}_t &= \eta^1{}_1 + \eta^2{}_2 - r\eta^3 \end{aligned}$$

which is now of the form (4.20) with

$$A^1 = \begin{pmatrix} 0 & 0 & 1 \\ 0 & 0 & 0 \\ 1 & 0 & 0 \end{pmatrix} \qquad A^2 = \begin{pmatrix} 0 & 0 & 0 \\ 0 & 0 & 1 \\ 0 & 1 & 0 \end{pmatrix} \qquad B = \begin{pmatrix} 0 & 0 & 0 \\ 0 & 0 & 0 \\ 0 & 0 & -r \end{pmatrix}$$

In this case,

$$D = \begin{pmatrix} 0 & 0 & 0 \\ 0 & 0 & 0 \\ 0 & 0 & -2r \end{pmatrix}$$

which is $\leq \Theta$ if $r \geq 0$, as we shall assume.

The potential energy plus the kinetic energy at time $t$ for the amplitude function of Eq. (4.22) is given by

$$\text{Energy} = \tfrac{1}{2} \int_G (|u_1|^2 + |u_2|^2)\, d\chi + \tfrac{1}{2} \int_G |u_t|^2\, d\chi = \tfrac{1}{2} \int_G \langle y,y \rangle\, d\chi$$

Extending this to the system (4.20), we take, as the energy at time $t$ of the amplitude function $y$, the quantity

$$E = \tfrac{1}{2} \int_G \langle y,y \rangle\, d\chi$$

This suggests the Hilbert space $H$ with inner product

$$(y,z) = \int_G \langle y,z \rangle\, d\chi$$

as the natural setting for our problem. If we assume that the solution $S(t)f$ is energy-nonincreasing, then we have

$$\|S(t)f\| \leq \|f\|$$

Thus the associated semigroup operator is contractional, as desired.

By differentiating $E$ with respect to $t$ and making use of an integration by parts, we obtain, at least formally,

$$\begin{aligned} 2E_t &= (y_t,y) + (y,y_t) \\ &= \left(\left\{\sum_{i=1}^{m} (A^i y)_i + By\right\}, y\right) + \left(y, \left\{\sum_{i=1}^{m} (A^i y)_i + By\right\}\right) \\ &= \int_G \left\{\sum_{i=1}^{m} \langle A^i y,y \rangle_i + \langle Dy,y \rangle\right\} d\chi \\ &= \int_{\dot{G}} \left\{\sum_{i=1}^{m} \langle A^i y,y \rangle n^i\right\} d\sigma + (Dy,y) \end{aligned} \tag{4.23}$$

where $\dot{G}$ is the boundary of $G$, $n = (n^1, n^2, \ldots, n^m)$ the outer normal to $\dot{G}$, and $d\sigma$ the surface element on $\dot{G}$. The first term on the right in Eq. (4.23), namely,

$$Q(y,y) = \int_{\dot{G}} \left\{\sum_{i=1}^{m} \langle A^i y,y \rangle n^i\right\} d\sigma$$

represents the rate at which energy enters the system through the boundary, the $m$-tuple with components $\langle A^i y,y \rangle$ being a generalization of the "Poynting vector." The other term, $(Dy,y)$, represents the rate at which energy enters the system from energy sources within $G$. In particular, if we assume that $E_t \leq 0$ for all smooth initial-value functions that vanish near $\dot{G}$, then for $t = 0$ the surface integral vanishes and we are left with the condition $(Dy,y) \leq 0$ for all such functions. This is easily seen to imply the dissipative condition (4.21), which we assume to be satisfied by the coefficients of our system. Finally, when we set

$$Ly = \sum_{i=1}^{m} (A^i y)_i + By \tag{4.24}$$

the computation (4.23) shows that

$$(Ly,y) + (y,Ly) = Q(y,y) + (Dy,y) \tag{4.25}$$

and hence the condition $E_t \leq 0$ for all solutions of the initial-value problem is equivalent to the condition that the domain of $L$ is so defined as to make this operator dissipative in the sense of Sec. 4.4. This comes as no surprise, since we have already shown in Secs. 4.1 and 4.4 that a necessary and sufficient condition for the problem to be well formulated is that $L$ be maximal dissipative.

Now if the domain of $L$ is such that $Q(y,y) \leq 0$ for all functions in this domain, then $L$ is obviously dissipative. It turns out that this is the case if and only if the domain of $L$ is suitably restricted by what may be called *boundary conditions.* In particular, this will be the case if the functions $y$ in $\mathfrak{D}(L)$ are subject to homogeneous constraints requiring that

$$Q^{\chi}(y,y) = \sum_{i=1}^{m} \langle A^i y, y \rangle n^i$$

be nonpositive at each point $\chi$ of $\dot{G}$. Boundary conditions of this sort will be called *local boundary conditions;* K. O. Friedrichs[4] has characterized boundary conditions of this kind that define $L$ as a maximal dissipative operator. There are also a wide variety of boundary conditions defining $L$ as maximal dissipative in which the boundary values at more than one point of $\dot{G}$ are related; for instance, periodic-type boundary conditions can be of this sort.

Since most of the present discussion deals with the question of boundary conditions, it will be helpful to elaborate on the description of local boundary conditions as given by Friedrichs. For fixed $\chi \in \dot{G}$, it is clear that the matrix

$$A^{\chi} = \sum_{i=1}^{m} A^i(\chi) n^i$$

is symmetric, say of rank $r$ with $n$ negative and $p$ positive eigenvalues.

Relative to the local quadratic form

$$Q^{x}(y,y) = \langle A^{x}y,y\rangle$$

a subspace $N$ of $Z_k$ will be called negative [positive] if

$$Q^{x}(y,y) \leq 0 \qquad [Q^{x}(y,y) \geq 0]$$

for all $y \in N$, and maximal negative [maximal positive] if $N$ is negative [positive] but not the proper subspace of some other negative [positive] subspace. It is easy to show that a negative [positive] subspace is maximal negative [maximal positive] if and only if it is of dimension $n + (k - r)$ [of dimension $p + (k - r)$]. It follows from Eq. (4.25) that $L$ will be dissipative if the functions in its domain lie in negative subspaces at each point of $\dot{G}$, and it is clear that $L$ cannot be maximal dissipative unless these subspaces are maximal negative. Friedrichs has shown that this condition is also sufficient. More precisely, Friedrichs assumes that $A^{x}$ is of constant rank throughout $\dot{G}$ and chooses a continuous family of maximal negative subspaces on $\dot{G}$; the set of all smooth functions having boundary values in the chosen subspaces determines a dissipative operator $L$ for which the smallest closed extension is maximal dissipative.

On the other hand, it is clear from Eqs. (4.23) and (4.25) that the condition $E_t \leq 0$, which is equivalent to $L$ being dissipative, does not require the surface integral $Q(y,y)$ to be nonpositive for all $y$ in the domain of $L$. If $D \not\equiv \Theta$, then $E_t \leq 0$ even if energy enters through the boundary, *provided that it is compensated for by the energy dissipated in the interior.* In order to ensure such a state of affairs, the domain of the operator $L$ must be restricted by *global lateral conditions* that relate the boundary values of a function in $\mathfrak{D}(L)$ to its values at the sinks in $G$. This is a new phenomenon that can occur only when $D \not\equiv \Theta$ or, what amounts to the same thing (as we shall see), only in the non-self-adjoint problem. Such lateral conditions are reminiscent of lateral conditions found by W. Feller[2] in describing certain return processes in diffusion theory. There is also a dual process, no longer governed by a pure partial differential operator, in which energy leaks out through the boundary and is in part transported back into the interior.

Returning now to our main problem, namely, that of finding all the dissipative solutions of the initial-value problem (4.20), we see by Theorem 4.3 that this is equivalent to the problem of characterizing all the maximal dissipative operators that can be associated with the differential operator $L$ of Eq. (4.24). There is still a certain amount of arbitrariness as to the operators that can be associated with $L$. We shall take this to mean either extensions of a *minimal operator* $L_0$ defined by Eq. (4.24), with domain containing in essence only the smooth functions that vanish near $\dot{G}$, or restrictions of a *maximal operator* $L_1$ also defined by

Eq. (4.24), but with domain containing in essence all smooth functions on $\bar{G}$.

More precisely, $L_0$ and $L_1$ are defined as the least closed extensions of $L_{00}$ and $L_{10}$, respectively, which are in turn defined as follows:

$$\mathfrak{D}(L_{00}) = \{y;\ y \text{ continuously differentiable and vanishing outside a compact subset of } G\}$$

$$L_{00}y = \sum_{i=1}^{m} (A^i y)_i + By$$

and

$$\mathfrak{D}(L_{10}) = \{y;\ y \text{ continuously differentiable in } \bar{G}\}$$

$$L_{10}y = \sum_{i=1}^{m} (A^i y)_i + By$$

The maximal operator $L_1$ can also be defined in terms of the formal adjoint to $L$, namely,

$$Mz = -\sum_{i=1}^{m} (A^i z)_i + \left(B^* + \sum_{i=1}^{m} A^i{}_i\right) z \tag{4.26}$$

We note that the differential operator $M$ is readily shown to be dissipative in the sense that its coefficients satisfy the analogue of Eq. (4.21); in fact, $D$ is the same for both $L$ and $M$. We proceed as above to define $M_0$ and $M_1$ by means of $M_{00}$ and $M_{10}$, respectively, where

$$\mathfrak{D}(M_{00}) = \mathfrak{D}(L_{00}) \qquad \mathfrak{D}(M_{10}) = \mathfrak{D}(L_{10})$$

It can be shown[3] that $L_1$ is the adjoint of $M_0$; in other words, $\mathfrak{D}(L_1)$ consists of all $y$ in $H$ for which there exists a $y'$ in $H$ such that

$$(y, M_0 z) = (y', z) \qquad \text{for all } z \text{ in } \mathfrak{D}(M_0) \tag{4.27}$$

in which case we have $L_1 y = y'$.

### 4.6 Maximal Dissipative Operators

Our problem, then, is to characterize both the maximal dissipative extensions of $L_0$ and the restrictions of $L_1$. The extensions of $L_0$ need not be of the form (4.24) for $y$ not in $\mathfrak{D}(L_0)$, whereas the restrictions of $L_1$ will in general have domains determined by global lateral conditions. The maximal dissipative operators that are both extensions of $L_0$ and restrictions of $L_1$ have domains determined by boundary conditions, and the elements of these domains satisfy the condition $Q(y,y) \leq 0$; it is with these operators that we shall be most concerned.

Before proceeding further with our characterization of the maximal dissipative operators that lie between $L_0$ and $L_1$, we shall illustrate the foregoing discussion with a simple example, in which $G$ is the interval $0 < x < 1$ and we take $m = 1$, $k = 1$, $H = L_2(0,1)$, and

$$\begin{aligned} y_t &= y_x - y \qquad 0 < x < 1 \quad t > 0 \\ y(x,0+) &= f(x) \end{aligned}$$

In this case we find that $A = 1$, $B = -1$, and $D = -2 \leq 0$. It can readily be verified that

$$\begin{aligned} \mathfrak{D}(L_1) &= \{y;\, y \text{ absolutely continuous on } (0,1) \text{ with } y \text{ and } y_x \text{ in } L_2(0,1)\} \\ \mathfrak{D}(L_0) &= \{y;\, y \text{ in } \mathfrak{D}(L_1) \text{ with } y(0) = 0 = y(1)\} \end{aligned}$$

Moreover, we have

$$Q^0(y,z) = -y(0)\overline{z(0)} \qquad Q^1(y,z) = y(1)\overline{z(1)}$$

The only local boundary condition defining the domain of a maximal dissipative operator is $y(1) = 0$. For the general boundary condition, we consider the form

$$Q(y,z) = Q^1(y,z) + Q^0(y,z) = y(1)\overline{z(1)} - y(0)\overline{z(0)}$$

for which $r = 2$ = dimension of the boundary space, and $n = 1 = p$. Each maximal negative subspace for this form is characterized as $y(1) = \alpha y(0)$ for some $\alpha$ of absolute value less than or equal to 1. Each such subspace defines the domain of a maximal dissipative operator $L_\alpha$, $L_0 \subset L_\alpha \subset L_1$, and conversely, each maximal dissipative operator between $L_0$ and $L_1$ is characterized in this way.

On the other hand, it is easy to construct extensions of $L_0$ that are maximal dissipative but not of the above type; for instance, consider

$$\begin{aligned} Uy &= L_1 y + ay(0) \\ \mathfrak{D}(U) &= \{y;\, y \text{ in } \mathfrak{D}(L_1) \text{ and } y(1) = 0\} \end{aligned}$$

Here $a = a(x)$ belongs to $L_2(0,1)$ and we assume that $(a,a) \leq 2$. In order to see that $U$ is dissipative, we observe that

$$\begin{aligned} (Uy,y) + (y,Uy) &= (L_1y,y) + (y,L_1y) + (a,y)y(0) + (y,a)\overline{y(0)} \\ &= -|y(0)|^2 - 2(y,y) + (a,y)y(0) + (y,a)\overline{y(0)} \\ &= -|y(0) - (y,a)|^2 - [2(y,y) - |(y,a)|^2] \leq 0 \end{aligned}$$

where the inequality follows from the Schwarz inequality:

$$|(y,a)| \leq \|y\|\,\|a\|$$

It is clear that $\mathfrak{D}(U)$ is dense in $L_2(0,1)$; hence, according to Theorem 4.2, we shall prove that $U$ is maximal dissipative if we show that

$$\mathfrak{R}(I - U) = H$$

This requires that we solve the equation

$$y - Uy = f$$

for arbitrary $f \in H$. It is readily verified that this is satisfied by the function

$$y(\chi) = y(0)e^{2\chi}\left(1 - \int_0^\chi e^{-2\sigma}a(\sigma)\,d\sigma\right) - e^{2\chi}\int_0^\chi e^{-2\sigma}f(\sigma)\,d\sigma$$

We remark that the term $ay(0)$ in the expression for $U$ corresponds physically to transporting energy into the interior in amounts having linear density $a(\chi)|y(0)|^2$.

The dual situation of a maximal dissipative operator $V$ that is a restriction of $L_1$ but not an extension of $L_0$ can be illustrated by the operator

$$Vy = L_1y$$
$$\mathfrak{D}(V) = \{y;\ y \text{ in } \mathfrak{D}(L_1) \text{ with } y(1) = (y,a)\}$$

where again it is assumed that $(a,a) \leq 2$. In this case,

$$\begin{aligned}(Vy,y) + (y,Vy) &= (L_1y,y) + (y,L_1y)\\ &= [|y(1)|^2 - |y(0)|^2] - 2(y,y)\\ &= -|y(0)|^2 - [2(y,y) - |(y,a)|^2] \leq 0\end{aligned}$$

so that $V$ is dissipative. One verifies as before that $V$ is maximal dissipative by showing that

$$\mathfrak{R}(I - V) = H$$

This amounts to solving the equation

$$y - Vy = g$$

and one can easily find that this equation is satisfied by

$$y(\chi) = e^{2\chi}\left[k - \int_0^\chi e^{-2\sigma}g(\sigma)\,d\sigma\right]$$

where

$$k\left(e^2 - \int_0^1 e^{2\sigma}\overline{a(\sigma)}\,d\sigma\right) = e^2\int_0^1 e^{-2\sigma}g(\sigma)\,d\sigma - \int_0^1\left[e^{2\chi}\overline{a(\chi)}\int_0^\chi e^{-2\sigma}g(\sigma)\,d\sigma\right]d\chi$$

The condition $(a,a) \leq 2$ is sufficient to make the coefficient of $k$ different from zero. It can be shown that for some $y$ in $\mathfrak{D}(V)$ the boundary integral $Q(y,y)$ is positive, so that energy can enter the system through the boundary; however, the lateral condition $y(1) = (y,a)$, which determines $\mathfrak{D}(V)$, requires that the rate of energy being dissipated in the interior, namely, $-2(y,y)$, is always at least as large as the flow in through the boundary, namely, $|y(1)|^2$.

We return now to the general problem of characterizing the boundary conditions for the maximal dissipative operators $L$ between $L_0$ and $L_1$. An integration by parts as in Eq. (4.25) shows that

$$Q(y,z) = (L_1 y,z) + (y,L_1 z) - (Dy,z) \tag{4.28}$$

at least for $y$ and $z$ in $\mathfrak{D}(L_{10})$; hence, by considering sequences in the graph topology we see that Eq. (4.28) gives us a means of defining the boundary integral $Q(y,z)$ for all $y$ and $z$ in $\mathfrak{D}(L_1)$. For functions $y$ in $\mathfrak{D}(L_{00})$, which by definition vanish near $\dot{G}$, and functions $z$ in $\mathfrak{D}(L_{10})$, it is clear that $Q(y,z) = 0$; upon again taking sequences in the graph topology, we see that this remains true for all $y$ in $\mathfrak{D}(L_0)$ and $z$ in $\mathfrak{D}(L_1)$. Thus, as far as the boundary integral is concerned, the functions in $\mathfrak{D}(L_0)$ have absolutely no influence and can be said to exhibit zero-like boundary behavior. An argument similar to that used for Eq. (4.27) shows that $\mathfrak{D}(L_0)$ is the largest class of functions in $\mathfrak{D}(L_1)$ with this property.

Suppose now we consider the residue classes of cosets

$$\dot{H} = \frac{\mathfrak{D}(L_1)}{\mathfrak{D}(L_0)} \tag{4.29}$$

Since each coset consists of functions that differ from each other by functions in $\mathfrak{D}(L_0)$, it follows that all the functions in such a coset exhibit the same boundary behavior. On the other hand, functions in different cosets do not differ by an element of $\mathfrak{D}(L_0)$ and hence exhibit different boundary behavior. Thus, there is a one-to-one correspondence between the cosets in $\dot{H}$ and the types of boundary behavior displayed by the functions in $\mathfrak{D}(L_1)$. Moreover, suppose that $y_1$ and $y_2$ belong to one coset and $z_1$ and $z_2$ belong to another. Then $y_1 - y_2$ and $z_1 - z_2$ both lie in $\mathfrak{D}(L_0)$ and, according to our previous remarks,

$$Q(y_1,z_1) = Q([y_2 + (y_1 - y_2)], [z_2 + (z_1 - z_2)]) = Q(y_2,z_2)$$

so that $Q(y,z)$ depends only on the cosets to which $y$ and $z$ belong. Thus, $Q$ induces a bilinear form on the residue classes; we denote this form by $\dot{Q}$.

We now have a space $\dot{H}$ of boundary data $\dot{y}$ together with a bilinear form $\dot{Q}$ that is in essence the boundary integral associated with the differential operator (4.24). A subset of $\dot{H}$ can be thought of as a boundary condition on a restriction of $L_1$ and, in particular, a linear subspace of $\dot{H}$ can be considered as a homogeneous boundary condition on such an operator; the corresponding domain of this operator would consist of the totality of functions in the cosets of such a subset or subspace.

The maximal dissipative operators between $L_0$ and $L_1$ can be described as follows. Relative to the quadratic form $\dot{Q}(\dot{y},\dot{y})$ defined on the cosets of $\dot{H}$, we can again define the notion of a negative [positive] subspace of $\dot{H}$ as well as that of a maximal negative [maximal positive] subspace of $\dot{H}$.

These maximal negative subspaces determine the boundary conditions for the maximal dissipative operators in question. More precisely, we have the following result.

THEOREM 4.4. *There is a one-to-one correspondence between the maximal negative subspaces* $\{\dot{N}\}$ *of* $H$ *and the maximal dissipative operators* $L$ *that lie between* $L_0$ *and* $L_1$*; this correspondence is given by*

$$\mathfrak{D}(L) = \{y;\, y \text{ belongs to } \mathfrak{D}(L_1) \text{ and lies in a coset of } \dot{N}\}$$

In the case of one spatial variable, $\dot{H}$ is finite-dimensional and the problem of determining the maximal negative subspaces of $\dot{H}$ relative to $\dot{Q}$ is purely algebraic. For instance, in the case of our previous example the cosets of $\dot{H}$ are determined by the values of the functions in $\mathfrak{D}(L_1)$ at 0 and 1, so that $\dot{H}$ is two-dimensional. Thus $\dot{y} = [a,b]$, and $y$ lies in the coset $\dot{y}$ if and only if

$$y(0) = a \qquad \text{and} \qquad y(1) = b$$

Moreover,

$$\dot{Q}(\dot{y},\dot{y}) = |b|^2 - |a|^2$$

and each maximal negative subspace of $\dot{H}$ is characterized by a number $\alpha$, $|\alpha| \leq 1$, as $b = \alpha a$.

In the many-spatial-variable case, $\dot{H}$ is no longer finite-dimensional and the problem is quite complex, as we shall see. Nevertheless, at each point $\chi$ of $\dot{G}$ we do have the bilinear form $Q^{\chi}$ that acts on the $k$-dimensional vector space $Z_k$. In effect, Friedrichs[4] has characterized the maximal negative subspaces of $\dot{H}$ that correspond to local boundary conditions as the direct sum of the maximal negative subspaces relative to $Q^{\chi}$ at each point $\chi$ of $\dot{G}$.

By way of illustration, we consider the initial-value problem for the circular membrane

$$\begin{gathered} u_{tt} = u_{11} + u_{22} \qquad (\chi^1)^2 + (\chi^2)^2 < 1 \quad t > 0 \\ u(\chi,0+) = f(\chi) \qquad u_t(\chi,0+) = g(\chi) \end{gathered}$$

which we rewrite as before with $\eta^1 = u_1$, $\eta^2 = u_2$, $\eta^3 = u_t$,

$$A^1 = \begin{pmatrix} 0 & 0 & 1 \\ 0 & 0 & 0 \\ 1 & 0 & 0 \end{pmatrix} \qquad A^2 = \begin{pmatrix} 0 & 0 & 0 \\ 0 & 0 & 1 \\ 0 & 1 & 0 \end{pmatrix} \qquad B = \Theta = D$$

If we parameterize the boundary point $\chi^1 = \cos\sigma$, $\chi^2 = \sin\sigma$ by $\sigma$, $0 \leq \sigma < 2\pi$, then the local quadratic form can be written as

$$Q^{\sigma}(y,y) = 2\Re[(\eta^1 \cos\sigma + \eta^2 \sin\sigma)\bar{\eta}^3]$$

where $\Re[\ \ ]$ denotes the real part of $[\ \ ]$. Each maximal negative subspace relative to $Q^{\sigma}$ is characterized by a number $\alpha(\sigma)$ with $\Re\alpha(\sigma) \geq 0$

and described by the relationship

$$\eta^1 \cos \sigma + \eta^2 \sin \sigma = -\alpha(\sigma)\eta^3$$

The maximal negative subspaces of Friedrichs are obtained by choosing $\alpha(\sigma)$ as a continuous function of $\sigma$ with

$$\alpha(2\pi-) = \alpha(0)$$

It is instructive actually to characterize $\dot{H}$ for this example and then exhibit certain maximal negative subspaces of $\dot{H}$. To this end, we note that the mapping

$$y \to [y, L_i y]$$

carries $\mathfrak{D}(L_i)$ in a one-to-one linear fashion onto the graph $\mathcal{G}(L_i)$ for $i = 0$ and 1. Consequently, Eq. (4.29) can be written equivalently as

$$\dot{H} = \frac{\mathcal{G}(L_1)}{\mathcal{G}(L_0)} \tag{4.30}$$

The advantage of Eq. (4.30) over Eq. (4.29) lies in the fact that the graphs of $L_1$ and $L_0$ are closed subspaces of $H \times H$, and it is therefore possible to take the orthogonal complement of $\mathcal{G}(L_0)$ with respect to $\mathcal{G}(L_1)$, in symbols $\mathcal{G}(L_1) \cap \mathcal{G}(L_0)^{\perp}$. Now each element in $\mathcal{G}(L_1)$ has a unique decomposition as an element in $\mathcal{G}(L_0)$ plus an element in $\mathcal{G}(L_1) \cap \mathcal{G}(L_0)^{\perp}$, and from this it follows that the cosets of $\dot{H}$ are in one-to-one correspondence with the elements of $\mathcal{G}(L_1) \cap \mathcal{G}(L_0)^{\perp}$. The elements of $\mathcal{G}(L_1) \cap \mathcal{G}(L_0)^{\perp}$ are just the pairs $[z^1,z^2]$ in $\mathcal{G}(L_1)$ for which

$$(z^1,y) + (z^2,L_0 y) = 0 \qquad \text{for all } y \text{ in } \mathfrak{D}(L_0)$$

In this example we have $D = \Theta$, so that $M_0 = -L_0$; see Eq. (4.26). It follows from Eq. (4.27) that $z^2$ lies in $\mathfrak{D}(L_1)$ and $z^1 = L_1 z^2$. On the other hand, $[z^1,z^2]$ being in $\mathcal{G}(L_1)$ means that $z^2 = L_1 z^1$, and hence it follows that

$$L_1^2 z^1 = z^1 \tag{4.31}$$

Accordingly, there is a one-to-one correspondence between the solutions of Eq. (4.31) and the elements of $\dot{H}$.

For the circular membrane, Eq. (4.31) is of the form

$$\begin{aligned} \zeta^1{}_{11} + \zeta^2{}_{21} &= \zeta^1 \\ \zeta^1{}_{12} + \zeta^2{}_{22} &= \zeta^2 \\ \zeta^3{}_{11} + \zeta^3{}_{22} &= \zeta^3 \end{aligned} \tag{4.32}$$

The first two of Eqs. (4.32) show that $\zeta^1{}_2 = \zeta^2{}_1$, and hence there is a function $\phi$ such that

$$\zeta^1 = \phi_1 \qquad \text{and} \qquad \zeta^2 = \phi_2$$

By combining this with Eqs. (4.32), we see that

$$\phi_{11} + \phi_{22} = \phi \tag{4.33}$$

Thus each element of $\dot{H}$ is characterized by two solutions of Eq. (4.33), say $\phi$ and $\zeta^3$, such that $[\phi_1,\phi_2,\zeta^3]$ lies in $\mathfrak{D}(L_1)$. Now $[\phi_1,\phi_2,\zeta^3]$ lies in $\mathfrak{D}(L_1)$ if and only if $\phi_1$, $\phi_2$, $\phi_{11} + \phi_{22} = \phi$, $\zeta^3$, $\zeta^3{}_1$, and $\zeta^3{}_2$ are in $\mathfrak{L}_2(G)$. If $\phi$ satisfies Eq. (4.33), then by Green's theorem we have

$$\int_G (|\phi|^2 + |\phi_1|^2 + |\phi_2|^2)\, d\chi = \int_{\dot{G}} \phi \frac{\overline{\partial \phi}}{\partial r}\, d\sigma$$

where

$$r^2 = (\chi^1)^2 + (\chi^2)^2$$

Hence the integrability conditions on $\phi$ and $\zeta^3$ are

$$\int_{\dot{G}} \phi \frac{\overline{\partial \phi}}{\partial r}\, d\sigma < \infty \qquad \text{and} \qquad \int_{\dot{G}} \zeta^3 \frac{\overline{\partial \zeta^3}}{\partial r}\, d\sigma < \infty$$

Further, in the present case,

$$Q(z,z) = \int_{\dot{G}} Q^\sigma(z,z)\, d\sigma = \int_{\dot{G}} \left( \zeta^3 \frac{\overline{\partial \phi}}{\partial r} + \overline{\zeta^3} \frac{\partial \phi}{\partial r} \right) d\sigma$$

Thus only the boundary values of $\partial\phi/\partial r$ and $\zeta^3$ enter into the quadratic form $Q$. Let us represent these boundary values by the Fourier series

$$\frac{\partial \phi}{\partial r} \sim \frac{1}{\sqrt{2\pi}} \sum_{k=-\infty}^{\infty} a_k e^{ik\sigma}$$

$$\zeta^3 \sim \frac{1}{\sqrt{2\pi}} \sum_{k=-\infty}^{\infty} b_k e^{ik\sigma}$$

Then, since both $\phi$ and $\zeta^3$ satisfy Eq. (4.33), the functions $\partial\phi/\partial r$ and $\zeta^3$ can be represented by means of Bessel-function expansions in $r < 1$; these expansions are, respectively,

$$\frac{\partial \phi}{\partial r} = \frac{1}{\sqrt{2\pi}} \sum_{k=-\infty}^{\infty} \frac{a_k I_k'(r) e^{ik\sigma}}{I_k'(1)}$$

$$\zeta^3 = \frac{1}{\sqrt{2\pi}} \sum_{k=-\infty}^{\infty} \frac{b_k I_k(r) e^{ik\sigma}}{I_k(1)}$$

The integrability conditions become simply

$$\sum_{k=-\infty}^{\infty} \rho_k^{-1} |a_k|^2 < \infty \qquad \text{and} \qquad \sum_{k=-\infty}^{\infty} \rho_k |b_k|^2 < \infty$$

where $\rho_k = I'_k(1)/I_k(1)$, which is of order $|k|$. Thus $\dot{H}$ can be represented as the direct product of two sequence spaces $\dot{h}_1$ and $\dot{h}_2$ with elements

$$\dot{y} = [\dot{\eta}^1, \dot{\eta}^2]$$

and inner product

$$(\dot{y}, \dot{z}) = (\dot{\eta}^1, \dot{\zeta}^1) + (\dot{\eta}^2, \dot{\zeta}^2)$$

where

$$(\dot{\eta}^1, \dot{\zeta}^1) = \sum_{k=-\infty}^{\infty} \rho_k^{-1} a_k \bar{c}_k \qquad \dot{\eta}^1 \sim \{a_k\} \quad \dot{\zeta}^1 \sim \{c_k\}$$

and

$$(\dot{\eta}^2, \dot{\zeta}^2) = \sum_{k=-\infty}^{\infty} \rho_k b_k \bar{d}_k \qquad \dot{\eta}^2 \sim \{b_k\} \quad \dot{\zeta}^2 \sim \{d_k\}$$

Here, of course, $\dot{\eta}^1$ corresponds to the boundary values of

$$\frac{\partial \phi}{\partial r} = \zeta^1 \cos \sigma + \zeta^2 \sin \sigma$$

and $\dot{\eta}^2$ corresponds to the boundary values of $\zeta^3$. The quadratic form $\dot{Q}$ is given by

$$\dot{Q}(\dot{y}, \dot{y}) = \sum_{k=-\infty}^{\infty} (a_k \bar{b}_k + \bar{a}_k b_k)$$

Let $\alpha(\sigma)$ be a bounded measurable function on $[0,2\pi)$ with nonnegative real part. We now show that the local boundary condition

$$\frac{\partial \phi}{\partial r} + \alpha(\sigma)\zeta^3 = 0 \tag{4.34}$$

defines the domain of a maximal dissipative operator $L$ between $L_0$ and $L_1$. We note that a sequence $\{b_k\}$ in $\dot{h}_2$ corresponds to a function

$$f(\sigma) \sim \frac{1}{\sqrt{2\pi}} \sum_{k=-\infty}^{\infty} b_k e^{jk\sigma}$$

in $L_2(0,2\pi)$. By imposing the usual norm in $L_2(0,2\pi)$, namely,

$$\|f\| = \left[\int_0^{2\pi} |f(\sigma)|^2 \, d\sigma\right]^{\frac{1}{2}} = \left[\sum_{k=-\infty}^{\infty} |b_k|^2\right]^{\frac{1}{2}}$$

we see that the mapping

$$\{b_k\} \to f$$

is continuous on $\dot{h}_2$ to $L_2(0,2\pi)$, in fact of norm $\rho_0^{-\frac{1}{2}}$, and that the mapping

$$[f(\sigma) \text{ in } L_2(0,2\pi)] \to [\alpha(\sigma)f(\sigma) \text{ in } L_2(0,2\pi)]$$

is also continuous and of norm

$$\|\alpha\| = \text{essential sup } |\alpha(\sigma)|$$

Finally, the mapping

$$\left[\alpha(\sigma)f(\sigma) \sim \frac{1}{\sqrt{2\pi}} \sum_{k=-\infty}^{\infty} a_k e^{jk\sigma} \text{ in } L_2(0,2\pi)\right] \to [\{a_k\} \text{ in } \dot{h}_1]$$

is also continuous and of norm $\rho_0^{-1/2}$, so that the composite mapping

$$[\{b_k\} \text{ in } \dot{h}_2] \to [\{a_k\} \equiv \alpha \cdot \{b_k\} \text{ in } \dot{h}_1]$$

is of norm $\rho_0^{-1}\|\alpha\|$. Consequently, the relationship (4.34) makes sense and defines a subspace $\dot{N}_\alpha$ in $\dot{H}$. It is clear that $\dot{N}_\alpha$ is closed, and since

$$\dot{Q}(\dot{y},\dot{y}) = -\int_0^{2\pi} [\alpha(\sigma) + \overline{\alpha(\sigma)}]|f(\sigma)|^2 \, d\sigma \leq 0 \qquad \text{for } \dot{y} \text{ in } \dot{N}_\alpha$$

it follows that $\dot{N}_\alpha$ is a negative subspace of $\dot{H}$. If $\dot{N}_\alpha$ were not maximal negative, then $\dot{N}_\alpha$ would be properly contained in another negative subspace $\dot{N}$, and there would be a $[\{c_k\}, \{d_k\}]$ in $\dot{N} - \dot{N}_\alpha$ and hence an element

$$[\{e_k\} = \{c_k\} - \alpha \cdot \{d_k\}, \{0\}]$$

in $\dot{N}$. Choose $b_k = e_k/\rho_k$ so that $\{b_k\}$ lies in $\dot{h}_2$ and

$$\sum_{k=-\infty}^{\infty} e_k \bar{b}_k \neq 0$$

In this case,

$$\dot{y} = \dot{u} + c\dot{v} = [\alpha \cdot \{b_k\}, \{b_k\}] + c[\{e_k\}, \{0\}]$$

lies in $\dot{N}$ and therefore

$$0 \geq \dot{Q}(\dot{y},\dot{y}) = \dot{Q}(\dot{u},\dot{u}) + c\dot{Q}(\dot{v},\dot{u}) + c\dot{Q}(\dot{u},\dot{v}) + |c|^2\dot{Q}(\dot{v},\dot{v})$$

But

$$\dot{Q}(\dot{v},\dot{v}) = 0 \qquad \text{and} \qquad \dot{Q}(\dot{u},\dot{v}) = \sum_{k=-\infty}^{\infty} b_k \bar{e}_k \neq 0$$

so that the above inequality cannot hold for all $c$. The supposition that $\dot{N}_\alpha$ is not maximal negative has thus led to a contradiction. It now follows from Theorem 4.4 that the boundary condition (4.34) defines a maximal dissipative operator $L$ between $L_0$ and $L_1$. The condition (4.34) is more general than Friedrichs' local boundary conditions in that continuity of $\alpha(\sigma)$ has been replaced by bounded measurability.

It is also easy to exhibit maximal negative subspaces of $\dot{H}$ that do not correspond to local boundary conditions. For example, if $\partial\phi/\partial r$ and $\zeta^3$ are represented as above by the Fourier coefficients $\{a_k\}$ and $\{b_k\}$, respectively, then for each sequence $\{\omega_k\}$ satisfying the conditions

$$|\omega_k| \leq M\rho_k \qquad \Re(\omega_k) \geq 0$$

for some $M > 0$, the subspace $\dot{N}_\omega$ of $\dot{H}$ defined by the relationships

$$a_k = -\omega_k b_k \qquad \text{for all integers } k \tag{4.35}$$

is closed and negative since the mapping

$$[\{b_k\} \text{ in } \dot{h}_2] \to [\{-\omega_k b_k\} \text{ in } \dot{h}_1]$$

is obviously continuous, in fact of norm $\leq M$, and since for

$$\dot{y} = [\{-\omega_k b_k\}, \{b_k\}]$$

we have

$$\dot{Q}(\dot{y},\dot{y}) = -\sum_{k=-\infty}^{\infty} (\omega_k + \bar{\omega}_k)|b_k|^2 \leq 0$$

Again, suppose that $\dot{N}_\omega$ were properly contained in a negative subspace $\dot{N}$; then, as above, there would be an element

$$\dot{v} = [\{e_k\}, \{0\}]$$

in $\dot{N} - \dot{N}_\omega$. Choose $b_k = \rho_k^{-1}e_k$; then the element

$$\dot{u} = [\{-\omega_k b_k\}, \{b_k\}]$$

lies in $\dot{N}$,

$$\dot{Q}(\dot{v},\dot{v}) = 0 \qquad \dot{Q}(\dot{u},\dot{v}) = \sum_{k=-\infty}^{\infty} b_k \bar{e}_k \neq 0$$

and again it is impossible to have

$$\dot{Q}(\dot{u} + c\dot{v}, \dot{u} + c\dot{v}) \leq 0$$

for all $c$. This shows that $\dot{N}_\omega$ is actually maximal negative and therefore defines the domain of a maximal dissipative operator between $L_0$ and $L_1$ as in Theorem 4.4.

## 4.7 Parabolic Partial Differential Equations

The Cauchy problem for the diffusion equation is associated with an entirely new set of concepts. Energy no longer plays a role; instead, we have to do with density and positivity, both of which are somewhat foreign to Hilbert-space theory. Consequently, one cannot expect the foregoing material to be especially suitable for the parabolic case. Indeed, the best results in this case have been obtained by W. Feller,[2] E. Hille,[5] and K. Yosida[10] with both the space of continuous functions and the space of Lebesgue-integrable functions as settings. But the previous development can be adapted[8] to take care of the diffusion equation, and this results in new information about the boundary-value problem when the domain is in a euclidean space of more than one dimension.

For notational convenience, we shall consider the initial-value problem with domain $G$ in $E_2$:

$$\begin{aligned} u_t &= (au_1)_1 + (bu_2)_2 + cu \qquad \chi \text{ in } G \quad t > 0 \\ u(\chi,0+) &= f(\chi) \end{aligned} \tag{4.36}$$

where $a$ and $b$ are positive continuously differentiable functions of $\chi$ in $\bar{G}$, while $c$ is merely nonpositive and continuous in $\bar{G}$. When we set

$$\eta^0 = u \qquad \eta^1 = u_1 \qquad \eta^2 = u_2$$

we can write Eq. (4.36) as

$$\begin{aligned} \eta^0{}_t &= (a\eta^1)_1 + (b\eta^2)_2 + c\eta^0 \\ 0 &= (a\eta^0)_1 - a_1\eta^0 - a\eta^1 \\ 0 &= (b\eta^0)_2 - b_2\eta^0 - b\eta^2 \end{aligned} \tag{4.37}$$

It is clear that this is not of the form (4.20). The right-hand member of the system (4.37) is again of the form (4.24), however, with

$$A^1 = \begin{pmatrix} 0 & a & 0 \\ a & 0 & 0 \\ 0 & 0 & 0 \end{pmatrix} \qquad A^2 = \begin{pmatrix} 0 & 0 & b \\ 0 & 0 & 0 \\ b & 0 & 0 \end{pmatrix} \qquad B = \begin{pmatrix} c & 0 & 0 \\ -a_1 & -a & 0 \\ -b_2 & 0 & -b \end{pmatrix}$$

Furthermore,

$$D = \begin{pmatrix} 2c & 0 & 0 \\ 0 & -2a & 0 \\ 0 & 0 & -2b \end{pmatrix} \leq \Theta$$

and hence the differential operator $L$ can be treated as before. We proceed, therefore, to define the minimal operator $L_0$ and the maximal operator $L_1$ as in Sec. 4.5; Theorem 4.4 then furnishes us with a characterization of all the maximal dissipative operators $L$ that lie between $L_0$ and $L_1$.

When we have obtained such a maximal dissipative operator $L$, our next step is somehow to restrict $L$ to the subspace of first components, namely,

$$H_1 = [\eta^0,0,0]$$

in order to recover an operator of the form

$$K\eta^0 = (a\eta^0{}_1)_1 + (b\eta^0{}_2)_2 + c\eta^0$$

To this end, we first define the restriction $L' \subset L$ by

$$\mathfrak{D}(L') = [y;\, y \text{ in } \mathfrak{D}(L) \text{ and } Ly \text{ in } H_1]$$

For $y$ in $\mathfrak{D}(L')$, we see that

$$\eta^1 = \eta^0{}_1 \qquad \eta^2 = \eta^0{}_2$$

and hence that

$$L'y = [K\eta^0,0,0]$$

There is still the objection that $L'$ acts on $H$ and not simply on $H_1$. With this in mind, we set $P_1$ equal to the projection

$$P_1[\eta^0,\eta^1,\eta^2] = [\eta^0,0,0]$$

and define the *retraction* of $L$ to $H_1$ as

$$\mathfrak{D}(L'') = [P_1y;\ y \text{ in } \mathfrak{D}(L')] \qquad L''P_1y = L'y$$

It is easily shown that $L''$ is uniquely determined by $P_1y$ alone anp hence is well defined. In fact, suppose that $y$ and $z$ lie in $\mathfrak{D}(L')$ and that

$$P_1y = P_1z$$

Then

$$w = y - z = [\omega^0,\omega^1,\omega^2]$$

lies in $\mathfrak{D}(L')$, and

$$P_1w = [\omega^0,0,0] = 0$$

Thus, $w$ is orthogonal to $L'w$ and

$$0 = (L'w,w) + (w,L'w) = Q(w,w) + (Dw,w)$$

by Eq. (4.25). Since $w$ also belongs to $\mathfrak{D}(L)$, and as such has nonpositive boundary conditions [that is, $Q(w,w) \leq 0$], we see that

$$0 \leq (Dw,w) = -2 \int_G (a|\omega^1|^2 + b|\omega^2|^2)\, d\chi$$

from which we can conclude that

$$\omega^1 = 0 = \omega^2$$

and hence that

$$w = 0 \qquad L'w = 0$$

It follows that $L''$ is a well-defined operator on $H_1$ to itself and of the form $K$ if we write the argument

$$P_1y = [\eta^0,0,0]$$

as simply $\eta^0$. In this case, for

$$\eta^0 = P_1y$$

and $y$ in $\mathfrak{D}(L')$, we have

$$(L''\eta^0,\eta^0) + (\eta^0,L''\eta^0) = (L'y,y) + (y,L'y) \leq 0$$

so that $L''$ is dissipative. Actually, it can be shown that $L''$ is maximal dissipative with a dense domain in $H_1$, so that $L''$ generates a semigroup of contraction operators $[S_1(t)]$ on $H_1$. Thus for $f \in \mathfrak{D}(L'')$, the function

$$u(\cdot,t) = S_1(t)f$$

solves the initial-value problem (4.36). Moreover, there is a one-to-one correspondence between the maximal dissipative operators $L$ that lie between $L_0$ and $L_1$ and the retractions defined in this way.

We note that the above procedure does not furnish us with all of the maximal dissipative operators that lie between the retractions of $L_0$ and $L_1$. We do, however, obtain all the usual operators associated with the system (4.36) and then some.

In the special case

$$a = 1 \qquad b = 1 \qquad G = [(\chi^1,\chi^2);\, (\chi^1)^2 + (\chi^2)^2 < 1]$$

the resulting operators $L$ are those considered in the example at the end of Sec. 4.6, with $\eta^3$ now replaced by $\eta^0$. For the restricted operator $L''$, we have

$$\eta^1 = \eta^0{}_1 \qquad \eta^2 = \eta^0{}_2$$

so that the function $\phi$ of the example can now be replaced by $\eta^0$; the boundary data are therefore determined in the present case by $\eta^0$ and $\partial\eta^0/\partial r$. The relationship (4.34) becomes

$$\frac{\partial \eta^0}{\partial r} + \alpha(\sigma)\eta^0 = 0$$

where again $\alpha(\sigma)$ is a bounded measurable function with nonnegative real part, and this relationship now defines the domain of a maximal dissipative operator $L''$. Likewise, the equations (4.35) furnish a nonlocal relationship between $\eta^0$ and $\partial\eta^0/\partial r$ and also define the domain of a maximal dissipative operator of type $L''$.

## REFERENCES


1. Beckenbach, E. F. (ed.), "Modern Mathematics for the Engineer," First Series, McGraw-Hill Book Company, Inc., New York, 1956.
2. Feller, W., The Parabolic Differential Equations and the Associated Semi-groups of Transformations, *Ann. of Math.*, ser. 2, vol. 55, pp. 468–519, 1952.
3. Friedrichs, K. O., Symmetric Hyperbolic Linear Differential Equations, *Comm. Pure Appl. Math.*, vol. 7, pp. 345–392, 1954.
4. ———, Symmetric Positive Linear Differential Equations, *Comm. Pure Appl. Math.*, vol. 11, pp. 333–418, 1958.
5. Hille, Einar, The Abstract Cauchy Problem and Cauchy's Problem for Parabolic Differential Equations, *J. Analyse Math.*, vol. 3, pp. 81–196, 1954.
6. ——— and R. S. Phillips, "Functional Analysis and Semi-groups," Amer. Math. Soc. Colloquium Publ., vol. 31, American Mathematical Society, New York, 1957.
7. Phillips, R. S., Dissipative Operators and Hyperbolic Systems of Partial Differential Equations, *Trans. Amer. Math. Soc.*, vol. 90, pp. 193–254, 1959.
8. ———, Dissipative Operators and Parabolic Partial Differential Equations, *Comm. Pure Appl. Math.*, vol. 12, pp. 249–276, 1959.

9. Riesz, F., and B. Sz.-Nagy, "Functional Analysis," Frederick Ungar Publishing Co., New York, 1955.
10. Yosida, K., Semi-group Theory and the Integration Problem of Diffusion Equations, *Proc. Internat. Cong. Math.*, vol. 2, pp. 1–16, 1954.
11. ———, "Lectures on Semi-group Theory and Its Applications to Cauchy's Problem in Partial Differential Equations," Tata Institute of Fundamental Research, Bombay, 1957.

# 5

# Asymptotic Formulas and Series

J. BARKLEY ROSSER
PROFESSOR OF MATHEMATICS
CORNELL UNIVERSITY

## 5.1 Introduction

Asymptotic formulas and series are useful primarily when one is concerned with the behavior of a function as the value of the independent variable becomes large. This sort of thing arises frequently in practical problems. Thus, in considering a physical system, one may wish to investigate its behavior when the number of particles is large, as in the kinetic theory of gases; or when the time is large, as in considerations of how rapidly equilibrium will be approached; or at a considerable distance from the main center of activity, as in the dispersion of waves. In the WKBJ method (see Ref. 8, pages 1092 to 1105), one determines atomic-interaction phenomena—such as phase shift for scattering—by relating the asymptotic formula for long wavelength with that for short wavelength (i.e., for large frequency). One could cite many other instances concerned with situations in which one of the quantities is large. In a considerable number of these, it would be very difficult to investigate the behavior without using asymptotic formulas and series.

In addition to indicating general behavior as the independent variable becomes large, asymptotic series can often be used to get quite accurate numerical estimates with relatively minor calculations. This is usually the case in those regions where computation by ordinary means becomes so laborious as to overtax even a large modern computer.

We shall not attempt to indicate in detail how asymptotic formulas and series are actually used in practical problems. Some such illustrations are to be found in the book by Morse and Feshbach[8] and in the one by Jeffreys and Jeffreys.[6] We shall give a general summary of the mathematical aspects of the question. Fuller treatments of these aspects have been given by Erdélyi[3] and de Bruijn;[2] the first of these two authors has compiled extensive bibliographical references,[3] which can be consulted if further information is needed.

## 5.2 Definitions

One can see that

$$\frac{z+1}{z+2} = 1 - \frac{1}{z} + \frac{2}{z^2} - \frac{4}{z^3} + \frac{8}{z^4} - \frac{16}{z^5} + \cdots \tag{5.1}$$

for $|z| > 2$ by noting that the expression on the right, after the first term, is a geometric series. On replacing the geometric series by its sum, one readily verifies Eq. (5.1). According to the definition that we shall shortly give, the right-hand side of Eq. (5.1) is an asymptotic series for the left-hand side.

One can have asymptotic series that diverge at all points. Note, for example, that $n$ successive integrations by parts will give

$$ze^z \int_z^\infty \frac{e^{-t}}{t}\,dt = 1 - \frac{1}{z} + \frac{2!}{z^2} - \frac{3!}{z^3} + \cdots + \frac{(-1)^{n-1}(n-1)!}{z^{n-1}} + (-1)^n n! z e^z \int_z^\infty \frac{e^{-t}}{t^{n+1}}\,dt \tag{5.2}$$

One is tempted to let $n$ go to infinity and replace the right-hand side by

$$\sum_{n=0}^{\infty} \frac{(-1)^n n!}{z^n} \tag{5.3}$$

Unfortunately, this series diverges for each value of $z$, so that such a replacement would be invalid.

In spite of this, the series (5.3) is quite useful. We note that the integral term in the right-hand side of Eq. (5.2) is the error that one makes if one uses $n$ terms of the series (5.3) as an approximation for the left-hand side of Eq. (5.2). If $z$ is positive, then clearly the error is positive or negative according as $n$ is even or odd. That is, if we compute the sum of an odd number of terms of the series (5.3), we get a larger value than the left-hand side of Eq. (5.2), but if we compute the sum of an even number of terms, we get a smaller value.

Suppose we desire an estimate for the left-hand side of Eq. (5.2) for $z = 10$. With $z = 10$, the sum of the first nine terms of the series (5.3) is 0.91582, while the sum of ten terms is 0.91546. Thus the true value must lie between these two numbers, and we have obtained an estimate of the true value correct to three significant figures.

There is a convergent series for the left-hand side of Eq. (5.2), namely,

$$ze^z\left(-\gamma - \log z + z - \frac{z^2}{2!2} + \frac{z^3}{3!3} - \frac{z^4}{4!4} + \cdots\right)$$

where $\gamma$ is the Euler-Mascheroni constant. To get three-significant-figure accuracy by means of this series would require using more than 40 terms, as against only 10 terms of the series (5.3). As $z$ becomes larger, the superior effectiveness of the series (5.3) becomes even more marked.

The property of the divergent series (5.3) that makes it useful for computation is a somewhat strengthened version of a general property that a series

$$\sum_{n=0}^{\infty} \frac{A_n}{z^n} \tag{5.4}$$

can have relative to a function $f(z)$. This property is that

$$\lim_{z\to\infty} z^N \left[ f(z) - \sum_{n=0}^{N} \frac{A_n}{z^n} \right] = 0 \tag{5.5}$$

for each fixed positive value of $N$. When the condition (5.5) holds, we say that the series (5.4) is an *asymptotic series* for $f(z)$ and write

$$f(z) \sim \sum_{n=0}^{\infty} \frac{A_n}{z^n}$$

Note that the relationship (5.5) expresses the fact that for $|z|$ large a finite number of terms of the series (5.4) is a good approximation for $f(z)$. Because only a finite number of terms of this series is used, the question of convergence need not arise.

Usually the condition (5.5) will hold not merely for $z$ going to infinity along the positive real axis, but for $|z|$ going to infinity subject to a restriction of the form $a \leq \arg z \leq b$, where $a$ and $b$ are a couple of real numbers with $a < b$. When this happens, we say that the series (5.4) is an asymptotic series for $f(z)$ in the angle $a \leq \arg z \leq b$. In this sense, the right-hand side of Eq. (5.1) is an asymptotic series for the left-hand side with no restriction on angle, i.e., for $-\pi \leq \arg z \leq \pi$.

For complex $z$ not at the origin or on the negative real axis, we can give a meaning to Eq. (5.2) by interpreting the integrals as line integrals along a straight line from $z$ to 1 and thence to infinity along the positive real axis. With this interpretation, Eq. (5.2) continues to hold. Moreover, by making a careful estimate of the size of the integral on the right, one can verify that the series (5.3) is an asymptotic series for the left-hand side of Eq. (5.2) in the angle

$$-\pi + \epsilon \leq \arg z \leq \pi - \epsilon \tag{5.6}$$

for each positive $\epsilon$. Indeed, if $z$ is in the right-hand half plane, including the imaginary axis except the origin, it can be shown that the error in using a finite number of terms of the series (5.3) is less in absolute value than the first term omitted. Thus, in this domain, the series is very effective for computation with $|z|$ large. The left-hand side of Eq. (5.2) appears in numerous practical applications. In particular, the case in which $z$ is a pure imaginary arises in various optical problems.

Although strictly speaking an asymptotic series must have the form (5.4), one commonly uses more general forms. Thus, instead of using explicitly the series (5.3) one would often write

$$\int_z^\infty \frac{e^{-t}}{t}\,dt \sim e^{-z} \sum_{n=0}^{\infty} \frac{(-1)^n n!}{z^{n+1}}$$

Equation (5.2) is given in Whittaker and Watson, as are the two following generalized series:

$$\Gamma(z) \sim \left(\frac{z}{e}\right)^z \sqrt{\frac{2\pi}{z}} \left(1 + \frac{1}{12z} + \frac{1}{288z^2} - \frac{139}{51{,}840z^3} - \frac{571}{2{,}488{,}320z^4} + \cdots\right) \tag{5.7}$$

$$J_n(z) \sim \sqrt{\frac{2}{\pi z}} \left[P \cos\left(z - \frac{\pi n}{2} - \frac{\pi}{4}\right) + Q \sin\left(z - \frac{\pi n}{2} - \frac{\pi}{4}\right)\right] \tag{5.8}$$

where

$$P = 1 + \sum_{r=1}^{\infty} \frac{(-1)^r(4n^2 - 1^2)(4n^2 - 3^2) \cdots [4n^2 - (4r - 1)^2]}{(2r)!2^{6r}z^{2r}}$$

and

$$Q = \sum_{r=1}^{\infty} \frac{(-1)^r(4n^2 - 1^2)(4n^2 - 3^2) \cdots [4n^2 - (4r - 3)^2]}{(2r - 1)!2^{6r-3}z^{2r-1}}$$

Here $\Gamma(z)$ is the gamma function and $J_n(z)$ is the Bessel function of order $n$. Both expressions (5.7) and (5.8) are valid in the angle (5.6).

Although the sense in which the representation (5.8) holds is a considerable generalization of that for the representation (5.5), one can use the series (5.8) very effectively for computation. Indeed, for large $z$, it is much more useful than the convergent series

$$J_n(z) = \sum_{r=0}^{\infty} \frac{(-1)^r z^{n+2r}}{2^{n+2r} r!\Gamma(n + r + 1)} \tag{5.9}$$

Thus for $n = 0$ and $z$ as small as 6, we can get five-decimal-place accuracy by using no more than seven terms of the series (5.8) as against at least twelve terms of the series (5.9), despite the remarkably fast convergence of the latter series. As $z$ increases, the superiority of the representation (5.8) over (5.9) improves.

We say that $g(z)$ is an *asymptotic formula* for $f(z)$ in the angle $a \leq \arg z \leq b$ if

$$\lim_{|z| \to \infty} \frac{f(z)}{g(z)} = 1 \tag{5.10}$$

for $a \leq \arg z \leq b$. In such a case, we write

$$f(z) \sim g(z)$$

From Eq. (5.2) we get

$$\int_z^\infty \frac{e^{-t}}{t}\, dt \sim \frac{e^{-z}}{z}$$

and from the representation (5.7) we obtain

$$\Gamma(z) \sim \left(\frac{z}{e}\right)^z \sqrt{\frac{2\pi}{z}}$$

in the angle (5.6). Since $n! = n\Gamma(n)$ when $n$ is a positive integer, we deduce the well-known Stirling approximation

$$n! \sim \frac{n^n \sqrt{2\pi n}}{e^n}$$

One might jump to the conclusion that the first term of an asymptotic series always gives an asymptotic formula. This is not so, as one can see by looking at the series (5.8). Although

$$\sqrt{\frac{2}{\pi z}} \cos\left(z - \frac{\pi n}{2} - \frac{\pi}{4}\right)$$

is a very useful approximation to $J_n(z)$ for large $z$, one does not have

$$\lim_{z \to \infty} J_n(z) \sqrt{\frac{\pi z}{2}} \left[\cos\left(z - \frac{\pi n}{2} - \frac{\pi}{4}\right)\right]^{-1} = 1$$

since the left-hand side goes to infinity at each of the points

$$z = \frac{\pi n}{2} + \frac{3\pi}{4} + \pi N$$

One can have asymptotic formulas and series that hold in a smaller angle than (5.6). In particular, we have

$$\cos\left(z - \frac{\pi n}{2} - \frac{\pi}{4}\right) \sim \frac{1}{2} \exp\left[-j\left(z - \frac{\pi n}{2} - \frac{\pi}{4}\right)\right]$$

$$\sin\left(z - \frac{\pi n}{2} - \frac{\pi}{4}\right) \sim \frac{j}{2} \exp\left[-j\left(z - \frac{\pi n}{2} - \frac{\pi}{4}\right)\right]$$

for

$$\epsilon \leq \arg z \leq \pi - \epsilon \tag{5.11}$$

as one can easily show. By substituting these approximations into the series (5.8), for $z$ satisfying (5.11) we get

$$J_n(z) \sim \sqrt{\frac{1}{2\pi z}}\, PQ \tag{5.12}$$

where

$$P = 1 + \sum_{r=1}^{\infty} \frac{(-j)^r(4n^2 - 1^2)(4n^2 - 3^2) \cdots [4n^2 - (2r - 1)^2]}{r!(8z)^r}$$

and

$$Q = \exp\left[-j\left(z - \frac{\pi n}{2} - \frac{\pi}{4}\right)\right]$$

The problem of determining the angle in which a given asymptotic series represents a function is quite tricky. One can even have a given function represented by different asymptotic series in different angles. In a given angle, however, an asymptotic series in the strict sense, of the form (5.4), must be unique.

For asymptotic series in the strict sense, of the form (5.4), one can perform the standard operations of adding, subtracting, multiplying, and dividing in the same manner as if the series were convergent. One can also integrate term by term. Differentiation is not always possible. Indeed, there are functions that have asymptotic series of the form (5.4) even though their derivatives do not possess such asymptotic series. Since term-by-term integration is possible, however, one can conclude that, if a function and its derivative both have asymptotic series of the form (5.4), then one can find the series for the derivative by differentiating term by term in the series for the function.

All these items, and similar matters, are discussed in considerable detail by Erdélyi[3] and de Bruijn.[2]

## 5.3 Integration by Parts

The method of integration by parts that we used in deriving Eq. (5.2) can be generalized. Consider

$$\int_A^B f(x) e^{zg(x)}\, dx \tag{5.13}$$

where $g'(x) \neq 0$ for $A \leq x \leq B$. We can rewrite this integral as

$$\int_A^B \frac{f(x)}{g'(x)}\, g'(x) e^{zg(x)}\, dx$$

whereupon an integration by parts changes it to

$$\frac{1}{z}\left\{\frac{f(B)}{g'(B)}e^{zg(B)} - \frac{f(A)}{g'(A)}e^{zg(A)} - \int_A^B \left[\frac{d}{dx}\frac{f(x)}{g'(x)}\right] e^{zg(x)}\,dx\right\}$$

Often one of the explicit terms is an asymptotic formula. In any case, the new integral has the same form as the integral (5.13), so that one can repeat the process. Continued repetition commonly produces an asymptotic series.

Thus, starting with

$$\int_0^A e^{-zx^2}\,dx = \frac{\sqrt{\pi}}{2\sqrt{z}} - \int_A^\infty e^{-zx^2}\,dx \tag{5.14}$$

we can take $B = \infty$, $f(x) = 1$, and $g(x) = -x^2$ to get

$$\int_0^A e^{-zx^2}\,dx = \frac{\sqrt{\pi}}{2\sqrt{z}} - Ae^{-A^2z}\left[\frac{1}{2A^2z} - \frac{1}{(2A^2z)^2} + \frac{1\cdot 3}{(2A^2z)^3} - \frac{1\cdot 3\cdot 5}{(2A^2z)^4} + \cdots + \frac{(-1)^n 1\cdot 3\cdot 5\cdots(2n-1)}{(2A^2z)^{n+1}}\right] - (-1)^n 1\cdot 3\cdot 5\cdots(2n+1)\int_A^\infty \frac{e^{-zx^2}}{(2x^2z)^{n+1}}\,dx$$

From this, one can conclude that

$$\int_0^A e^{-zx^2}\,dx \sim \frac{\sqrt{\pi}}{2\sqrt{z}} + \frac{Ae^{-A^2z}}{2\sqrt{\pi}}\sum_{n=1}^{\infty}\frac{\Gamma[(2n-1)/2]}{(-A^2z)^n} \tag{5.15}$$

holds for

$$-\frac{\pi}{2} \le \arg z \le \frac{\pi}{2}$$

if $A$ is positive. Actually, the representation (5.15) holds for complex $A$ and for a wider angle than stated, but the exact specifications are a bit involved, and we shall not give them.

If $A^2z$ is at all large in absolute value, then the series (5.15) is excellent for computing numerical approximations. Indeed, it is far superior to either of the following convergent series:

$$\int_0^A e^{-zx^2}\,dx = A\sum_{n=0}^{\infty}\frac{(-A^2z)^n}{n!(2n+1)} = Ae^{-A^2z}\left[1 + \frac{2A^2z}{1\cdot 3} + \frac{(2A^2z)^2}{1\cdot 3\cdot 5} + \frac{(2A^2z)^3}{1\cdot 3\cdot 5\cdot 7} + \cdots\right]$$

## 5.4 The Generalized Watson's Lemma

If we wish to evaluate

$$\int_0^\infty f(x)e^{-zx}\,dx$$

we can proceed formally as follows. Expand $f(x)$ in a Maclaurin series and integrate term by term. The result is

$$\sum_{n=0}^{\infty} \frac{f^{(n)}(0)}{z^{n+1}}$$

Sometimes this or similar devices are legitimate. Thus, if

$$f(x) = \frac{\cos\sqrt{x}}{\sqrt{x}}$$

we can expand $f(x)$ in a Laurent series in $\sqrt{x}$ and integrate term by term to get

$$\int_0^\infty \frac{\cos\sqrt{x}}{\sqrt{x}}\,e^{-zx}\,dx = \frac{1}{\sqrt{z}}\sum_{n=0}^{\infty}\frac{\Gamma[(2n+1)/2]}{(2n)!(-z)^n} = \sqrt{\frac{\pi}{z}}\,e^{-1/4z}$$

which is a known result. Usually the procedure is not permissible, however, since the resulting series diverges for all $z$. Even in these cases, the method commonly is legitimate if one interprets the resulting series as an asymptotic series for the original integral. Indeed, if we take $A = 0$, $B = \infty$, and $g(x) = -x$ in the integral (5.13), then under very modest assumptions as to the behavior of $f(x)$ and its derivatives, we can infer from repeated integration by parts that

$$\int_0^\infty f(x)e^{-zx}\,dx \sim \sum_{n=0}^{\infty}\frac{f^{(n)}(0)}{z^{n+1}} \tag{5.16}$$

for

$$-\frac{\pi}{2} + \epsilon \le \arg z \le \frac{\pi}{2} - \epsilon \tag{5.17}$$

To illustrate the use of the foregoing procedure, take

$$x = \frac{y}{\sqrt{z}} + A$$

in the right-hand side of Eq. (5.14) to get

$$\int_0^A e^{-zx^2}\,dx = \frac{\sqrt{\pi}}{2\sqrt{z}} - \frac{e^{-A^2z}}{\sqrt{z}}\int_0^\infty e^{-y^2}e^{-2Ay\sqrt{z}}\,dy$$

Now, by using the series (5.16) with exp $(-x^2)$ for $f(x)$, we get

$$\int_0^A e^{-zx^2}\,dx \sim \frac{\sqrt{\pi}}{2\sqrt{z}} - \frac{e^{-A^2z}}{\sqrt{z}} \sum_{n=0}^{\infty} \frac{(-1)^n(2n)!}{n!(2A\sqrt{z})^{2n+1}}$$

which is the same as (5.15), except that according to the inequalities (5.17) it holds for

$$-\pi + \epsilon \leq \arg\,(A^2z) \leq \pi - \epsilon$$

The result embodied in the relationships (5.16) and (5.17) is usually called Watson's lemma. There is a generalized form of it that is useful for getting asymptotic series in a wide variety of cases. This says that

$$\int_0^K f(x^s)x^r e^{-zx}\,dx \sim \sum_{n=0}^{\infty} \frac{\Gamma(r+ns+1)f^{(n)}(0)}{z^{r+ns+1}n!} \tag{5.18}$$

in the angle (5.17). The representation (5.18) holds for each positive $K$ or for $K = \infty$, subject only to some very mild hypotheses such as that, for some positive $z$,

$$\int_0^K |f(x^s)|x^r e^{-zx}\,dx$$

exists and that $f(w)$ is analytic in the neighborhood of $w = 0$. One can also let $K$ be complex, usually at the expense of having the representation (5.18) hold in a smaller angle than (5.17).

To prove the validity of the series (5.18), one can generalize the proof given in Sec. 17.03 of the book by Jeffreys and Jeffreys.[6] Note that the right-hand side of (5.18) is what one would get by expanding $f(x^s)$ as a Maclaurin series in $x^s$ and integrating term by term from zero to infinity. Indeed, the proof of (5.18) depends on this fact.

For the special cases $s = 1$ and $s = \frac{1}{2}$, the representation (5.18) has been established by Rosser,[10] who used alternative procedures involving integration by parts. This has the advantage of simplifying the estimation of the error that one would commit in using a given finite number of terms of the series.

## 5.5 Asymptotic Solution of Differential Equations

The method of solution of differential equations by series is well known. To refresh our memories, let us attempt to solve the differential equation

$$\frac{d^2y}{dz^2} + \frac{1}{z}\frac{dy}{dz} + \left(1 - \frac{n^2}{z^2}\right)y = 0 \tag{5.19}$$

by a series of the form

$$y = \sum_{r=0}^{\infty} A_r z^{r+s} \tag{5.20}$$

If we differentiate term by term in Eq. (5.20) and substitute into Eq. (5.19), then we get the relationship

$$A_{r-2} = [n^2 - (r + s)^2]A_r \tag{5.21}$$

by equating to zero the coefficient of $z^{r+s}$ in the left-hand member. If we are to have $A_0 \neq 0$, then we must have $n^2 = s^2$. When we take $s = n$, Eq. (5.21) reduces to

$$A_{r-2} = -r(2n + r)A_r \tag{5.22}$$

Since any solution of Eq. (5.19) remains a solution if multiplied by a constant, the choice of $A_0$ is arbitrary. If we choose

$$A_0 = \frac{1}{2^n\Gamma(n + 1)}$$

we verify by Eq. (5.22) that the series on the right in Eq. (5.9) is a solution of Eq. (5.19). If we take $s = -n$, we get a second solution of Eq. (5.19) unless $n$ is an integer, in which case the relationship (5.21) fails at $r = 2n$.

Because Eq. (5.19) is the differential equation for the Bessel function of order $n$, these results are consistent with Eq. (5.9).

One might expect that in a similar way one could get an asymptotic series for $J_n(z)$ by differentiating term by term in the series (5.4), substituting into Eq. (5.19), and equating to zero the coefficient of each power of $z$ in the left-hand member. Actually, this is a rather dubious procedure, since, as we remarked earlier, term-by-term differentiation is not always permissible with an asymptotic series. In the present case, there is a more serious difficulty in that $J_n(z)$ does not have an asymptotic series of the form (5.4). Thus it is not surprising that the attempt to find a solution for Eq. (5.19) of the form (5.4) fails.

Upon referring to the representation (5.12), we see that $J_n(z)$ has an asymptotic series of a different form. This suggests that we write

$$J_n(z) = z^{-1/2}F(z) \exp\left[-j\left(z - \frac{\pi n}{2} - \frac{\pi}{4}\right)\right]$$

By substituting this into Eq. (5.19), we find that $F(z)$ must satisfy the differential equation

$$F''(z) - 2jF'(z) + \frac{1 - 4n^2}{4z^2}F(z) = 0$$

In this we can substitute

$$F(z) \sim \sum_{r=0}^{\infty} \frac{A_r}{z^r}$$

differentiate term by term, and equate the coefficient of $z^{-r}$ on the left to zero. This gives

$$A_{r+1} = -j \frac{4n^2 - (2r+1)^2}{8(r+1)} A_r$$

which immediately verifies the series given in (5.12), except for a constant factor. As remarked earlier, one can determine a solution of Eq. (5.19) only to within a multiplicative constant.

In summary, given a differential equation of the form (5.19), one may seek a solution of the form

$$CK(z) \sum_{r=0}^{\infty} \frac{A_r}{z^r}$$

where $C$ is a multiplicative constant and $K(z)$ is a multiplicative function. One cannot determine $C$ from the differential equation. Because $C$ is arbitrary, one can take $A_0 = 1$. Then, if $K(z)$ is known, one can determine the other $A$'s from the differential equation. There is a procedure for determining $K(z)$ also from the differential equation in special cases. It is intricate, and we shall not attempt to describe it. It is discussed briefly in Sec. 17.12 of the book by Jeffreys and Jeffreys,[6] and more extensively beginning with Sec. 7.3 of the text by Ince.[5]

As another example, we shall later consider a certain function $G(w)$ that satisfies the differential equation

$$wG'''(w) + 2G''(w) + 2G(w) = 0$$

We shall show that it has an asymptotic expansion of the form

$$G(w) \sim \frac{e^{\beta^2}}{\beta\sqrt{\pi}} \left[ 1 + \sum_{r=1}^{\infty} \frac{A_r}{r!(12\beta^2)^r} \right]$$

with

$$\beta = \left( \frac{-27w^2}{4} \right)^{1/6}$$

As it happens, our method for showing this is such that it is extremely laborious to compute $A_r$ for $r$ larger than 1 or 2. It is fairly easy, however, to compute $A_r$ for larger $r$ by using the differential equation. We take $\beta$ as the independent variable, and set

$$G(w) = \frac{e^{\beta^2}}{\beta\sqrt{\pi}} H(\beta)$$

Then $H(\beta)$ satisfies the differential equation

$$\beta^3 H'''(\beta) + 3(2\beta^4 - \beta^2)H''(\beta) \\ + 2(6\beta^5 - 3\beta^3 + 2\beta)H'(\beta) + 2(\beta^2 - 2)H(\beta) = 0$$

Into this we substitute

$$H(\beta) \sim \sum_{r=0}^{\infty} \frac{A_r}{r!(12\beta^2)^r}$$

differentiate term by term, and equate the coefficient of $\beta^{-2r}$ on the left to zero. This gives

$$A_1 = A_0$$

$$A_{r+2} = (12r^2 + 36r + 25)A_{r+1} - 12(2r + 1)^2(r + 1)(r + 2)A_r$$

Starting with the known value of unity for $A_0$, we compute

$$\begin{array}{lll} A_0 = 1 & A_2 = 2 & A_4 = -86{,}975 \\ A_1 = 1 & A_3 = -575 & \text{etc.} \end{array}$$

## 5.6 Other Methods of Deriving Asymptotic Series

In Chap. 3 of the book by de Bruijn[2] are listed a number of special means for deriving asymptotic series. We shall not attempt even to list them but shall note one that is fairly widely known. It involves use of the Euler-Maclaurin sum formula. It is applicable only in certain special cases, but when it is applicable, it generally gives very effective series. The Euler-Maclaurin sum formula is described in the book by Whittaker and Watson[13] and also in the one by Jeffreys and Jeffreys.[6] In the latter reference, it is used in deriving several asymptotic series, including the well-known one

$$\log \Gamma(z) \sim (z - \tfrac{1}{2}) \log z - z + \tfrac{1}{2} \log (2\pi) - \sum_{r=1}^{\infty} \frac{(-1)^r B_r}{2r(2r - 1)z^{2r-1}}$$

which is valid in the angle (5.6). Here $B_r$ is the $r$th Bernoulli number, with

$$B_1 = \tfrac{1}{6} \qquad B_2 = \tfrac{1}{30} \qquad B_3 = \tfrac{1}{42} \qquad \text{etc.}$$

Many other uses of the Euler-Maclaurin sum formula to derive asymptotic series are given by Rosser.[9]

## 5.7 Eulerizing

In Sec. 5.2, we get a three-decimal estimate for the left-hand side of Eq. (5.2) with $z = 10$ by adding together nine or ten terms of the series

(5.3). If one wishes more accuracy, one cannot get it merely by directly adding together more terms. The terms after the tenth get numerically larger and larger so that, while the partial sums are alternately too large and too small, they get farther away from the true value.

Quite a great deal more accuracy can be obtained, however, if one transforms the remainder of the series instead of merely trying to add its terms. The Euler transformation

$$\sum_{n=0}^{\infty} a_n x^n = \frac{1}{1-x} \sum_{n=0}^{\infty} (\Delta^n a_0) \left(\frac{x}{1-x}\right)^n \tag{5.23}$$

is commonly useful for this purpose. We note that the coefficients on the right are defined by

$$\begin{aligned} \Delta^0 a_r &= a_r \\ \Delta^{n+1} a_r &= (\Delta^n a_{r+1}) - (\Delta^n a_r) \end{aligned}$$

so that the computation of these coefficients can be readily accomplished by repeatedly taking differences.

A discussion of Eq. (5.23) together with many illustrations of its use in connection with asymptotic series is given by Rosser.[9] Such use of Eq. (5.23) is often called "eulerizing."

Let us now see how eulerizing can be employed to give more accuracy in the case we considered above. The sum of the first nine terms was exactly 0.9158192. With $z = 10$, the next terms are

$$-\left(\frac{9!}{10^9} - \frac{10!}{10^{10}} + \frac{11!}{10^{11}} - \frac{12!}{10^{12}} + \cdots\right)$$

Let us now apply Eq. (5.23) to give an estimate for $10^8$ times the quantity written in parentheses. That is, we try to identify

$$\frac{9!}{10} - \frac{10!}{10^2} + \frac{11!}{10^3} - \frac{12!}{10^4} + \cdots$$

with the left-hand side of Eq. (5.23). For this, we take $x = -1$ and

$$a_0 = \frac{9!}{10} = 36{,}288 \qquad a_2 = \frac{11!}{10^3} = 39{,}917 \text{ (approx)}$$

$$a_1 = \frac{10!}{10^2} = 36{,}288 \qquad a_3 = \frac{12!}{10^4} = 47{,}900 \text{ (approx)}$$

$$a_4 = \frac{13!}{10^5} = 62{,}270 \text{ (approx)}$$

etc.

By using the approximations listed above, we need to use only simple differencing operations to compute the following approximations:

$$\begin{array}{llll}
\Delta a_0 = 0 & \Delta^2 a_0 = 3{,}629 & \Delta^3 a_0 = 725 & \Delta^4 a_0 = 1{,}308 \\
\Delta a_1 = 3{,}629 & \Delta^2 a_1 = 4{,}354 & \Delta^3 a_1 = 2{,}033 & \text{etc.} \\
\Delta a_2 = 7{,}983 & \Delta^2 a_2 = 6{,}387 & \text{etc.} & \\
\Delta a_3 = 14{,}370 & \text{etc.} & & \\
\text{etc.} & & &
\end{array}$$

Then the right-hand side of Eq. (5.23) takes the form

$$\frac{1}{2}\left(a_0 - \frac{\Delta a_0}{2} + \frac{\Delta^2 a_0}{4} - \frac{\Delta^3 a_0}{8} + \cdots\right)$$
$$= 18{,}144 - 0 + 454 - 45 + 41 + \cdots = 18{,}594 + \cdots$$

So our revised estimate for the left-hand side of Eq. (5.2) is

$$0.91581920 - 0.00018594 = 0.91563326$$

An accurate value, computed by using a large number of terms of the convergent series, is 0.91563334, so that our estimate is in error only by eight units in the eighth decimal place.

This agreement is all the more surprising when we note that, strictly speaking, Eq. (5.23) is valid only when applied to convergent series, and not always then. When Eq. (5.23) is applied to an asymptotic series, however, it generally yields another asymptotic series, although an instance has been given by Rosser[9] in which this does not happen. In cases when the use of Eq. (5.23) yields an asymptotic series, one can often apply Eq. (5.23) a second time, to the transformed series, to get still greater accuracy.

## 5.8 Continued Fractions

A considerable discussion of means for obtaining more accurate numerical values by getting sharp numerical estimates of the error after summing several terms of an asymptotic series has been given by Rosser.[9,10] The simplest of these, often highly effective, is eulerizing, which we have just discussed. The most powerful and widely useful method involves the use of continued fractions. Unfortunately, the theory is quite involved, so that it is not a method that can be readily applied without considerable preliminary study. We shall not attempt an explanation; indeed, we shall not even say what a continued fraction is. For some discussion and illustrations, the reader may consult Ref. 9. A full and extensive treatment is given by Wall.[12]

We shall, however, illustrate one of the key ideas in the theory of continued fractions. Note that, as in Eq. (5.1), the quotient of two polynomials gives a series of the form (5.4) that converges for all sufficiently large $z$. If we choose two polynomials such that the first $n$ terms of their quotient exactly equal the first $n$ terms of the series (5.4), then the quotient should be a good approximation. Indeed, it usually is.

As an example, note that the series in (5.1) agrees exactly with the series (5.3) for the first three terms. Taking $z = 10$ in the left-hand side of Eq. (5.1) gives

$$^{11}\!/_{12} = 0.9167$$

This is a fair approximation to the true value, which is 0.9156 to four decimal places.

To get a quotient of two quadratic polynomials, we write formally

$$z^2 + az + b = (z^2 + cz + d)\left(1 - \frac{1}{z} + \frac{2!}{z^2} - \frac{3!}{z^3} + \cdots\right)$$

multiply out on the right, and equate coefficients of the first five powers of $z$ on the two sides. Solving the resulting equations gives $a = 5$, $b = 2$, $c = 6$, and $d = 6$. We have, in fact,

$$\frac{z^2 + 5z + 2}{z^2 + 6z + 6} = 1 - \frac{1}{z} + \frac{2}{z^2} - \frac{6}{z^3} + \frac{24}{z^4} - \frac{108}{z^5} + \cdots$$

for $|z| > 3 + \sqrt{3}$. The first five terms on the right agree exactly with the first five terms of the series (5.3), and the sixth terms are similar. If we take $z = 10$ in the left-hand side above, we get

$$^{152}\!/_{166} = 0.91566$$

as compared with the more accurate five-decimal-place approximation 0.91563. This is considerably closer than one could come directly with any partial sum of the series (5.3).

Quotients of polynomials such as we have illustrated arise automatically as convergents of continued-fraction expansions. One of the tremendous advantages of continued-fraction expansions is that in a large number of cases a divergent asymptotic series will transform into a convergent continued fraction. In such a case, one can get as accurate an approximation as desired by going to a quotient of polynomials of sufficiently high degree.

## 5.9 Laplace's Method

One often wishes to determine an asymptotic formula or asymptotic series for

$$\int_{-\infty}^{\infty} e^{-zt^2} g(t)\, dt \tag{5.24}$$

or

$$\int_0^{\infty} e^{-zt^2} f(t)\, dt \tag{5.25}$$

Of these integrals, (5.24) is actually a special case of (5.25), as can be seen by taking

$$f(t) = g(t) + g(-t)$$

in (5.25). If $f(t)$ is analytic in the neighborhood of the origin, one can get an asymptotic series in $\sqrt{z}$ for the integral (5.25) by expanding $f(t)$ as a Maclaurin series in $t$ and integrating term by term. One can get the same expansion by using the generalized Watson's lemma; specifically, substitute $x = t^2$ in the integral (5.25) and then take $s = \frac{1}{2}$ and $r = -\frac{1}{2}$ in the expansion (5.18).

Various classical procedures such as Laplace's method, the method of steepest descent, and the saddle-point method amount to nothing more than applying various ingenious devices to reduce the function in question to one of the two forms (5.24) or (5.25). A common instance of this arises when one is considering the integral (5.13) and $g'(x) = 0$ for a unique $x$ with $A \leq x \leq B$, so that one cannot use integration by parts as discussed in Sec. 5.3.

We shall illustrate by showing how to derive the expansion (5.7). We start with the Euler formula

$$\Gamma(z) = \frac{1}{z} \int_0^\infty t^z e^{-t}\, dt$$

By putting $t = zx$, we get

$$\Gamma(z) = z^z \int_0^\infty x^z e^{-zx}\, dx = \left(\frac{z}{e}\right)^z \int_0^\infty e^{z(1-x+\log x)}\, dx$$

We now let

$$t = \sqrt{x - 1 - \log x}$$

It turns out that $x$ is an analytic function of $t$ near the origin, and we have

$$\int_0^\infty e^{z(1-x+\log x)}\, dx = \int_{-\infty}^\infty e^{-zt^2} \frac{dx}{dt}\, dt$$

This has the form (5.24). If we expand $dx/dt$ as a Maclaurin series in $t$ and integrate term by term, we get the expansion (5.7), valid in the angle (5.17).

A detailed discussion of this particular expansion is given by Rosser.[10] Its validity in the larger angle (5.6) is established by Whittaker and Watson.[13]

If $g'(x) = 0$ for several points in the interval $(A,B)$, we break the integral (5.13) into several integrals, with $g'(x) = 0$ at most once in each interval. Then we treat the various integrals separately.

The procedure illustrated above can be employed to devise a proof for results such as the following, use of which is commonly called Laplace's method.

THEOREM. *Let $A$, $B$, and $C$ be real numbers satisfying $A < C < B$ Let $f(x)$ be a function, not necessarily real, that is continuous for $A \leq x \leq B$.*

*and analytic at $x = C$. Let $g(x)$ be a real function, analytic at $x = C$, that has a continuous second derivative for $A < x < B$. Let $g'(C) = 0$, $g''(C) < 0, f(C) \neq 0$; let also $g'(x) \neq 0$ for $A < x < C$ and for $C < x < B$. Then an asymptotic formula for*

$$\int_A^B f(x)e^{zg(x)}\,dx$$

*is*

$$\frac{f(C)\sqrt{2\pi}\,e^{zg(C)}}{\sqrt{-zg''(C)}}$$

*valid for*

$$-\frac{\pi}{2} + \epsilon \leq \arg z \leq \frac{\pi}{2} - \epsilon$$

*while in the same angle each of*

$$\int_A^C f(x)e^{zg(x)}\,dx \qquad \text{and} \qquad \int_C^B f(x)e^{zg(x)}\,dx$$

*has the asymptotic formula*

$$\frac{f(C)\sqrt{\pi}\,e^{zg(C)}}{\sqrt{-2zg''(C)}}$$

For example, if we take $A = 0$, $B = \infty$, $C = 1$,

$$f(x) = 1 \qquad g(x) = 1 - x + \log x$$

in the theorem, we get

$$\int_0^\infty e^{z(1-x+\log x)}\,dx \sim \sqrt{\frac{2\pi}{z}}$$

Certain intuitive considerations are quite helpful in understanding the above theorem. The hypotheses of the theorem ensure that $g(x)$ takes its maximum at $x = C$. Thus, since $0 < \Re(z)$, we see that

$$|e^{zg(x)}|$$

also takes its maximum at $x = C$. Furthermore, since $g(x)$ is in the exponent, this maximum will be very sharp at $x = C$. Consequently, most of the value of the integral comes from the values of $x$ near $C$. So we replace $f(x)$ and $g(x)$ by their values near $x = C$, specifically

$$f(C) \quad \text{for } f(x) \qquad \text{and} \qquad g(C) + \tfrac{1}{2}g''(C)(x - C)^2 \quad \text{for } g(x)$$

Then our integral assumes the form

$$f(C)e^{zg(C)} \int_A^B \exp\left[\frac{z}{2}\,g''(C)(x - C)^2\right] dx$$

For large $z$, the value of this integral is changed only slightly if we replace $A$ and $B$ by $-\infty$ and $+\infty$, respectively. This gives the asymptotic formula stated in the theorem.

By means of a proof based on these ideas, the theorem can be strengthened in various ways. For instance, the hypotheses that $f(x)$ and $g(x)$ are analytic at $x = C$ are quite unnecessary. One can strengthen the theorem still further by weakening the hypotheses of continuity and existence of derivatives. Various results of this sort are given by Rosser.[11]

Of much more practical use is the fact that the foregoing theorem continues to hold even if $f$ and $g$ are allowed to be functions of both $x$ and $z$ instead of merely $x$, provided the dependence on $z$ is rather mild. For details, the reader should consult Ref. 11, but we shall cite one useful result. If there are constants $k$, $a$, and $b$ such that

$$\lim_{x\to C} f(x,z) = k \qquad \lim_{x\to C} g(x,z) = a \qquad \lim_{x\to C} g''(x,y) = b$$

uniformly in the region

$$-\frac{\pi}{2} + \epsilon \leq \arg z \leq \frac{\pi}{2} - \epsilon$$

and $|z|$ sufficiently large, then the above theorem continues to hold with $k$ in place of $f(C)$, $a$ in place of $g(C)$, and $b$ in place of $g''(C)$.

We shall illustrate with a simple example. Consider

$$I(z) = \int_0^\infty \frac{y^z e^{-y}}{1 + y}\, dy$$

Take $y = zx$ and get

$$I(z) = \frac{z^{z+1}}{1+z} \int_0^\infty \frac{1+z}{1+zx} x^z e^{-zx}\, dx$$

Now take $A = 0$, $B = \infty$, $C = 1$,

$$f(x,z) = \frac{1+z}{1+zx} \qquad g(x,z) = \log x - x$$

and get $k = 1$, $a = -1$, and $b = -1$. So an asymptotic formula for $I(z)$ is

$$J(z) = \frac{z^z \sqrt{2\pi z}}{e^z(1+z)}$$

The efficiency of this formula is illustrated in Table 5.1.

**Table 5.1** Efficiency of the Asymptotic Formula $J(z)$ for the Integral $I(z)$

| $z$ | $I(z)$ | $J(z)$ | Per cent error |
|---|---|---|---|
| 1 | 0.40366 | 0.461 | 14 |
| 2 | 0.59634 | 0.640 | 7 |
| 5 | 19.404 | 19.67 | 1.4 |
| 10 | $3.2698 \times 10^5$ | $3.2717 \times 10^5$ | 0.06 |

## 5.10 The Method of Stationary Phase

If $A$ and $B$ are both finite, then the theorem of the previous section holds even for $z$ on the imaginary axis, and indeed without the hypotheses that $f(x)$ and $g(x)$ are analytic at $C$; see the discussion by Erdélyi[3] under the heading "The method of stationary phase." This is the name given to the employment of the theorem when $z$ is a pure imaginary. Specifically, let the hypotheses of the theorem hold, except perhaps those concerning analyticity. Also let $A$ and $B$ both be finite. Then, for $t$ real and positive,

$$\lim_{t\to\infty} \sqrt{-g''(C)t}\, e^{-jtg(C)} \int_A^B f(x)e^{jtg(x)}\, dx = f(C)\sqrt{\pi}\,(1-j)$$

and

$$\lim_{t\to\infty} \sqrt{-g''(C)t}\, e^{jtg(C)} \int_A^B f(x)e^{-jtg(x)}\, dx = f(C)\sqrt{\pi}\,(1+j)$$

In the method of stationary phase, one can let $g''(C) > 0$, since it suffices to change the signs of $j$, $g(x)$, $g(C)$, and $g''(C)$ simultaneously in the above results to recover the case $g''(C) < 0$.

As in the Laplace method, each of the integrals, from $A$ to $C$ and from $C$ to $B$, has as an asymptotic formula one-half the formula given for the integral from $A$ to $B$.

The intuitive argument for the method of stationary phase goes as follows: At $x \neq C$, $g(x)$ is changing. Then for large $t$, $tg(x)$ is changing very rapidly. Thus $e^{jtg(x)}$ is oscillating very rapidly. So when we evaluate

$$\int f(x)e^{jtg(x)}\, dx$$

the alternate plus and minus values produced by the oscillation tend to cancel out. At $x = C$, $g(x)$ is stationary, and $e^{jtg(x)}$ does not oscillate. Thus there is not the cancellation that there is at other values of $x$. So the value of the integral depends primarily on what $f(x)$ and $g(x)$ do near $x = C$. Specifically, we make the same replacements for $f(x)$ and $g(x)$ as in our intuitive discussion of the previous section, and then replace $A$ and $B$ by $-\infty$ and $+\infty$. This gives the results stated.

If $f(x)$ is positive and $g(x)$ and $t$ are real, then $f(x)e^{jtg(x)}$ represents an oscillation of amplitude $f(x)$ and phase $tg(x)$. Hence the name "stationary phase," since $x = C$ is the point at which $g(x)$ is stationary.

We illustrate the method of stationary phase by finding the asymptotic behavior of $J_n(t)$ for $t$ positive and $n$ a nonnegative integer. It is shown by Whittaker and Watson[13] that in this case

$$J_n(t) = \frac{1}{\pi}\int_0^\pi \cos(nx - t\sin x)\, dx = \Re\frac{1}{\pi}\int_0^\pi e^{-jnx}e^{jt\sin x}\, dx$$

We take $A = 0$, $B = \pi$, $C = \pi/2$, $f(x) = e^{-jnx}$, and $g(x) = \sin x$. This gives

$$\frac{1}{\pi}\int_0^\pi e^{-jnx}e^{jt \sin x}\,dx \sim \frac{e^{-jn\pi/2}\sqrt{\pi}\,(1-j)e^{jt}}{\pi\sqrt{t}}$$

By rewriting the latter as

$$\sqrt{\frac{2}{\pi t}}\exp\left[j\left(t - \frac{\pi n}{2} - \frac{\pi}{4}\right)\right]$$

we see that $J_n(t)$ behaves like

$$\sqrt{\frac{2}{\pi t}}\cos\left(t - \frac{\pi n}{2} - \frac{\pi}{4}\right)$$

for large positive $t$. In this connection, we refer to the expansion (5.8).

## 5.11 The Method of Steepest Descent

The method of steepest descent and the saddle-point method are essentially the same, and they amount to reducing the quantity to be studied to one of the forms (5.24) or (5.25). Unfortunately, there is no general procedure for discovering how to make the required reduction. Illustrations, with little motivation, are given by Jeffreys and Jeffreys[6] and by Erdélyi.[3] In Chap. 5 of the book by de Bruijn,[2] there is a serious discussion of heuristic considerations that would be helpful in undertaking to carry out reductions to one of the foregoing forms. Some slightly different considerations are put forth by Chester and Friedman.[1] Anyone who is faced with the need for expanding a given function in an asymptotic series, and suspects that the method of steepest descent might be of value in doing it, would do well to consult one of the sources cited, or indeed several of them. He will still find that the method requires considerable resourcefulness and ingenuity on his own part.

We shall not attempt any motivation, but shall remark only that if one has an integral of the form (5.13) to deal with, one attempts to transform the path of integration in the complex plane so that it passes through a "saddle point." One feature of saddle points that helps in finding them is that $g'(z) = 0$ at a saddle point. We shall give a particularly simple example.

The Airy integral is useful in a number of practical applications. For real $z$, it is defined by

$$\text{Ai } z = \frac{1}{\pi}\int_0^\infty \cos\left(tz + \frac{t^3}{3}\right)dt$$

This is clearly equivalent to

$$\text{Ai } z = \frac{1}{2\pi j}\int_{L_1} \exp\left(tz - \frac{t^3}{3}\right)dt$$

where $L_1$ is the imaginary axis. If $z$ is positive, we can justify changing the path of integration to $L_2$, a vertical line through $-\sqrt{z}$. On this line, we have $t = -\sqrt{z} + jy$, and so get

$$\begin{aligned}\operatorname{Ai} z &= \frac{1}{2\pi j}\int_{L_2} \exp\left(tz - \frac{t^3}{3}\right) dt \\ &= \frac{1}{2\pi}\int_{-\infty}^{\infty} \exp\left(-\frac{2z^{3/2}}{3} - \sqrt{z}\,y^2 + \frac{jy^3}{3}\right) dy \\ &= \frac{\exp(-2z^{3/2}/3)}{\pi}\int_0^{\infty} e^{-\sqrt{z}y^2}\cos\frac{y^3}{3}\,dy\end{aligned}$$

Expanding the cosine term in a Maclaurin series and integrating term by term gives

$$\operatorname{Ai} z \sim \frac{\exp(-2z^{3/2}/3)}{2\sqrt{\pi}\,z^{1/4}}\left[1 - \frac{3\cdot 5}{144z^{3/2}} + \frac{5\cdot 7\cdot 9\cdot 11}{2!(144z^{3/2})^2} - \frac{7\cdot 9\cdot 11\cdot 13\cdot 15\cdot 17}{3!(144z^{3/2})^3} + \cdots\right]$$

Note that the point through which we passed the new path, namely $t = -\sqrt{z}$, is one of the two points at which the derivative of

$$g(z) = tz - \frac{t^3}{3}$$

is zero. Why we chose this zero rather than the other, why we chose the particular path through $-\sqrt{z}$ that we did rather than some other path, and similar questions are illuminated, but not entirely answered, in de Bruijn's discussion.[2]

We mention in passing that one can extend the definition of the Airy integral into the complex plane. When one does, the asymptotic series given above is valid in the angle (5.6).

We saw above that one can get an asymptotic formula for the integral (5.13) even when $f$ and $g$ are functions of both $x$ and $z$, provided the dependence on $z$ is slight. One can get asymptotic series also in this case. There is no well-developed theory for this general case, and one often proceeds by improvising through analogy with simpler situations. A fairly intricate example is given at the end of Ref. 10.

In the method of steepest descent, in which one is trying to reduce a given expression to one of the form (5.24) or (5.25) by deforming the path of integration to pass through a saddle point, one can often greatly simplify the choice of a path by permitting $g$ in the integral (5.24) or $f$ in the integral (5.25) to depend on $z$ as well as $t$. This is discussed in Sec. 4.4 of the book by de Bruijn[2] and in Ref. 11. In the latter, a proof is sketched for the following principle.

Let us consider

$$\int h(w)e^{zg(w)}\,dw \tag{5.26}$$

where $z$ and $w$ are complex variables, and the integral is taken along some path $P$ of finite length. Let the path begin at a point $A$ such that

$$g'(A) = 0$$

and never return to $A$. Let $h(w)$ and $g(w)$ be analytic in the neighborhood of $A$. Let $h(w)$ be bounded on $P$. Let there be real constants $a$, $b$, and $c$, with $a \leq b$ and $c$ positive, such that

$$\Re\{z[g(w) - g(A)]\} \leq -c|z(w - A)^2| \tag{5.27}$$

for $w$ on $P$ and $a \leq \arg z \leq b$. Let $P$ have the inclination $\theta$ at the point $A$, in the sense that the half ray defined by $w = A + xe^{j\theta}$, with $0 \leq x$, shall be tangent to $P$ at $A$. Introduce $t$ as a new variable of integration by the relationship

$$w = A + te^{j\theta} \tag{5.28}$$

Take $\alpha$ to be the constant such that the Maclaurin series for $g(w) - g(A)$ in terms of $t$ starts with the term $-\alpha t^2$. Under the substitution (5.28), the integral (5.26) takes the form

$$e^{zg(A)} \int_0^\infty e^{-\alpha z t^2} f(t)\,dt$$

with

$$f(t) = e^{j\theta}h(w)e^{z[\alpha t^2 + g(w) - g(A)]}$$

We now have a form analogous to the integral (5.25) and can proceed as suggested for dealing with (5.25). Specifically, form the Maclaurin expansion

$$e^{j\theta}h(w)e^{z[\alpha t^2 + g(w) - g(A)]} = \sum_{n=0}^{\infty} K_n t^n \tag{5.29}$$

Each $K_n$ is a polynomial in $z$ of degree not more than $n/3$. If we now formally integrate term by term in

$$\int_0^\infty e^{-\alpha z t^2} \sum_{n=0}^{\infty} K_n t^n\,dt \tag{5.30}$$

and collect terms in like powers of $\sqrt{z}$, we get a series

$$\sum_{n=1}^{\infty} \frac{B_n}{z^{n/2}}$$

with the $B_n$ constant, such that

$$e^{zg(A)} \sum_{n=1}^{\infty} \frac{B_n}{z^{n/2}}$$

is an asymptotic series for the integral (5.26) in the angle $a \leq \arg z \leq b$.

It may appear that our conditions on the path $P$ have been rather stringent. Actually, we can encompass a wide variety of paths by minor adaptations. For example, if the path ends at $A$ instead of beginning at $A$, we can reverse the order of integration; then the principle above can be applied and will give an asymptotic series for the negative of our function. If the path passes several times through $A$, we cut it into several parts; some will begin at $A$ and others will end at $A$, but we can handle both cases. In case the path passes once through $A$ with no change in direction at $A$ and does not begin or end at $A$, we could cut the path into two pieces, as proposed above, but it can be shown that the same result would be achieved if we left the path in one piece and merely replaced the integral (5.30) by

$$\int_{-\infty}^{\infty} e^{-azt^2} \sum_{n=0}^{\infty} K_n t^n \, dt \tag{5.31}$$

If the path passes through several points at which $g'(w) = 0$, it may be necessary to cut the path into several pieces and treat these pieces separately. Finally, if one has a case in which $P$ is infinite in length, one will have to make some estimate to justify throwing away all but a finite portion; this is usually very easy.

We shall now give an illustration. We should warn the reader that the above principle is advantageous primarily in that it often permits a simple choice of path. The penalty that one pays for this simplicity in the choice of the path is that commonly the calculations become extremely difficult if one wishes more than one or two terms of the series. One can often, however, find a differential equation that is satisfied by the function. Then one can usually get the higher coefficients with comparative ease by applying the method of Sec. 5.5. Our illustration will display these features, and it will also show how one can get easier calculations by using more sophistication in the choice of a path.

Suppose we write

$$F(x) = \sum_{n=0}^{\infty} \frac{(-x^2)^n}{n!(2n)!} \tag{5.32}$$

and require the asymptotic behavior of $F(x)$ for large positive $x$. We easily obtain

$$F(x) = \sum_{n=0}^{\infty} \frac{(-x^2)^n}{n!} \frac{1}{2\pi j} \oint \frac{e^t}{t^{2n+1}}\, dt$$
$$= \frac{1}{2\pi j} \oint \frac{e^t}{t} \exp\left(-\frac{x^2}{t^2}\right) dt$$

We put

$$t = (2x^2)^{1/3} w$$

and get

$$F(x) = \frac{1}{2\pi j} \oint \frac{1}{w} \exp\left[\left(\frac{x}{2}\right)^{2/3} \left(2w - \frac{1}{w^2}\right)\right] dw \tag{5.33}$$

This has the form (5.26) with

$$h(w) = \frac{1}{w} \qquad g(w) = 2w - \frac{1}{w^2} \qquad z = \left(\frac{x}{2}\right)^{2/3}$$

We can easily take the path of integration to be the unit circle, whereupon it will pass through all three zeros of $g'(w)$, to wit, the three cube roots of $-1$. Of these, $-1$ itself causes no trouble, but the other two interfere with each other. If we attempt to take either root as $A$, then condition (5.27) fails because of the presence of the other root.

One can proceed to cut the path into two parts, namely, the upper and lower halves of the unit circle. Thus let us take

$$I_k(z) = \frac{1}{2\pi j} \int_{P_k} \frac{1}{w} \exp\left[z\left(2w - \frac{1}{w^2}\right)\right] dw$$

where $P_1$ is the upper half of the unit circle and $P_2$ is the lower half.

On $P_1$, we take

$$A = e^{\pi j/3} \qquad \theta = \frac{5\pi}{6} \qquad w = e^{\pi j/3}(1 + jt)$$

We can satisfy the condition (5.27) with $a = -\pi/2$, $b = 0$, and any reasonably small value of $c$, say $c = 0.2$. We get

$$g(w) - g(A) = -3e^{\pi j/3}t^2 + 4e^{5\pi j/6}t^3 + \cdots$$

so that

$$\alpha = 3e^{\pi j/3}$$

Then (5.29) takes the form

$$\frac{j}{1 + jt} \exp\left[e^{\pi j/3}z(4jt^3 + 5t^4 + \cdots)\right]$$
$$= (j + t - jt^2 - t^3 + jt^4 + \cdots)\left[1 + e^{\pi j/3}z(4jt^3 + 5t^4 + \cdots) + \frac{e^{2\pi j/3}z^2}{2!}(-16t^6 + \cdots) + \cdots\right]$$
$$= j + t - jt^2 - (4e^{\pi j/3}z + 1)t^3 + (9e^{\pi j/3}z + 1)jt^4 + (15e^{\pi j/3}z + 1)t^5 - (8e^{2\pi j/3}z^2 + 22e^{\pi j/3}z + 1)jt^6 + \cdots$$

Substituting this into the integral (5.31) gives

$$\int_{-\infty}^{\infty} \exp(-3e^{\pi j/3}zt^2)(j + t - \cdot \cdot \cdot)\, dt$$
$$= \frac{\sqrt{\pi}}{e^{\pi j/6}\sqrt{3z}}\left[j - \frac{j}{6e^{\pi j/3}z} + (9e^{\pi j/3}z + 1)\frac{j}{12e^{2\pi j/3}z^2} - (8e^{2\pi j/3}z^2 + 22e^{\pi j/3}z + 1)\frac{5j}{72e^{\pi j}z^3} + \cdot \cdot \cdot\right]$$

From this we can get the first two terms of the desired asymptotic series, as well as some of the constituents of the third and fourth terms. To get all the constituents of the third term of the asymptotic series, we would have to carry the integration as far as the term in $t^{12}$. In general, to get the $(n + 1)$st term of the asymptotic series, we would have to carry the integration as far as the term in $t^{6n}$. We have verified, however, that

$$I_1(z) \sim \frac{e^{\beta^2}}{2\beta\sqrt{\pi}}\left(1 + \frac{1}{12\beta^2} + \cdot \cdot \cdot\right)$$

for

$$-\frac{\pi}{2} \le \arg z \le 0$$

with

$$\beta = e^{\pi j/6}(3z)^{1/2} = e^{\pi j/6}\left(\frac{27x^2}{4}\right)^{1/6} \tag{5.34}$$

We can proceed similarly with $I_2(z)$, taking

$$A = e^{-\pi j/3} \qquad \theta = -\frac{5\pi}{6} \qquad \text{etc.}$$

Since we are interested only in real $x$, however, we can proceed more expeditiously by noting that in this case $I_1(z)$ and $I_2(z)$ are conjugate complexes of each other.

It will be noted that we were able to use the very simplest sort of path, but that the calculations were so complicated as to discourage us from attempting to compute even the third term of the series. We shall see shortly that this is not a serious matter, as the later coefficients can be obtained fairly readily by means of a differential equation. We digress temporarily, however, to show how one could simplify the calculations by using more subtlety in connection with the path of integration. Putting $w = e^{jv}$ in Eq. (5.33) and recalling that $x$ is real gives

$$F(x) = \frac{1}{2\pi}\int_{-\pi}^{\pi} \exp\left[\left(\frac{x}{2}\right)^{2/3}(2e^{jv} - e^{-2jv})\right] dv$$
$$= \frac{1}{\pi}\Re\int_{0}^{\pi} \exp\left[\left(\frac{x}{2}\right)^{2/3}(2e^{jv} - e^{-2jv})\right] dv$$

We can apply the principle stated earlier to the integral above, with

$$h(v) = 1 \qquad g(v) = 2e^{jv} - e^{-2jv} \qquad z = \left(\frac{x}{2}\right)^{2/3}$$

$$A = \frac{\pi}{3} \qquad a = -\frac{\pi}{2} \qquad b = 0 \qquad c = 0.2$$

$$\theta = 0 \qquad v = t + \frac{\pi}{3} \qquad \alpha = 3e^{\pi j/3}$$

Then

$$\begin{aligned}\alpha t^2 + g(v) - g(A) &= e^{\pi j/3}(3t^2 + 2e^{jt} + e^{-2jt} - 3)\\ &= e^{\pi j/3}\left(jt^3 + \frac{3t^4}{4} - \frac{jt^5}{4} - \frac{11t^6}{120} + \cdots\right)\end{aligned}$$

Next, writing only those terms that will be needed to get three terms of the asymptotic series, we get

$$\begin{aligned}e^{j\theta}h(v)e^{z[\alpha t^2+g(v)-g(A)]} &= 1 + ze^{\pi j/3}\left(jt^3 + \frac{3t^4}{4} - \frac{jt^5}{4} - \frac{11t^6}{120} + \cdots\right)\\ &\quad + \frac{z^2e^{2\pi j/3}}{2!}\left(-t^6 + \frac{3jt^7}{2} + \frac{17t^8}{16} - \cdots\right)\\ &\quad + \frac{z^3e^{\pi j}}{3!}\left(-jt^9 - \frac{9t^{10}}{4} + \cdots\right) + \frac{z^4e^{4\pi j/3}}{4!}(t^{12} - \cdots) + \cdots\end{aligned}$$

Substituting into the integral (5.31) and integrating term by term gives

$$F(x) = \Re\left[\pi^{-1/2}e^{\beta^2}\left(\frac{1}{\beta} + \frac{1}{12\beta^3} + \frac{1}{288\beta^5} + \cdots\right)\right]$$

with $\beta$ as in (5.34).

We now show how to get more terms of the series by use of a differential equation. Starting from the series definition (5.32), we easily discover a differential equation

$$xF'''(x) + 2F''(x) + 2F(x) = 0$$

that $F(x)$ satisfies. Unfortunately, $F(x)$ equals not an explicit asymptotic series, but the real part of one, so that we have to work not with $F(x)$ directly, but with a closely related function. We define

$$G(x) = \frac{1}{\pi j}\int_L \frac{1}{w}\exp\left[\left(\frac{x}{2}\right)^{2/3}\left(2w - \frac{1}{w}\right)\right]dw$$

where $L$ is the path in the $w$ plane that starts at the origin, proceeds right along the real axis to $w = 1$, then counterclockwise halfway around the unit circle to $w = -1$, and finally left along the negative real axis to $w = -\infty$. Clearly, $G(x)$ is closely related to

$$2I_1\left[\left(\frac{x}{2}\right)^{2/3}\right]$$

Indeed, an elementary estimate shows that the difference $G - 2I_1$ is exponentially small compared with the asymptotic value of $I_1$ that we discovered above. This lets us infer that $G$ and $2I_1$ have the same asymptotic series, and so furnishes an instance of the case for which one has a path of infinite length, for example $L$, and justifies throwing away all but a finite portion in order to be able to apply the principle given above. We noted above that the real part of the asymptotic series for $G$ and $2I_1$ is the asymptotic series for $F(x)$.

By putting

$$w = (2x^2)^{-1/3}t$$

and deforming the contour of integration, we conclude that

$$G(x) = \frac{1}{\pi j}\int_L \exp\left(t - \frac{x^2}{t^2}\right)\frac{dt}{t}$$

Then

$$\begin{aligned} xG'''(x) + 2G''(x) + 2G(x) \\ &= \frac{1}{\pi j}\int_L \left(\frac{2}{t} - \frac{4}{t^3} + \frac{20x^2}{t^5} - \frac{8x^4}{t^7}\right)\exp\left(t - \frac{x^2}{t^2}\right)dt \\ &= \frac{1}{\pi j}\int_L \frac{\partial}{\partial t}\left[\left(\frac{2}{t} + \frac{2}{t^2} - \frac{4x^2}{t^4}\right)\exp\left(t - \frac{x^2}{t^2}\right)\right]dt \\ &= 0 \end{aligned}$$

Thus we can proceed as in Sec. 5.5 to get the higher terms of the asymptotic series for $G(x)$, with the result

$$F(x) \sim \Re\left\{\pi^{-1/2}\beta^{-1}e^{\beta^2}\left[1 + \frac{1}{12\beta^2} + \frac{1}{2!(12\beta^2)^2} - \frac{575}{3!(12\beta^2)^3} - \frac{86{,}975}{4!(12\beta^2)^4} + \cdots\right]\right\}$$

## 5.12 Further Use of Integration by Parts

Suppose that we are considering the integral (5.13) and that the hypotheses of the theorem of Sec. 5.9 are satisfied. As we noted, most of the value of the integral (5.13) comes from the values of $x$ near $C$. This suggests that

$$f(C)\int_A^B e^{zg(x)}\,dx \tag{5.35}$$

should be a good approximation for the integral (5.13) for large $z$. Indeed, it usually is. One commonly cannot get a closed form for the integral (5.35), so that this result is not particularly useful. When one does wish to use (5.35) as an approximation for (5.13), it is useful to have an estimate of the error, which is

$$\int_A^B [f(x) - f(C)]e^{zg(x)}\,dx = \int_A^B \frac{f(x) - f(C)}{g'(x)}\, g'(x)e^{zg(x)}\,dx$$

Since our hypotheses ensure that $g'(x)$ has a simple zero at $x = C$, we see that

$$\frac{f(x) - f(C)}{g'(x)}$$

is a well-behaved function for $A \leq x \leq B$. Thus we can integrate by parts in the right-hand integral above, getting

$$\frac{1}{z}\left\{\frac{f(B) - f(C)}{g'(B)}\, e^{zg(B)} - \frac{f(A) - f(C)}{g'(A)}\, e^{zg(A)} - \int_A^B e^{zg(x)} \frac{d}{dx}\left[\frac{f(x) - f(C)}{g'(x)}\right] dx\right\}$$

as an expression for the error incurred in using the integral (5.35) as an approximation for the integral (5.13). The integral on the right above has the same form as (5.13), so that one could repeat the process, usually getting an even better estimate for (5.13).

Instances of useful expansions obtained by repeated applications of the above process are given by Rosser.[10] In the report by Franklin and Friedman,[4] this process is studied for $f$ and $g$ of a special form, and it is shown that the process leads in this case to a convergent asymptotic series.

To give an example, let us consider again the function $I(z)$ introduced in Sec. 5.9. We have

$$I(z) = z^{z+1} \int_0^\infty \frac{x^2}{1 + zx}\, e^{-zx}\,dx$$

and so take $A = 0$, $B = \infty$, $C = 1$,

$$f(x) = \frac{1}{1 + zx} \qquad g(x) = \log x - x$$

This gives

$$\begin{aligned} I(z) &= \frac{z^{z+1}}{1+z} \int_0^\infty x^z e^{-zx}\,dx + z^{z+1} \int_0^\infty \left(\frac{1}{1 + zx} - \frac{1}{1+z}\right) x^z e^{-zx}\,dx \\ &= \frac{\Gamma(z+1)}{1+z} + \frac{z^{z+2}}{1+z} \int_0^\infty \frac{x}{1+zx}\left(\frac{1}{x} - 1\right) x^z e^{-zx}\,dx \end{aligned}$$

We integrate by parts on the right, getting

$$I(z) = \frac{\Gamma(z+1)}{1+z} - \frac{z^{z+1}}{1+z} \int_0^\infty \frac{1}{(1+zx)^2}\, x^z e^{-zx}\,dx$$

We can repeat the process on the right-hand integral, getting

$$
\begin{aligned}
I(z) &= \frac{\Gamma(z+1)}{1+z} - \frac{z^{z+1}}{(1+z)^3}\int_0^\infty x^z e^{-zx}\,dx \\
&\qquad - \frac{z^{z+1}}{1+z}\int_0^\infty \left[\frac{1}{(1+zx)^2} - \frac{1}{(1+z)^2}\right] x^z e^{-zx}\,dx \\
&= \frac{\Gamma(z+1)}{1+z} - \frac{\Gamma(z+1)}{(1+z)^3} \\
&\qquad - \frac{z^{z+2}}{(1+z)^3}\int_0^\infty \frac{2x+zx+zx^2}{(1+zx)^2}\left(\frac{1}{x}-1\right) x^z e^{-zx}\,dx \\
&= \frac{\Gamma(z+1)}{1+z} - \frac{\Gamma(z+1)}{(1+z)^3} + \frac{z^{z+1}}{(1+z)^3}\int_0^\infty \frac{2+z-z^2x}{(1+zx)^3} x^z e^{-zx}\,dx
\end{aligned}
$$

One could repeat the process indefinitely. If we start the next repetition, but do not carry through the integration by parts, we get three terms of an approximate series for $I(z)$ as follows:

$$
J(z) = \Gamma(z+1)\left[\frac{1}{1+z} - \frac{1}{(1+z)^3} + \frac{2-z}{(1+z)^5} + \cdots\right]
$$

The effectiveness of this series is illustrated in Table 5.2.

**Table 5.2** Efficiency of the Asymptotic Series $J(z)$ for the Integral $I(z)$

| $z$ | $I(z)$ | $J(z)$, one term | $J(z)$, two terms | $J(z)$, three terms |
|---|---|---|---|---|
| 1 | 0.40366 | 0.50 | 0.375 | 0.40625 |
| 2 | 0.59634 | 0.67 | 0.5926 | 0.59259 |
| 5 | 19.404 | 20 | 19.44 | 19.398 |
| 10 | $3.2698 \times 10^5$ | $3.3 \times 10^5$ | $3.2716 \times 10^5$ | $3.2698 \times 10^5$ |

## EXERCISES

Most of the exercises given below are intended to illustrate points raised in this chapter and to afford the reader an opportunity to test his mastery of these points. The final exercise (No. 8) is a problem that arose in the theory of boundary layers in aerodynamics, and it is included as an illustration of the kind of problem that can arise in practical applications.

**1.** For each of the following integrals, find an asymptotic formula or series as requested and state in what angle in the $z$ plane it is valid.

*a.* Asymptotic series for

$$
\int_0^\infty \frac{\sin x}{x+z}\,dx \tag{5.36}
$$

*b.* Asymptotic formula for

$$
\int_0^\infty e^{-z(\theta\cos\beta+\sinh\theta)} \qquad 0 \le \beta \le \frac{\pi}{2} \tag{5.37}
$$

*c.* Asymptotic series for

$$\int_0^\infty \frac{e^{-zx}}{1+x^2}\,dx \tag{5.38}$$

*d.* Asymptotic formula for

$$\int_0^\infty x^z e^{-zx}\,dx \tag{5.39}$$

*e.* Asymptotic formula for

$$\int_0^\infty \frac{x^z e^{-zx}}{1+x}\,dx \tag{5.40}$$

*f.* Asymptotic formula for

$$\int_0^1 \left(\frac{x}{1+x^2}\right)^z dx \tag{5.41}$$

**2.** Derive an asymptotic formula for

$$\int_0^\infty \frac{(x-1)x^z e^{-zx}}{1+x}\,dx \tag{5.42}$$

*Hint:* First perform an integration by parts in the manner suggested for handling the integral (5.13) when $g'(x) \neq 0$; note that, since the factor $x - 1$ is present, this integration by parts will succeed even though $g'(x) = 0$ in the present case.

**3.** Note that the integral (5.42) plus twice the integral (5.40) equals the integral (5.39), but the same does not hold for their asymptotic formulas. Explain this.

**4.** Find an asymptotic formula for

$$\frac{1}{2\pi j}\int_C w^{-z}e^{zw}\,dw$$

where $C$ is the path given by

$$w = \frac{\theta}{\sin\theta}e^{j\theta} \qquad -\pi < \theta < \pi$$

**5.** By using the principle set forth in Sec. 5.11, get two nonvanishing terms of the asymptotic series for

$$\int_0^{\pi/2} \theta\cos^z\theta\,d\theta$$

**6.** Compute a four-decimal value of $\Gamma(j)$. *Hint:* For some positive integer $N$, compute a sufficiently accurate value for $\Gamma(N+1+j)$ by the series (5.7). Then use

$$\Gamma(N+1+j) = (N+j)(N-1+j)\cdots(1+j)j\Gamma(j)$$

**7.** Prove that

$$2^{2z-1}\Gamma(z)\Gamma(z+\tfrac{1}{2}) = \sqrt{\pi}\,\Gamma(2z)$$

*Hint:* Define

$$G(z) = \frac{2^{2z-1}\Gamma(z)\Gamma(z+\frac{1}{2})}{\Gamma(2z)}$$

By the result $\Gamma(w+1) = w\Gamma(w)$, prove that $G(z) = G(z+1)$, and so conclude that $G(z) = G(z+N)$, where $N$ is a large integer. Now use the series (5.7) to show that

$$G(z) = \lim_{N\to\infty} G(z+N) = \sqrt{\pi}$$

**8.** Determine the two leading terms of a series that gives the asymptotic behavior of

$$\int_0^1 e^{a\theta^2} \cos b(\theta^3 - \theta)\, d\theta \tag{5.43}$$

for $a$ and $b$ both large and real. *Hint:* Clearly the integral (5.43) is the real part of

$$\int_0^1 \exp\,[a\theta^2 + jb(\theta^3 - \theta)]\, d\theta$$

Now proceed as suggested under the integral (5.13).

## REFERENCES


1. Chester, C., and B. Friedman, "Uniform Asymptotic Expansions," Report No. IMM-NYU 219, New York University Institute of Mathematical Sciences, July, 1955.
2. de Bruijn, N. G., "Asymptotic Methods in Analysis," Interscience Publishers, Inc., New York, 1958.
3. Erdélyi, A., "Asymptotic Expansions," Dover Publications, New York, 1956.
4. Franklin, J., and B. Friedman, "A Convergent Asymptotic Representation for Integrals," Research Report No. BR-9, New York University Institute of Mathematical Sciences, December, 1954.
5. Ince, E. L., "Ordinary Differential Equations," Longmans, Green & Co., Ltd., London, 1926. Reprinted by Dover Publications, New York, 1956.
6. Jeffreys, H., and B. S. Jeffreys, "Methods of Mathematical Physics," 2d ed., Cambridge University Press, New York, 1950.
7. Morrey, Charles B., Nonlinear Methods, chap. 16 in "Modern Mathematics for the Engineer," First Series, edited by E. F. Beckenbach, McGraw-Hill Book Company, Inc., New York, 1956.
8. Morse, P. M., and H. Feshbach, "Methods of Theoretical Physics," McGraw-Hill Book Company, Inc., New York, 1953.
9. Rosser, J. B., Transformations to Speed the Convergence of Series, *J. Res. Natl. Bur. Standards*, vol. 46, pp. 56–64, 1951.
10. ———, Explicit Remainder Terms for Some Asymptotic Series, *J. Rational Mech. Anal.*, vol. 4, pp. 595–626, 1955.
11. ———, Some Sufficient Conditions for the Existence of an Asymptotic Formula or an Asymptotic Expansion, a chapter (pp. 371–387) in "On Numerical Approxition—Proceedings of a Symposium Conducted by the Mathematics Research Center, U.S. Army, at the University of Wisconsin, Madison, April 21–23, 1958," University of Wisconsin Press, Madison, Wis., 1959.
12. Wall, H. S., "Analytic Theory of Continued Fractions," D. Van Nostrand Company, Inc., Princeton, N.J., 1948.
13. Whittaker, E. T., and G. N. Watson, "A Course of Modern Analysis," American ed., The Macmillan Company, New York, 1946.

# PART 2
# Statistical and Scheduling Studies

# 6

# Chance Processes and Fluctuations

WILLIAM FELLER
PROFESSOR OF MATHEMATICS
PRINCETON UNIVERSITY

## 6.1 Introduction

Chance processes play a role in almost all phases of modern life: The efficient organization of traffic and communication systems, the provision of an adequate power supply and of storage and emergency facilities, the setting of tolerance limits, etc., all depend on an understanding of the phenomena of chance fluctuations and on an estimation of their future magnitude. Accordingly, probability theory has to cope with a large variety of problems and methods, and it would obviously be impossible here to present a fair sample of them. We shall therefore limit our attention essentially to two simple but methodologically interesting problems.

First we move in the direction of the classical central limit theorem of probability theory. It is well known that the normal (or Gaussian) distribution appears almost every time when we have to deal with the summation of many small chance effects. We shall attempt to explain this phenomenon, approaching it by the method of random walks.[1] This method is useful in many connections, and it has the advantage of showing that the normal distribution enters the theory in its role as fundamental solution of the heat equation. This leads us naturally to the more general Fokker-Planck, or diffusion, equation and permits us to deal with important variations on the theme, such as the ruin problem.

In the second part, we use the random-walk approach to discuss the simplest model for fluctuations in a waiting line or queue.[3,4] We discuss processes characterized by a lack of memory, or aftereffect, and the role of the exponential law for decay and for holding times. An effort is made to elucidate the operational meaning of the so-called steady state (or state of statistical equilibrium), which, together with the law of large numbers,[1] is the subject of widespread and dangerous misconceptions.

## SUMS OF RANDOM VARIABLES

### 6.2 Cumulative Effects

We begin with an analysis of the probable fluctuations of a quantity that may be considered as the resultant of a huge number of small—i.e., individually negligible—chance effects. For example, the electricity consumption in a city at any given time is the sum of the demands of a great number of individual households, offices, factories, etc. Similarly, the supply and the demand in a variety of economic processes are due to many contributing sources, and certain heritable characteristics are the cumulative effect of many individually small genetic contributions. The error of measurement in a physical experiment is composed of many unobservably small errors, and the whole theory to be presented traces its origin to this example. The thickness of washers produced in a factory and the total damage due to fire accidents may serve as further familiar examples.

In each case, we deal with a large (not necessarily known) number of random variables $X_1, X_2, \ldots, X_n$ and their sum

$$S_n = X_1 + X_2 + \cdots + X_n \tag{6.1}$$

Perhaps the most important and intuitively simplest case arises in connection with stochastic processes, i.e., processes with a time-dependent random variable $S(t)$. Typical examples are the amount of water in a reservoir, the total damage due to accidents up to time $t$, and the electricity consumption in a city. These quantities may be the cumulative effect of many contributing factors, but there is another way in which we may treat them as a sum of the form (6.1). In fact, we may split the time interval from 0 to $t$ into $n$ intervals of length $\tau = t/n$ and consider the increment $S(t) - S(0)$ as the sum of $n$ corresponding increments, $S(k\tau + \tau) - S(k\tau)$. Here the number $n$ of components is arbitrarily large, and by letting $n \to \infty$ we obtain correct laws rather than approximations.

### 6.3 The Simplest Random-walk Model

For simplicity, we consider only sums of the form (6.1) in which the components $X_k$ are statistically independent. In the case of a stochastic process $S(t)$, this means that the increment $S(t'') - S(t')$ in any time interval $(t',t'')$ is in no way influenced by the past of the process. ("Processes with independent increments." A more general scheme of Markovian processes will be considered in Sec. 6.6.)

The next restriction will at first appear much harder, but will turn out to be rather harmless (see Sec. 6.6). For simplicity of exposition, we shall restrict our attention to the case where the components $X_k$ can assume only three values, namely 0, $h$, and $-h$, where $h > 0$ is a constant. The probabilities that $X_k$ equals $h$, $-h$, and 0 are $p$, $q$, and $r$, respectively, with $p + q + r = 1$.

In the tritest example, $X_k$ represents the gain or loss of a gambler at the $k$th trial (with $r$ the probability of a tie), and $S_n$ the accumulated gain in the first $n$ trials. It is convenient to imagine that the trials take place at a uniform rate at times $\tau$, $2\tau$, $3\tau$, . . . , so that the index $n$ may be interpreted as a time variable, $t = n\tau$.

In *Brownian* motion, or *diffusion*, one considers a particle exposed to random molecular collisions. These, of course, are neither uniformly spaced in time nor of equal magnitude. In first approximation, however, we can work with averages and assume that the shocks can occur only at times $\tau$, $2\tau$, $3\tau$, . . . and that the resulting displacement is always 0, $h$, or $-h$. In this case, $S_n$ represents the position of the moving particle at time $t = n\tau$, and if at some time $t$ the position happens to be $x = kh$, then the position of the particle at the next moment $t + \tau$ is $x + h$, $x - h$, or $x$, with probabilities $p$, $q$, and $r$, respectively.

In other words, we are concerned with a simple random walk on the $x$ axis starting from the origin and such that all steps have the same length $h$. This terminology is so convenient that it is used in general for sums $S_n = X_1 + \cdots + X_n$ of the type described, whatever the empirical meaning (if any) of the random variables $X_k$. The "particle," of course, becomes a symbolic indicator for $S_n$.

Our first task is to calculate the probability

$$u_{n,k} = \text{prob }\{S_n = kh\}$$

that at time $t = n\tau$ the particle occupies the position $x = kh$; here $n = 1, 2, 3, \ldots$, and $k = 0, \pm 1, \pm 2, \ldots$. Now $S_{n+1}$ can equal $kh$ in only three mutually exclusive cases: either $S_n = (k - 1)h$ and $X_{n+1} = h$; or $S_n = (k + 1)h$ and $X_{n+1} = -h$; or, finally, $S_n = kh$ and $X_{n+1} = 0$. By adding the corresponding probabilities, we find that, for $n \geq 1$,

$$u_{n+1,k} = pu_{n,k-1} + qu_{n,k+1} + ru_{n,k} \tag{6.2}$$

This equation remains valid also for $n = 0$ if we put

$$u_{0,0} = 1 \qquad u_{0,k} = 0 \qquad \text{for } k \neq 0 \tag{6.3}$$

From Eq. (6.2), we may calculate recursively all the required probabilities $u_{n,k}$ by substituting successively $n = 0, 1, 2, \ldots$. Thus the *difference equation* (6.2) together with the initial conditions (6.3) furnishes the solution to our problem. Instead of deriving the rather unwieldy exact solution, however, we prefer to approximate Eq. (6.2) by a *differ-*

*ential equation.* This method has the advantage of leading to an important class of more general diffusion processes.

## 6.4 The Fokker-Planck Equation

It is natural, for our problem, to introduce such units of measurement that the probable fluctuations of $S_n$ will be numerically neither very small nor very large. Since we are interested in large $n$, this means that $h$ and $\tau$ will be numerically very small. One would then expect that for a fixed time $t = n\tau$ the probability distribution $\{u_{n,k}\}$ may be interpolated by a probability density $u(t,x)$. This means that, for any integers $\nu < \mu$, the probability

$$\text{prob}\,\{\nu h < S_n \leq \mu h\} = u_{n,\nu+1} + u_{n,\nu+2} + \cdots + u_{n,\mu}$$

may be approximated by the integral of $u(t,x)$ over the interval $\nu h < x \leq \mu h$. Then $u_{n,k}$ corresponds to the integral of $u(t,x)$ over an interval of length $h$ centered at $x = kh$; that is, we should have

$$u_{n,k} \approx u(n,kh)h$$

This leads us simply to put

$$u(n\tau,kh) = h^{-1}u_{n,k}$$

and to assume that $u(t,x)$ may be defined for all $t > 0$ and $x$ as a function with two continuous derivatives. This step can be justified rigorously, but we shall proceed purely heuristically.

The difference equation (6.2) now takes on the form

$$u(t + \tau, x) = pu(t, x - h) + qu(t, x + h) + ru(t,x)$$

or

$$u(t + \tau, x) - u(t,x) = -p[u(t,x) - u(t, x - h)] + q[u(t, x + h) - u(t,x)] \tag{6.4}$$

By using Taylor's formula, we conclude that

$$\frac{\partial u(t,x)}{\partial t} = -\frac{(p - q)h}{\tau}\frac{\partial u(t,x)}{\partial x} + \frac{1}{2}\frac{(p + q)h^2}{\tau}\frac{\partial^2 u(t,x)}{\partial x^2} + \cdots \tag{6.5}$$

The coefficients occurring on the right have a simple physical significance. In fact, $(p - q)h = E(X_k)$ is the mean displacement of the particle at each individual step. Since each step takes a time $\tau$, the ratio

$$b = \frac{(p - q)h}{\tau}$$

is the *mean displacement per time unit.* Similarly,

$$a = \frac{(p + q)h^2}{\tau}$$

is the *mean-square displacement per time unit.* By keeping these quantities fixed, we obtain in the limit as $h \to 0$ the Fokker-Planck (or general diffusion, or heat) equation

$$\frac{\partial u}{\partial t} = -b\frac{\partial u}{\partial x} + \frac{1}{2}a\frac{\partial^2 u}{\partial x^2} \tag{6.6}$$

For the initial conditions (6.3), we require the particular solution of Eq. (6.6) that, as $t \to 0$, approaches the so-called *Dirac function,* that is, $u(t,x) \to 0$ for each $x \neq 0$ and $u(t,0) \to \infty$ in such a way that the integral of $u$ equals 1 for all $t > 0$ (see Chap. 1). With reasonable coefficients $a$ and $b$, there exists exactly one solution of Eq. (6.6) satisfying this condition, and we may use it as an approximation to the probability distribution of $S_n$ for large $n$.

The advantage of this method is that our problem has been reduced to the analysis of an equation that plays an important role in many theories. In Sec. 6.6, we shall indicate how it may be derived under less restrictive conditions and in a more general form. It will be seen in Sec. 6.7 that the same equation with appropriate boundary conditions serves several additional purposes.

## 6.5 Example

Consider a coin-tossing game in which at each trial the gambler wins or loses the amount $h > 0$ with probabilities $\frac{1}{2}$. Here $p = q = \frac{1}{2}$, $r = 0$, and hence $b = 0$. The Fokker-Planck equation reduces to the ordinary heat equation

$$\frac{\partial u}{\partial t} = \frac{1}{2}a\frac{\partial^2 u}{\partial x^2} \tag{6.7}$$

where $a$ is a nonzero constant that may be chosen arbitrarily since this amounts only to a choice of the scale.† The required solution of Eq. (6.7) is given by the famous *normal* (or Gaussian) probability density

$$u(t,x) = \frac{1}{(2\pi a t)^{1/2}} \exp\left(-\frac{x^2}{2at}\right)$$

To apply our approximation formula, we have to put $\tau = h^2/a$. To calculate approximately the probability distribution of the gambler's accumulated gain after $n$ trials, we must put $t = n\tau$, that is, $at = nh^2$. Thus,

$$P\{x_1 < S_n < x_2\} \approx \frac{1}{(2\pi n)^{1/2}h}\int_{x_1}^{x_2} \exp\left(-\frac{x^2}{2nh^2}\right)dx \tag{6.8}$$

† We have refrained from the natural norming $a = 1$ in order to exhibit the general form (6.9) of the Fokker-Planck equation.

The quantity $S_n/h$ is, of course, independent of $h$, and it is more natural to rewrite the relationship (6.8) in the form

$$P\left\{y_1 < \frac{1}{h} S_n < y_2\right\} \approx \frac{1}{(2\pi n)^{1/2}} \int_{y_1}^{y_2} \exp\left(-\frac{y^2}{2n}\right) dy$$

This approximation formula was given by De Moivre in 1718 and is a special case of the *central limit theorem* of probability theory.[1,2] It shows in particular that the probable fluctuations of $S_n$ will be of the order of magnitude $hn^{1/2}$. For example, for large $n$ the probability that $|S_n| > 3hn^{1/2}$ is about 0.0027, and the probability that $|S_n| > 4hn^{1/2}$ is smaller than 0.000065.

If $a > 0$ and $b$ are arbitrary constants, the required solution of the Fokker-Planck equation (6.5) becomes

$$u(t,x) = \frac{1}{(2\pi at)^{1/2}} \exp\left[-\frac{(x-bt)^2}{2at}\right]$$

which is simply a normal distribution with a moving center (modal point) at $x = bt$. This explains why so many different phenomena are subject to the normal law.

## 6.6 Generalizations

The Fokker-Planck equation (6.6) actually may be derived from much more general assumptions than those introduced in Sec. 6.3, and its validity is much wider than is indicated by the above discussion. For example, the restriction that the $X_k$ assume only the values 0 and $\pm h$ was introduced only for notational convenience. To illustrate this point, it will suffice to consider another special case, namely, that each $X_k$ ranges over the five possible values $-2h$, $-h$, 0, $h$, $2h$, which are assumed, respectively, with probabilities $p_{-2}$, $p_{-1}$, $p_0$, $p_1$, $p_2$. In this case, the difference equation (6.4) takes the form

$$\begin{aligned} u(t+\tau, x) - u(t,x) = {} & -p_2[u(t,x) - u(t, x-2h)] \\ & - p_1[u(t,x) - u(t, x-h)] + p_{-1}[u(t, x+h) - u(t,x)] \\ & + p_{-2}[u(t, x+2h) - u(t,x)] \end{aligned}$$

and the use of Taylor's formula leads to a differential equation of the form (6.5) except that the coefficients on the right are replaced by

$$\frac{(2p_2 + p_1 - p_{-1} - 2p_{-2})h}{\tau} \quad \text{and} \quad \frac{(4p_2 + p_1 + p_{-1} + 4p_{-2})h^2}{\tau}$$

Now these coefficients again represent, respectively, the mean displacement and mean-square displacement per time unit, and so the Fokker-Planck equation (6.6) remains valid without change in the physical (or probabilistic) meaning of the coefficients.

This consideration should render plausible the fact that the differential equation (6.6) represents a rather universal law governing the fluctuation of random variables obtained by a summation of many independent components. In fact, it applies with little change to the much more general case of random walks in which the probability distribution of each step depends on the actual position of the particle, i.e., instead of independent and equally distributed $X_k$ we may consider the case in which $X_{n+1}$ and its distribution depend on $S_n$. The essential new feature is that the coefficients $a$ and $b$ are no longer constants but may depend on $x$. The derivation outlined above leads in this case to the *general Fokker-Planck equation*

$$\frac{\partial u(t,x)}{\partial t} = -\frac{\partial b(x)u(t,x)}{\partial x} + \frac{1}{2}\frac{\partial^2 a(x)u(t,x)}{\partial x^2} \tag{6.9}$$

## 6.7 The "Ruin" Problem

In many applications, the problem of estimating the probable fluctuations of $S_n$ or $S(t)$ for a fixed $n$ or $t$ is of a secondary interest; what we really require is the probability that $S_n$ or $S(t)$ will remain within preassigned bounds $\alpha$ and $\beta$ for all times, or at least for all $t$ of a certain period. In technical applications, the setting of tolerance limits and the provision of adequate power supplies, safety measures, or storage facilities are likely to lead to questions of this type. The estimation of the probable extremes of various chance variables in the past or future provides another example. Even in ordinary gambling, the De Moivre formula (6.8) does not give the pertinent information. In fact, if the two players have respectively the finite capitals $\alpha$ and $\beta$, then the first gambler's gain cannot fall outside the interval $(-\alpha,\beta)$. Once one of the limits is reached, one of the gamblers is "ruined" and the game necessarily ends. A picturesque terminology is desirable, and it has therefore become customary to refer to the event that $S_n$ reaches a preassigned limit as "ruin." The example of Sec. 6.11 will show, however, that in technical applications the "ruin" may stand for the most desirable event.

The random-walk method remains applicable with almost no change. With the conventions of Sec. 3.3, we now let $u_{n,k}$ be the probability that $S_n = k$ under the additional restriction that none of the variables $S_1$, $S_2$, . . . , $S_n$ has reached either the level $-\alpha$ or the level $\beta$. The probability $\rho_n$ that such "ruin" does not occur within the first $n$ trials is then given by

$$\rho_n = u_{n,-\alpha+1} + u_{n,-\alpha+2} + \cdots + u_{n,\beta-1}$$

The quantities $u_{n,k}$ are, of course, undefined for $k \leq -\alpha$ and $k \geq \beta$. The difference equation (6.2) obviously remains valid when $-\alpha + 1 < k < \beta - 1$, but it must be modified for the extreme values $k = -\alpha + 1$ and

$k = \beta - 1$. For example, when $k = \beta - 1$, the event $S_{n+1} = kh$ can occur only if $S_n$ equals either $kh$ or $(k - 1)h$, and so Eq. (6.2) in this case reduces to $u_{n+1,\beta-1} = pu_{n,\beta-2} + ru_{n,\beta-1}$; that is, the term $qu_{n,p}$ is missing. Since $u_{n,\beta}$ is undefined, we can take care of this exception by defining $u_{n,\beta} = 0$. The required probabilities $u_{n,k}$ will then *satisfy the difference equations* (6.2) *for all k in the pertinent interval* $-\alpha < k < \beta$, *but we have to impose the two boundary conditions*

$$u_{n,-\alpha} = 0 \qquad u_{n,\beta} = 0$$

The argument leading to the Fokker-Planck equation (6.6) remains valid, but this equation will now be restricted to the given interval and supplemented by the boundary condition that $u$ vanish at the two end points for all times.

The general applicability and usefulness of this method will be illustrated in Sec. 6.11.

## QUEUEING PROBLEMS

### 6.8 Holding and Waiting Times; Discipline

Consider a counter, telephone trunk line, port facility, or any other device that at any given moment either is "free" or else is "serving a customer." If a new customer arrives when the counter is free, he is served without delay. If the counter is occupied, the new customer joins a waiting line or queue. The number $W(t)$ of customers in the waiting line (always including the customer being served) is a time-dependent random variable of a nature totally different from the random variables studied above.

In focusing our attention on the queue length, we are neglecting features of the process that from other points of view may be of great importance. If I am standing in a queue, the point of interest to me is my *waiting time*, and this depends only on the number of customers to be served before me, not on the total number of customers in the queue. Here the *queue discipline* determines the precedence. Under the *first come, first served* rule, the queue is ordered and I am interested only in the number of customers ahead of me. Unfortunately, this order of things is not as natural or as common as one might expect. Thus in an automatic telephone exchange, a (usually short) queue may be formed with every dialing. Here, as in other technical devices of a similar nature, the prevailing system is based on *random* choice: at the termination of each service time, in case of an existing queue, one of its members is selected at random irrespective of the order of arrivals. Finally, for tracking- and servomechanisms using information that is fed to them continuously, the latest information is naturally most important, and accordingly their queue discipline is based on the *last come, first served* rule.

As long as we are interested only in the length of the queue, the queue discipline plays no role, but we still have to specify the laws governing the duration of the *service time* and the *incoming traffic.* A great variety of models is of theoretical interest and of importance in applications, but we shall limit our discussion to the simplest and most important case of a "Markovian process." It is characterized by a complete *lack of memory.*

To introduce this notion in the simplest way, let us begin by considering a *waiting time T;* this random variable may stand for the lifetime of a radioactive atom, a piece of mechanical equipment, or an individual. Alternatively, $T$ may stand for the duration of a service time in our queue, for the length of a telephone conversation, or for the time between the arrivals of two customers.

In principle, this random variable $T$ can have an arbitrary probability distribution, but we consider only the case in which *no aging* occurs: given that the waiting time $T$, starting at 0, has lasted for a time $t$, the probability that it ends during the time interval $(t, t + \tau)$ is independent of the present age $t$ and depends only on $\tau$. A typical example is a radioactive atom since, according to accepted and tested theories, the probability of its disintegrating during the next minute remains constant throughout its lifetime. The law of *exponential decay* for radioactive matter is well known, and we now proceed to show that it is a simple consequence of the assumed lack of aging or memory.

Let us first discretize time; i.e., let us assume that the waiting time $T$ ranges only over integral multiples of a fixed time, say $\tau$ sec. Let us denote by $q_n$ the probability that $T$ exceeds $n\tau$, that is,

$$q_n = \text{prob }\{T > n\tau\}$$

Then $1 - q_1$ is the probability that $T$ expires at time $\tau$; therefore, given that $T$ has surpassed $n\tau$, the probability that $T$ will expire at time $(n + 1)\tau$ is again $1 - q_1$. This means that $q_{n+1} = q_n q_1$, and hence by induction that $q_n = q_1^n$. The probability that the waiting time $T$ exceeds $n\tau$ is accordingly subject to the *geometric distribution* $\{q_1^n\}$, and the latter characterizes the lack of aging. Of course, $q_1$ is an arbitrary constant, $0 < q_1 < 1$. For example, in throwing a die the probability that the waiting time for the first ace exceeds $n$ trials is $(5/6)^n$. For a waiting time subject to the geometric distribution $\{q^n\}$, the probability that $T$ lasts exactly $n$ time units is $q^{n-1} - q^n$, and therefore the *expected* (or *mean*) *duration* is given by

$$\begin{aligned} E(T) &= [(1 - q) + 2(q - q^2) + 3(q^2 - q^3) + \cdots]\tau \\ &= (1 + q + q^2 + q^3 + \cdots)\tau = \frac{\tau}{1 - q} \end{aligned}$$

To obtain the corresponding formulas for a continuous time parameter, we have only to pass to the limit as $\tau \to 0$. As we shorten the intervals of subdivision, the probability $q$ will vary in such a way that the life expectancy $E(T)$ remains constant, say

$$E(T) = a^{-1}$$

This means that

$$q = 1 - a^{-1}\tau$$

and hence the probability prob $\{T > t\}$ that $T$ lasts longer then time $t = n\tau$ becomes

$$q^n = (1 - a^{-1}\tau)^{t/\tau}$$

This quantity approaches exp $(-at)$ as $\tau \to 0$, and we see thus that our assumption of absence of aging leads to the probability distribution

$$\text{prob } \{T > t\} = \exp(-at)$$

which is the law of exponential decay.

In many applications there exists a theoretical justification for the assumption that the waiting and service times are subject to this law, but frequently this assumption is introduced as an approximation merely because without it the theory becomes unmanageable. Such a procedure may appear rough, but it is accompanied by an astounding success. Practically the whole theory of telephone communication uses this assumption even though it cannot be correct; for example, long-distance conversations are more likely to last merely 3, 6, 9, . . . min, which contradicts the assumption that the past has no aftereffect. For further details and applications, see Ref. 2.

## 6.9 Random-walk Model; the Differential Equations

We begin by describing our queueing model with a discretized time; i.e., we assume that changes can occur only at times $\tau$, $2\tau$, $3\tau$, . . . , where $\tau$ is a fixed small fraction of the time unit. For a continuous model, we shall then pass to the limit as $\tau \to 0$.

Our model is based on the assumption that, independently of the past history of the system, at any time $t = n\tau$ the probability of an arrival of a new customer is $\alpha\tau$, and, independently, if at that instant the counter is occupied, the probability that the service time ends at the next moment $t + \tau$ is $\beta\tau$. We do not consider the possibility that several customers arrive simultaneously. It will be noted that both the service times and the waiting times between consecutive arrivals of customers lack memory and therefore have geometric distributions. The *expected duration* of a service time is $\beta^{-1}$, and the expected duration between two consecutive arrivals is $\alpha^{-1}$. It follows that the average *number of arrivals* per time unit is $\alpha$.

We may describe the process in random-walk terminology, with the position of the "particle" at any time indicating the number of customers in the queue (including the one actually being served). The possible positions are now restricted to the integers $k \geq 0$. If at any time there are $k \geq 1$ customers in the line, the next step takes the particle to $k + 1$ if a new customer arrives before the service time ends; the step leads to the left if the service time ends before a new customer arrives; in the remaining cases the particle stays fixed. In terms of the notations of Sec. 6.3, the probabilities $p$, $q$, and $r$ of a step to the right, to the left, or no change, respectively, are therefore given by

$$p = \alpha\tau(1 - \beta\tau) \qquad q = \beta\tau(1 - \alpha\tau) \qquad r = 1 - p - q \tag{6.10}$$

An exceptional situation arises for $k = 0$, since no service time can end when none is going on. For $k = 0$, we therefore have $p = \alpha\tau$, $q = 0$.

If $u_{n,k}$ stands for the probability that at time $t = n\tau$ there are exactly $k$ customers in the line, we have again the difference equation (6.2), valid for $k \geq 2$; for $k = 0, 1$, the equation takes the forms

$$\begin{aligned} u_{n+1,1} &= \alpha\tau u_{n,0} + qu_{n,2} + ru_{n,1} \\ u_{n+1,0} &= (1 - \alpha\tau)u_{n,0} + qu_{n,1} \end{aligned}$$

If the counter is free at time 0, we again have the initial conditions (6.3).

For a queueing model with *continuous time*, we have only to pass to the limit as $\tau \to 0$. Let $u_k(t)$ be the probability that at time $t$ the line contains $k \geq 0$ customers. If in our difference equations we subtract $u_{n,k}$ from both sides and divide by $\tau$, we obtain on the left the difference ratios $[u_k(t + \tau) - u_k(t)]/\tau$. It follows that the required probabilities $u_k(t)$ satisfy the *differential equations*

$$\begin{aligned} u_0'(t) &= -\alpha u_0(t) + \beta u_1(t) \\ u_k'(t) &= -(\alpha + \beta)u_k(t) + \alpha u_{k-1}(t) + \beta u_{k+1}(t) \qquad k \geq 1 \end{aligned} \tag{6.11}$$

This is a system of infinitely many differential equations, and its explicit solution requires the use of Laplace transforms or other tricks. However, much information can be obtained directly from Eqs. (6.11). As an example, in the next section we shall consider the so-called "steady state."

## 6.10 Steady State

It is not difficult to show that for the solutions $u_k(t)$ of Eqs. (6.11), and for all similar systems, the limits

$$p_k = \lim_{t \to \infty} u_k(t) \tag{6.12}$$

always exist. Moreover, only two possibilities arise: either (*a*) $p_k = 0$

for all $k$ or (b) the $p_k$ are a probability distribution—that is,

$$p_0 + p_1 + p_2 + \cdots = 1$$

In the first case, the probability of finding fewer than $N$ customers in the line tends to 0, and hence *the waiting line is bound to increase indefinitely* as time goes on.

In the second case, we obtain a solution of Eqs. (6.11) if we put

$$u_k(t) = p_k$$

for all times. This means that if initially, at time 0, the number of customers in the line is randomly distributed in accordance with the probability distribution $\{p_k\}$, then the same probability distribution will prevail at all times: we have a "*steady state*," or a state of "*statistical equilibrium*." Much harm has been done by this misleading term; its true significance will be discussed in Sec. 6.12.

The probabilities $p_k$ must obviously satisfy the system of equations obtained on putting $u_k'(t) = 0$ in Eqs. (6.11), that is,

$$\alpha p_0 = \beta p_1$$
$$(\alpha + \beta)p_k = \alpha p_{k-1} + \beta p_{k+1} \qquad k \geq 1$$

Thus $p_1 = p_0(\alpha/\beta)$; by substituting into the second equation for $k = 1$, we find $p_2 = p_0(\alpha/\beta)^2$; and by continuing in this manner by recursion, we get $p_k = p_0(\alpha/\beta)^k$. But the series $\Sigma p_k$ must converge and its sum is unity whenever the $p_k$ represent steady-state probabilities. Hence we have the result that

$$p_k = 0 \quad \text{if } \alpha \geq \beta, \qquad \text{and} \qquad p_k = \left(1 - \frac{\alpha}{\beta}\right)\left(\frac{\alpha}{\beta}\right)^k \quad \text{if } \alpha < \beta$$

In other words, a "steady state" exists only if $\alpha < \beta$. The *average queue length in the steady state is*

$$p_1 + 2p_2 + 3p_3 + \cdots = \frac{\alpha}{\beta - \alpha} \tag{6.13}$$

Before discussing (see Sec. 6.12) the significance of this concept and the frequent misconceptions concerning it, we turn to a related notion.

## 6.11 Busy Periods

Suppose that at some time $t$, which we choose as $t = 0$, there are $i$ customers in the waiting line. How long will it take for the counter to become free for the first time? How will the queue length vary in the meantime? Many important questions of this type may be raised, and they are all related to the *ruin problem* as discussed in Sec. 6.7. Indeed, if we restrict our consideration to a time during which the counter remains constantly busy, the queue length 0 does not occur and thus 0 becomes the "ruin": the busy period ends when 0 is reached. For example, consider first the model with discrete time and let $v_{n,k}$ be the probability at time $t = n\tau$, during a busy period, of finding $k$ customers in the line. That is, we have $v_{0,i} = 1$, and $v_{n,k}$ is the probability of the compound event that the counter is not free at times $\tau, 2\tau, \ldots, (n-1)\tau$ and that there are exactly $k$ customers at time $n\tau$. As we have seen in Sec. 6.7, for $k \geq 1$ these $v_{n,k}$ will satisfy the same difference equations as the $u_{n,k}$, namely,

$$v_{n+1,k} = pv_{n,k-1} + qv_{n,k+1} + rv_{n,k}$$

But we have to impose the *boundary condition* $v_{n,0} = 0$. Similarly, in the continuous model, the probability $Q_k(t)$ that a busy period starting with $i$ customers lasts longer than $t$, and that at time $t$ there are exactly $k$ customers in the line, satisfies the differential equations

$$Q'_k(t) = -(\alpha + \beta)Q_k(t) + \alpha Q_{k-1}(t) + \beta Q_{k+1}(t)$$

for $k \geq 1$, with the initial and boundary conditions

$$Q_i(0) = 1 \qquad Q_0(t) = 0$$

These equations uniquely determine the probabilities $Q_k(t)$. The sum $\Sigma Q_k(t)$ is the probability that the busy period extends beyond $t$.

We shall here calculate only the *expected* or *average duration* $L_i$ of this busy period. For that purpose, let $t_1, t_2, t_3, \ldots, t_i$ be the moments when the queue length (initially of size $i$) first decreases to $i-1, i-2, i-3, \ldots, 0$. These moments break up the busy period into $i$ subintervals of length $t_1, t_2 - t_1, t_3 - t_2, \ldots, t_i - t_{i-1}$, each of which is characterized by the condition that at the end (and only at the end) the queue is shorter than at the beginning. It follows that the durations of the several subintervals are random variables with a common distribution and the common expected value $L$; therefore, $L_i = iL$. To calculate $L = L_1$, consider the discrete-time model at an instant when the queue length is 1. At the next moment the queue length is 0, 1, or 2, with respective probabilities $q$, $r$, and $p$. In the first case, the busy period lasted one trial (twice $\tau$). In the second case, the situation remained unchanged and, on the average, it will take another $L$ trials before 0 is reached. Finally, in the last case, the expected *continued* duration of the busy period is $2L$. Thus,

$$L = \tau + q \cdot 0 + r \cdot L + p \cdot 2L$$

or

$$L = \frac{\tau}{1 - 2p - r} = \frac{\tau}{q - p}$$

This, of course, makes sense only if $q > p$; if $q \leq p$, there is a chance that the busy period never ends, and we cannot speak of its expected duration.

By using the values (6.10) and letting $\tau \to 0$, we find that in the continuous-time model, corresponding to the differential equations (6.11), *the expected duration of a busy period starting with i customers is* $i/(\beta - \alpha)$, provided, of course, that $\beta > \alpha$.

For example, let the average service time be 5 min and suppose that the average frequency of arrivals is 10 per hr. This corresponds to $\alpha = 1/6$, $\beta = 1/5$. The average queue length (6.13) in the "steady state" is 5; the average duration of a busy period starting with $i$ customers is $30i$ min; the steady-state probability $p_0$ of finding the counter unoccupied is $1/30$.

## 6.12 Fluctuations in the Individual Process vs. Ensemble Averages

What are the practical or operational implications of the "steady state," i.e., of the existence of the limits (6.12)? It is a widespread but dangerous fallacy to assume that these relationships refer to the fluctuations in an individual process or the successive stages of the queue at any one particular counter. In reality, the relationships (6.12) and the "statistical equilibrium" refer to a large ensemble of similar counters. If we had a large number $N$ of counters operating independently for a long time, as described by our differential equations (6.11), then it would be reasonable to expect that, at any particular moment, $Np_0$ of the counters would be found free, $Np_1$ with exactly one customer, etc. Averaging over this ensemble of counters shows that the average number of customers is approximately $\alpha/(\beta - \alpha)$, etc. Once this steady-state distribution $\{p_k\}$ is attained, it will automatically maintain itself—at least ideally in an infinitely large ensemble.

At each individual counter, however, the fluctuations will continue unabatedly without the slightest attenuation. There is no tendency of the queue length to converge in any sense of the word or to settle down near the average value. In our last example, the average queue length was 5. If at any time the queue length happens to be 5, however, with probability about 0.287 that it will reach the level of 10 before the counter ever gets free again, it will take a very long time for the queue to reach 10, and from that moment on it will, on the average, take another 5 hr to find the counter free for the first time. In more than 10 per cent of all situations the queue will, starting from a length 5, reach the length 15 before disappearing, and in such cases the busy-time period will be extraordinarily long.

We saw that a busy period (lifetime of a queue) starting with one customer is $1/(\beta - \alpha)$. Unfortunately, this statement has almost no practical significance because of the exceedingly large fluctuations. One common measure of the probable magnitude of such fluctuations is the variance. For the busy period, this variance is $(\alpha + \beta)/(\beta - \alpha)^3$, which can be much larger than the mean. In our example the mean is 30 while the variance is 9,900. In practice, our counter would be "jammed" at frequent intervals.

It is very dangerous to rely—as is usually done—on expectations as indication for what will really happen and to believe that the individual process shows similarity to the whole ensemble. This point may be illustrated by the example given in the next section.

## 6.13 The Example of D. G. Kendall's Taxicab Stand[3]

At a perfectly balanced taxicab stand, either customers wait for taxis or taxis wait for customers. Customers and taxis arrive with equal frequencies. If customers are counted $+$ and taxis $-$, the queue length may be any integer $0, \pm 1, \pm 2, \ldots$. From a queue of length $k$, the next change leads with equal probabilities to $k + 1$ or $k - 1$. Thus the successive changes are represented by a symmetric random walk. *The expected queue length is* 0, but it is easily seen that at each individual stand the queue length is *bound to grow to* $+\infty$ *or* $-\infty$. The zero expectation says nothing about the fluctuations at an individual stand; it assures us merely that, in a large ensemble, for any stand with thousands of taxis waiting in despair for customers there is somewhere a stand with equally many customers waiting vainly for a taxi.

It should be borne in mind that in building taxi stands, elevators, etc., we are interested in the fluctuations in time at *one particular* counter, not in large ensembles balanced in the manner described. Statistical equilibrium is good where it is really meaningful—e.g., in an ensemble of many telephone trunk lines. But little satisfaction can be derived from a judicial statistical equilibrium where for each innocently condemned person we find a felon running free.†

## REFERENCES

1. Brown, George W., Monte Carlo Methods, chap. 12 in "Modern Mathematics for the Engineer," First Series, edited by E. F. Beckenbach, McGraw-Hill Book Company, Inc., New York, 1956.
2. Feller, William, "An Introduction to Probability Theory and Its Applications," vol. 1, 2d ed., John Wiley & Sons, Inc., New York, 1957.
3. Kendall, D. G., Some Problems in the Theory of Queues, *J. Roy. Statist. Soc.*, ser. B, vol. 13, pp. 151–185, 1951.
4. Morse, P. M., "Queues, Inventories, and Maintenance," John Wiley & Sons, Inc., New York, 1958.

† See Ref. 2, Chap. III, for a discussion and experiment concerning the accumulated gains $S_1, S_2, S_3, \ldots, S_n$ in an *individual* coin-tossing game. A majority of coins are what our psychologists and sociologists would call maladjusted; i.e., they behave quite differently from the ensemble of coins. For example, the sums $S_k$ are likely to remain of the same sign for unbelievably long periods: they show what many economists would call a trend.

# 7

# Information Theory†

DAVID BLACKWELL
PROFESSOR OF STATISTICS
UNIVERSITY OF CALIFORNIA, BERKELEY

## 7.1 Introduction

Information theory was founded by C. E. Shannon in 1948. In his fundamental paper,[3] Shannon introduced the concepts of entropy and channel capacity and related them by the coding theorem. For an excellent treatment of Shannon's work and its later development, the reader is referred to the book by Amiel Feinstein.[2]

## 7.2 An Example

Suppose that a man, called the *sender*, is given a 10-digit number and is required to communicate it to a second man, called the *receiver*. Suppose further that their only means of communication is a set of 100 postcards, labeled serially 1, 2, . . . , 100. These cards are in the possession of the sender and are addressed to the receiver. On each card is a space in which a single number may be written; this number is required to be 0 or 1. The sender, after being told the 10-digit number, will write a number, 0 or 1, on each of the 100 cards and mail the entire set to the receiver. The latter, on receiving the set of cards and noting the number written on each card, will try to deduce the 10-digit number. Can the sender and receiver agree in advance on a code with which the receiver can make a correct deduction?

We can easily answer this question by counting. The number of 10-digit numbers is $10^{10}$. The number of ways of filling out the cards is $2^{100} = 10^{30.103} > 10^{10}$. Thus to each 10-digit number there may be assigned a corresponding sequence of 100 0s and 1s in such a way that no

† This chapter was prepared with the partial support of the Office of Naval Research (Nonr-222-53); it may be reproduced in whole or in part for any purpose of the United States government.

two 10-digit numbers are assigned the same sequence. By agreeing in advance on any particular such code, the sender and receiver can communicate perfectly.

We now modify the problem by supposing imperfect postal service, so that each card has a probability of only 0.5 that it will be delivered, with the deliveries of the separate cards being independent events. Perfect communication in this case is impossible, since every card could be lost in transit. We now ask, can the sender and receiver agree in advance on a code with which the probability of an error is small, no matter which 10-digit number is selected?

In general, we formulate the problem as follows. There is given a number $k$, the number of possible messages, of which exactly 1 is to be selected for transmission. In our example, we have $k = 10^{10}$. There is given also a *channel*, i.e., a means of communication. This channel is specified by ($a$) an *input set* $A$, from which the sender selects any one element, ($b$) an *output set* $B$, of which the receiver observes one element, and ($c$) the *probability law* $p$ of the channel, which specifies, for each input element $a$ and output element $b$, the probability $p(b|a)$ that the receiver will observe element $b$ when the sender selects element $a$. In our example, $A$ consists of the $2^{100}$ sequences of length 100 that can be formed by using the two symbols 0 and 1; $B$ consists of the $3^{100}$ sequences of length 100 that can be formed by using the three symbols 0, 1, and $X$, where $X$ represents nondelivery of a card; $p(b|a) = 2^{-100}$ for any $b$ that has 0 or $X$ wherever $a$ has 0 and 1 or $X$ wherever $a$ has 1; and $p(b|a) = 0$ otherwise.

A $k$ *code* consists of ($a$) a sequence of $k$ input elements $a_1, \ldots, a_k$ and ($b$) a division of the set $B$ of outputs into $k$ disjoint parts $E_1, \ldots, E_k$. The code is used as follows. When the sender wants to transmit the $i$th message, he chooses input $a_i$. When the receiver observes output $b$, he locates the set $E_j$ to which $b$ belongs, and he concludes that message $j$ was sent. For message $i$, the probability of error is then

$$\text{prob error } (i) = \sum_{b \text{ not in } E_i} p(b|a_i) \tag{7.1}$$

The maximum, over all $i$, of this error probability is called the *maximum error probability* for the given code. Our problem is this: For a given channel and a given $k$, how small can we make the maximum error probability, by proper choice of a $k$ code? We shall see that, in our example, there is a $10^{10}$ code for which the maximum probability of error is less than 0.03.

## 7.3 Entropy

If there are $k$ possible messages, we can encode each message by a binary sequence of length $\log_2 k$. Thus a channel is adequate for transmitting any message from a set of $k$ possible messages only if it is adequate for transmitting any binary sequence of length $\log_2 k$. Now suppose that each of the $k$ messages has a certain probability of being presented for transmission and that there is a relatively small number $l$ of these messages such that the total probability of the remaining $k - l$ messages is small. It will then be possible to encode all messages except a set of small total probability by using only $\log_2 l$ binary digits. The entropy of a set of messages with associated probabilities is defined, roughly, as the number of binary digits per message required to encode all long sequences of messages, except for a set of small total probability. Before making this notion precise, let us consider an example.

Suppose that there are two possible messages, 0 and 1, with probabilities $\alpha$ and $1 - \alpha$, respectively. In a long series of $N$ such messages, the law of large numbers[1] tells us that there will very probably be about $N\alpha$ 0s and $N(1 - \alpha)$ 1s. Thus the message sequence that occurs will very likely be one with probability approximately equal to

$$\alpha^{N\alpha}(1 - \alpha)^{N(1-\alpha)} = 2^{-Nh(\alpha)}$$

where
$$h(\alpha) = -[\alpha \log_2 \alpha + (1 - \alpha) \log_2 (1 - \alpha)] \tag{7.2}$$

There are then about $2^{Nh(\alpha)}$ of these "likely" message sequences, and it is very unlikely that the actual message sequence will not be one of these. Note that, unless $\alpha = \frac{1}{2}$, we have $h(\alpha) < 1$, so that the likely messages are an extremely small fraction of the possible messages. Instead of the $N$ binary digits required to encode all possible message sequences of length $N$, we need only $Nh(\alpha)$ binary digits to encode all sequences of length $N$, except for a set of small total probability. The number $h(\alpha)$ is then the entropy of the pair of messages 0 and 1 with associated probabilities $\alpha$ and $1 - \alpha$.

In general, if a random variable $X$ has $k$ different possible values, with probabilities $p_1, p_2, \ldots, p_k$, respectively, the entropy of the random variable is defined by

$$H(X) = -\sum_{i=1}^{k} p_i \log_2 p_i \tag{7.3}$$

An extension of the above argument shows that we can encode all long sequences of messages $X$, except for a set of small total probability, with $NH(X)$ binary digits, where $N$ is the number of messages in the sequence. The following properties of $H$ are easily established.

*a.* The function $H(X)$ satisfies the inequality

$$0 \leq H(X) \leq \log_2 k$$

with $H(X) = \log_2 k$ if and only if all values have the same probability $1/k$.

*b*. For any two random variables $X$, $Y$, if $(X,Y)$ denotes the cartesian product set of values $X$, $Y$, then the function $H(X,Y)$ satisfies the inequality

$$H(X,Y) \leq H(X) + H(Y)$$

with equality only if $X$ and $Y$ are independent.

The number $R(X,Y) = H(X) + H(Y) - H(X,Y)$ is called the *mutual information* in $X$ and $Y$. It represents the number of bits (binary digits) of information that we obtain about $X$ by observing $Y$. This can be seen as follows: If we observe $Y$, we observe $H(Y)$ bits of information about the pair $(X,Y)$, so that $H(X,Y) - H(Y)$ additional bits are needed to identify $(X,Y)$, that is, to identify $X$. Had we not observed $Y$, we would have needed $H(X)$ bits to identify $X$. Thus, observing $Y$ reduces the number of bits required to identify $X$ by

$$H(X) - [H(X,Y) - H(Y)] = R(X,Y)$$

bits. The number

$$H(X|Y) = H(X,Y) - H(Y)$$

is called the *conditional entropy* of $X$, given $Y$; it is, as we have just seen, the additional number of bits required to identify $X$, once we are given $Y$.

As an example, suppose a fair coin is tossed 100 times. Let $X$ describe the result of the first 60 tosses and $Y$ the result of the last 70. Then

$$H(X) = 60 \qquad H(Y) = 70 \qquad H(X,Y) = 100 \qquad R(X,Y) = 30$$
$$H(X|Y) = 30 \qquad H(Y|X) = 40$$

## 7.4 Capacity of a Channel

Consider a channel with input set $A$, output set $B$, and probability law $p(b|a)$. If we select an input $a$ according to some probability distribution $q$ on $A$, then the probability of the input-output pair $(a,b)$ is

$$r(a,b) = q(a)p(b|a)$$

and the probability of the output $b$ is

$$s(b) = \sum_a r(a,b) = \sum_a q(a)p(b|a) \tag{7.4}$$

If $X$ is the input variable and $Y$ is the output variable, we have

$$\begin{aligned} H(X) &= \sum_a q(a) \log_2 q(a) \\ H(Y) &= \sum_b s(b) \log_2 s(b) \\ H(X,Y) &= \sum_{a,b} r(a,b) \log_2 r(a,b) \end{aligned} \tag{7.5}$$

With our interpretation of $R(X,Y)$ as the number of bits of information that we get about $X$ from observing $Y$, we may hope that our channel is capable of transmitting $R(X,Y)$ bits of information, i.e., that a message set with not more than $2^{R(X,Y)}$ messages can be transmitted with small error probability. We shall see that this is approximately true, provided that we are allowed to encode long sequences of messages rather than individual ones.

For the present, we note that, by varying $q$, we obtain different numbers $R(X,Y)$. The maximum of $R(X,Y)$ over all input distributions $q$ on $A$ is called the *capacity* of the channel.

As an example, consider the channel consisting of a single postcard of the type considered in Sec. 7.2. We have $A = (0,1)$, $B = (0,1,X)$, and $p(b|a)$ is as shown below:

| A \ B | 0 | 1 | X |
|---|---|---|---|
| 0 | 0.5 | 0 | 0.5 |
| 1 | 0 | 0.5 | 0.5 |

For $q(0) = \alpha$, $q(1) = 1 - \alpha$, we obtain

$$\begin{aligned} H(X) &= -[\alpha \log_2 \alpha + (1 - \alpha) \log_2 (1 - \alpha)] = h(\alpha) \\ H(Y) &= -\{0.5\alpha \log_2 (0.5\alpha) + 0.5(1 - \alpha) \log_2 [0.5(1 - \alpha)] \\ &\qquad + 0.5 \log_2 (0.5)\} \\ H(X,Y) &= -\{\alpha \log_2 (0.5\alpha) + (1 - \alpha) \log_2 [0.5(1 - \alpha)]\} \\ R(X,Y) &= 0.5h(\alpha) \end{aligned}$$

Thus $R(X,Y)$ is maximized for $\alpha = 0.5$, and its maximum value, which is the channel capacity, is 0.5. This is exactly what we should expect: A channel that transmits one bit of information when it works, and that works half the time, should be able to transmit half as much information as a similar channel that works every time.

## 7.5 The Fundamental Theorem

The main result asserts that, if a channel has capacity $C$ and, if we are allowed to use the channel $N$ times, where $N$ is large, then we can transmit any one of about $2^{CN}$ messages with small error probability. Note that, since our postcard channel has $C = \frac{1}{2}$, we should be able to transmit, with our 100 postcards, any one of about $2^{50}$ messages with small error probability. Since the message set of interest has only $10^{10} = 2^{30.103}$ messages, the theorem implies that we should be able to find a code with small maximum error probability, since $N = 100$ is large.

The use of a channel $N$ times may be considered as the single use of a large channel—the so-called $N$ extension of the original channel. Thus our 100 postcards constitute the 100 extension of the single-card channel. In general, for a channel with input set $A$, output set $B$, and probability function $p(b|a)$, the $N$ extension has input set $U$ consisting of all sequences of length $N$ of elements of $A$, output set $V$ consisting of all sequences of length $N$ of elements of $B$, and probability function $f(v|u)$ defined by

$$f(b_1, \ldots, b_N|a_1, \ldots, a_N) = p(b_1|a_1) \cdots p(b_N|a_N) \qquad (7.6)$$

We can now state the following result.

FUNDAMENTAL THEOREM. *Given a channel of capacity $C$ and any number $\epsilon > 0$, there is an integer $N_0$, depending on $\epsilon$ and the channel, such that, for any $N \geq N_0$ and for the $N$ extension of the channel, there is a $2^{N(C-\epsilon)}$ code having the property that the maximum error probability of the code is at most $\epsilon$.*

The theorem can be proved easily from a simple inequality, which will also enable us to say something useful about how large $N$ must be; the proof actually describes the construction of a particular code.

INEQUALITY. *Given a channel having input set $U$, output set $V$, and probability law $f(v|u)$, a probability distribution $q$ on $U$, a positive number $c$, and a positive number $k$, there is a $k$ code for the channel such that the maximum error probability of the code is at most*

$$\text{prob}\,\{J(u,v) < c\} + \frac{k}{c} \qquad (7.7)$$

*where*

$$J(u,v) = \frac{\text{prob}\,(u,v)}{[\text{prob}\,(u)][\text{prob}\,(v)]} \qquad (7.8)$$

*and where the probability of a pair $(u,v)$ is $q(u)f(v|u)$.*

We shall not prove the fundamental theorem from our inequality but shall instead illustrate its use on our original postcard problem. We have $U$ consisting of all sequences of 100 0s and 1s; $V$ consisting of all sequences of 100 0s, 1s, and $X$'s; and $f(v|u) = 2^{-100}$ if $v$ has 0 or $X$ wherever $u$ has 0 and 1 or $X$ wherever $u$ has 1. We are interested in $k = 10^{10}$. We may choose any distribution $q$ on $U$ and any number $c$. We shall choose $q(u) = 2^{-100}$ for all $u$ and $c = 2^{40}$. The second term in our error bound is

$$\frac{k}{c} = \frac{10^{10}}{2^{40}} = 0.0091$$

Let us evaluate the first term, that is, prob $\{J(u,v) < c\}$.

For any possible pair $(u,v)$, we have

$$\text{prob}\,(u,v) = g(u,v) = q(u)f(v|u) = 2^{-100} \cdot 2^{-100} = 2^{-200}$$

Also for any $v$, we have

$$\text{prob}\ (v) = e(v) = 2^{-200} \times \text{number of } u\text{'s that could produce } v$$

so that

$$e(v) = 2^{-200} \times 2^{x}$$

where $x$ is the number of $X$'s in $v$, that is, the number of cards that fail to arrive. Thus we obtain

$$J(u,v) = \frac{g(u,v)}{q(u)e(v)} = 2^{-200} \times 2^{100} \times 2^{200-x} = 2^{100-x}$$

The probability that $J(u,v) < c$ (i.e., that $2^{100-x} < 2^{40}$, or that $x > 60$) is the probability that more than 60 cards are lost in transit. This probability is found from tables of the binomial distribution to be 0.0176. Thus our inequality asserts the existence of a $10^{10}$ code with maximum error probability at most $0.0176 + 0.0091 = 0.0267$.

*Proof of the Inequality.* Write

$$g(u,v) = q(u)f(v|u) \qquad e(v) = \sum_{u} g(u,v) \tag{7.9}$$

and let

$$\epsilon = \text{prob}\ \{J(u,v) < c\} + \frac{k}{c}$$

Let $S$ consist of all pairs $(u,v)$ for which $J(u,v) \geq c$. We proceed to construct a code, i.e., a series $u_1, \ldots, u_k$ of elements of $U$ and a series $E_1, \ldots, E_k$ of disjoint subsets of $V$, such that

$$\sum_{v \text{ not in } E_i} f(v|u_i) \leq \epsilon \tag{7.10}$$

Choose, if possible, for $u_1$ any element of $U$ for which there is a set $E_1$ in $V$ such that

*a.* $\sum_{v \text{ in } E_1} f(v|u_1) \geq 1 - \epsilon.$

*b.* All pairs $(u_1,v)$ with $v$ in $E_1$ are in $S$.

Having found $u_1$ and $E_1$, choose, if possible, for $u_2$ any element of $U$ for which there is a set $E_2$ in $V$ such that

*a'.* $\sum_{v \text{ in } E_2} f(v|u_2) \geq 1 - \epsilon.$

*b'.* All pairs $(u_2,v)$ with $v$ in $E_2$ are in $S$.

*c'.* $E_2$ is disjoint from $E_1$.

Having continued in this manner, and thus having found $u_1, \ldots, u_i$ and $E_1, \ldots, E_i$, choose, if possible, for $u_{i+1}$ any element of $U$ for which there is a set $E_{i+1}$ in $V$ such that

$a''$. $\sum_{v \text{ in } E_{i+1}} f(v|u_{i+1}) \geq 1 - \epsilon$.

$b''$. All pairs $(u_{i+1},v)$ with $v$ in $E_{i+1}$ are in $S$.

$c''$. $E_{i+1}$ is disjoint from $E_1, E_2, \ldots, E_i$.

Clearly, if $u_1, \ldots, u_i$ and $E_1, \ldots, E_i$ can be found, they are an $i$ code with maximum error probability at most $\epsilon$. It remains to be shown that they can be found for $i = k$. Suppose, then, that we have found $u_1, \ldots, u_i$ and $E_1, \ldots, E_i$ and are unable to find $u_{i+1}$ and $E_{i+1}$ satisfying the conditions $a''$, $b''$, and $c''$. We must show that $i \geq k$. For any $u$, let $F_u$ consist of all $v$ for which $(u,v)$ is in $S$ and $v$ is not in $E_1$ or $E_2$ or . . . or $E_i$. Then, for every $u$, we have

$$\sum_{v \text{ in } F_u} f(v|u) < 1 - \epsilon \tag{7.11}$$

since otherwise $u$ and $F_u$ would be a possible choice for $u_{i+1}$ and $E_{i+1}$. Multiplying this inequality by $q(u)$ and summing on $u$ yields

$$\sum_{(u,v) \text{ in } F} g(u,v) < 1 - \epsilon \tag{7.12}$$

where $F$ consists of all pairs $(u,v)$ for which $(u,v)$ is in $S$ and $v$ is not in $E_1$ or $E_2$ or . . . or $E_i$. Also, for $v$ in $E_j$, we have

$$e(v) \leq \frac{f(v|u_j)}{c}$$

so that

$$\sum_{v \text{ in } E_j} e(v) \leq \sum_{v \text{ in } E_j} \frac{f(v|u_j)}{c} \leq \frac{1}{c}$$

whence

$$\sum_{v \text{ in } E_1 \text{ or } \ldots \text{ or } E_i} e(v) \leq \frac{i}{c} \tag{7.13}$$

But

$$\sum_{(u,v) \text{ in } S} g(u,v) \leq \sum_{(u,v) \text{ in } F} g(u,v) + \sum_{v \text{ in } E_1 \text{ or } \ldots \text{ or } E_i} e(v) \leq 1 - \epsilon + \frac{i}{c}$$

that is,

$$\epsilon \leq \text{prob}\,\{J(u,v) < c\} + \frac{i}{c} \tag{7.14}$$

Since we have

$$\epsilon = \text{prob}\ \{J(u,v) < c\} + \frac{k}{c}$$

it follows that $i \geq k$ and the proof is complete.

## 7.6 Multistate Channels

In constructing the $N$ extension of a given channel, we defined the probability function for the $N$ extension by

$$f(b_1, \ldots, b_N | a_1, \ldots, a_N) = p(b_1|a_1) \cdots p(b_N|a_N)$$

This is correct for a single channel used $N$ successive times only if the probability function of the channel does not depend on the history of the channel. Channels of this type are called memoryless or single-state channels. More generally, a finite-state channel is specified by an input set $A$, an output set $B$, a finite set $S$ of states, and a probability function $p(b,s'|a,s)$, which specifies the probability that, if the channel is currently in state $s$ and receives input $a$, it will produce output $b$ and move to state $s'$.

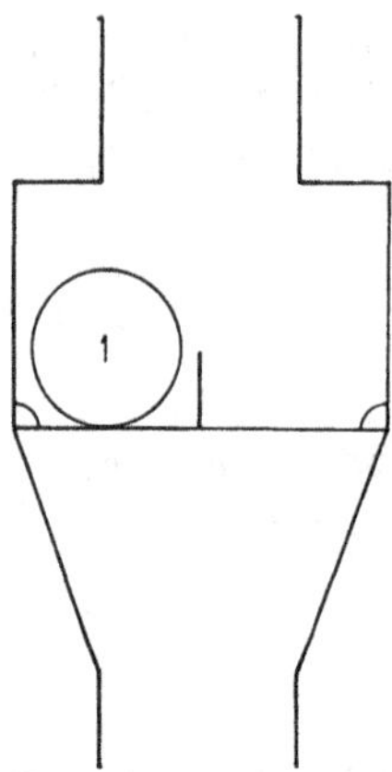


**Fig. 7.1** A simple two-state channel.

A simple two-state channel is illustrated in Fig. 7.1. At the beginning of a cycle, the channel contains a single ball, numbered 0 or 1, and occupying either compartment. An input consists of a ball, labeled 0 or 1, dropped into the channel from the top. This input ball descends and occupies the remaining compartment. One of the two trap doors, selected at random, then opens, allowing one ball to descend. This ball is the output of the channel, and the cycle is complete. We have $A = (0,1)$, $B = (0,1)$, $S = (0,1)$, and

$$\begin{aligned} p(b,s'|a,s) &= \tfrac{1}{2} && \text{if } a \neq s \quad b = a \quad s' = s \\ &= \tfrac{1}{2} && \text{if } a \neq s \quad b = s \quad s' = a \\ &= 1 && \text{if } b = s' = a = s \\ &= 0 && \text{otherwise} \end{aligned}$$

For a finite-state channel, we can again consider its $N$ extension, which will be a finite-state channel. For a given code, the probability of error will be a function of the initial state of the channel as well as of the particular message being transmitted. Again each finite-state channel has a capacity $C$, and there is a $2^{(C-\epsilon)N}$ code for the $N$ extension such that the maximum error probability of the code, over initial states and messages, is small. For these channels, however, $C$ is defined in a more complicated way, as described below, and cannot easily be evaluated. For instance, the capacity of the channel described above is not known.

Indecomposability of a finite-state channel is defined as follows. Let $s$ be any state and let $u$ be any finite sequence $a_1, \ldots, a_k$ of input letters.

A state $s'$ is *u-attainable* from $s$ if, when the channel is initially in state $s$ and receives the infinite sequence of inputs $a_1, \ldots, a_k; a_1, \ldots, a_k; a_1, \ldots, a_k; \ldots$, it is possible that at some time it enters state $s'$. A channel is indecomposable if for every two states $s_1$, $s_2$, and every $u$, there is a state $s$ that is $u$-attainable from either $s_1$ or $s_2$.

Our two-state channel illustrated above is clearly indecomposable. For if $u$ contains at least one 1, the state 1 is $u$-attainable from either 0 or 1, and similarly for 0.

## 7.7 Entropy of a Process; Capacity of Finite-state Channels

In defining the entropy of a random variable $X$, we noted that if $X_1, X_2, \ldots, X_N$ are independent variables with the same distribution as $X$, then the sequence of values $a_1, \ldots, a_N$ that occurs is practically certain to be one having probability about $2^{-NH(X)}$. More generally, let $(A,B,S,p)$ be any indecomposable finite-state channel, let the initial state of the channel be chosen according to some probability distribution $\lambda$, let a specified input $a^*$ be used again and again, and let $Y_1, Y_2, \ldots$ be the resulting sequence of outputs. Sequences $Y_1, Y_2, \ldots$ defined in this way are called *finitary processes*. There is then a number $H$, depending on the channel and on $a^*$, but not on $\lambda$, such that the sequence $b_1, \ldots, b_N$ that occurs is practically certain to have probability about $2^{-HN}$. This number $H$ is called the *entropy* of the process $Y_1, Y_2, \ldots$.

For instance, if the output $b$ is identical with the state of the channel, so that $Y_1, Y_2, \ldots$ are the successive states of the channel, the probability of the sequence $s_1, \ldots, s_N$ is

$$\sum_{s_0} \lambda(s_0)q(s_1|s_0) \cdots q(s_N|s_{N-1}) \tag{7.15}$$

where $q(s'|s)$ is the probability of moving to state $s'$, when the channel is in state $s$ and receives input $a^*$. Among the pairs $(s_0,s_1), (s_1,s_2), \ldots, (s_{N-1},s_N)$, let $f(s,s')$ be the proportion that are equal to the pair $(s,s')$. Then, approximately,

$$\sum_{s'} f(s,s') = \sum_{s'} f(s',s) = g(s) \tag{7.16}$$

say, since both sums represent approximately the relative frequency of state $s$ among $s_0, s_1, \ldots, s_N$. Now it is very likely that, among all occurrences of $s$, a proportion equal to about $q(s'|s)$ are followed by $s'$; i.e., we have, approximately,

$$g(s)q(s'|s) = f(s,s') \tag{7.17}$$

Summing over $s$ yields the system

$$\sum_{s} g(s)q(s'|s) = g(s') \tag{7.18}$$

of linear equations in the proportions $g(s)$. This system, for indecomposable channels, will have a unique solution $g^*(s)$. Thus it is very probable that the sequence $s_0, s_1, \ldots, s_N$ that occurs is one for which $f(s,s')$ is approximately $g^*(s)q(s'|s)$, so that its probability is about

$$\prod_{s,s'} q(s'|s)^{Ng^*(s)q(s'|s)} = 2^{-NH} \tag{7.19}$$

where
$$H = \sum_{s} g^*(s) \sum_{s'} q(s'|s) \log_2 q(s'|s) \tag{7.20}$$

Thus, for the case in which the output is identical with the state, the entropy of the process is given by the above formula. No similarly tractable formula is known for the entropy of finitary processes in general.

In terms of the entropy of finitary processes, it is easy to define the capacity of an indecomposable finite-state channel. If $(A,B,S,p)$ is such a channel, let $\{X_1,X_2, \ldots\}$ be any finitary process such that the value of each $X_n$ is one of the input letters of the channel. This process has an entropy $H_X$. If we start the channel in an arbitrary state and drive it with the inputs $X_1, X_2, \ldots$, the resulting outputs $Y_1, Y_2, \ldots$ will form a finitary process, with entropy $H_Y$, and the pairs $(X_1,Y_1)$, $(X_2,Y_2), \ldots$ will form a third finitary process, with entropy $H_{X,Y}$. The capacity $C$ of the channel is defined as the upper bound over all finitary sources $X_1, X_2, \ldots$ of the numbers

$$R_{X,Y} = H_X + H_Y - H_{X,Y} \tag{7.21}$$

## EXERCISES

**1.** Find the capacity of each of the following channels:

*a.* $A = (0,1)$ $B = (0,1)$

$p$:

| A \ B | 0 | 1 |
|---|---|---|
| 0 | $\alpha$ | $1 - \alpha$ |
| 1 | $1 - \alpha$ | $\alpha$ |

(Binary symmetric channel)

*b.* $A = (0,1)$ $B = (0,1)$

$p$:

| A \ B | 0 | 1 |
|---|---|---|
| 0 | 1 | 0 |
| 1 | ½ | ½ |

**2.** Under what conditions will a memoryless channel have capacity zero?

**3.** Let $C$ be any memoryless channel $(A,B,p)$ and let $C^*$ be the channel $(A,B^*,p^*)$, where $B^*$ consists of $B$ and a single new symbol $x$, and

$$p^*(b|a) = \tfrac{1}{2}p(b|a) \qquad p^*(x|a) = \tfrac{1}{2}$$

for all $a$. Show that the capacity of $C^*$ is one-half the capacity of $C$.

**4.** A channel has input set (0,1) and output set (0,1). For input 0, outputs 0 and 1 are equally probable. For input 1, the output is 0 if the previous input and output agree and 1 if they differ. Describe this as a two-state channel.

**5.** Let $C = (A,B,p)$ and $C^* = (B,D,q)$ be two memoryless channels, and let $C^{**} = (A,D,r)$, where

$$r(d|a) = \sum_b p(b|a)q(d|b)$$

Show that the capacity of $C^{**}$ does not exceed that of $C$ or that of $C^*$.

## REFERENCES


1. Brown, George W., Monte Carlo Methods, chap. 12 in "Modern Mathematics for the Engineer," First Series, edited by E. F. Beckenbach, McGraw-Hill Book Company, Inc., New York, 1956.
2. Feinstein, Amiel, "Foundations of Information Theory," McGraw-Hill Book Company, Inc., New York, 1958.
3. Shannon, C. E., A Mathematical Theory of Communication, *Bell System Tech. J.*, vol. 27, pp. 379–423, 623–658, 1948.

# 8

# The Mathematical Theory of Control Processes

RICHARD BELLMAN
RESEARCH MATHEMATICIAN
THE RAND CORPORATION

## 8.1 Introduction

Let us begin by stating precisely what we mean by a *control process*.† Consider a physical system $S$ that we suppose to be described at any time $t$ by a finite-dimensional vector

$$\mathbf{x}(t) \equiv [x_1(t), x_2(t), \ldots, x_n(t)]$$

which we shall call the *state vector*. Initially, let us assume that this vector is determined as a function of time by means of a vector differential equation

$$\frac{d\mathbf{x}}{dt} = \mathbf{g}(\mathbf{x},t) \qquad \mathbf{x}(0) = \mathbf{c} \tag{8.1}$$

where the vector $\mathbf{c}$ represents the initial state.

A large part of the theory of differential equations is devoted to the study of the existence and uniqueness of the solutions of Eq. (8.1) and to the qualitative and quantitative analysis of the solution. Let us assume here that we are able in one way or another to predict the future behavior of the system.

On comparing this actual behavior with some idealized or desired behavior, we may find that the extent of agreement is not satisfactory. If, as is often the case, it is impossible, or too expensive in time or resources, to construct a totally new system, we must devise an efficient method for modifying the behavior of the system we possess.

† A detailed discussion of the ideas and techniques of this article will be found in the author's book "Adaptive Control Processes: A Guided Tour," Princeton University Press, Princeton, N.J., 1960.

One ingenious idea is to use the very deviation of the system from desired behavior as a restoring force that will influence its behavior in the right way. This is the concept of *feedback control*, an idea that goes back at least as far as the governor on the Watt steam engine.

In many cases the mathematical version of this fruitful concept is the following: Suppose that the vector $\mathbf{z}(t)$ is the desired state of the system at any particular time, and let $\mathbf{h}(\mathbf{x} - \mathbf{z})$ represent a restoring force applied to the system at time $t$. Assume that this is done in such a way that the new vector differential equation describing the system is

$$\frac{d\mathbf{x}}{dt} = \mathbf{g}(\mathbf{x},t) + \mathbf{h}(\mathbf{x} - \mathbf{z}) \qquad \mathbf{x}(0) = \mathbf{c}$$

In general, there will be a certain cost involved in doing this over a time interval $[0,T]$, measured by a functional $J_1(\mathbf{x},\mathbf{z},\mathbf{h})$, and a cost due to the deviation of $\mathbf{x}$ from $\mathbf{z}$. This latter may be a tractable functional such as

$$J_2(\mathbf{x},\mathbf{z}) = \int_0^T (\mathbf{x} - \mathbf{z}, \mathbf{x} - \mathbf{z})\, dt$$

where $(\mathbf{x},\mathbf{y})$ is the inner product

$$\sum_{i=1}^{n} x_i y_i$$

Or, more realistically, the functional may be a measure such as

$$J_2(\mathbf{x},\mathbf{z}) = \max_{0 \le t \le T} \max_{1 \le i \le n} |x_i - z_i|$$

where $x_i$ and $z_i$ are the $i$th components of $\mathbf{x}$ and $\mathbf{z}$, respectively.

We thus arrive at the problem of choosing the vector function, subject perhaps to certain physical constraints that we shall discuss in detail below, so as to minimize the sum $J_1 + J_2$.

Analytically, many of these problems can be subsumed under the problem of choosing a vector function $\mathbf{y}$ to minimize the functional

$$J(\mathbf{y}) = \int_0^T h(\mathbf{x},\mathbf{y})\, dG(t)$$

subject to the condition that $\mathbf{x}$ and $\mathbf{y}$ are related by a vector differential equation of the form

$$\frac{d\mathbf{x}}{dt} = \mathbf{g}(\mathbf{x},\mathbf{y}) \qquad \mathbf{x}(0) = \mathbf{c}$$

and perhaps subject to additional constraints of the type

$$r_i(\mathbf{x},\mathbf{y}) \le 0 \qquad i = 1, 2, \ldots, M \qquad 0 \le t \le T$$

A process of this type we shall call a *control process.*

Two particularly important cases are those in which $T$ is fixed and either $dG(t) = dt$ or $G(t)$ is a jump function with a single jump at $t = T$, so that

$$J(\mathbf{y}) = h(\mathbf{x}(T),\mathbf{y}(T))$$

In this latter case, we often speak of *terminal control.*

The case in which $T$ is variable and itself dependent on $\mathbf{x}$ and $\mathbf{y}$ is of great current interest. Particular cases are the "bang-bang" control problem[27,41,48,53-55] and various trajectory problems.[42] We shall call problems of this nature *implicit variational problems.* Little has been done on these problems, and they offer a fascinating field of research.

The role of the mathematician in this challenging study is threefold. In the first place, he has the task of formulating questions of this nature—which arise in economic, industrial, engineering, and biological context—in precise mathematical terms. He must convert a well-posed scientific problem into a well-posed mathematical problem. The two types of problems are not at all equivalent; see Ref. 21 for a further discussion of this point.

The foregoing consideration necessitates a study of the existence and uniqueness of solution of the variational problems that arise. Even more, it requires a choice of formulation. One of the main points of this survey will be the fact that in studying control processes we have two approaches: the classical calculus of variations,[49] with modern amplifications, and the theory of dynamic programming.[9,32,49]

A second part of the general problem is that of determining the analytic structure of the solution. Infrequently, this can be done by means of an explicit solution, but most often it must be done without this luxury. We shall present, below, many examples of what we mean here.

A third part of the problem, and one that is required before we can consider the problem to be resolved in any sense, is that of producing algorithms that can be used for numerical solution.

The three parts are completely intertwined and interdependent. Clearly there is no point in attempting to find the numerical solution to a problem that does not possess a solution. Conversely, a knowledge of the uniqueness or multiplicity of a solution is an essential feature of a computational approach.

Finally, the very formulation of the control process in analytic terms is dependent on the numerical information that is available and the numerical results that are desired.

## DETERMINISTIC CONTROL PROCESSES

### 8.2 The Calculus of Variations

A powerful tool for the investigation of problems of the type posed in Sec. 8.1 is the classical calculus of variations. Let us, for the sake of simplicity, consider the scalar variational problem of maximizing the functional

$$J = \int_0^T h(u,v)\, dt$$

over all scalar functions $v(t)$, where $u$ and $v$ are related by means of the relationship

$$\frac{du}{dt} = g(u,v) \qquad u(0) = c$$

By proceeding in the usual way, we obtain the Euler equation

$$\frac{g_u h_v - g_v h_u}{h_v} - \frac{d}{dt}\left(\frac{g_v}{h_v}\right) = 0 \tag{8.2}$$

and the boundary conditions

$$u(0) = c \qquad \left.\frac{h_v}{g_v}\right]_{t=T} = 0 \tag{8.3}$$

The first condition is one with which we started, while the second arises from the variational process.

There is a temptation to regard Eq. (8.2) as a solution of the original variational problem. Actually, the problem has only begun at this point. Often, the computational solution is so difficult that the original variational problem is used to furnish a solution to Eqs. (8.2) and (8.3).

## 8.3 A Catalogue of Catastrophes

Let us now indicate briefly some of the difficulties encountered in applying the variational technique outlined above. Further discussion will be found in Ref. 32.

*a. Two-point Boundary-value Problems.* The computational solution of a differential equation with initial conditions is today a rather simple matter with the aid of digital computers. Systems involving 100 simultaneous differential equations in 100 unknowns are readily treated.

The situation is quite different when two-point conditions are imposed. For a discussion of these matters and the methods that can be used to treat these more complex matters, see Ref. 17.

*b. Relative Extrema.* As in the case of ordinary calculus, the variational equation determines the position of the absolute maximum, but also the positions of relative maxima, relative minima, and critical points of more complex nature.

*c. Constraints.* If, in the problem of Sec. 8.2, we introduce the constraint

$$|v| \leq m$$

then the process of solution becomes very much more intricate, and in some cases it eludes us completely. This is due to the fact that there is no longer one variational condition; rather, there is a set of Euler equations and inequalities. Furthermore, the number and order of these relationships depend on the solution itself.

*d. Implicit Functionals.* The Euler equation presupposes the existence of an analytic functional. As we shall see, some quite important problems cannot be posed in these terms.

*e. Linearity.* Far on the other extreme, variational problems posed in the simplest of analytic terms, even in linear terms, cannot be treated by means of classical variational techniques.

In view of these briefly sketched facts (a more detailed account of which will be found in Ref. 32) our program will be the following: We shall first discuss various classes of problems that can be treated by conventional techniques and modern extensions. Then we shall present a new technique, the theory of dynamic programming,[9] which will enable us to treat deterministic, stochastic, and adaptive control processes in a uniform fashion.

## 8.4 Quadratic Criteria and Linear Equations

Let us begin by indicating one class of variational problems that can be analyzed rigorously and completely. Problems of this nature, regardless of their apparent simplicity and lack of rigor, are always valuable as stepping stones, via successive approximations, to the solution of more significant problems.

Consider the problem of minimizing the quadratic functional

$$J(\mathbf{y}) = \int_0^T (\mathbf{x}, C(t)\mathbf{x})\, dG(t) + \int_0^T (\mathbf{y}, D(t)\mathbf{y})\, dH(t)$$

over all vector functions $\mathbf{y}(t)$, where $\mathbf{x}$ and $\mathbf{y}$ are related by means of the linear equation

$$\frac{d\mathbf{x}}{dt} = A(t)\mathbf{x} + \mathbf{y} \qquad \mathbf{x}(0) = \mathbf{c} \tag{8.4}$$

Here, $A(t)$, $C(t)$, and $D(t)$ are matrix functions.

It is easy to verify that the basic variational equation is linear in this case. Since, however, the boundary conditions are of the two-point type, there are still matters of existence and uniqueness to settle.

In place of deciding these questions by means of the theory of differential equations, it is better to take a more general point of view that simultaneously resolves a large class of problems of similar form. By using the solution $X(t)$ of the matrix equation

$$\frac{dX}{dt} = A(t)X \qquad X(0) = I$$

where $I$ denotes the unit matrix, we can write the solution of Eq. (8.4) in the form

$$\mathbf{x} = X\mathbf{c} + \int_0^t X(t)X(s)^{-1}\mathbf{y}(s)\,ds$$

Let us then write

$$\mathbf{x} = \mathbf{L}(\mathbf{y})$$

where $\mathbf{L}$ is an inhomogeneous linear transformation of the vector $\mathbf{y}$.

The variational problem is now that of minimizing

$$J(\mathbf{y}) = \int_0^T (\mathbf{L}(\mathbf{y}), C(t)\mathbf{L}(\mathbf{y}))\,dG(t) + \int_0^T (\mathbf{y}, D(t)\mathbf{y})\,dH(t)$$

over all admissible vectors $\mathbf{y}$. The advantage of this formulation lies in the fact that a large class of problems arising from differential-difference equations, partial differential equations, difference equations, integral equations, and so on, can all be subsumed under this general form.

Once the problem has been posed in this manner, we can employ quite general Hilbert-space techniques to establish existence and uniqueness of the solution, to derive the variational equations, and to deduce the dependence of the solution on basic parameters such as $T$. For detailed discussions see Refs. 26 and 29.

## 8.5 Linear Criteria and Linear Constraints

A problem of a type that arises frequently in mathematical economics and occasionally in engineering work is that of maximizing the linear functional

$$J(\mathbf{y}) = \int_0^T [(\mathbf{x}(t), \mathbf{a}(t)) + (\mathbf{y}(t), \mathbf{b}(t))]\,dG(t)$$

over all vector functions $\mathbf{y}(t)$ satisfying the constraints

$$\mathbf{c}(t) \leq \mathbf{y}(t) \leq \mathbf{d}(t) \qquad 0 \leq t \leq T$$

where the vectors $\mathbf{x}$ and $\mathbf{y}$ are connected by the linear equation

$$\frac{d\mathbf{x}}{dt} = A(t)\mathbf{x} + \mathbf{y} \qquad \mathbf{x}(0) = \mathbf{c}$$

Here, we face an excess of linearity! The correct approach to variational problems of this nature is by means of the famous Neyman-Pearson lemma of mathematical statistics.[26,29] See Refs. 26, 36, and 56 for the solution of particular problems.

## 8.6 Nonlinear Criteria and Constraints

The addition of constraints renders nontrivial even the simplest-appearing problems. As mentioned above, the reason for this lies in the fact that, in place of one Euler equation over the entire interval, one has a set of equalities and inequalities. Thus, if the problem is that of maximizing

$$J(v) = \int_0^T h(u,v)\,dt$$

over all scalar functions $v$, where $u$ and $v$ are related by means of the differential equation

$$\frac{du}{dt} = g(u,v) \qquad u(0) = c \qquad \text{and} \qquad |v| \leq k$$

then the solution can take the form shown in Fig. 8.1 (see page 238).

This means that the Euler equation is satisfied for $0 < t < T_1$, an Euler inequality for $T_1 < t < T_2$, and so on.

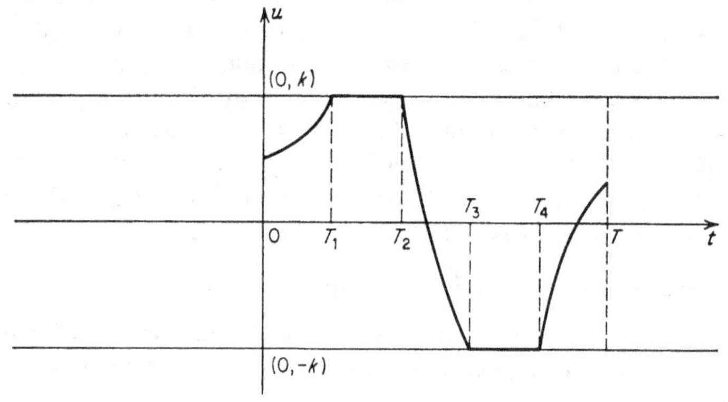


**Fig. 8.1** Possible solution of variational problem with inequality constraint.

At the present time, there exists no systematic technique for determining the number of transition points, or their location. For a detailed discussion of particular problems under various assumptions, see Refs. 25, 26, and 57.

## 8.7 Implicit Functionals

An interesting class of control processes that has attracted a great deal of attention in recent years arises from the following considerations. Let the state of the system be determined by the linear vector differential equation

$$\frac{d\mathbf{x}}{dt} = A(t)\mathbf{x} + \mathbf{y} \qquad \mathbf{x}(0) = \mathbf{c}$$

and let it be required to determine the vector $\mathbf{y}$, subject to the constraint

$$|y_i(t)| \leq m_i \qquad i = 1, 2, \ldots, n \qquad 0 \leq t$$

so as to minimize the time required to drive the system into a preassigned state. This problem, under the name of the "bang-bang" control problem, has been extensively investigated in the United States; see Refs. 27, 41, 54, and 55, where further references may be found. Under the name of "optimal time systems," it has been extensively studied also in Russia; see Refs. 48 and 53, where references to the work of Pontrjagin, Gamkrelidze, and others are given.

The methods that have been used are a combination of classical analytical techniques and modern analytical-topological procedures based on moment-space concepts. Dynamic programming can also be used for both analytic and computational purposes.

## 8.8 Dynamic Programming

Let us now briefly outline the application of the theory of dynamic programming to control processes, referring the reader to Refs. 9 and 32 for more detailed discussions. In a number of cases, the techniques of dynamic programming furnish the only feasible computational solution of the variational problems that arise. We shall indicate the limitation of these methods below.

The basic idea of the dynamic-programming approach is to regard a control process of the type described, say, in Sec. 8.2, as a multistage decision process of continuous type. If we write

$$f(c,T) = \max \int_0^T h(u,v)\, dt$$

then the principle of optimality[5,9] yields the nonlinear partial differential equation

$$\frac{\partial f}{\partial T} = \max_v \left[ h(c,v) + g(c,v) \frac{\partial f}{\partial c} \right] \qquad T > 0 \quad f(c,0) = 0$$

If a constraint such as $|v| \leq k$ is added, then this becomes

$$\frac{\partial f}{\partial T} = \max_{|v| \leq k} \left[ h(c,v) + g(c,v) \frac{\partial f}{\partial c} \right] \tag{8.5}$$

We can use Eq. (8.5) for computational purposes, invoking standard difference techniques, or we can use the approximation equation

$$f(c, T + \Delta) = \max_v \left[ h(c,v)\Delta + f(c + g(c,v)\Delta,\, T) \right]$$

arising from a discrete version of the original process.

A discussion of the computational aspects will be found in the author's book on adaptive control processes.[19]

The advantage of this approach is that it handles linearities or nonlinearities, constraints or not, implicit or explicit functionals, in precisely the same manner, and with the same basic equation. The disadvantages arise from the requirement that the computer tabulate the function

$f(c,T)$. For the case of scalar variational problems, this is easily done. But in the general case, we face a function of the $n$ variables $c_i$, the components of the vector $\mathbf{c}$.

With present computers, this limits routine solution to variational problems in which the dimension $n$ of $\mathbf{x}$ is at most 2; and it limits feasible solution by various coarsenings of grid, and so on, to cases in which the dimension is at most 4. For problems of higher dimensions, we must use analytic devices,[6,12] the method of successive approximations,[13] and techniques of analytic representation.[24]

## 8.9 Trajectories

A very interesting type of control process arises from the determination of optimal trajectories for space rockets and satellites. A great deal of effort has been devoted to this field, using both conventional variational techniques[47,49,57] and dynamic-programming techniques.[37,42]

Generalized trajectory problems arise in a number of different fields. Abstractly, we wish to transform a system from one point in phase space to another point in phase space at minimum cost. Here, the cost may be measured in terms of time, fuel, or other quantities.

Particular versions of this general problem are the actual trajectory problems, the bang-bang control problem mentioned above, shortest-route problems,[14,31] some problems in chemical engineering,[2] and a variety of problems in the field of sequential machines.[18,30] All of these can be treated in a uniform fashion by means of the theory of dynamic programming.

## 8.10 Computational Aspects

Those readers who are interested in the actual numerical solutions of variational problems by means of functional-equation techniques may wish to refer to the author's book in collaboration with S. Dreyfus[23] and to Refs. 22 and 42.

It is interesting to note that even in cases of linear equations and quadratic criteria, for which the variational equations are linear and hence capable of being solved explicitly, the functional-equation approach may still yield a simpler and quicker computational treatment; see Refs. 10 and 11.

# STOCHASTIC CONTROL PROCESSES AND GAME THEORY

## 8.11 Stochastic Effects

A mathematical theory of stochastic processes can be introduced either on the grounds of expediency, as a mathematical device for circumventing analytic or conceptual difficulties, or on the grounds of realism, as a more accurate picture of the universe. The reader can take his choice of the philosophical basis he prefers, since the mathematical consequences are the same.

At the present time, although there is great interest in stochastic control processes, the problems that have been treated fall into only two categories, those specified by linear equations and quadratic criteria and those that have been approached by means of dynamic programming.

In place of the deterministic differential equations governing the process, we have stochastic equations of the form

$$\frac{du}{dt} = g(u, v, r(t)) \qquad u(0) = c$$

where $r(t)$ is a random variable with given distribution. In place of maximizing a prescribed functional, we maximize the expected value of a functional of the form

$$J(v) = \int_0^T h(u, v, r(t))\, dt$$

Since the mathematical theory of variational problems of this nature is in its infancy, it is well to begin with discrete versions that can be treated rigorously.

For the case of linear equations and quadratic criteria, however, a well-established theory exists; it is discussed in a number of readily available books.[3,63]

For the treatment of more general stochastic control processes by means of dynamic-programming techniques, see Ref. 15.

## 8.12 Games against Nature

Game theory enters into the consideration of control processes in several different ways: as a substitute for knowledge, in the study of pursuit processes, and as an analytic technique for treating some important classes of variational problems. As in the case of stochastic processes, the theory has barely been conceived and many fascinating and significant problems are as yet untouched. A number of interesting results may be found in Ref. 64.

Let us introduce the concept of a "game against Nature," to begin our discussion. One of the objections that can be leveled against the theory of stochastic control processes, as outlined in the preceding section, is that the distribution of the random function may not be known. This is a question we shall examine in much greater detail in the following part on adaptive control processes. Here, we overcome the foregoing objection by supposing that the distribution function for $r(t)$ will be chosen by someone completely opposed to our interests. This "someone" we call Nature. In this way, the mathematical theory of games enters the theory of control processes.

There are now two approaches, as in the theory of deterministic control processes, one based on an extension of the classical calculus of variations,[38,43,64] and one based on the theory of dynamic programming. A game against Nature of this type is viewed as a multistage game, and accordingly it is susceptible to the functional-equation approach.[7,20,35,58,59]

All aspects of the problem are now very much more difficult. This is the price that we must pay for ignorance of the precise disturbing effects in the system that we wish to control.

### 8.13 Pursuit Processes

In various applications of control processes, we encounter situations in which one person or object is chasing another person or object. One is trying to effect capture, and the other is trying to evade capture. These processes may quite legitimately be regarded as two-person games of multistage type, and a great deal of effort has been devoted to their study. The problems are quite difficult to pin down precisely, and most of the significant problems remain not only open, but not even precisely formulated.[32,50]

### 8.14 Analytic Techniques

It sometimes happens that variational problems that are originally maximization or minimization problems of rather complex type become quite simple when viewed as min-max problems. They can then be regarded as arising from the theory of games. Not only does this simplify our solution of the problem, but it enormously simplifies the proof of the extremum property. The technique is used in Ref. 28, and the idea of replacing an ordinary equation by one containing a maximization or minimization in order to obtain a stronger foothold is extensively discussed in Refs. 4 and 52.

## ADAPTIVE CONTROL PROCESSES

### 8.15 Adaptive Systems

Let us now, again quite briefly, glance into one of the newest and most exciting domains of science: the theory of adaptive control processes. In treating stochastic control processes and multistage games, we have skirted the unknown; in discussing adaptive processes, we must boldly grapple with it.

The general problem that faces us is that of exerting control in some efficient way in situations in which we do not possess enough information for a classical formulation. This can occur if we do not know *a.* The distribution of the random variables involved

*b.* The analytic form of the function $g(u, v, r(t))$ appearing in the describing equation

*c.* The state of the system at any particular time

*d.* The analytic form of the criterion function $h(u, v, r(t))$

*e.* The duration of the process

The type of process that confronts us is then a learning process, in which we must obtain the missing information, to the best of our ability, as the control process unravels. The decisions at any stage may then consist of control operations and information probes.

The first theory of this type to be developed, from a statistical and economic background, was the theory of sequential analysis,[65,66] although the first analytic efforts were in the field of biological control processes.[61,62] For subsequent papers in this field, see Refs. 8 and 51. From the psychological side, the subject has been extensively studied; see Ref. 40, where many references are given. For a discussion of some experimental work, see Refs. 44 and 45.

In the following section, we shall present an approach to these problems based on dynamic-programming ideas.

## 8.16 Functional-equation Approach

We now characterize the state of a system in which uncertainties occur in terms of two quantities:

$$\begin{aligned} p &= \text{the state in phase space} \\ P &= \text{the information pattern} \end{aligned}$$

The problem of making precise what we mean by the information pattern is a formidable one, and it is one of the principal obstacles to progress. One of the aspects of these problems that is so frustrating is that the very formulation of a problem is a task of sometimes major nature. The concept is best elucidated by particular examples. Thus, $P$ may consist of a priori estimates for the distribution function of a random variable, or of parameters specifying the particular distribution function; or it may merely be a catalogue of past events of importance.

Our assumption will be that a decision modifies $p$ and $P$ in known fashion, dependent on the actual outcome of the decision. Thus

$$\begin{aligned} p_1 &= T_1(p,P;\mathbf{y};r) \\ P_1 &= T_2(p,P;\mathbf{y};r;p_1) \end{aligned}$$

Here $r$ represents a random variable, which we suppose to be governed by our a priori probability distributions. Thus the distribution function for $r$ has the form

$$dG(r;p,P)$$

Let us now consider a typical terminal-control problem in which, starting in state $(p,P)$, we wish to minimize the expected value of $\phi(p_N)$. Write

$$f_N(p,P) = \min [E\ \phi(P_N)]$$

where $E$ denotes expected value.

The expected value E is to be computed according to the a priori probability distributions. Then, by the principle of optimality,[5,9] the recurrence relationship we obtain is

$$f_N(p,P) = \min_{y} [\mathrm{E}_r\, f_{N-1}(T_1(p,P;y,r),T_2(p,P;y,r;p_1))]$$

for $N = 2, 3, \ldots$, with

$$f_1(p,P) = \min_{y} [\textstyle\int \phi(p)\ dG(r,p:P)] \tag{8.6}$$

## 8.17 Computational Aspects

The foregoing relationships provide a conceptual basis for a theory of adaptive control processes. More detailed discussions, and applications to feedback control, communication processes, and medical testing, will be found in Refs. 1, 16, 33, 34, 46, and 8.

A successful computational approach is dependent on characterizing $P$ by means of numbers rather than functions. For this, one can draw upon a number of devices used by statisticians, devices that may be lumped under the heading of "sufficient statistics." Applications of these ideas are given in the references cited in the first paragraph of this section.

# AN ILLUSTRATIVE EXAMPLE

## 8.18 Formulation

In order to make some of the foregoing concepts precise, let us consider a particular control problem in three versions: deterministic, stochastic, and adaptive. As our differential equation, we shall take the van der Pol equation,

$$u'' + \lambda(u^2 - 1)u' + u = r(t) \qquad u(0) = c_1 \qquad u'(0) = c_2 \tag{8.7}$$

where $r(t)$ is a forcing term. For the purposes of digital computation, we shall use a discrete version of this equation, namely,

$$\begin{aligned} u_{n+1} &= u_n + v_n\Delta \qquad u_0 = c_1 \\ v_{n+1} &= v_n + \Delta[r_n - \lambda(u_n^2 - 1)v_n - u_n] \qquad v_0 = c_2 \end{aligned} \tag{8.8}$$

obtained from the system equivalent to Eq. (8.7),

$$\begin{aligned} u' &= v \qquad u(0) = c_1 \\ v' &= r(t) - \lambda(u^2 - 1)v - u \qquad v(0) = c_2 \end{aligned} \tag{8.9}$$

We shall suppose that our aim in introducing feedback control is to keep the system ruled by Eq. (8.7) as close to the equilibrium state

$$u = u' = 0$$

as possible over a time interval $[0,T]$. For the sake of simplicity, we shall use the criterion

$$\int_0^T (|u| + |u'|)\, dt \tag{8.10}$$

as a measure of the deviation from equilibrium, or, in discrete terms, the criterion

$$\sum_{n=0}^{N} (|u_n| + |v_n|) \tag{8.11}$$

Let us represent the effect of feedback control by a forcing term $w_n$, so that Eq. (8.8) takes the form

$$\begin{aligned} u_{n+1} &= u_n + v_n\Delta \qquad u_0 = c_1 \\ v_{n+1} &= v_n + \Delta[r_n - \lambda(u_n^2 - 1)v_n - u_n + w_n] \qquad v_0 = c_2 \end{aligned} \tag{8.12}$$

The cost of control will be measured by a function $g(w_n)$, so that the total measure of the control process will be given by the function

$$J_N(w) = \sum_{n=0}^{N} [|u_n| + |v_n| + g(w_n)] \tag{8.13}$$

## 8.19 Deterministic Case

Let us take $r_n$ to be independent of $n$, for the sake of simplicity, and suppose that we wish to choose the $w_n$ to minimize the function $J_N(w)$ appearing above. Write

$$f_N(c_1,c_2) = \min_w J_N(w) \tag{8.14}$$

Then we readily obtain the recurrence relationship

$$\begin{aligned} f_N(c_1,c_2) = \min_{w_0} (&|c_1| + |c_2| + g(w_0) \\ &+ f_{N-1}\{c_1 + c_2\Delta,\ c_2 + \Delta[r - \lambda(c_1^2 - 1)c_2 - c_1 + w_0]\}) \end{aligned} \tag{8.15}$$

for $N \geq 1$, with

$$f_0(c_1,c_2) = |c_1| + |c_2| \tag{8.16}$$

## 8.20 Stochastic Case

Let us now consider the case in which each $r_n$ is a random variable drawn from a common distribution $dG(r)$. We then wish to choose the

$w_n$, in sequential fashion, to minimize the expected value of $J_N(w)$. Write

$$f_N(c_1,c_2) = \min_w E_r\, J_N(w) \tag{8.17}$$

Then the recurrence relationship is very much as above,

$$f_N(c_1,c_2) = \min_{w_0} \Big( |c_1| + |c_2| + g(w_0) + \int f_{N-1}\{c_1 + c_2\Delta,\, c_2 + \Delta[r - \lambda(c_1^2 - 1)c_2 - c_1 + w_0]\}\, dG(r) \Big) \tag{8.18}$$

for $N \geq 1$, with

$$f_0(c_1,c_2) = |c_1| + |c_2| \tag{8.19}$$

For computational purposes, we replace $dG(r)$ by a discrete probability distribution $(p_1,p_2, \ldots ,p_M)$. For example, if each $r_n$ can assume only two values, 0 and 1, let $p_1$ be the probability that $r_n = 0$ and $p_2$ the probability that it is 1. Then Eq. (8.18) takes the form

$$\begin{aligned} f_N(c_1,c_2) = \min_{w_0} \Big( &|c_1| + |c_2| + g(w_0) \\ &+ p_1 f_{N-1}\{c_1 + c_2\Delta,\, c_2 + \Delta[-\lambda(c_1^2 - 1)c_2 - c_1 + w_0]\} \\ &+ p_2 f_{N-1}\{c_1 + c_2\Delta,\, c_2 + \Delta[1 - \lambda(c_1^2 - 1)c_2 - c_1 + w_0]\}\Big) \end{aligned} \tag{8.20}$$

## 8.21 Adaptive Case

Consider now the case in which $p_1$ is not known, but in which we do have an a priori estimate for the distribution function for $p_1$, $dH(p_1)$. We agree to transform $dH(p_1)$ into

$$dH_{m,n}(p_1) = \frac{p_1^m(1 - p_1)^n\, dH(p_1)}{\int_0^1 p_1^m(1 - p_1)^n\, dH(p_1)} \tag{8.21}$$

if we observe $m$ zero values and $n$ unit values for the $r_k$ over the previous $m + n$ stages. Write

$$\begin{aligned} p_{m,n} &= \frac{\int_0^1 p_1^{m+1}(1 - p_1)^n\, dH(p_1)}{\int_0^1 p_1^m(1 - p_1)^n\, dH(p_1)} \\ &= \int_0^1 p_1\, dH_{m,n}(p_1) \end{aligned} \tag{8.22}$$

the expected probability at the end of $m + n$ stages.

The theory of sufficient statistics tells us that it is not necessary to keep track of the actual sequence of 0s and 1s, merely the number of each. Let us then introduce the function

$f_N(c_1,c_2,m,n)$ = the minimum, over sequential choice of the $w_k$, of the expected value of $J_N(w)$ over the $r_k$, having observed $m$ 0s and $n$ 1s (8.23)

Then, quite analogous to the recurrence relation of the foregoing section,

$$\begin{aligned} f_N(c_1,c_2,m,n) = \min_{w_0} \big( &|c_1| + |c_2| + g(w_0) \\ &+ p_{m,n} f_{N-1}\{c_1 + c_2\Delta,\ c_2 + \Delta[-\lambda(c_1^2 - 1)c_2 - c_1 + w_0],\ m + 1,\ n\} \\ &+ (1 - p_{m,n}) f_{N-1}\{c_1 + c_2\Delta,\ c_2 + \Delta[1 - \lambda(c_1^2 - 1)c_2 \\ &\qquad - c_1 + w_0],\ m,\ n + 1\}\big) \end{aligned} \tag{8.24}$$

for $N \geq 1$, with

$$f_0(c_1,c_2,m,n) = |c_1| + |c_2|$$

The computational difficulties are, of course, much more severe in this case, and various special techniques, which we shall not enter into here, must be employed to obtain results, even when using the largest and most versatile currently existing computers.

## REFERENCES


1. Aoki, M., "Adaptive Control for Nonlinear Systems," Ph.D. Thesis, University of California, Los Angeles, 1959.
2. Aris, G., R. Bellman, and R. Kalaba, Dynamic Programming and Chemical Process Control (to appear).
3. Battin, R. H., and J. H. Laning, "Random Processes in Automatic Control," McGraw-Hill Book Company, Inc., New York, 1956.
4. Bellman, R., Functional Equations in the Theory of Dynamic Programming—V: Positivity and Quasi-linearity, *Proc. Natl. Acad. Sci. U.S.A.*, vol. 41, pp. 743–746, 1955.
5. ———, The Theory of Dynamic Programming, chap. 11 in "Modern Mathematics for the Engineer," First Series, edited by E. F. Beckenbach, McGraw-Hill Book Company, Inc., New York, 1956.
6. ———, Dynamic Programming and Lagrange Multipliers, *Proc. Natl. Acad. Sci. U.S.A.*, vol. 42, pp. 767–769, 1956.
7. ———, Functional Equations in the Theory of Dynamic Programming—III, *Rend. Circ. Mat. Palermo*, ser. 2, vol. 5, pp. 1–23, 1956.
8. ———, A Problem in the Sequential Design of Experiments, *Sankhyā*, vol. 16, pp. 221–229, 1956.
9. ———, "Dynamic Programming," Princeton University Press, Princeton, N.J., 1957.
10. ———, On Some Applications of Dynamic Programming to Matrix Theory, *Illinois J. Math.*, vol. 1, pp. 297–301, 1957.
11. ———, On the Computational Solution of Linear Programming Problems Involving Almost Block Diagonal Matrices, *Management Sci.*, vol. 3, pp. 403–406, 1957.
12. ———, Some New Techniques in the Dynamic Programming Solution of Variational Problems, *Quart. Appl. Math.*, vol. 16, pp. 295–305, 1958.
13. ———, Dynamic Programming, Successive Approximations and Monotone Convergence, *Proc. Natl. Acad. Sci. U.S.A.*, vol. 44, pp. 578–580, 1958.
14. ———, On a Routing Problem, *Quart. Appl. Math.*, vol. 16, pp. 87–90, 1958.
15. ———, Dynamic Programming and Stochastic Control Processes, *Information and Control*, vol. 1, pp. 228–239, 1958.
16. ———, Note on Matrix Theory—Multiplicative Properties from Additive Properties, *Amer. Math. Monthly*, vol. 65, pp. 693–694, 1958.

17. ———, Dynamic Programming, Invariant Imbedding and Two-point Boundary-value Problems, *Proc. Symposium on Two-point Boundary-value Problems*, University of Wisconsin, April, 1959.
18. ———, "Sequential Machines, Ambiguity and Dynamic Programming," Paper P-1597, The RAND Corporation, Santa Monica, Calif., 1959.
19. ———, "Adaptive Control Processes: A Guided Tour," Princeton University Press, Princeton, N.J., 1960.
20. ——— and D. Blackwell, "On a Particular Non-zero-sum Game," Research Memorandum RM-250, The RAND Corporation, Santa Monica, Calif., 1949.
21. ——— and P. Brock, On the Concepts of a Problem and Problem-solving, *Amer. Math. Monthly*, vol. 67, pp. 119–134, 1960.
22. ——— and S. Dreyfus, On the Computational Solution of Dynamic Programming Processes—I: On a Tactical Air Warfare Model of Mengel, *J. Operations Res. Soc. Amer.*, vol. 6, pp. 65–78, 1958.
23. ——— and ———, Computational Aspects of Dynamic Programming (to appear).
24. ——— and ———, "Functional Approximations and Dynamic Programming," Paper P-1176, The RAND Corporation, Santa Monica, Calif., 1959.
25. ———, W. Fleming, and D. V. Widder, Variational Problems with Constraints, *Ann. Mat. Pura Appl.*, ser. 4, vol. 41, pp. 301–323, 1956.
26. ———, I. Glicksberg, and O. Gross, On Some Variational Problems Occurring in the Theory of Dynamic Programming, *Rend. Circ. Mat. Palermo*, ser. 2, vol. 3, pp. 1–35, 1954.
27. ———, ———, and ———, On the "Bang-bang" Control Problem, *Quart. Appl. Math.*, vol. 14, pp. 11–18, 1956.
28. ———, ———, and ———, Some Non-classical Problems in the Calculus of Variations, *Proc. Amer. Math. Soc.*, vol. 7, pp. 87–94, 1956.
29. ———, ———, and ———, "Some Aspects of the Mathematical Theory of Control Processes," Report R-313, The RAND Corporation, Santa Monica, Calif., 1958.
30. ———, J. Holland, and R. Kalaba, "On an Application of Dynamic Programming to the Synthesis of Logical Systems," Paper P-1551, The RAND Corporation, Santa Monica, Calif., 1958.
31. ——— and R. Kalaba, "On $k$th Best Policies," Paper P-1417, The RAND Corporation, Santa Monica, Calif., 1958.
32. ——— and ———, "A Mathematical Theory of Adaptive Control Processes," Paper P-1699, The RAND Corporation, Santa Monica, Calif., 1959.
33. ——— and ———, On Adaptive Control Processes, *IRE National Convention Record*, 1959.
34. ——— and ———, Dynamic Programming and Adaptive Processes—I: Mathematical Foundation, *Proc. IRE* (to appear).
35. ——— and J. P. LaSalle, "On Non-zero-sum Games and Stochastic Processes," Research Memorandum RM-212, The RAND Corporation, Santa Monica, Calif., 1949.
36. ——— and S. Lehman, Studies in Bottleneck Processes (unpublished).
37. ——— and J. M. Richardson, "On the Application of Dynamic Programming to a Class of Implicit Variational Problems," Paper P-1374, The RAND Corporation, Santa Monica, Calif., 1958.
38. Berkovitz, L. D., and W. H. Fleming, "On Differential Games with Integral Payoff," Paper P-717, The RAND Corporation, Santa Monica, Calif., 1955.
39. Bohnenblust, H. Frederic, The Theory of Games, chap. 9 in "Modern Mathematics for the Engineer," First Series, edited by E. F. Beckenbach, McGraw-Hill Book Company, Inc., New York, 1956.

40. Bush, R. R., and F. Mosteller, "Stochastic Models for Learning," John Wiley & Sons, Inc., New York, 1955.
41. Bushaw, D. W., "Differential Equations with a Discontinuous Forcing Term, Experimental Towing Tank," Ph.D. Thesis, Princeton University, 1952, Stevens Institute of Technology Report No. 469, 1953; Optimal Discontinuous Forcing Terms, in "Contributions to the Theory of Nonlinear Oscillations," vol. 4, edited by S. Lefschetz, Annals of Mathematics Studies, study 24, Princeton University Press, Princeton, N.J., 1958.
42. Cartaino, T. F., and S. Dreyfus, "Applications of Dynamic Programming to the Airplane Minimum Time-to-climb Problem," Paper P-834, The RAND Corporation, Santa Monica, Calif., 1956.
43. Fleming, W. H., "On a Class of Games over Function Space and Related Variational Problems," Paper P-405, The RAND Corporation, Santa Monica, Calif., 1953.
44. Flood, M. M., "On Game-learning Theory and Some Decision-making Experiments," Paper P-346, The RAND Corporation, Santa Monica, Calif., 1952.
45. ———, "On Stochastic Learning Theory," Paper P-353, The RAND Corporation, Santa Monica, Calif., 1952.
46. Freimer, M., "A Dynamic Programming Approach to Adaptive Control Processes," Lincoln Laboratories, Cambridge, Mass., 1959.
47. Fried, B. D., General Formulation of Powered Flight Optimization Problems, *J. Appl. Physics*, vol. 29, pp. 1203–1209, 1958.
48. Gamkrelidze, R. V., Theory of Time-optimal Processes for Linear Systems, Russian, *Izv. Akad. Nauk SSSR*, Ser. Mat., vol. 22, pp. 449–474, 1958.
49. Hestenes, M. R., Elements of the Calculus of Variations, chap. 4 in "Modern Mathematics for the Engineer," First Series, edited by E. F. Beckenbach, McGraw-Hill Book Company, Inc., New York, 1956.
50. Isaacs, R. P., "Differential Games I: Introduction," Research Memorandum RM-1391; "Differential Games II: The Definition and Formulation," Research Memorandum RM-1399; "Differential Games III: The Basic Principles of the Solution Process," Research Memorandum RM-1411; "Differential Games IV: Mainly Examples," Research Memorandum RM-1486; The RAND Corporation, Santa Monica, Calif., 1954–1955.
51. Johnson, S. M., and S. Karlin, "A Bayes Model in Sequential Design," Paper P-328, The RAND Corporation, Santa Monica, Calif., 1954.
52. Kalaba, R., On Nonlinear Differential Equations, the Maximum Operation, and Monotone Convergence, *J. Math. Mech.*, vol. 8, pp. 519–574, 1959.
53. Krasovskii, N. N., Concerning the Theory of Optimum Control, Russian, *Avtomat. i Telemeh.*, vol. 18, pp. 960–970, 1957.
54. LaSalle, J. P., Abstract 247t, *Bull. Amer. Math. Soc.*, vol. 60, p. 154, 1954; "Study of the Basic Principle Underlying the Bang-bang Servo," Report GER-5518, Goodyear Aircraft Corporation, 1953.
55. ———, Time Optimal Control Systems, *Proc. Natl. Acad. Sci. U.S.A.*, vol. 45, pp. 573–577, 1959.
56. Lehman, S., "On the Continuous Simplex Method," Research Memorandum RM-1386, The RAND Corporation, Santa Monica, Calif., 1954.
57. Miele, A., and J. O. Cappellari, "Topics in Dynamic Programming," Purdue University, Lafayette, Ind., 1958.
58. Milnor, J. W., and L. S. Shapley, "On Games of Survival," Paper P-622, The RAND Corporation, Santa Monica, Calif., 1955.
59. Peisakoff, M. P., "More on Games of Survival," Research Memorandum RM-884, The RAND Corporation, Santa Monica, Calif., 1952.

60. Robbins, H., Some Aspects of the Sequential Design of Experiments, *Bull. Amer. Math. Soc.*, vol. 58, pp. 527–535, 1952.
61. Thompson, W. R., On the Likelihood That One Unknown Probability Exceeds Another in View of the Evidence of Two Samples, *Biometrika*, vol. 25, pp. 285–294, 1933.
62. ———, On the Theory of Apportionment, *Amer. J. Math.*, vol. 57, pp. 450–456, 1935.
63. Truxal, J. G., "Automatic Feedback Control System Synthesis," McGraw-Hill Book Company, Inc., New York, 1955.
64. Tucker, A. W., et al. (eds.), "Contributions to the Theory of Games," vols. 1–4, Annals of Mathematics Studies, studies 24, 28, 39, and 40, Princeton University Press, Princeton, N.J., 1950–1959.
65. Wald, A., "Sequential Analysis," John Wiley & Sons, Inc., New York, 1947.
66. ———, "Statistical Decision Functions," John Wiley & Sons, Inc., New York, 1950.

# 9

# Formulating and Solving Linear Programs

GEORGE B. DANTZIG
RESEARCH MATHEMATICIAN
THE RAND CORPORATION

## 9.1 Introduction

One of the reasons why the programming tool has assumed importance, both in industry and in the military establishment, is that *it is a method for studying the behavior of systems.*† In philosophy, it is close to what some describe as the distinguishing feature of management science or operations research,[5] to wit: "Operations are considered as an entity. The subject matter studied is not the equipment used, nor the morale of the participants, nor the physical properties of the output, it is the combination of these in total as an economic process."[4]

To many, the term "linear programming" refers to mathematical methods for solving linear inequality systems. While the solution of such systems may be the *central* mathematical problem, it is not the definition of linear programming. Linear programming is a technique for building a model describing the interrelations of the components of a system. As such, it is probably the simplest mathematical model that can be constructed that is of any value for broad programming problems of industry and government. Thus the importance of the linear-programming model is that it has wide applicability.

## 9.2 Formulating a Linear-programming Model

Suppose that the system under study, which may be one actually in existence or one that we wish to design, is a complex of machines, people, facilities, and supplies. There are certain over-all reasons for its exist-

† Material of the first four sections of this chapter is drawn from Refs. 1 and 3.

ence. For the military, the purpose may be to provide a striking force; for industry, it may be to make certain types of products.

The linear-programming approach is first to consider the entire system as decomposable into a number of elementary functions called *activities;* each type of activity is abstracted to be a kind of "black box" into which flow tangible things called *items,* such as supply and money, and out of which may flow other items such as the products of manufacture or trained crews for the military. What goes on inside the "box" is the concern of the engineer or the educator; but to the programmer, only the rates of flow in and out are of interest.

The next step in building a model is to select some unit for measuring the quantity of each activity. For a production-type activity, it is natural to measure the quantity of the activity by the amount of some product produced by it. This quantity is called the *activity level.* To increase the activity level, it will be necessary, of course, to increase the flows into and out of the activity.

*In the linear-programming model, the quantities of flow of various items into and out of the activity are always proportional to the activity level.*

*While any positive multiple of an activity is possible, negative quantities of activities are not possible.*

On the basis of these two assumptions, we see that it is sufficient to know the flows for the *unit activity level.* If we wish to *double the activity level,* we simply *double* all the corresponding flows for the unit activity level.

One of the items in our system is regarded as *precious,* in the sense that the total quantity of it that is produced by the system measures the costs. The contribution of each activity to the total cost is the amount of the precious item that flows into or out of each activity. Thus if the objective is to minimize total cost, activities that *require money* contribute positively and those that *produce money* contribute negatively to the total cost.

Finally, it is required that the system of activities be *complete,* in the sense that *a complete accounting of each item can be made by the system of activities.* To be precise, for each item it is required that the total amount on hand is equal to the amount flowing into the various activities minus the amount flowing out. Thus, in our abstract system, each item is characterized by a *material balance equation,* the various terms of which represent the flows into or out of the various activities.

To review, the basic assumptions underlying the linear-programming model are:

*a.* Input-output flows are *proportional* to the activity level.
*b.* Activity levels are *nonnegative.*

*c*. Item flows are *additive*.

*d*. The net flow of one selected item, the linear objective function, is to be minimized.

## 9.3 Building the Model

We consider now the following steps in *model building*.

*a*. Decompose the entire system under study into its elementary functions, the *activities*, and choose a unit for each activity in terms of which its quantity, or *level*, can be measured.

*b*. Determine the classes of objects, the *items*, that are consumed or produced by the activities and choose a unit for measuring each item. Select one item such that the net quantity of it that is required by the system as a whole measures the *cost* (or such that its negative measures the *profit*) of the entire system.

*c*. Determine the quantity of each item consumed or produced by the operation of each activity at its unit level. These numbers, the *input-output coefficients*, are the factors of proportionality between activity levels and item flows.

*d*. Determine the *exogenous flows*—the net inputs or outputs of the items between the system, taken as a whole, and the outside.

*e*. Assign unknown *nonnegative* activity levels $x_1, x_2, \ldots$ to all the activities; then, for each item, write the *material balance equation*, which asserts that the algebraic sum of the flows of that item into each activity, given as the product of the activity level by the appropriate input-output coefficient, is equal to the exogenous flow of the item.

The result of the model building is thus the collection of mathematical relationships characterizing all the feasible programs of the system. This collection is the *linear-programming model.*

Once the model has been built, the *programming problem* has been formulated in mathematical terms. The solution of the mathematical problem is then the ultimate aim of the particular application of linear programming:

The Linear-programming Problem

For all the activities of the system, determine levels that

*a*. Are nonnegative

*b*. Satisfy the material balance equations

*c*. Collectively minimize the total cost

## 9.4 The Linear-programming Model Illustrated

To illustrate the foregoing principles of the linear-programming approach to model building, let us turn to an application in the petroleum industry, where linear-programming methods have been very successful.

The complicated piece of plumbing of Fig. 9.1 is a simplified flow diagram of an oil refinery.† The problem facing management is this: By turning valves, setting temperatures and pressures, and starting pumps, crude oil is drawn from one or several oil fields under the control of the

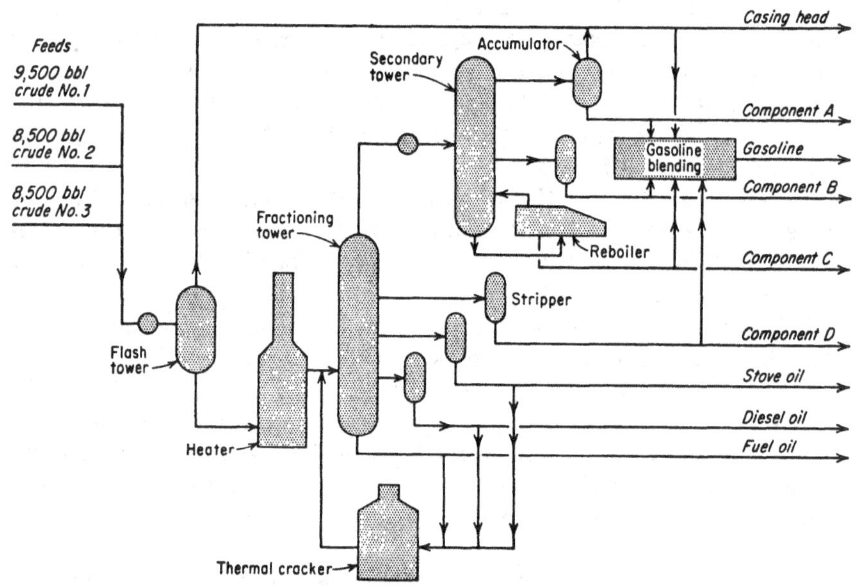


**Fig. 9.1** Refinery flow.

refinery, as shown on the left in the figure. As in the song about the music, the crude oil "goes 'round and 'round" and comes out as several streams of pure oils, shown on the right. The latter can be marketed at varying prices. By changes in the controls, the quantities in the various streams of pure oils can be altered. This will change both the costs of operating the equipment and the revenues from the sales of the final products. The various components are interrelated, however, in such a complicated manner that it is not obvious what is the *best way* to operate the equipment to *maximize profits*. In spite of these complex interrelationships, when this system is decomposed into elementary functions as the first step in building a model, it turns out that there are essentially

† This refinery example was taken from a term paper of R. J. Ullman.

only three main kinds of activities taking place: distilling, cracking, and blending.

*a. Distilling.* The net effect of the flash tower, heater, fractionating towers, strippers, etc., is to separate the crude into the varying amounts of pure oils of which it is composed. Crudes drawn from different oil fields will have different decompositions. Hence, there must be a separate distillation activity developed for each type of crude. The maximum amount of crude that can be distilled depends on which one of the pieces

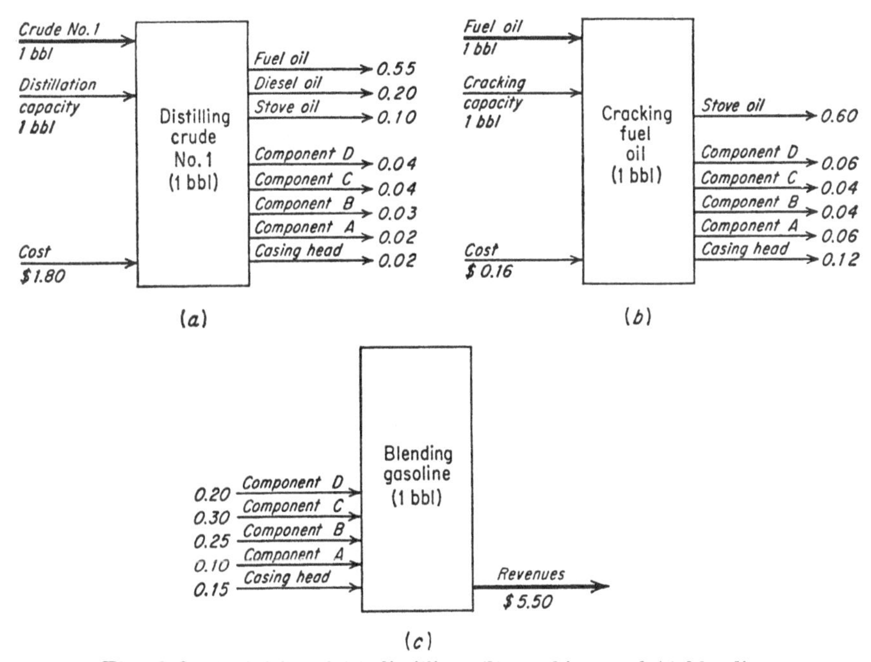


**Fig. 9.2** Activities of (*a*) distilling, (*b*) cracking, and (*c*) blending.

of equipment it passes through is the bottleneck. In our case, let us suppose that it is the heater and that it has a fixed capacity of 14,000 bbl/day independent of the type of crude processed.

From this description it is evident that, if the level of distillation activity is measured in number of barrels of crude input, then a unit level of activity can be pictured as in Fig. 9.2*a*. It is seen that 1 bbl of crude 1 will use 1 bbl of distillation capacity and will cost $1.80 to purchase and to distill; the outputs will be streams of pure oils in the amounts shown. These outputs are principally the heavier oils—fuel, diesel, and stove—and smaller amounts of the lighter types used in making gasoline. If instead of 1 bbl, it is desired to distill 10 or $x$ bbl of crude, all input and output quantities of Fig. 9.2*a* would have to be multiplied by 10 or $x$.

*b. Cracking.* The net effect of the cracking equipment is to take one of the heavier types of oil and to cause it to be broken down into lighter types of oil. In the case of fuel oil, it will produce a small amount of the lighter types and a larger amount of stove oil; the latter, if desired, can in turn be recycled back into the cracker and made into lighter oils. It is seen from Fig. 9.2*b* that 1 unit of fuel oil requires 1 unit of cracking capacity, will cost $0.16, and will produce pure oils in the amounts shown on the right. A separate type of activity must be set up for cracking of fuel, diesel, and stove oils.

*c. Blending.* Gasoline is not a pure oil but is a blend of several of the lighter types of pure oil; see Fig. 9.2*c*. It will be noted that the only output shown is the *net revenue* from marketing 1 bbl of gasoline. The latter is assumed to be the sales price at the refinery less the cost of the blending operation.

Once the flows for these major activities have been determined on a per-barrel basis, it is a simple matter to set up the linear-programming model by means of which the managers can determine the best manner of operating the refinery in order to maximize profits. In Table 9.1, each row represents an activity. The input and output quantities per unit level of activity are shown in the row; to distinguish outputs from inputs, outputs are shown with a minus sign. For example, in Table 9.1 the data of Fig. 9.2*a* are shown in the row captioned "Distilling—Crude 1"; the data of Fig. 9.2*b* are denoted "Cracking—Fuel Oil"; the data of Fig. 9.2*c* are labeled "Marketing—Gasoline." The other activity rows are self-explanatory. The amounts of various items available to the system are shown at the bottom of the table.

The unknown activity levels to be determined are denoted by $x_1, x_2, \ldots, x_{20}$. When these unknowns are multiplied by the corresponding numbers found in any column and the terms are summed, the total obtained should be equal to the availability shown at the bottom.

For example, the first material-balance equation reads

$$1x_1 + 1x_4 = 9{,}500$$

which means the amount of crude 1 available, 9,500 bbl, is completely accounted for by the amount left in the ground, $x_1$, plus the amount distilled, $x_4$.

The fourth material-balance equation, referring to distillation capacity, reads simply

$$1x_4 + 1x_5 + 1x_6 + 1x_7 = 14{,}000$$

which means that the distillation capacity of 14,000 bbl is completely accounted for by the amount used in distilling the various types of crude plus any excess capacity not used.

**Table 9.1** Linear-programming Model of an Oil Refinery

| Activity ($x$) | Crude 1 | Crude 2 | Crude 3 | Distill. capac. | Fuel oil | Diesel oil | Stove oil | Crack. capac. | Comp. D | Comp. C | Comp. B | Comp. A | Casing head | Profit |
|---|---|---|---|---|---|---|---|---|---|---|---|---|---|---|
| **Unused** | | | | | | | | | | | | | | |
| Crude 1 ($x_1$) | 1 | | | | | | | | | | | | | |
| Crude 2 ($x_2$) | | 1 | | | | | | | | | | | | |
| Crude 3 ($x_3$) | | | 1 | | | | | | | | | | | |
| **Distilling** | | | | | | | | | | | | | | |
| Crude 1 ($x_4$) | 1 | | | 1 | −0.55 | −0.20 | −0.10 | | −0.04 | −0.04 | −0.03 | −0.02 | −0.02 | −1.8 |
| Crude 2 ($x_5$) | | 1 | | 1 | −0.61 | −0.12 | −0.07 | | −0.06 | −0.05 | −0.04 | −0.02 | −0.03 | −1.9 |
| Crude 3 ($x_6$) | | | 1 | 1 | −0.50 | −0.11 | −0.14 | | −0.05 | −0.08 | −0.05 | −0.03 | −0.04 | −2.0 |
| Unused capac. ($x_7$) | | | | 1 | | | | | | | | | | |
| **Cracking** | | | | | | | | | | | | | | |
| Fuel oil ($x_8$) | | | | | 1 | | −0.60 | 1 | −0.06 | −0.04 | −0.04 | −0.06 | −0.12 | −0.16 |
| Diesel oil ($x_9$) | | | | | | 1 | −0.20 | 1 | −0.28 | −0.20 | −0.04 | −0.12 | −0.16 | −0.21 |
| Stove oil ($x_{10}$) | | | | | | | 1 | 1 | −0.30 | −0.30 | −0.04 | −0.10 | −0.14 | −0.21 |
| Unused capac. ($x_{11}$) | | | | | | | | 1 | | | | | | |
| **Marketing** | | | | | | | | | | | | | | |
| Fuel oil ($x_{12}$) | | | | | 1 | | | | | | | | | 1.8 |
| Diesel oil ($x_{13}$) | | | | | | 1 | | | | | | | | 4.0 |
| Stove oil ($x_{14}$) | | | | | | | 1 | | | | | | | 4.2 |
| Gasoline ($x_{15}$) | | | | | | | | | 0.20 | 0.30 | 0.25 | 0.10 | 0.15 | 5.5 |
| Component D ($x_{16}$) | | | | | | | | | 1 | | | | | 4.0 |
| Component C ($x_{17}$) | | | | | | | | | | 1 | | | | 4.1 |
| Component B ($x_{18}$) | | | | | | | | | | | 1 | | | 4.2 |
| Component A ($x_{19}$) | | | | | | | | | | | | 1 | | 4.3 |
| Casinghead ($x_{20}$) | | | | | | | | | | | | | 1 | 3.3 |
| **Available (barrels in 1 day)** | | | | | | | | | | | | | | |
| Available | 9.500 | 8,500 | 8,500 | 14,000 | 0 | 0 | 0 | 3,500 | 0 | 0 | 0 | 0 | 0 | Maximum |

Finally, the profit equation states that the revenue obtained from marketing various products,

$$1.8x_{12} + 4.0x_{13} + 4.2x_{14} + 5.5x_{15} + 4.0x_{16} + 4.1x_{17} + 4.2x_{18} + 4.3x_{19} + 3.3x_{20}$$

less the cost of distilling and crude purchases,

$$1.8x_4 + 1.9x_5 + 2.0x_6$$

less the cost of cracking,

$$0.16x_8 + 0.21x_9 + 0.21x_{10}$$

is the amount of profit. The problem, of course, is to choose the program of activity levels in such a way that the material balance equations are satisfied and the profits are maximized.

## 9.5 Algebraic Statement of the Linear-programming Problem

The minimization of a linear form subject to linear inequality constraints has been called the central mathematical problem of linear programming.† The *standard form* for such problems, because it arises naturally in many applications, is taken to be the following: Find a solution of a system of linear equations, in nonnegative variables, that minimizes a linear form. We shall see in a moment why this particular form is chosen as standard. At the same time, we shall formalize in mathematical terms our remarks regarding linear-programming models.

If the subscript $k$ denotes the $k$th type of activity, $k = 1, 2, \ldots, n$, and $x_k$ its quantity, or *activity level*, then usually $x_k \geq 0$. If, for example, $x_k$ represents the quantity of a stockpile allocated for the $k$th use, it does not, as a rule, make sense to allocate a negative quantity. In certain cases, however, one may wish to interpret a negative quantity as meaning taking stock from the $k$th use. Here some care must be exercised; for example, there may be costs, such as transportation charges, that are positive regardless of the direction of flow of the stock. One must also be careful not to overdraw the stock of the activity. For these reasons, it is better in formulating models to distinguish two activities, each with a nonnegative range, for their respective $x_k$, rather than to try incorporating them into a single range.

The interdependencies between various activities arise because all practical programming problems are circumscribed by commodity limitations of one kind or another. The limiting commodities may be raw materials, manpower, facilities, or funds; these are referred to by the general term *item*. In chemical equilibrium problems, where *molecules* of

† Material in this section is taken from Ref. 2.

different types play the role of activities, the different kinds of *atoms* in the mixture are the items. The different types of items are denoted by a subscript $i$, where $i = 1, 2, \ldots, m$.

In linear-programming work, the quantity of an item required as input or produced as output by an activity is assumed to be *proportional* to the quantity of activity level; the coefficient of proportionality is denoted by $a_{ik}$. The sign of $a_{ik}$ depends on whether the item is required or produced by the activity. The sign convention used will be + if required and − if produced, as shown in the diagram. Finally, let $b_i$, if positive, denote the

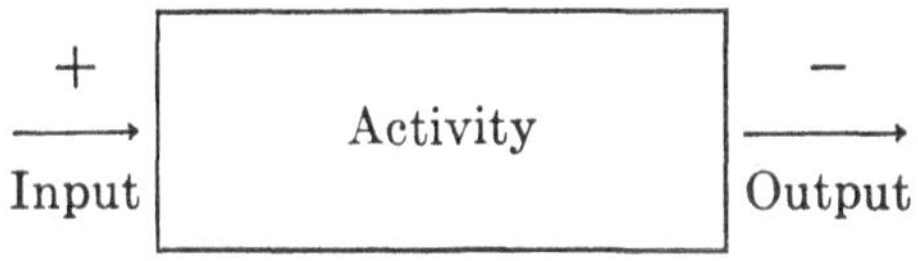

| + → Input | Activity | − → Output |
|---|---|---|

quantity of the $i$th item made available as input to the system of activities from outside (or exogenous) sources, and let $b_i$, if negative, denote the quantity required as output by the system. Then the interdependencies between the $x_k$ can be expressed as a set of $m$ linear equations; the $i$th such equation gives a complete accounting of the $i$th item. Thus,

$$\begin{aligned} a_{11}x_1 + a_{12}x_2 + \cdots + a_{1n}x_n &= b_1 \\ a_{21}x_1 + a_{22}x_2 + \cdots + a_{2n}x_n &= b_2 \\ \cdots\cdots\cdots\cdots\cdots\cdots\cdots\cdots & \\ a_{m1}x_1 + a_{m2}x_2 + \cdots + a_{mn}x_n &= b_m \end{aligned} \tag{9.1}$$

where

$$x_k \geq 0 \qquad k = 1, 2, \ldots, n \tag{9.2}$$

Any set of values $x_k$ satisfying Eqs. (9.1) and (9.2) is called a *feasible solution* because the corresponding schedule is possible or feasible.

The objective of a program, in practice, often is most difficult to express in mathematical terms. There are many historical reasons for this; they go beyond the scope of the present discussion. In many problems, however, the objective is simply one of carrying out the requirements, expressed by those $b_i$ that are negative, in such a manner *that total costs are minimized.* Costs may be measured in dollars, or in number of people involved, or in the quantity of a scarce commodity used. In linear programming, the total cost, denoted by $z$, is assumed to be a *linear function of the activity levels:*

$$c_1x_1 + c_2x_2 + \cdots + c_nx_n = z$$

The linear form $z$ is called the *objective function.*

In some problems, the linear objective form, as originally presented, is to be *maximized* rather than minimized. For example, the problem may be to produce the maximum dollar value of products under a fixed budget, fixed machine capacity, and fixed labor supply. Suppose the linear form expressing total profits to be maximized is

$$p_1x_1 + p_2x_2 + \cdots + p_nx_n$$

Maximizing this form obviously is mathematically equivalent to minimizing

$$-p_1x_1 - p_2x_2 - \cdots - p_nx_n$$

For these reasons, *the standard form of the linear-programming problem is taken to be the problem of determining a solution of a given system of linear equations in nonnegative variables in such a way as to minimize a given linear function of those variables.*

### 9.6 Outline of the Simplex Method

The standard procedure for solving the central mathematical problem of linear programming is the simplex method. It makes use of two important characteristics of solutions:

*a.* If a feasible solution exists, so that Eqs. (9.1) and (9.2) are satisfied, then one exists with at least $n - m$ variables equal to zero.

*b.* If feasible solutions exist and the objective form $z$ has a finite lower bound, then a minimal feasible solution exists with at least $n - m$ variables equal to zero.

Suppose that there is at hand a feasible solution with the property that $n - m$ variables are equal to zero, say $x_{m+1} = x_{m+2} = \cdots = x_n = 0$. We wish to test whether or not the solution is optimal. The procedure is first to eliminate each of the remaining variables, $x_1, x_2, \ldots, x_m$, from all but one of the equations. This will be possible if the determinant of the coefficients of the selected set of $m$ variables does not vanish; in this case, the selected set is referred to as a *basic set of variables.* We shall refer to this process as reduction to *canonical form.* If $x_1, x_2, \ldots, x_m$ are the variables selected for elimination, then the canonical form for the simplex method relative to these basic nonnegative variables is as follows:

$$\begin{array}{llllll}
x_1 & + \bar{a}_{1m+1}x_{m+1} & + \cdots & + \bar{a}_{1n}x_n & = \bar{b}_1 \\
\quad x_2 & + \bar{a}_{2m+2}x_{m+1} & + \cdots & + \bar{a}_{2n}x_n & = \bar{b}_2 \\
\quad\cdot & & & & \cdot \\
\quad\;\cdot & & & & \cdot \\
\quad\;\;\cdot & & & & \cdot \\
\qquad x_m & + \bar{a}_{mm+1}x_{m+1} & + \cdots & + \bar{a}_{mn}x_n & = \bar{b}_m \\
 & \quad\bar{c}_{m+1}x_{m+1} & + \cdots & + \bar{c}_nx_n & = z - \bar{z}_0
\end{array}$$

where $\bar{a}_{ik}$, $\bar{b}_i$, $\bar{c}_k$, and $\bar{z}_0$ are constants.

It is important to note that this system is *equivalent* to the original system of Eqs. (9.1) and may be used in its place for any further discussions. In this form, it is easy to obtain a particular solution to the equations and to test whether or not it is feasible and optimal.

## 9.7 Test for Optimal Feasible Solution

THEOREM 9.1. *If, in the canonical form, the values of the constant terms $\bar{b}_i$ and the coefficients $\bar{c}_k$ in the cost equation are nonnegative, then the basic solution obtained by setting nonbasic variables to zero and solving for values of basic variables, namely,*

$$\begin{aligned} x_k &= 0 \qquad \text{for } k = m+1, \ldots, n \\ x_i &= \bar{b}_i \qquad \text{for } i = 1, 2, \ldots, m \\ z &= \bar{z}_0 \end{aligned} \tag{9.3}$$

*is optimal.*

*Proof.* Since $x_k$ must be nonnegative and $\bar{c}_k \geq 0$ *whatever the feasible solution*, it follows from the last equation of the canonical form that the values of $z$ for the class of feasible solutions must always exceed $\bar{z}_0$ by the sum of the nonnegative terms on the left. Hence any particular feasible solution with $z = \bar{z}_0$, such as (9.3), would be a minimizing solution.

For example, consider the problem of finding min $z$ and nonnegative $x_1$, $x_2$, $x_3$, $x_4$ satisfying

$$\begin{aligned} \textcircled{x_1} - 2x_2 + 2x_3 - 4x_4 &= 1 \\ -x_1 + 3x_2 - 4x_3 + 5x_4 &= 2 \qquad x_k \geq 0 \\ x_1 + 2x_2 + 3x_3 + 4x_4 &= z \text{ (min)} \end{aligned}$$

Suppose we would like to check if the particular solution obtained by setting $x_3 = x_4 = 0$ and solving for $x_1$, $x_2$, $z$ is feasible and optimal. If we pivot on $x_1$ in the first equation by dividing through by its coefficient, if necessary, so that this coefficient becomes unity, and then eliminating $x_1$ from the other equations, we obtain

$$\begin{aligned} x_1 - 2x_2 + 2x_3 - 4x_4 &= 1 \\ \textcircled{x_2} - 2x_3 + x_4 &= 3 \\ 4x_2 + x_3 + 8x_4 &= z - 1 \end{aligned}$$

If next we pivot on the $x_2$ term in the second equation, eliminating $x_2$ from all the other equations including the first, we obtain the equivalent canonical form with respect to $x_1$ and $x_2$:

$$\begin{aligned} x_1 \qquad - 2x_3 - 2x_4 &= 7 \\ x_2 - 2x_3 + x_4 &= 3 \\ 9x_3 + 4x_4 &= z - 13 \end{aligned} \tag{9.4}$$

Note that Eqs. (9.4) fulfill the conditions of Theorem 9.1; hence, an optimal solution is obtained by setting the nonbasic variables $x_3$ and $x_4$

equal to 0 and solving for $x_1$, $x_2$, $z$:

$$x_1 = 7 \quad x_2 = 3 \quad x_3 = x_4 = 0 \quad z = 13$$

This solution is also *the unique optimal solution*, as will become clear by applying the following result:

THEOREM 9.2. *If in the canonical system the basic solution is feasible and the coefficients $\bar{c}_k$ in the cost form are all positive for nonbasic variables, then the solution is the unique feasible optimal solution.*

COROLLARY. *If in the canonical system the basic solution is feasible and the coefficients $\bar{c}_k$ satisfy $c_k \geq 0$ for all $k$, then no other feasible solution can also be optimal unless the values $x_k$ satisfy $x_k = 0$ whenever $\bar{c}_k > 0$.*

*Proof.* Consider another *feasible solution* and suppose that the value of some $x_k$ is positive for a positive $\bar{c}_k$. Then the value of $z$ must be *greater* than $\bar{z}_0$, since the term $\bar{c}_k x_k$ in the last equation of the canonical form satisfies $\bar{c}_k x_k > 0$. Hence, any feasible solution that is also minimal must have $x_k = 0$ for all variables with corresponding $\bar{c}_k > 0$, establishing the corollary.

Under the hypothesis of Theorem 9.2 that *all* $\bar{c}_k$ corresponding to nonbasic $x_k$ satisfy $\bar{c}_k > 0$, all nonbasic $x_k$ must satisfy $x_k = 0$, but upon substitution this leaves only the values $x_i = \bar{b}_i$ for basic variables, and accordingly $z = z_0$.

## 9.8 Improving a Nonoptimal Basic Feasible Solution

The standard simplex method is applicable only to basic solutions that are feasible; i.e., the $\bar{b}_i$ satisfy $\bar{b}_i \geq 0$. To understand the underlying principle, let us suppose that one or more coefficients $\bar{c}_k$ are negative in the canonical form for nonbasic variables. In this case, the test for optimality fails.

If all $\bar{b}_i$ satisfy $\bar{b}_i > 0$ (note that zero is excluded), we can immediately improve the basic feasible solution by increasing the value of one of the corresponding nonbasic variables $x_k$ for which $\bar{c}_k < 0$, keeping the other nonbasic variables at value zero. As a good empirical rule, we might choose to increase that variable $x_s$ for which

$$\bar{c}_s = \min \bar{c}_k < 0 \tag{9.5}$$

For example, suppose the canonical form is

$$\begin{aligned} x_1 \qquad - 2x_3 - 2x_4 &= 7 \\ x_2 - 2x_3 + \textcircled{x_4} &= 3 \\ + 9x_3 - 4x_4 &= z - 13 \end{aligned} \tag{9.6}$$

which is the same as our earlier example except that the sign of the coefficient $\bar{c}_4$ has been reversed. The basic solution formed by setting nonbasic variables $x_3$ and $x_4$ equal to 0 and solving for $x_1$, $x_2$, and $z$ yields $x_1 = 7$, $x_2 = 3$, $z = 13$, which is feasible.

Since $\bar{c}_4 = -4 = \min \bar{c}_k$, we now choose to increase $x_4$, maintaining all other nonbasic variables (in this case $x_3$) at zero value. The values of the basic variables and $z$ will now depend on the value chosen for $x_4$. For example, the value of $z$ is given by

$$z = 13 - 4x_4$$

so that the larger we take the value $x_4$, the lower will be the value $z$. On the other hand, the values of the basic variables become

$$\begin{aligned} x_1 &= 7 - (-2x_4) \\ x_2 &= 3 - (x_4) \end{aligned}$$

so that the largest value that $x_4$ can take, without destroying the feasibility of the solution, is $x_4 = 3$. Above this value, $x_2$ would be negative. By setting $x_4 = 3$, we see that the value of $z$ is lowered by $4x_4$, or 12 units, and the new solution becomes

$$x_1 = 13 \qquad x_4 = 3 \qquad x_2 = 0 \qquad x_3 = 0 \qquad z = 1$$

Notice that, at the critical value $x_4 = 3$, $x_2$ has vanished. Accordingly, *in the next iteration* we *interchange* the roles of $x_2$ and $x_4$ as basic and nonbasic variables by pivoting on the circled term $x_4$ in Eqs. (9.6) to eliminate $x_4$ from the other equations. The pivot term is always in the column of the variable to be introduced into the basic set and the row of the basic variable to be dropped (in this case $x_2$). This yields

$$\begin{aligned} x_1 + 2x_2 + 2x_3 \qquad &= 13 \\ x_2 - 2x_3 + x_4 &= 3 \\ 4x_2 + x_3 \qquad &= z - 1 \end{aligned}$$

Notice that this is now in canonical form relative to $x_1$ and $x_4$, except that we have not bothered to rearrange terms.

The new basic solution is obtained by setting the new nonbasic variables $x_2$ and $x_3$ equal to 0 and solving for the remaining variables: $x_1 = 13$, $x_4 = 3$, $z = -23$. Upon applying our test for optimality, we see that the solution is *optimal* because all $\bar{b}_i$ and $\bar{c}_k$ satisfy $\bar{b}_i \geq 0$ and $\bar{c}_k \geq 0$. Moreover, it is *unique*, since all $\bar{c}_k$ satisfy $\bar{c}_k > 0$ (not zero) for nonbasic variables.

## 9.9 General Iterative Procedure

As we have seen in the numerical example, the canonical form provides in general an immediate criterion for testing optimality of a basic feasible solution. Furthermore, if the solution is not optimal, and if all $\bar{b}_i$ satisfy

$\bar{b}_i > 0$, then it leads to another solution that reduces the value of the cost or objective function.

The new basic feasible solution can again be tested for optimality by seeing if all $\bar{c}_k$ satisfy $\bar{c}_k \geq 0$ (Theorem 9.1). If it is not optimal, then by criterion (9.5) one may choose a new variable $x_s$ to increase and proceed to construct either (1) a class of solutions in which there is no finite lower bound for $z$ (if all $\bar{a}_{is}$ satisfy $\bar{a}_{is} \leq 0$) or (2) a new basic feasible solution in which the cost $z$ is lower than the previous one (provided that the values of the basic variables for the latter are strictly positive; otherwise the new value of $z$ may be equal to the previous value).

The simplex algorithm consists in repeating this cycle again and again, terminating only when there has been constructed one of the following:

*a.* A class of feasible solutions in which $z \to -\infty$

*b.* An optimal basic feasible solution

If all $\bar{b}_i$ satisfy $\bar{b}_i > 0$ at each cycle, *the entire process will terminate in a finite number of cycles.* The reason for this is that there are only a finite number of ways of choosing a set of $m$ basic variables out of $n$ variables. If the algorithm were to continue indefinitely, it could do so only by repeating the same basic set of variables—hence the same cost $z$. The latter is prevented from happening because on each cycle the value of $z$ decreases a positive amount.

## 9.10 Finding an Initial Basic Feasible Solution

Up to the present, we have been assuming that it is possible to specify a basic set of variables that could be used to perform the necessary initial eliminations and reduce the problem to canonical form. It has also been assumed that the associated basic solution is feasible.

We shall now give a simple device that uses the simplex algorithm itself to provide (if it exists) a starting basic feasible solution. This part of the process is referred to as *Phase I.* The second part of the process, obtaining an optimal basic feasible solution, is then referred to as *Phase II.* The device has several important features that should be noted:

*a.* No assumptions are made regarding the original system; it may be redundant, inconsistent, or not solvable in nonnegative numbers.

*b.* No eliminations are required in order to obtain an initial solution in canonical form for Phase I.

*c.* The end product of Phase I is a basic feasible solution, if it exists, in canonical form ready to initiate Phase II. The procedure for Phase I is this:

*a. Arrange the original system of equations so that all constant terms $b_i$ are positive or zero by changing, where necessary, the signs on both sides of any of the equations.*

*b. Augment the system to include a basic set of "artificial" variables $x_{n+1} \geq 0$, $x_{n+2} \geq 0$, . . . , $x_{n+m} \geq 0$ so that it becomes*

$$\begin{aligned} a_{11}x_1 + a_{12}x_2 + \cdots + a_{1n}x_n + x_{n+1} \qquad\qquad &= b_1 \\ a_{21}x_1 + a_{22}x_2 + \cdots + a_{2n}x_n \qquad + x_{n+2} \qquad &= b_2 \\ a_{m1}x_1 + a_{m2}x_2 + \cdots + a_{mn}x_n \qquad\qquad + x_{n+m} &= b_m \end{aligned} \tag{9.7}$$

*where all $b_i$ satisfy $b_i \geq 0$ and for $k = 1, \ldots, n, n+1, \ldots, n+m$, all $x_k$ satisfy*

$$x_k \geq 0 \tag{9.8}$$

*c. By using the simplex algorithm, find a solution to Eqs.* (9.7) *and* (9.8) *that minimizes the sum of the artificial variables, denoted by $w$:*

$$x_{n+1} + x_{n+2} + \cdots + x_{n+m} = w$$

We call this the *infeasibility form.* Its coefficients, after elimination of basic variables, are referred to as $\bar{d}_k$; these are the analogue of $\bar{c}_k$ for the *cost form.*

*d. If min $w > 0$, then no feasible solution exists and the procedure is terminated. On the other hand, if min $w = 0$, initiate Phase II of the simplex algorithm as follows:*

*i. Drop from further consideration all nonbasic variables $x_k$ for which the corresponding coefficients $d_k$ are positive (not zero) in the final modified $w$ equation.*

*ii. Replace the linear form $w$, as modified through various eliminations, by the linear form $z$, after first eliminating from $z$ all basic variables.*

In practical computational work, the elimination of the basic variables from the $z$ form is usually done on each cycle of Phase I.

## REFERENCES


1. Dantzig, George B., "Thoughts on Linear Programming and Automation," Paper P-824, The RAND Corporation, Santa Monica, Calif., 1956.
2. ———, "The Central Mathematical Problem," Paper P-892, The RAND Corporation, Santa Monica, Calif., 1956.
3. ———, "Formulating a Linear Programming Model," Paper P-893, The RAND Corporation, Santa Monica, Calif., 1956.
4. Hermann, C. C., and J. F. Magee, Operations Research for Management, *Harvard Bus. Rev.*, July, 1953.
5. King, Gilbert W., Applied Mathematics in Operations Research, chap. 10 in "Modern Mathematics for the Engineer," First Series, edited by E. F. Beckenbach, McGraw-Hill Book Company, Inc., New York, 1956.

# 10

# The Mathematical Theory of Inventory Processes

SAMUEL KARLIN
PROFESSOR OF MATHEMATICS AND STATISTICS
STANFORD UNIVERSITY

## 10.1 Introduction

The "inventory process" can be considered as a general title for a broad class of sequential decision activities. The mathematics of sequential decision problems embodies techniques of stochastic processes, statistical inference, and decision analysis. Statistical inference is used to determine the parameters of the inventory model. The theory of stochastic processes is needed in describing the workings and the structure of the model. Finally, decision analysis is exploited to optimize preassigned measures of efficiency from among all available inventory policies. This combination underlies most sequential decision problems.

It is not essential for our purposes to offer a precise definition of the inventory process. Only through a study of typical models can one acquire a sense of the scope of the subject. In this chapter, our approach to the inventory process will be to set up and describe the analysis of a series of inventory models. The models chosen for discussion are intended to exemplify the methods of analysis, in addition to the formulation of meaningful inventory situations.

An inventory process might, for example, involve deciding how many pen-and-pencil sets to stock each quarterly period. When production is involved, the inventory process might require determining how much wheat to plant per year or how much gasoline of a certain variety to blend. How much water to release from a dam for electricity and irrigation purposes is an inventory problem; how many workers to hire for a given labor force is another. Inventory models may involve scheduling production, determining efficient distribution of commodities in certain markets, finding proper replacement policies for old equipment, determining proper prices for goods produced, or a combination of these elements.

## 10.2 Factors of the Inventory Process

In general, there are four main factors contributing to the structure of the inventory process:

*a.* Cost factors
*b.* The nature of the demand of the inventory commodity
*c.* The nature of the supply of the inventory commodity
*d.* Mechanism of the inventory process

In the next four sections, we shall briefly elaborate on the interpretation and significance of these factors. For simplicity of discussion we assume, throughout this chapter, consideration of a single commodity to be subject to inventory control.

## 10.3 Cost Factors

The selection of an inventory policy for implementation is governed mainly by consideration of relative profitability. The choice of keeping inventory is influenced by a desire to balance ordering costs, holding costs, shortage costs, etc. In the general formulation of the inventory model, seven such cost factors are recognized:

*a. The Cost of Ordering or Producing.* Various assumptions concerning the cost function $c(z)$, where $c(z)$ = the cost of ordering or producing a given amount $z$, appear reasonable in different circumstances. A common assumption is that $c(z)$ is composed of a cost proportional to the amount ordered plus a constant set-up cost independent of the amount ordered. We shall describe in Sec. 10.11 how the qualitative nature of the optimal inventory policies appears to be very sensitive to the form of the function $c(z)$.

*b. Storage Cost.* This cost is associated with the stock of inventories on hand (the cumulative excess of supply over demand). Storage or handling costs may be incurred by actual maintenance of stocks or by the renting of storage space; or, in a more generalized form, they may be a measure of obsolescence or spoilage.

*c. Discount Rate.* The discount rate can be construed as a holding cost, since money tied into inventory stock could be earning interest in government bonds or acquiring higher rates of return on other investments. The discount factor also permits us to compare future costs in terms of present costs. It expresses the time preference for money.

*d. Shortage or Penalty Costs.* These describe the costs due to insufficient stocks to meet demand. Such costs are difficult to evaluate, for they embrace concepts like essentiality, program forecast, compensability, and similar considerations. The importance of these costs as part of the inventory model was recognized only in recent years.

*e. Revenues.* These are self-evident negative costs.

*f. Costs of Changes in the Rate of Production.* In some cases, a change in the rate of production leads to costs that are different from those associated with continued production at the new rate. They are associated with the derivative of the rate of production and not with the rate itself.

*g. Salvage Cost.* This is a negative cost. It measures the value of stock left over at the end of a decision period. For most purposes, it is appropriate to assign to the remaining inventory a salvage value that reflects its utility to the future operations of the firm. In some cases, from the point of view of method, this cost may be combined with the holding cost.

## 10.4 The Nature of Demand

Inventories are held for the ultimate purpose of satisfying demand. The nature of the demand can assume different forms. If the demand in the subsequent time periods can be perfectly forecast, then we have a case of deterministic demand. We thus have a sequence of future time periods for each of which the demand is assumed known, though it will in general vary from period to period.

If the demand is not known ahead of time, it is sometimes justifiable to assume that the demand in each future period is a random variable with a known probability distribution. The natural extension of these ideas is to consider the demand flow as a general stochastic process. An even more realistic hypothesis recognizes that even the form of the stochastic process need not be known. It is usually necessary to apply procedures of statistical inference in order to determine the parameters describing the demand process. Even under the restrictive assumption that the demands in successive periods are independent identically distributed random variables, the problem of determining optimal policies is difficult and intrinsically complicated.

## 10.5 The Nature of Supply

We have tacitly assumed in the above discussion that demand is independent of the firm's control, although other assumptions are certainly reasonable in some circumstances. The situation for supply is different. In most cases, the entrepreneur can order any quantity desired. In contrast, there may exist inventory situations in which a decision to order means the delivery of a fixed quantity of goods. For example, in the petroleum industry a blend of gasoline is automatically produced by the kettleful, the reason being the high set-up cost. There are also situations, notably in agricultural production, in which a decision to order (produce) does not imply complete control over the amount delivered. The actual amount produced may be a random variable with known probability distribution function. Thus, different inventory models arise under varying hypotheses as to the nature of and possibilities for supply.

## 10.6 The Structure of the Inventory Process

*a. Discrete vs. Continuous Time.* In order to describe the inventory model, it is necessary first to specify whether decisions are to be made once in each fixed time period or whether the model is best viewed continuously in time and admits the possibility of replenishment decisions at any instant of time.

For the case in which the firm examines its inventory only at certain fixed time intervals (e.g., every 3 months), consideration of time in discrete units is well suited as an idealization of the inventory process. On the other hand, when demand arises continuously in time, then the expression of the model in terms of continuous time is the appropriate framework.

*b. Lags.* Another basic element in the formulation of the inventory model involves the nature of the lags in delivery and production. Time lags, which generally represent an inability of the process to respond instantly, occur in a number of different forms. There are situations in which the lag factor is sufficiently small that it may be disregarded. In many practical cases, however, the lag is of fundamental significance. The lag period could be deterministic or random, each appropriate under different conditions. In Sec. 10.17, we shall describe an inventory model in which the lag between placing the order and its delivery is random.

*c. The Inventory Process.* The analytical formulation of the inventory model is based on the stock-flow identities, in conjunction with a description of the natural constraints of the model. Let $x_t$ be the initial stock level in the $t$th period, $y_t$ the stock level after ordering, $z_t$ the amount ordered, and $\xi_t$ (which may be random) the amount demanded. If the commodity dealt with is not in any way perishable, then obviously

$$z_t = y_t - x_t \qquad x_{t+1} = y_t - \xi_t \tag{10.1}$$

If time is taken to be continuous, then this identity is to be replaced by

$$\frac{dy}{dt} = z(t) - \xi(t) \tag{10.2}$$

Physical considerations require that all the variables are nonnegative. This is not absolutely essential, since a negative value of $x$ can be interpreted as stock owed to consumption; $y_t - \xi_t$ is then a shortage, which is backlogged for future delivery.

In the case of deterministic demand, it is usual to insist on inventory policies for which demands are constantly met, so that $y_t \geq \xi_t$. However, if the demand has a stochastic form, then it is not possible to guarantee with certainty the fulfillment of all demands. In this event, a shortage is suffered.

There are several assumptions about the firm's behavior in dealing with the shortage. One way is to order the necessary goods for immediate delivery, paying a premium over the usual price. In this case, Eq. (10.1) is modified to

$$x_{t+1} = \max\,(0, y_t - \xi_t) \tag{10.3}$$

A second possibility is that unsatisfied demand is never met, and thus again Eq. (10.3) applies. The third possibility is to permit backlogging as mentioned previously and to satisfy demand as soon as possible (again a penalty is suffered). Now Eq. (10.1) is valid.

In the description of Eq. (10.3), we tacitly assumed the presence of no lags in delivery of ordered goods. The modification of these formulas, when the lag factor is relevant, is left to the reader.

The inventory process can be represented by the process $x_t$.

## 10.7 Classification of Inventory Models

There are several types of principles that can be used as a basis for classifying inventory models.

*a.* Inventory problems can be divided into two types—deterministic and stochastic.[12] Deterministic models imply full knowledge of the future demands and complete control over the workings of the inventory policy effected. In this case, the problem of finding the optimal policy reduces to a problem recognized as one in the calculus of variations. The standard methods of this discipline are not applicable, however, since the constraints do not appear in linear form. Nonnegativity and general inequality boundary conditions are typical. New approaches must be developed.

In the stochastic case (i.e., in case demand is not predictable with certainty), it is necessary, in analyzing the inventory process, to appeal to the extensive theory of stochastic processes. The present development of this aspect of the subject indicates that we are at the very beginning of the problem.

*b.* A second breakdown of inventory models is along the lines of static (one-stage) models, contrasted with dynamic models. The study of one-stage inventory models is of interest for two reasons. First, there are practical circumstances in which an inventory decision is essentially made only once (e.g., some types of military spending). Secondly, a study of one-stage inventory problems is preliminary to any detailed investigation of the dynamic models.

The justification and significance of the dynamic inventory models are obvious, since demand flow and supply flow take place in time. The essence of the analysis of the dynamic problem seems to be the exploitation of the recursive, time-dependent nature of the problem and the infinite symmetries made available by the extended horizon of time.

*c.* Inventory models may also be classified according to the method of analysis and the nature of the objective function. In most cases, with respect to a given objective, we aim to find the optimal policy from among the class of all policies. Unfortunately, many simple inventory problems lead to formidable mathematical problems, and the greater the realism and complexity inserted into the formulation of the model, the more likely the problem is unsolvable in analytic terms. The second-best alternative is to perform the optimization from among a restricted range of policies. This has been the principal point of view of the analysis in the bulk of the literature on inventory studies.

Instead of optimizing, we may turn the problem around and ask what the effects of a given inventory policy will be. We refer to this as the procedure of calculating the operating characteristics of the policy. A given policy, together with a given specification of random demands, determines a stochastic process involving inventories and production. One can study the probabilistic structure of this inventory process independently of any costs. In investigating the inventory process, one can concentrate on the stationary or equilibrium distribution of the process (if such exists) or one can direct attention to the time-dependent behavior of the process. Both kinds of studies have relevance to the analysis of inventory policies. After determination of the distribution theory of the process, it is then routine to superimpose a cost structure and to choose the parameters of the policy to minimize expected costs.

## 10.8 Historical Inventory Models

In order to place the recent developments in inventory analysis in the proper perspective, it seems appropriate to examine briefly several more well-known classical inventory models.

*a. Lot-size Law.* The most popular and mathematically the most sophisticated formula that is being used in practical control of the size of inventory is the famous "square-root law." The justification of this law is based on the following inventory model. Suppose a firm decides to stock $Q$ units of a commodity and repeatedly to replenish the stock up to the original size whenever all available stock is consumed. The demand is assumed to flow at a uniform rate, say $\lambda$ stock units per unit time. The cost factors recognized include a unit holding cost $h$ per unit time and a fixed set-up cost $k$ associated with each ordering (replenishment). The inventory decision problem is to choose the level $Q$ to minimize total costs per unit time. The whole process may be regarded as stationary in time. The desired policy, correspondingly, will have a steady-state optimality characteristic.

The variation in stock over time, according to the policy just described, can be diagramed as shown in Fig. 10.1. The slope of the slanted lines is $-\lambda$. The number of reorders (= number of triangles) in a time period $T$ is asymptotically equal to $\lambda T/Q$. The holding cost per time unit can be computed as the sum of the areas occurring in the time period of

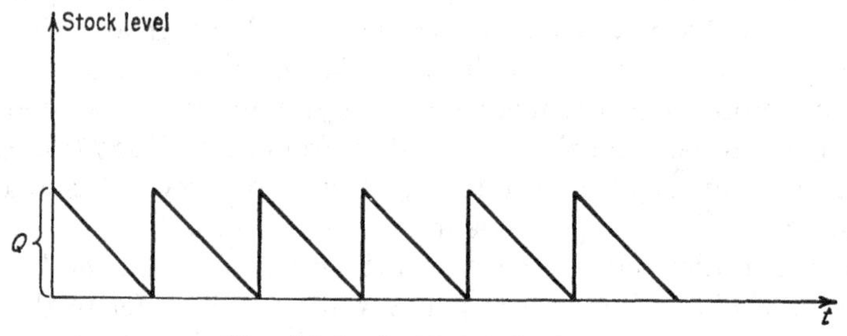


**Fig. 10.1** Stock flow in time.

length $T$, multiplied by the unit holding cost. Explicitly, the holding cost in the time duration $T$ is asymptotically equal to $hTQ/2$. The total cost over this time period divided by $T$ tends to

$$\frac{k\lambda}{Q} + \frac{hQ}{2}$$

which is minimized for

$$Q = \sqrt{\frac{2k\lambda}{h}} \tag{10.4}$$

Equation (10.4) is the famous square-root formula. It states that the optimal stock level is directly and inversely proportional to the square root of the set-up cost and holding cost, respectively.

Various generalizations of this model, and alternatives to it, have been developed. For example, it is trivial to see that the introduction of a cost proportional to the quantity ordered does not alter the solution in any way.

Another alternative is as follows: Instead of a holding cost, we recognize a unit purchase cost $c$ with value discounted in time by an interest-rate factor $r$; that is, the value of $d$ dollars $t$ time units later in terms of present value is $de^{-rt}$ dollars. Suppose also that demand occurs at a uniform rate of $\lambda$ units per unit time.

We consider the same class of policies as before described by a single critical number $Q$. The total cost function over the infinite future, properly discounted, reduces to $(k + cQ)/(1 - e^{-Qr/\lambda})$. The minimum with respect to the variable $Q$ occurs at the solution of the equation

$$(1 - e^{-Qr/\lambda})c - \frac{(k + cQ)r}{\lambda e^{-Qr/\lambda}} = 0$$

When $Qr$ is small, the optimal value of $Q$ is approximately

$$Q^* \sim \sqrt{\frac{k\lambda}{rc}} \tag{10.5}$$

Comparison of Eqs. (10.4) and (10.5) shows that $rc$ acts here like a holding cost.

*b. Econometric Inventory Model.* Another inventory model that has received considerable attention in the economic literature is as follows: Suppose a firm wants to decide how large a quantity $z$ of a commodity to make available on the market. The price received per unit of the commodity is a function $p(z)$ of the stock made available. This situation may exist in the case of monopolies or duopolies that can influence the price of the commodity by the quantity produced. The same thing sometimes applies to commodities regulated by government agencies.

The profit, in offering a quantity $z$ of the commodity, is clearly

$$P = zp(z)$$

The decision problem is to choose $z$ to maximize $P$. When

$$p(z) = \begin{cases} a - bz & \text{for } z \leq \dfrac{a}{b} \\ 0 & \text{for } z \geq \dfrac{a}{b} \end{cases}$$

the optimal choice of $z$ becomes

$$z = \frac{a}{2b}$$

This inventory problem is called the econometric model since, generally, $p(z)$ is assumed to be a random variable depending on the parameter $z$ in the form $a - bz + u$, where $a$ and $b$ are constants to be estimated and $u$ is a random variable normally distributed with mean zero.

The model has been extended in several directions. Thus, the problem can be made more realistic by having the variable $z$ limited through capacity restrictions. Also, we may consider an obvious dynamic version. The corresponding econometric model then involves time-series analysis. Finally, we may examine the many-commodity analogue. The bulk of the economics literature regarding inventory analysis is in fact concerned with such ramifications.

*c. Linear Decision Rules for Economic Models Based on Quadratic Costs.* Another point of view in treating inventory problems is to approximate all cost factors by quadratic functions.[10] The advantage of this rests in the result established by H. Simon[18] to the effect that an optimal decision rule based on a quadratic cost function depends on future demands only through their expected values. An example of an application of this criterion is the following: Consider a scheduling problem in which at regular time intervals the production rate $P$ and work force $W$ must be adjusted for the following time period. A decision rule that will minimize expected costs over $N$ future periods is required.

The costs to be minimized are represented by the following function of work force $W_t$, aggregate production $P_t$, net inventory $I_t$, and ordered shipments $O_t$, where the subscript $t$ designates the time period. The total cost function breaks down per period as follows:

$$\begin{aligned}
\text{Regular payroll cost} &= c_1 W_t \\
\text{Cost of hirings and layoffs} &= c_2(W_t - W_{t-1})^2 \\
\text{Expected cost of overtime} &= c_3(P_t - c_4 W_t)^2 + c_5 P_t - c_6 W_t \\
\text{Expected inventory, back order,}& \\
\text{and set-up costs} &= c_7[I_t - (c_8 + c_9 O_t)]^2
\end{aligned}$$

The solution, namely, the determination of $W_t$ and $P_t$ for all $t$ to minimize total costs, can be obtained by applying generating-function techniques.[10]

*d. Other Inventory Models.* Another method in inventory analysis consists in applying linear-programming methods. To the extent that costs are not additive and to the extent that the objective function is not linear, the linear-programming approach is severely limited[6] (see Chap. 9).

Another attack on inventory models is by exploiting an analogy between the inventory process and a servomechanism.[17] The analogy appears usually to be on the surface, and thus not very applicable.

## 10.9 The Literature of Inventory Theory

The main results prior to 1950 concerning inventory control procedures are summarized in the book of Whitin.[19] He devotes his attention largely to a discussion of the lot-size formula, its historic development, and some of its ramifications and justifications. On the other hand, the importance of the dynamic and uncertain character of demand is not made manifest. The relevance of shortage costs likewise plays an inessential role, throughout this book, in governing the choice of inventory policy.

The significance of these two factors was first emphasized in the fundamental paper of Arrow, Harris, and Marschak.[1] They also were the first to formulate correctly a dynamic inventory model. Dvoretzky, Kiefer, and Wolfowitz,[7] in a series of papers, greatly generalized this formulation, but they did not concern themselves with the specific analysis of solutions. The first actual solution of the simplest nontrivial version of this dynamic inventory model was uncovered by Bellman, Glicksberg, and Gross.[5] These results were extended in many directions in a series of papers appearing in the book by Arrow, Karlin, and Scarf.[2]

The above-mentioned book[2] examines several of the models proposed in the earlier papers, and with some success the optimal policies are determined for additional models. Moreover, several chapters of the book display and emphasize the relationships between inventory processes and stochastic processes originating in engineering, biology, economics, and other domains. In this connection, we also mention the book of Morse[14] devoted to the study of queueing systems. He shows how special models in inventory theory can be identified with queueing processes. We shall describe one such example in Sec. 10.17.

In the remainder of this chapter, we borrow from the book by Arrow, Karlin, and Scarf[2] and try to highlight the principal results of this book and its underlying philosophy.

## 10.10 Deterministic Inventory Models

In this section, we formulate three examples of deterministic inventory models. We describe typical solutions in each case.

EXAMPLE 10.1. *Production Scheduling.* We are interested in scheduling the production of a commodity so as to minimize costs. We shall assume that planning is for $n$ periods (the horizon). The requirements for each period $r_i$, $i = 1, \ldots, n$, can be forecast exactly, and we shall insist only on those policies that involve producing a quantity $z_i$ for the $i$th period that guarantees the fulfillment of all requirements. These constraints in mathematical form are

$$z_i \geq 0 \text{ and } z_0 + Z_i = z_0 + \sum_{j=1}^{i} z_j \geq R_i = \sum_{j=1}^{i} r_j \tag{10.6}$$

for $i = 1, 2, \ldots, n$, where $z_0$ represents initial stock on hand.

In this model, we recognize three costs:

$$\begin{aligned} c(z_i) &= \text{cost per period of producing } z_i \text{ units} \\ g(z_i - z_{i-1}) &= \text{cost per period of changing the rate of production} \\ h(y_i) &= \text{cost per period of holding inventory stock} \\ y_i &= z_0 + Z_i - R_i \end{aligned}$$

The total costs discounted at a rate $\alpha$ per period taken over the entire horizon is

$$C(z) = \sum_{i=1}^{n} [c(z_i) + g(z_i - z_{i-1}) + h(z_0 + Z_i - R_i)]\alpha^{i-1} \tag{10.7}$$

subject to the constraints (10.6). The problem is to determine $z_i$ by minimizing $C(z)$.

The continuous-time analogue involves replacing the sum in Eq. (10.7) by an integral. The difference as the argument of $g$ becomes a derivative of $z(t)$ is to be interpreted as a rate of production. We have

$$C[z(t)] = \int_0^T \left\{ c[z(t)] + g\left(\frac{dz}{dt}\right) + h[y(t)] \right\} e^{-\rho t}\,dt$$

where $$y(t) = y(0) + \int_0^t z(\xi)\,d\xi - \int_0^t r(\xi)\,d\xi \tag{10.8}$$

The function $r(t)$ is the requirement schedule, $T$ is the time horizon, and $\rho$ represents the rate of interest. The constraints assume the form

$$z(t) \geq 0 \qquad \text{and} \qquad y(t) \geq 0 \tag{10.9}$$

Analysis of Eq. (10.7) or Eq. (10.8) in general is very complex, and the problem cannot yet be given a constructive solution. It has thus far been found necessary to make special assumptions concerning the functions $c$, $g$, and $h$.

The problem can be regarded as one in the calculus of variations.[9] The tools of this discipline, however, are not directly applicable, because of the inequality constraints. It is therefore necessary to build up a solution by composing it partly from the solution provided by the calculus of variations when ignoring the inequality constraints, and elsewhere fitting the solution to the constraints, as indicated on pages 199 and 200.

We describe as typical the form of the solution of Eq. (10.8) for the special case in which $c(z) = z^2/2$, $r(t) = r$ $(0 \leq t \leq T)$, $\rho = 0$, and $h(z)$ is a linear function with proportionality constant $h$. In order to make the solution nontrivial, we assume available an initial stock $y(0) > 0$. The function $y(t)$ in this case is evaluated as

$$y(t) = y(0) + \int_0^t [z(\xi) - r(\xi)]\,d\xi \tag{10.10}$$

The solution divides into six cases as shown in Table 10.1. The detailed enumeration of cases and the complex character of the solutions even in this special example testify to the intrinsic difficulties of the general problem.

**Table 10.1** Solution when $c(z) = z^2/2$ and $r(t) = r$

| Case | Solution $z_0(t)$ |
| --- | --- |
| $2y(0)h - r^2 > 0 \quad \frac{r}{2h} + \frac{y(0)}{r} \le T$<br>$y(0) < rT$ | $z_0(t) = \begin{cases} 0 & 0 \le t \le v^0 \\ h(t - v^0) & v^0 \le t \le v_0 + \frac{r}{h} = t_0 \\ r & t_0 \le t \le T \end{cases}$<br>where<br>$v^0 = \frac{2hy(0) - r^2}{2hr}$ |
| $2y(0)h - r^2 > 0 \quad \frac{r}{2h} + \frac{y(0)}{r} > T$<br>$y(0) < rT$ | $z_0(t) = \begin{cases} 0 & 0 \le t \le v^* \\ h(t - v^*) & v^* \le t \le T \end{cases}$<br>where<br>$v^* = T - \sqrt{\frac{2}{h}[rT - y(0)]}$ |
| $y(0) > rT$ | $z_0(t) = 0$ |
| $2y(0)h - r^2 \le 0 \quad \sqrt{\frac{2y(0)}{h}} \le T$<br>$y(0) < rT$ | $z_0(t) = \begin{cases} r - \sqrt{2hy(0)} + ht & v \le t \le \sqrt{\frac{2y(0)}{h}} \\ r & \sqrt{\frac{2y(0)}{h}} \le t \le T \end{cases}$ |
| $2y(0)h - r^2 \le 0 \quad \sqrt{\frac{2y(0)}{h}} > T$<br>$rT - y(0) - \frac{hT^2}{2} \ge 0$ | $z_0(t) = ht + \frac{rT - y(0) - hT^2/2}{T} \qquad 0 \le t \le T$ |
| $2y(0)h - r^2 \le 0 \quad \sqrt{\frac{2y(0)}{h}} \ge T$<br>$rT - y(0) - \frac{hT^2}{2} < 0$ | $z_0(t) = \begin{cases} 0 & 0 \le t \le v^* \\ h(t - v^*) & v^* \le t \le T \end{cases}$ |

The derivation of these solutions as well as the solution of other cases of Eq. (10.8) can be found in Chaps. 4 to 6 of Ref. 2.

Example 10.2. *Optimal Expansion of the Capacity of a Firm.* Let $\xi(t)$ be the known rate of demand for a firm's product over the horizon of time $[0,T]$. Let $y(t)$ be the capacity of the firm at time $t$. Suppose for simplicity that the operating profit (price minus current cost) per unit of output is 1. Then the operating profit at time $t$ is $\min[\xi(t), y(t)]$. Let $c$ be the cost of increasing capacity by one unit per unit time and $r$ the rate of interest. The cost of capacity expansion at time $t$ is $c\dot{y}(t)$, and the present value of the income stream over the period $T$ is

$$\int_0^T \{\min[\xi(t), y(t)] - c\dot{y}(t)\}e^{-rt}\,dt \tag{10.11}$$

The problem is to choose $y(t)$ in such a way as to maximize the expression

(10.11), with $y(0)$ and $T$ given and $0 \leq \dot{y}(t) \leq M$, where $M$ is the maximum permissible rate of expansion.

The complete solution can be obtained for arbitrary $\xi(t)$ (Chap. 7 of Ref. 2). We describe the form of the solution for one special case. Assume $\xi(t)$ is monotone increasing. We define $t^*$ by the equation

$$e^{-rt^*} = \frac{e^{-rT}}{1 - rc}$$

provided $1 > rc$. (In the contrary case, $1 \leq rc$, the solution is trivial and entails no expansion.) If $\xi(0) = y(0)$, then the solution is $y_0(t) = \xi(t)$ until $t^*$ is reached, and afterward there is no expansion. In this case, and only in this case, the solution is independent of the exact nature of the function $\xi(t)$.

If $\xi(0) < y(0)$, there is an opening interval $[0,\bar{t}]$ in which no expansion takes place; $\bar{t}$ is characterized by the equation $\xi(\bar{t}) = y(0)$ provided the value of $\bar{t}$ does not exceed $t^*$. If it does, then set $y_0(t) \equiv 0$.

If $\xi(0) > y(0)$, then there is an opening interval $[0,t']$ in which

$$\dot{y}_0(t) = M$$

The point $t'$ is the solution of the equation $y(0) + Mt' = \xi(t')$ provided the solution does not exceed $t^*$. In that case, $\dot{y}_0(t) = \xi(t)$ in $[t',t^*]$ and 0 thereafter. Otherwise, $\dot{y}_0(t)$ becomes zero after $t^*$.

The algorithm of the solution is considerably more complicated if $\xi(t)$ is not monotonic.[2]

EXAMPLE 10.3. *Optimal Hydroelectric Operation.* The following model of hydroelectric operation is proposed in order to show the flexibility of inventory problems. In Sec. 10.13, the optimal policy for a dynamic version of this model is discussed.

Water (due to rainfall and melting snows) flows into a reservoir at a rate $r(t)$ $(0 \leq t \leq T)$ units per unit time. The function

$$R(t) = \int_0^t r(s)\, ds$$

specifies the gross volume of water that has been added to the reservoir during the time period $(0,t)$. The demand for water in terms of the rate at which water must flow through the turbines in order to meet the hydroelectric demand is given by a function $z(t)$. The functions $r(t)$ and $z(t)$ are assumed to be continuously differentiable. A policy is described by the rate of flow of water discharged through the turbines and will be denoted by $u(t)$, $0 \leq t \leq T$.

The problem is to regulate the rate at which water is discharged from the turbines in such a way as to minimize the cost incurred in supplying the electric energy demanded during a period of time $[0,T]$. There are two sources of electric energy. One source is water power, which is assumed to be free. The amount of electric energy that can be supplied from this source is limited by turbine capacity $u_{max}$, the reservoir capacity $V$, the volume of water in the reservoir at time $t = 0$, and the rate of water flow into the reservoir. The amount of water in storage at time $t$ is denoted by $V(t)$.

The second source of electric energy might be, for example, steam power. The cost of electric energy from this source is an increasing convex function of the amount of power so generated. It is assumed that the demand for power $z(t)$ will be met by this secondary source if it is not met hydroelectrically.

The problem considered here can now be stated mathematically as follows. Find $u_0(t)$ and $w_0(t)$, defined on the interval $0 \leq t \leq T$, that minimize

$$C[u(t)] = \int_0^T c[z(s) - u(s)]\, ds$$

and satisfy the following constraints for $0 \leq t \leq T$:

$$0 \leq u(t) \leq y(t) = \min\,[u_{max}, z(t)]$$
$$0 \leq V(t) = V(0) + \int_0^t [r(s) - u(s) - w(s)]\, ds \leq V$$
$$0 \leq w(t)$$

The first constraint states that no more electricity need be generated than is demanded or is possible by virtue of the turbine capacity. The second condition asserts that the amount of water in storage is necessarily nonnegative and is limited by the capacity $V$. Here, the function $w(t)$ represents the overflow function, which indicates the rate at which water is released with no intent but to prevent overflow.

The fact that the overflow function $w(t)$ does not appear in the cost function, which is to be minimized, and does appear in the constraints with the same sign as $u(s)$ implies that nothing is gained by letting water out of the reservoir when the reservoir capacity $V$ is not exceeded. Taking account of this, we can restate the problem as follows:

Minimize the functional

$$C[u(t)] = \int_0^t c[z(s) - u(s)]\, ds \tag{10.12}$$

subject, for $0 \leq t \leq T$, to the constraints

$$\begin{aligned} &0 \leq u(t) \leq y(t) \\ &0 \leq V(0) + \int_0^t [r(s) - u(s)]\, ds = V^*(t) \\ &V \geq \max_{0 \leq t^1 \leq t} V(t^1) - V^*(t) \end{aligned} \tag{10.13}$$

The solution of Eq. (10.12) is intricate, and for details the reader should consult Refs. 8 and 13.

## 10.11 One-stage Stochastic Inventory Models

Much of the early work in inventory theory has dealt with models that involve one time period. Static models form a reasonable approximation when the item is perishable or when it becomes obsolete rapidly because of technological change. A production or ordering decision is to be made at the beginning of the period so that the inventory $y$ is determined. Then the demand occurs and the revenues and costs are computed as functions of the demand and the resulting inventory.

One general-type single-period stochastic inventory model can be formulated as follows: Let $c(\cdot)$, $h(\cdot)$, and $p(\cdot)$ represent ordering, holding, and penalty cost, respectively, as functions of the stock level. We suppose for definiteness that penalty and holding costs are evaluated at the end of the period. Let $r$ represent the unit revenue and suppose that the total sales value is proportional to the total demand, which is assumed to be a random variable $\xi$ with known density function $\phi(\xi)$. If $x$ is the initial stock, if ordering results in raising the stock level to $y$, and if the demand during the period is $\xi$, then the costs suffered are evidently

$$L_x(y) = c(y - x) + h(y - \xi) + p(\xi - y) - r \min(\xi, y)$$

Here, $h$ and $p$ are zero for negative arguments. By summing over the various contingencies for demand, we secure the evaluation of expected total cost as

$$c(y - x) + \int_0^y [h(y - \xi) - r\xi]\phi(\xi)\, d\xi + \int_y^\infty [p(\xi - y) - ry]\phi(\xi)\, d\xi \qquad (10.14)$$

The problem is to determine $y = y^*(x) \geq x$ as a function of $x$ in such a way as to minimize the expression (10.14).

The form of the solution $y^*(x)$ appears to be fairly sensitive to the growth behavior of $h$ and $p$ as well as to the nature of the demand distribution, and it depends decisively on whether $c(z)$ is convex, concave, or linear.

For example, if $c(z)$ is linear, i.e., if $c(z) = cz$, $p$ is concave, $r > c$, and $\phi$ is a Pólya frequency function,[2] then the optimal policy $y^*(x)$ is described as follows: There exists an $\bar{x}$ such that

$$y^*(x) = \max(x, \bar{x})$$

and $\bar{x}$ is the *unique* solution of the equation

$$c + \int_0^{\bar{x}} h'(\bar{x} - \xi)\phi(\xi)\, d\xi - \int_{\bar{x}}^{\infty} [p'(\xi - \bar{x}) + r]\phi(\xi)\, d\xi = 0$$

The fact that the solution is unique simplifies significantly the computation of $\bar{x}$. Also, in this case any statistical estimate of $\phi$ can be translated quickly into a reliable statistical estimate of the critical number $\bar{x}$.

The class of densities that are Pólya frequency functions includes a large number of the standard unimodal densities arising in statistical applications. If $\phi(\xi)$ is a density of a gamma distribution of integral order, a truncated normal distribution, or a Kolmogoroff-Smirnoff distribution, it is a Pólya frequency function. In the case of discrete distributions, the Poisson and the Pascal family (negative binomial) qualify as discrete Pólya frequency functions. For a detailed discussion of Pólya frequency functions, the reader is referred to Refs. 11, 15, and 16.

As another example, we discuss now the case in which $c(z)$ is strictly convex and increasing, with $h$ and $p$ also convex functions.

Let $y^*(x)$ denote the *unique* root of $c'(y - x) + G'(y) = 0$ (whenever it exists), where

$$G(y) = \int_0^y [h(y - \xi) - r\xi]\phi(\xi)\, d\xi + \int_y^{\infty} [p(\xi - y) - ry]\phi(\xi)\, d\xi$$

Again, uniqueness is important to guarantee computation facility as well as stability in the solution when the parameters of the model are subject to slight changes. It follows easily that $y^*(x)$ increases with $x$. If $c'(0) < -G'(0)$, then for $x$ sufficiently small the root $y^*(x) > x$ exists. Furthermore, inserting $y^*(x)$ into the expression

$$\frac{\partial[EL_x(y)]}{\partial y} = c'(y - x) + G'(y)$$

where $EL_x(y)$ represents the expected loss for the policy $y$, leads to the conclusion that $dy^*/dx < 1$. This means that, although the point to which we order, according to the policy $y^*$, increases as a function of the stock level, the actual quantity of stock ordered decreases with respect to $x$. Finally, it follows that, for $x$ beyond some critical value $\bar{x}$, no positive ordering should be made.

For the case in which $c(z)$ is concave, the analysis is more complicated. A special important case, in which $c(z)$ is assumed to be concave, is nevertheless of great practical interest and deserves particular mention. Specifically, we refer to the case for which

$$c(z) = cz + K(z) \qquad \text{and} \qquad K(z) = \begin{cases} K & \text{for } z > 0 \\ 0 & \text{for } z = 0 \end{cases}$$

The last term $K(z)$ is commonly referred to as the set-up cost. Provided the equation $G'(y) + c = 0$ has a unique solution in the variable $y$, the optimal policy is characterized by two critical numbers $s$ and $S > s$ and is effected as follows: When $x \leq s$, ordering should be done to the point $S$; and when $x > s$, no ordering should be done (see Chap. 8 of Ref. 2). In this same reference, other one-stage inventory models are discussed.

### 10.12 Optimal Policy for Dynamic Stochastic Inventory Problems

In an $n$-stage model, the ordering rule that is to be used during each of the last $n - 1$ periods can be determined recursively by the methods of backward induction. The general solution of the one-stage problem is obtained first; then the two-stage solution is determined, using the full answer to the one-stage problem, and so on.

In the full dynamic or infinite-stage model, the same situation is faced at the beginning of the $n$th period as was faced at the beginning of the first period. Hence, in this model the optimal ordering rule for the first opportunity indicates the course of action to be followed in all subsequent stages.

The basic relationship used in deriving the optimal ordering rule is expressed by a functional equation; it is a recursive relationship that expresses the symmetries and renewal properties of the dynamic model. The derivation of this functional equation goes as follows: Suppose $x$ is the size of the initial stock; then the expected loss for the initial period if we order up to the amount $y$ is

$$L(y;x) = c(y - x) + \int_0^y [h(y - \xi) - r\xi]\phi(\xi)\,d\xi + \int_y^\infty [p(\xi - y) - ry]\phi(\xi)\,d\xi \quad (10.15)$$

where the terms of this expression have the same meaning as set forth in the previous section.

Let $f(x)$ be the minimum discounted expected loss that will be incurred during an infinite time period if $x$ is the initial stock level and an optimal ordering rule is followed throughout the total future. If we order at the first stage up to the amount $y$ and thereafter follow an optimal ordering policy, then the expected loss from the second stage on, discounted in terms of the present value of money, is

$$\alpha\left[f(0)\int_y^\infty \phi(\xi)\,d\xi + \int_0^y f(y - \xi)\phi(\xi)\,d\xi\right]$$

where $\alpha$ is the discount factor. It is intuitively clear that the optimal policy must be such that $f(x)$ satisfies the functional equation

$$f(x) = \min_{y \geq x}\left\{L(y;x) + \alpha\left[f(0)\int_y^\infty \phi(\xi)\,d\xi + \int_0^y f(y - \xi)\phi(\xi)\,d\xi\right]\right\} \quad (10.16)$$

The process by which Eq. (10.16) has been derived is referred to by Bellman as the *principle of optimality*.[3,4]

Even though $f(x)$ can rarely be determined explicitly, it is sometimes possible to characterize the optimal policy without evaluating $f(x)$. We illustrate one example of this.

Consider the case in which $c$ is linear, that is, $c(z) = cz$, $p$ is convex or concave, $h$ is increasing, and $\phi$ is a Pólya frequency function. It can be shown that the optimal policy is characterized by a single critical number $S$, such that, if $x < S$, then the optimal policy calls for replenishment to the level $S$, and otherwise no ordering is done. The value $S$ is determined as the unique solution of the equation

$$c + \int_0^S [h'(S - \xi) - \alpha c]\phi(\xi)\,d\xi - \int_S^\infty [r + p'(\xi - S)]\phi(\xi)\,d\xi = 0 \quad (10.17)$$

Further results along these lines are described in Chaps. 8 and 9 of Ref. 2.

We close this section with a brief discussion of the dynamic inventory model in which the lag in delivery has to be taken into account. The basic model is as follows: The three principal cost factors as usual are ordering, handling, and shortage costs, which are all assumed to be linear. This simplification is introduced in order to expedite the analysis and is definitely not essential to the qualitative nature of the results. Between delivery and order, we assume that there is a lag of $\lambda$ periods of time, with $\lambda$ fixed. When delivery is made, it takes place at the start of a period.

Let $x$ represent current stock size and let $y_1, y_2, \ldots, y_{\lambda-1}$ represent the outstanding orders, such that $y_1$ is due at the start of the present period, $y_2$ is to be delivered at the start of the second period hence, etc. Define $z$ to be the quantity of stock to be ordered at the start of the present period. Finally, let $f(x,y_1,y_2, \ldots ,y_{\lambda-1})$ = minimum expected loss following an optimal policy, where $(x,y_1,y_2, \ldots ,y_{\lambda-1})$ expresses all the information about the current stock level as well as the amounts of goods for which orders have been submitted and are to be delivered during the following $\lambda - 1$ periods. In the model of Eq. (10.16), we implicitly assumed that excess demand is to be interpreted as lost sales or alternatively as sales satisfied through priority shipments at a penalty cost. A second possible way of treating excess demand is to allow for deferring this demand to a later period. To exhibit the wide scope of these models, we concentrate here on the latter interpretation. The corresponding functional equation for $f(x,y_1,y_2, \ldots , y_{\lambda-1})$ may be derived as previously, and we obtain

$$f(x,y_1,y_2, \ldots ,y_{\lambda-1}) = \min_{z \geq 0} \left[ cz + L(x) + \alpha \int_0^\infty f(x - \xi + y_1, y_2, \ldots , y_{\lambda-1}, z)\phi(\xi)\,d\xi \right] \quad (10.18)$$

where

$$L(x) = \begin{cases} \int_0^x h(x - \xi)\phi(\xi)\,d\xi + \int_x^\infty p(\xi - x)\phi(\xi)\,d\xi & \text{for } x \geq 0 \\ \int_0^\infty p(\xi - x)\phi(\xi)\,d\xi & \text{for } x < 0 \end{cases}$$

The differences in the form of Eq. (10.16) vs. Eq. (10.18) result from the differences in the manner in which excess demand is regarded. The analysis of Eq. (10.18) is generally intricate and we merely indicate the form of the solution for some special cases. For example, the optimal policy $z^*(x,y_1,y_2, \ldots ,y_{\lambda-1})$ reduces to a function of the sum $x + y_1 + \cdots + y_{\lambda-1}$ and is characterized by a single critical number $\bar{x}$ as follows:

$$z^*(x,y_1,y_2, \ldots ,y_{\lambda-1}) = \begin{cases} \bar{x} - (x + y_1 + y_2 + \cdots + y_{\lambda-1}) & \text{for } x + y_1 + \cdots + y_{\lambda-1} < \bar{x} \\ 0 & \text{otherwise} \end{cases}$$

where $\bar{x}$ is determined as the *unique* positive root of the equation

$$c(1 - \alpha) + \alpha^\lambda \int_0^\infty \cdots \int_0^\infty L'(\bar{x} - \xi_1 - \xi_2 - \cdots - \xi_\lambda) \phi(\xi_1)\phi(\xi_2) \cdots \phi(\xi_\lambda)\,d\xi_1 \cdots d\xi_\lambda = 0$$

Table 10.2 shows the critical values $\bar{x}$ for $\lambda = 1, 2$, where

$$\phi(\xi) = \frac{\mu^k \xi^{k-1} e^{-\mu\xi}}{k!}$$

Here $\alpha = 1$, $h = p$, and $m$ represents the mean of the distribution $\phi$.

**Table 10.2** Optimal $\bar{x}$

| $\phi(\xi)$ | Lag of one period | Lag of two periods |
|---|---|---|
| $\mu e^{-\mu\xi}$ | $(1.68)m$ | $(2.67)m$ |
| $\mu^2 \xi e^{-\mu\xi}$ | $(1.83)m$ | $(2.83)m$ |

In particular, we see that, if the demand density is an exponential with unit average demand, then the optimal level of stock to maintain is 1.68 when the lag in delivery involves one time period and is 2.67 when the lag involves two periods.

## 10.13 Model of Hydroelectric Generation with Stochastic Inflow

A deterministic model of hydroelectric generation was formulated as the third example of Sec. 10.10. In the present section we discuss a hydroelectric operation that is an inventory model, such that the amount added to stock during a unit of time is a random variable with known distribution function. The total demand of electrical energy during each of $N$ future time periods is assumed to be known. Let $z_i$ be the demand during the $i$th time period ($i = 1, 2, \ldots, N$). The inventory is referred to as the volume of water that can be used to generate electrical energy. The units of measurement are so adjusted that the number of kilowatthours generated in a period of time is assumed equal to the volume of water that flows through the turbines during the period. All cost quantities and demand units may be stated in terms of volume of water.

It is assumed that up to $K$ units of the electrical energy may be supplied from supplementary sources (e.g., generated from thermal sources) in a single time period at a cost of $c$ dollars per unit. A penalty cost of $p$ dollars per unit not supplied is incurred if the demand for electricity is not fully met. Hydroelectricity is assumed to be free. Finally, the inflow of water in the $i$th period is assumed to be a random variable $r$, with known density $\phi_i(r)$.

The discounted expected cost, if an optimal water-storage policy is followed during $i$ consecutive subintervals and the reservoir level at the beginning of these $i$ subintervals is $R$, is denoted by $f_i(R)$ for $i = 1, \ldots, T$. The following basic recurrence relationship can be deduced in the usual way:

$$f_i(R) = \min_{0 \leq u \leq \min(z_i, R)} \left\{ c \min(K, z_i - u) + p \max(0, z_i - u - K) + \alpha \int_0^\infty f_{i-1}(R + r - u)\phi_i(r)\, dr \right\}$$

Subject to very slight conditions, it is possible to show that the optimal policy is given as follows: Let $R$ be the volume of water in storage at the beginning of the $N$ time periods over which the expected cost is to be minimized. There is a critical value $R'_N$ of the following sort: If $0 < R < z_1 - K$, use all available water to generate power during the first time period; if $z_1 - K < R < R'_N$, use $z_1 - K$ units of water in the first time period; if $R'_N < R < R'_N + K$, use $z_1 - K + R - R'_N$ units of water; if $R'_N + K < R$, use $z_1$ units of water. The optimal policy thus depends solely on the single parameter $R'_N$, and the value of $R'_N$ can be determined from a straightforward calculation of the relationship between $R'_N$ and the expected cost.

The following explicit numerical example is of some interest. Let the time period over which costs are to be minimized be 4 weeks and let this period be divided into 2-week subintervals. Suppose the parameters of the model have the following values, where 1 kcfs is a volume equal to the accumulation of 1,000 cfs for a 24-hr period:

$$\frac{c}{p} = \frac{1}{2} \qquad z_1 = z_2 = 1{,}087 \text{ kcfs} \qquad K = 436 \text{ kcfs}$$

$$\phi(r) = \lambda e^{-\lambda r} \qquad \frac{1}{\lambda} = 560 \text{ kcfs} \qquad \alpha = 1$$

In this case, the optimal policy asserts that

$$R_2' = 1{,}514 \text{ kcfs} \qquad R_2' + K = 1{,}950 \text{ kcfs}$$

For the case in which

$$\phi(r) = \lambda^2 r e^{-\lambda r} \qquad \frac{2}{\lambda} = 560 \text{ kcfs}$$

and all other parameters are unchanged, we have

$$R_2' = 563 \text{ kcfs} \qquad R_2' + K = 999 \text{ kcfs}$$

Thus, the critical number $R_2'$ is sensitive to changes in the density of inflow even though the expected inflow is kept constant.

## 10.14 Steady-state Solution of Inventory Problems

Investigations of the problem of the control of inventories have followed two principal methodologies. The first viewpoint, normative in concept, has dealt with the problem of determining optimal inventory policies, optimal in the sense of maximizing (or minimizing) an appropriate objective function as described in Secs. 10.11 to 10.13. In general, the problem of characterizing optimal policies as well as evaluating them is exceedingly difficult.

A second approach to the study of inventory problems is to concentrate attention on policies of "simple form." The policies are those that can be characterized by one of two parameters, and the aim is so to choose values for these parameters that the average long-run cost is minimized. In order to achieve such a goal, the asymptotic properties of the inventory process induced by implementation of "simple policies" must be ascertained.

If $x_0, x_1, x_2, \ldots$ represent a possible realization of the stock levels at the various periods resulting from the choice of a particular inventory policy and $\mathcal{L}(x_n)$ denotes the expected costs for the $(n + 1)$st period, we shall be interested in minimizing $L$, where $L$ is given by

$$L = \lim_{n \to \infty} \frac{\mathcal{L}(x_0) + \cdots + \mathcal{L}(x_n)}{n + 1}$$

In the case in which the sequence of random variables $x_n$ converges in the sense of distributions to a limiting random variable $x$ (the stationary distribution of the process), $L$ can be evaluated as $L = E_x\mathcal{L}(x)$. Since the distribution of $x$ is a function only of the policy and the nature of the demand variability, the stationary distribution can be calculated independently of the cost functions.

In most cases, the sequence $x_0, x_1, x_2, \ldots$ constitutes a Markoff process. For this situation, the calculation of the stationary distribution reduces to a standard technique. It is necessary, first, to write out the one-step probability transition law $P(x,dy)$. Specifically, $P(x,dy)$ represents the probability that, if the initial stock level is $x$, then one period later the stock level has a value in the interval $(y, y + dy)$. Initially, if the stock size is a random variable with distribution $F$, then, following an elapse of one epoch, the distribution function $G$ of the stock is calculated by the integral formula

$$\int_{-\infty}^{\infty} P(x,dy)\, dF(x) = T(F) = dG(y) \tag{10.19}$$

In particular, a cumulative distribution is a stationary distribution if and only if it is a fixed point of the transformation $T$. The problem of resolving Eq. (10.19) corresponding to the case of $T(F) = F$ is in general a formidable task. We illustrate the method of the stationary distribution for two inventory examples.

## 10.15 Stationary Inventory Model

We consider the well-known $(s,S)$ policy with $0 \leq s < S$ (see Sec. 10.11, page 244), which is operated as follows: Whenever the stock level $x$ falls below $s$, the ordering rule calls for replenishing stock to the level $S$. When the quantity of goods in supply exceeds $s$, then no ordering is done. Delivery of goods when ordered is assumed to be immediate. Decisions as to whether or not to order are to be made at the start of successive periods (each month, year, or the like).

The state of the system $x$ at the start of each period is described by the current stock level. The model to be considered allows $x$ to take on positive or negative values, where a negative stock level for $x$ should be interpreted as the amount owed to consumption. The penalty cost could be interpreted as a measure of the cost associated with lost sales, or alternatively interpreted as the cost suffered because of the presence of unsatisfied demand. In the present formulation of the model, all demand is ultimately filled. It is therefore meaningful to refer to negative stock levels. The essence of the penalty cost in this circumstance refers to the extra costs involved in having to make up current demand as soon as possible by future supplies.

The principal costs that will be considered as effecting the optimal inventory policy are as follows:

| | |
|---|---|
| Handling or storage cost | $h(z) = hz$ |
| Penalty or shortage cost | $p(z) = pz$ |
| Ordering cost | $c(z) + K(z) = cz + K(z)$ |

with
$$K(z) = \begin{cases} K & \text{for } z > 0 \\ 0 & \text{for } z = 0 \end{cases}$$

Each of the costs is a function of stock on hand at the close of a typical period. All cost functions, disregarding the ordering cost, are assumed to be linear functions. Many of the results can easily be extended, however, to allow for more general assumptions.

The demand $\xi$ in each period is assumed to be a positive random variable with known distribution function. For convenience of exposition, we assume that the distribution function of $\xi$ has a continuous density denoted by $\phi(\xi)$.

Once the $(s,S)$ policy is prescribed and an initial distribution for stock level $x$ (which might be a fixed initial stock size) is specified, all future probability distributions of stock levels are determined. We denote the stock level at the $n$th stage (period) by $x_n$, with distribution function given by $F_n$. The change of $x_n$ from one period to the next is described by the relationship

$$x_{n+1} = \begin{cases} x_n - \xi_n & \text{for } s < x_n < S \\ S - \xi_n & \text{for } x_n \leq s \end{cases}$$

where $\xi_n$ is a random observation, according to the density $\phi(\xi)$, that represents the demand arising during the $n$th period. As $n \to \infty$, the distributions $F_n$ converge, in the sense of the convergence of distributions, to the equilibrium distribution $F$. It is not difficult to verify that $F$ possesses a density $f$. The random variable $x$ in its stationary state must obviously satisfy the relationship

$$x = \begin{cases} x - \xi & \text{if } s < x < S \\ S - \xi & \text{if } x \leq s \end{cases} \tag{10.20}$$

This is to be interpreted not as an algebraic equation, but rather as a relationship of probability distributions. By expressing the relationship (10.20) in terms of probability densities, we obtain

$$f(x) = \phi(S - x) \int_{-\infty}^{s} f(t)\, dt + \int_{x}^{S} \phi(t - x) f(t)\, dt \qquad s < x < S \tag{10.21}$$

and

$$f(x) = \phi(S - x) \int_{-\infty}^{s} f(t)\, dt + \int_{s}^{S} \phi(t - x) f(t)\, dt \qquad x \leq s \tag{10.22}$$

Inspection of Eqs. (10.21) and (10.22) shows that $f(x)$ is continuous at $x = s$.

Before describing the general case, we specialize the density $\phi$ to

$$\phi(\xi) = e^{-\xi}$$

Equations (10.21) and (10.22) may be solved, and we conclude that

$$f(x) = \begin{cases} \dfrac{1}{1 + S - s} & s \leq x \leq S \\ \dfrac{1}{1 + S - s} e^{-(s-x)} & x \leq s \end{cases} \tag{10.23}$$

Computing the total expected costs for a single period, with initial random stock level having distribution of density given by Eq. (10.23), yields the formula

$$L(S,s) = \frac{K}{1 + S - s} + c + h \frac{s - 1 + (S^2 - s^2)/2}{1 + S - s} + \frac{(h + p)e^{-s}}{1 + S - s} \tag{10.24}$$

This represents the stationary costs according to the prescribed policy $(s,S)$. If $s_0$ and $S_0$ are the values that minimize $L(S,s)$ in Eq. (10.24), and $\Delta_0$ is defined by $\Delta_0 = S_0 - s_0$, then the equations

$$\Delta_0 = \sqrt{\frac{2K}{h}} \qquad e^{-s_0} = \frac{h + \sqrt{2Kh}}{h + p} \tag{10.25}$$

are valid whenever $\sqrt{2Kh} \leq p$. In case $\sqrt{2Kh} > p$, $s_0$ is to be taken as zero. It is of course intuitively clear that if no storage costs were involved (that is, $h = 0$), the stock level should be infinite. This is indeed evident from the form of the solution. Moreover, practical considerations suggest that the optimal policy should in no way depend on the marginal ordering cost parameter $c$, since all demands are ultimately satisfied. This is indeed the case. Another striking feature of the solution for this example is that $\Delta_0$ does not depend on the marginal penalty factor $p$.

The reader should compare the famous square-root law for the optimal lot size in inventory control with that of Eq. (10.4) in Sec. 10.8.

The total expected costs can be calculated and the corresponding decision problem resolved for numerous other choices of cost functions besides the case of linear costs. We are able to do this because the fundamental stationary distribution of the process has been obtained, independently of the cost terms of the model.

Another specific example might be worth mentioning. Let

$$\phi(\xi) = \xi e^{-\xi}$$

Then the stationary distribution is obtained as above and we have

$$f(x) = \begin{cases} \dfrac{2}{3 + 2\Delta + e^{-2\Delta}} [1 - e^{-2(S-x)}] & s \leq x \leq S \\ \dfrac{2}{3 + 2\Delta + e^{-2\Delta}} [e^{-(s-x)}(1 - e^{-2\Delta}) + (s - x)e^{-(s-x)}(1 + e^{-2\Delta})] & \\ & x \leq s \end{cases}$$

where $\Delta = S - s$. The optimal choices $\Delta_0$ and $s_0$ are found to satisfy the two equations

$$h\left(2\Delta + 1 + \frac{\Delta^2}{2}\right) + 2K = (h + p)e^{-s}(1 + s + 2\Delta + s\Delta)$$

$$\frac{h}{2}(3 + 2\Delta + e^{-2\Delta}) = (h + p)e^{-s}(s - 2 + se^{-2\Delta})$$

These transcendental equations may be solved by numerical methods. Again $\Delta_0$ and $s_0$ are found to be independent of $c$, but each now definitely depends on all other cost parameters.

The solution of Eqs. (10.21) and (10.22) may be accomplished in general. The first of these equations is essentially the classical renewal equation, for which the solution is

$$f(u) = c \sum_{n=1}^{\infty} \Phi^{(n)}(S - u) \qquad s \leq u \leq S$$

where $\Phi(t)$ is the cumulative distribution of $\phi$ and $\Phi^{(n)}$ is the $n$-fold convolution of $\Phi$ with itself. The second equation has a solution that is a multiple of the so-called "excess distribution" attached to the renewal process induced by the random variables $\xi_i$ independently distributed with density $\phi(\xi)$. The reader is referred to Chap. 15 of Ref. 2, where details concerning the general theory and its applications are developed.

## 10.16 Inventory Model with a Random Supply

We shall now discuss an inventory model with the special feature that, whenever a decision to order is made, a random quantity of goods with known distribution is added to the stock level. The yield from a plot of ground planted in wheat, corn, beans, etc., would be an addition to stock of this sort. This model would also fit a stock of electronic tubes, the manufacture of which involves a high rate of rejection of product for nonconformance to specifications.

In each period, the demand is random with realizations occurring in accordance with an estimable known density $\phi(\xi)$. At the start of each period, two possible actions are available: ordering and not ordering. When ordering is made, the delivery time is assumed again to be negligible. The result of ordering is an addition of a random quantity of goods $\eta$, with known density $\psi(\eta)$, to the present stock.

We consider policies of the form described by a single critical number $x^*$. The policy is effected as follows: Whenever the present stock level $x$ is smaller than $x^*$, then the policy requirements call for ordering; whenever $x$ is at least as great as $x^*$, then no ordering is to be done. Again, we allow the values of $x$ to achieve any possible real number, with the obvious meaning assigned to negative values.

The change in the stock level, which results from the policy and the nature of demand variability, in two succeeding periods is expressed by means of the relationship

$$x_{n+1} = \begin{cases} x_n + \eta_n - \xi_n & \text{if } x_n < x^* \\ x_n - \xi_n & \text{if } x_n \geq x^* \end{cases}$$

where $\eta_n$ represents an observation of $\psi(\eta)$ and $\xi_n$ is an observation of $\phi(\xi)$ during the $n$th period.

To ensure some stability for the process, it becomes necessary to impose the equilibrium condition $E(\eta) > E(\xi)$ ($E$ stands for expected value). This obviously guarantees that the supply is sufficiently large that all possible demands may ultimately be satisfied with probability 1. When the equilibrium condition is satisfied, the stock level settles in time to a stationary state. The form of the stationary distribution may be obtained explicitly. We describe the solutions for two cases:

EXAMPLE 10.4. *Let* $\phi(\xi) = \lambda e^{-\lambda\xi}$ *and take* $\psi(\eta)$ *arbitrary but fixed.* The unique stationary density is found to be

$$f(x) = \begin{cases} Ce^{-\alpha(x^*-x)} & x < x^* \\ C\lambda \int_{x-x^*}^{\infty} dz \int_z^{\infty} e^{-\alpha(y-z)}\psi(y)\, dy & x \geq x^* \end{cases}$$

where $\alpha$ is the unique positive solution of the transcendental equation

$$\lambda - \alpha = \lambda \int_0^{\infty} e^{-\alpha t}\psi(t)\, dt \tag{10.26}$$

The constant $C$, which is a normalizing factor for the density $f$, is clearly independent of the parameter $x^*$.

In dealing with the inventory problem, one desires the parameter $x^*$ to be chosen in such a way as to minimize the stationary costs. We assume here, for simplicity, that only the holding and penalty costs are relevant and that they are linear functions. Then a lengthy calculation shows that the optimal choice of $x^*$ is the unique solution of the equation

$$e^{-\alpha x^*} = E(\eta)\,\frac{h}{h+p} \tag{10.27}$$

It is interesting to observe that the optimal $x^*$ depends on the supply distribution $\psi(\eta)$ only through its expected value $E(\eta)$ and the value $\alpha$ that is determined by Eq. (10.26).

Some numerical cases are given in Tables 10.3 to 10.5.

**Table 10.3** Values of $x^* = -\frac{\log \gamma E(\eta)}{\alpha}$, for $E(\eta) = a$, $\lambda = 1$, $\gamma = \frac{h}{h+p}$, $\psi(\eta) = \frac{1}{a} e^{-\eta/a}\, d\eta$

| $\gamma$ \ $a$ | 1.01 | 1.05 | 1.1 | 1.2 | 1.5 | 2 | 5 |
|---|---|---|---|---|---|---|---|
| 0.1 | 231.56 | 47.33 | 24.28 | 12.72 | 5.69 | 3.22 | 0.87 |
| 0.25 | 139.01 | 28.09 | 14.20 | 7.22 | 2.94 | 1.39 | −0.28 |
| 0 5 | 69.00 | 13.53 | 6.58 | 3.06 | 0.86 | 0 | −1.15 |
| 0.75 | 28.05 | 5.02 | 2.12 | 0.63 | −0.35 | −0.81 | −1.65 |
| 0.9 | 9.64 | 1.19 | 0.11 | −0.46 | −0.90 | −1.18 | −1.88 |

**Table 10.4** Values of $x^* = -\frac{\log \gamma E(\eta)}{\alpha}$, for $E(\eta) = a$, $\lambda = 1$, $\gamma = \frac{h}{h+p}$, $\psi(\eta) = \left(\frac{6}{a}\right)^6 \frac{\eta^5}{\Gamma(6)} e^{-6/(a\eta)}\, d\eta$

| $\gamma$ \ $a$ | 1.01 | 1.05 | 1.1 | 1.2 | 1.5 | 2 | 5 |
|---|---|---|---|---|---|---|---|
| 0.1 | 135.38 | 27.93 | 14.49 | 7.75 | 3.65 | 2.21 | 0.71 |
| 0.25 | 81.27 | 16.58 | 8.47 | 4.40 | 1.89 | 0.95 | −0.23 |
| 0.5 | 40.34 | 7.99 | 3.92 | 1.87 | 0.55 | 0 | −0.94 |
| 0.75 | 16.40 | 2.96 | 1.26 | 0.39 | −0.23 | −0.56 | −1.36 |
| 0.9 | 5.63 | 0.70 | 0.066 | −0.28 | −0.58 | −0.81 | −1.55 |

**Table 10.5** Values of $x^* = -\frac{\log \gamma E(\eta)}{\alpha}$, for $E(\eta) = a$, $\lambda = 1$, $\gamma = \frac{h}{h+p}$, with Distribution of $\eta$ Concentrated at $\eta = a$

| $\gamma$ \ $a$ | 1.01 | 1.05 | 1.1 | 1.2 | 1.5 | 2 | 5 |
|---|---|---|---|---|---|---|---|
| 0.1 | 116.16 | 24.05 | 12.53 | 6.76 | 3.26 | 2.02 | 0.70 |
| 0.25 | 69.73 | 14.27 | 7.33 | 3.84 | 1.68 | 0.87 | −0.22 |
| 0.5 | 34.62 | 6.88 | 3.39 | 1.63 | 0.49 | 0 | −0.92 |
| 0.75 | 14.07 | 2.55 | 1.09 | 0.34 | −0.20 | −0.51 | −1.33 |
| 0.9 | 4.83 | 0.60 | 0.057 | −0.25 | −0.51 | −0.74 | −1.51 |

Examination of Tables 10.3 to 10.5 indicates that a supply distribution having a larger extent of randomness with expected value $a$ has necessarily a larger critical value $x^*$ for the corresponding optimal policy. This may be explained as being due to the fact that, although all three distributions have the same expected value, the median value varies greatly. The probability in the case considered in Table 10.3 of actually obtaining a small supply, when ordering is done, is substantially larger than that in the case exhibited in Table 10.5.

EXAMPLE 10.5. *Let the demand distribution have a gamma density. That is, let*

$$\phi(\xi) = \frac{\lambda^{k+1}\xi^k e^{-\lambda\xi}}{\Gamma(k+1)}$$

*with k an integer. Assume $\psi(\eta)$ is arbitrary as before.*

The stationary distribution may be obtained and the optimal choice $x^*$ subsequently determined. It is found that $x^*$ is the unique positive solution (if such exists; otherwise set $x^* = 0$) of the equation

$$\sum_{r=1}^{k+1} \frac{A_r e^{-\alpha_r x^*}}{\alpha_r} = \frac{\lambda^{k+1}}{\alpha_1\alpha_2 \cdots \alpha_{k+1}} \frac{h}{h+p} E(\eta)$$

where the $\alpha_r$ are complex solutions of the transcendental equation

$$(\lambda - \alpha)^{k+1} = \lambda^{k+1} \int_0^\infty e^{-\alpha t}\psi(t)\, dt$$

lying strictly in the right-hand half of the complex plane. It may be demonstrated that there exist precisely $k + 1$ (counting multiplicities) such solutions. The constants $A_r$ turn out to be independent of $x^*$ and are evaluated as the unique solutions of an associated system of linear equations that we do not reproduce here. The reader is referred to Chap. 14 of Ref. 2, where these equations are explicitly written out.

A particular case is appended to illustrate the theory.

Suppose $\phi(\xi) = \xi e^{-\xi}$ and $\psi(\eta) = be^{-b\eta}$. It can be established that the stationary distribution, excluding a normalizing factor, has the form

$$f(x) = \begin{cases} \dfrac{b+\alpha_2}{\sqrt{b^2+4b}} e^{-\alpha_1(x^*-x)} - \dfrac{b+\alpha_1}{\sqrt{b^2+4b}} e^{-\alpha_2(x^*-x)} & x < x^* \\ e^{-b(x-x^*)} & x \geq x^* \end{cases}$$

where $\alpha_1 = \dfrac{2 - b - \sqrt{b^2+4b}}{2} \qquad \alpha_2 = \dfrac{2 - b + \sqrt{b^2+4b}}{2}$

The optimal $x^*$ satisfies the equation

$$\frac{b+\alpha_2}{\sqrt{b^2+4b}} \frac{e^{-\alpha_1 x^*}}{\alpha_1} - \frac{b+\alpha_1}{\sqrt{b^2+4b}} \frac{e^{-\alpha_2 x^*}}{\alpha_2} = \frac{h}{h+p} \frac{1}{b} \frac{1}{1-2b}$$

## 10.17 Stationary Distribution for a Model of Lagged Delivery

We imagine a central mail-order house that receives orders from all parts of the country. The time interval $t$ between successive orders is assumed to have a known distribution. For the purposes of the present discussion, we assume this distribution is

$$\phi(t) = \lambda e^{-\lambda t}$$

(The theory may be developed for the general case.) It is further assumed that individual orders have no connection with each other. Consequently, the various demands as they arise may be thought of as forming a "Poisson process" in time. Associated with each order is a random variable $s$ with known distribution function $H(s)$ representing the lag involved in filling a typical order. The lag may also be interpreted as an expression of the labor involved in dealing with any specific order. It is assumed that separate orders are handled independently.

One quantity of interest in this model is the number $N_t$ of unfilled orders at a given moment of time $t$, where initially the state of the system (number of unfilled orders) is zero. The second problem is the stationary analogue of the first problem; namely, it is the problem of determining the distribution of the number of unfilled orders

$$N^* = \lim_{t \to \infty} N_t$$

at an arbitrary instant of time, given that the process has been going on infinitely long in the past. The distribution theory of $N_t$ and $N^*$ is related to studies conducted by Takács on the stochastic process of the shot effect. One can prove generally that $N^*$ has a Poisson distribution with parameter

$$\lambda \int_0^\infty [1 - H(s)]\, ds = \lambda\mu$$

where $\mu = E(s)$. It is indeed interesting to observe that the distribution of $N^*$ depends on $H(s)$ only through its mean value. It can also be shown that the distribution of $N(t)$ is likewise Poisson, with parameter

$$\lambda \int_0^t [1 - H(s)]\, ds$$

From a knowledge of the stationary distribution, various quantities of interest with respect to the inventory model can be computed. For example, the probability of having no outstanding order at a random moment in time is $e^{-\lambda\mu}$, etc. Other possibilities of these kinds of investigations should be clear to the reader. Any superposition of the various cost elements on the preceding structure is routine.

It is possible to interpret the process $N(t)$ as a queueing process with an infinite number of servers. The interarrival distribution is exponential, and the service distribution is $H(s)$. Let us consider a modification of this model such that, whenever $N$ demands are being served, all additional demands are refused until at least one of the servers becomes free. It can be shown that in this case of limited facilities the random variable $N^*$ has a truncated Poisson distribution. See Chap. 6.

Another stochastic quantity of interest is $w_t$, which represents the time necessary for all of the servers who were busy at time $t$ to finish their current work. It is not difficult to prove that

$$\text{prob}\,\{w^* \leq x\} = e^{-\lambda \int_x^\infty [1 - H(\xi)]\,d\xi}$$

where

$$w^* = \lim_{t \to \infty} w_t$$

and $\lambda$ is the parameter of the Poisson process. The proofs of these results and further extensions are developed in Chaps. 16 and 17 of Ref. 2.

## REFERENCES


1. Arrow, K. J., T. E. Harris, and J. Marschak, Optimal Inventory Policy, *Econometrica*, vol. 19, pp. 250–272, 1951.
2. ———, S. Karlin, and H. Scarf, "Studies in the Mathematical Theory of Inventory and Production," Stanford University Press, Stanford, Calif., 1958.
3. Bellman, Richard, The Theory of Dynamic Programming, chap. 11 in "Modern Mathematics for the Engineer," First Series, edited by E. F. Beckenbach, McGraw-Hill Book Company, Inc., New York, 1956.
4. ———, "Dynamic Programming," Princeton University Press, Princeton, N.J., 1957.
5. ———, I. Glicksberg, and O. Gross, On the Optimal Inventory Equation, *Management Sci.*, vol. 2, pp. 83–104, 1955.
6. Dantzig, G. B., Optimal Solution of a Dynamic Leontief Model with Substitution, *Econometrica*, vol. 25, pp. 295–302, 1955.
7. Dvoretzky, A., J. Kiefer, and J. Wolfowitz, The Inventory Problem, *Econometrica*, vol. 20, pp. 187–222, 450–466, 1952.
8. Gessford, J., "The Use of Reservoir Water for Hydroelectric Power Generation," Dissertation, Stanford University, 1957.
9. Hestenes, Magnus R., Elements of the Calculus of Variations, chap. 4 in "Modern Mathematics for the Engineer," First Series, edited by E. F. Beckenbach, McGraw-Hill Book Company, Inc., New York, 1956.
10. Holt, C. C., F. Modigliani, and H. A. Simon, Linear Decision Rule for Production and Employment Scheduling, *Management Sci.*, vol. 2, pp. 1–31, 1955.
11. Karlin, S., Pólya Type Distributions, II, *Ann. Math. Statist.*, vol. 28, pp. 281–308, 1957.
12. King, Gilbert W., Applied Mathematics in Operations Research, chap. 10 in "Modern Mathematics for the Engineer," First Series, edited by E. F. Beckenbach, McGraw-Hill Book Company, Inc., New York, 1956.
13. Koopmans, T. C., Water Storage Policy in a Simplified Hydroelectric System, "Proceedings of the International Conference on Operational Research," John Wright and Sons, Bristol, England, 1958.
14. Morse, P., "Queues, Inventories and Maintenance," John Wiley & Sons, Inc., New York, 1958.
15. Schoenberg, I. S., On Pólya Frequency Functions, *Jour. Analyse Math.*, vol. 1, pp. 331–374, 1951.

16. ———, On Smoothing Operations and Their Generating Functions, *Bull. Am. Math. Soc.*, vol. 59, pp. 199–230, 1953.
17. Simon, H. A., On the Application of Servomechanism Theory in the Study of Production Control, *Econometrica*, vol. 20, pp. 247–268, 1952.
18. ———, Dynamic Programming under Uncertainty with a Quadratic Criterion Function, *Econometrica*, vol. 24, pp. 74–81, 1956.
19. Whitin, T., "The Theory of Inventory Management," Princeton University Press, Princeton, N.J., 1953.

# PART 3
# Physical Phenomena

# 11

# Monte Carlo Calculations in Problems of Mathematical Physics

STANISLAW M. ULAM
RESEARCH ADVISOR
LOS ALAMOS SCIENTIFIC LABORATORY

## 11.1 Introduction

In this chapter we shall outline the general idea and summarize certain applications of the procedures constituting the so-called Monte Carlo method, and we shall discuss briefly some mathematical problems connected with these procedures. The Monte Carlo method is used in problems for which explicit solutions cannot be obtained by methods of classical analysis and also in cases such that the numerical work in solving the corresponding differential or integrodifferential equations would take a prohibitively long time even on the fast modern electronic computing machines. The chapter by George W. Brown, Monte Carlo Methods, in the first volume of "Modern Mathematics for the Engineer,"[1] gives an excellent description of the scope of this approach and of the basic statistical concepts. Some of the different applications will be mentioned here, and the role of *branching processes* will be discussed by means of examples.

The development of electronic computers has made it possible to perform statistical experiments, not with the physical situations in question, but rather—so to say—"on paper," mocking up the physical problem rather directly on the computing machines themselves. Viewed rather broadly, when the problem involves many (often more than 2!) independent variables or when the combinatorial complexities and the geometrical complications of the problem seem to preclude obtaining solutions in closed form, the set of procedures that came to be called the Monte Carlo method may often be used advantageously.

The idea of this approach may be illustrated, in a very simple case, by the following example:

Suppose a region $R$ is defined in the $n$-dimensional euclidean space by the inequalities

$$\begin{array}{l} F_1(x_1,x_2, \ . \ . \ . \ ,x_n) < 0 \\ F_2(x_1,x_2, \ . \ . \ . \ ,x_n) < 0 \\ . \ . \ . \ . \ . \ . \ . \ . \ . \ . \ . \ . \ . \\ F_k(x_1,x_2, \ . \ . \ . \ ,x_n) < 0 \end{array} \tag{11.1}$$

The problem is to find the volume of the region $R$.

Suppose that $R$ is contained in the unit cube and that we know a priori that this volume is not very small compared with 1. If the computation of the definite integral cannot be performed explicitly, an idea of the volume of $R$ can be obtained, actually with arbitrary accuracy, from numerical work corresponding to the definition of the integral. That is to say, the unit cube is subdivided into small cubes, and by counting the number of those cubes that lie in the region $R$ we obtain an approximation to its volume. If $n$ is, say, 10, by subdividing each axis into $r$ intervals we obtain $r^{10}$ small cubes. The proportion of those among them that lie wholly in $R$ gives us an approximation to the volume. Even if the functions $F_i$ are analytically well behaved, in order to obtain an answer to our question within 1 per cent or so, we have to subdivide the interval into some 10 to 100 subintervals. Assume that $r$ is of the order of 100 and that the number of subcubes is therefore $100^{10}$. It is obviously impractical to consider all of these. Instead of employing the foregoing "deterministic" procedure, we could do the following: Suppose we select $N$ points at *random* in the unit cube in our ten-dimensional space, and count among them those that belong to $R$, that is to say, those points having coordinates that satisfy all the given inequalities (11.1). Suppose we know in advance that the volume of $R$ is of the order of 1, i.e., that it is about ½ or so. Then if $N$ is of the order of $100^2$, that is, 10,000, we could hope that the fraction of the number of points lying in $R$ will give us an idea of the value of the volume that *should* differ from the true value by less than 1 per cent or so. This is due to the fact that we have here a situation similar to the simple "Bernoulli case" problem in the elementary theory of probabilities. If the true value of the volume is $v$, we perform an experiment $N$ times by selecting a point in the unit cube, with the probability $v$ that it will lie in $R$. The theorem of Bernoulli states that, if $N \to \infty$, the probability that the frequency of successes will approach $v$ tends to 1. What is more, we can estimate for a finite $N$ the mean-square error of the difference between the frequency and $v$. If $N$ is now of the order of 10,000, the probability is close to 1 that this error will not surpass a small percentage. By making $N$ several times 10,000, we can, with high probability, obtain an answer to our question within 1 per cent. The number $N$ of experiments is obviously much smaller than the above number $100^{10}$ of subcubes. In this way, by performing a stochastic experiment, we obtain a much quicker idea of the true volume —to be sure not with certainty, but only with high probability.

It remains for us to explain how one selects points "at random," with uniform probability, in a unit cube. It is immediately apparent that it suffices to be able to select randomly with uniform probability real numbers between 0 and 1, independent of each other in succession. Ten such numbers will define a point in the unit cube. Therefore, in order to obtain 10,000 random points in space of 10 dimensions, it suffices to produce 100,000 real numbers between 0 and 1. We shall discuss briefly how one can, on a computing machine—by a mathematical operation—produce such sequences of real numbers possessing certain properties of randomness. One could think of producing such real numbers by a physical process corresponding somehow to "tossing a coin." Indeed, one could use counts of radioactive disintegrations in electronic tubes, etc., but the problem is to produce these numbers very quickly without any auxiliary physical equipment in the computing machine itself. Many methods have been devised for achieving this aim. One of them is to start with a suitably selected real number written in the binary expansion as $x_1 = 0.\alpha_1\alpha_2 \cdots \alpha_{40}$, where $\alpha_i = 0$ or 1. To obtain $x_2$, we square $x_1$. This number will have 80 binaries, and we select the middle 40 of them. That is to say, the $\alpha$'s in $x_1^2$ having indices from 21 to 60 are used to define $x_2$. By iteration of this procedure we obtain $x_3$ from $x_2$, and so on. Obviously, the sequence of numbers thus obtained will not be truly random, since for one thing it has to start repeating, the process being a finite one if we have only a finite number—40 in our case—of digits; however, very long sequences of the $x_i$, of the order of 1,000,000 terms or so, can be obtained. The properties of uniformity of position in the unit interval, the lack of correlation between each term and its immediate successors, and so on, can be tested. Once this is done ahead of time, these numbers can be used very quickly, since each step requires only a multiplication and shifting on the machine to produce our "random" sequences.

There are other methods for obtaining such sequences. An account of these is given, for example, in the article by O. Taussky and J. Todd.[13]

## 11.2 A Combinatorial Problem

To give an example of how a sampling method may be of value in obtaining deterministically defined quantities by a random procedure, we shall discuss the following combinatorial problem.

A well-known theorem asserts that, given $n^2 + 1$ integers in any order, it is always possible to find among them a subsequence of $n + 1$ that is either increasing or decreasing. So, for example, with $n = 10$, if we have an arbitrary permutation of 101 integers, it is possible to find among them a subsequence of 11 in increasing order or 11 in decreasing order. Suppose we want to find, however, not the minimum length but rather, considering *all possible* permutations, the average length of the maximal increasing or decreasing subsequences in the permutations. The average is meant with respect to all possible permutations of the $n^2 + 1$ integers. If $n = 10$, the number of permutations is $101!$, and it is obviously impossible to examine them all. Imagine now that we examine only a few thousand out of the $(n^2 + 1)!$ permutations, these few thousand being chosen "at random" with uniform probability among all possible ones. The average performed on these should give us a fair idea of the true value of the average. An electronic computing machine like the IBM 704 will process several thousand permutations of the 101 integers in a few minutes.

We should explain here how one may obtain random permutations of $N$ integers. One way might be the following: We produce, to obtain a "random" permutation of $N$ integers, a sequence of $N$ random real numbers on the interval (0,1). If we write 1 at the position determined by the smallest among them, 2 at the position occupied by the next smallest among them, and so on, finally writing $N$ at the position occupied by the largest among them, we obtain one permutation of the $N$ integers. To produce $K$ permutations, we perform this process $K$ times. For various values $n$, E. Neighbor examined the mean length of the maximal increasing or decreasing subsequence (using a few thousand permutations for each case) starting with $n = 4$ and going to $n = 10$. The theorem guarantees that the minimum length of a monotone subsequence for $n = 4$ is 5 (for a permutation of 17 integers). The average length of a maximal monotone subsequence in a permutation of 17 integers turned out for his sample to be 6.69. For $n = 5$ the average was 8.46, for $n = 8$ the average was 14.02, and for $n = 10$ the average came out 17.85. The averages formed a linear function of $n$ and were about 1.7 times $n$. Another question of interest would be to find the distribution of the length of the maximum monotone subsequence around this average. It turned out to have a Gaussian form starting at the guaranteed minimum, having its maximum at the average, and becoming vanishingly small at about 2.2 times the minimum. We mention this example to show how, in problems of combinatorics, a statistical approach may help to obtain an idea of regularities inherent in complicated situations. We shall mention later a Monte Carlo study in some problems of statistical mechanics.

### 11.3 Branching Processes

One of the largest areas of application of the Monte Carlo method lies in the problems involving branching or multiplicative processes. Schematically, these problems can be described as follows: Imagine a collection of particles lying in a space $E$. This could be the ordinary three-dimensional space, but we can consider a much more general case, the particles being rather abstract entities located in a space of many dimensions. Each of the particles, in addition to its position in the space $E$, may have some other characteristics described by indices running from 1 to $K$. Let us imagine further that we consider time proceeding in a discrete manner by generations, one "second" apart. The particles perform a random walk in space during each generation and, in addition, they may multiply or demultiply; in particular, they may disappear altogether by absorption. They may also reflect on certain given boundaries, etc. The random walk may be isotropic—that is to say, the particles go with equal probability in any direction—or some directions may be preferred. The multiplication of the particle may depend, in an explicit fashion, on its position in space and on its index.

The simplest illustration is the elementary random-walk problem with no multiplication. This is a possible interpretation of problems of diffusion.[1] If we consider the steps to be very short and the intervals of time to tend to zero, the analytical description of the density $u$ of such diffusing particles as a function of space and time is determined by the differential equation

$$\frac{\partial u(x,y,z;t)}{\partial t} = a\,\Delta u$$

where

$$\Delta u = \frac{\partial^2 u}{\partial x^2} + \frac{\partial^2 u}{\partial y^2} + \frac{\partial^2 u}{\partial z^2}$$

If the particles, in addition to performing a random walk, should multiply in number by a factor $V(x,y,z)$ at the place $(x,y,z)$, the equation would be

$$\frac{\partial u}{\partial t} = \Delta u + Vu$$

a Schrödinger equation. Here, $V$ is a given function. More generally, we think of equations of the type

$$\frac{\partial u}{\partial t} = \Phi(u)$$

where $\Phi(u)$ is an operator on the function $u$, not involving time explicitly. If we consider time as increasing by a constant amount from generation to generation, the difference version of such an equation is

$$u^{t+h} = u^t + h\Phi(u) = \Psi(u)$$

or

$$u^{t+nh} = \Psi^n(u)$$

where $\Psi^n$ is the $n$th *iterate* of the operator $\Psi$. A solution of the equation may be obtained approximately by numerical work, through iteration of a given operator, starting from an initial distribution $u(x,y,z;0)$. For numerical work, the space of the variables $x$, $y$, $z$ may be considered as being divided into a finite number of zones. The operator $\Phi$ will involve, in some problems, not only differential expressions as in the continuous model of diffusion or the random walk, but integral parts corresponding to "large" displacements, etc. A stochastic or Monte Carlo approach is often useful in problems of this sort. As we shall see throughout this chapter, the procedure involves an iteration, in a timelike variable, of a given process. We intend now to outline some properties of the iterates of such operators, restricting our attention to a discrete time and space.

Consider the branching or multiplicative features of the processes on a simple prototype as follows: We start with a particle in the zeroth generation. This particle may now do one of the following things. With probability $p_0$ it may disappear altogether; with probability $p_1$ it reproduces itself; with probability $p_2$ it produces two particles; . . . ; with probability $p_n$ it produces $n$ particles like itself. For the simplest case, assume that the new particles, if any, in the second generation are independently subject to the same rules. We may obtain particles in the third generation, and the process may continue. The problem is to determine the properties of such a process—in particular, to discuss the probability distribution of the population as a function of time, that is to say, the generation index.

A useful tool for the study of these questions is the generating function of Laplace. Let

$$f(x) = p_0 + p_1 x + p_2 x^2 + \cdots p_i x^i + \cdots$$

be a function of the real variable $x$. We have $f(1) = 1$, since

$$\sum_{i=0}^{\infty} p_i = 1$$

Further,

$$f'(x)_{x=1} = \nu$$

gives us the expected value of the number of offspring from one particle. The second and higher moments of the probability distribution for the number of offspring from each particle can be obtained from $f(x)$ by successive differentiations. So, for example,

$$\sum_{i=0}^{\infty} i^2 p_i = f''(x)_{x=1} + \nu \qquad \text{etc.}$$

It is very easily proved that, if $f(x)$ is the generating function of the known probability distribution for the number of particles in the first generation, then the generating function for the second generation is given by $f[f(x)]$. For the third generation the generating function is given by $f\{f[f(x)]\}$. For the $k$th generation it is $f^k(x)$, where

$$f^k(x) = f[f^{k-1}(x)]$$

Note that, for the generating functions at least, the solution to our problem is obtained by iterating a given function. From the generating function we can obtain explicitly the moments of the probability distribution for the number of particles in the $k$th generation. So, for example, the expected value is readily obtained through the equation

$$[f^k(x)]'_{x=1} = [f'(x)_{x=1}]^k$$

from the rule for differentiation of composite functions. The second derivative $[f^k(x)]''$ is simply

$$f''(x)\{f'(x) + [f'(x)]^2 + \cdots + [f'(x)]^k\} \qquad \text{at } x = 1$$

In general, by representing

$$\frac{d^r}{dx^r} f^k(x) = M_{k,r} \qquad M_{1,r} = M_r$$

we obtain difference equations

$$\begin{aligned} M_{k,1} &= M_1 M_{k-1,1} \\ M_{k,2} &= M_2 M_{k-1}^2 + M_1 M_{k-1,2} \\ M_{k,3} &= M_3 M_{k-1,1}^3 + 3M_2 M_{k-1,1} M_{k-1,2} + M_1 M_{k-1,3} \\ &\cdots\cdots\cdots\cdots\cdots\cdots\cdots\cdots \end{aligned}$$

Each is of the form

$$x_k = A_{k-1} + M_1 x_{k-1}$$

with general solution

$$x_k = \sum_{s=2}^{k} M_1^{k-s} A_{s-1} + M_1^{k-1} x_1$$

Solutions for the first three derivatives are

$$\begin{aligned} M_{k,1} &= M_1^k \\ M_{k,2} &= M_2 M_1^{k-1} \frac{1 - M_1^k}{1 - M_1} \\ M_{k,3} &= M_3 M_1^{k-1} \frac{1 - M_1^{2k}}{1 - M_1^2} + \frac{3M_2^2 M_1^{k-1}}{1 - M_1} \frac{1 - M_1^{k-1}}{1 - M_1} \\ &\qquad - \frac{3M_2^2}{1 - M_1} M_1^k \frac{1 - M_1^{2k-2}}{1 - M_1^2} \end{aligned}$$

The probability of having no particles left in the $k$th generation is given by the constant term of its generating function. It is easy to find the asymptotic probability for the "death" of the system, that is to say, the limit of the value of this constant term of the $k$th generation as $k \to \infty$. This is done as follows: In the first generation, $f(0) = p_0$. We want to know the asymptotic value of $f^k(0)$. The function $f(x)$ is monotone increasing. As we iterate $f$, the values at $x = 0$ increase and approach a limit that is equal to the first $x$ such that $f(x) = x$. Therefore, the probability of the system dying out at some time is given by the root of this equation. Three cases are to be distinguished: $\nu > 1$, $\nu = 1$, and $\nu < 1$. In the first case the function $f(x)$ has a slope at $x = 1$ that is greater than 1, and since all its coefficients in the power-series development are nonnegative, it is convex upward. Since, as we may assume, we have $p_0 > 0$, it must cross the diagonal *inside* the interval (0,1), and the smallest root of the equation $f(x) = x$ is less than 1. Therefore, the probability of the system dying out is less than 1, and there is a positive chance for immortality of the system. The same argument shows that in the other cases the first root of the equation is $x = 1$, and therefore the system will die out with probability 1.

The problem of finding the probability distribution itself and not merely its moments, as above, is much more difficult. It can be solved explicitly only in some simple cases. For example, if $f(x)$ is a power series of a form that can be written as

$$f(x) = \frac{ax + b}{cx + d}$$

where $a$, $b$, $c$, and $d$ are constants so chosen as to ensure that all the $p_i$ are nonnegative with their sum equal to 1, then we have a three-parameter family of functions that can be explicitly iterated. This is due to the fact that functions of this form constitute a group under composition. The iterates have the same form and the constants can be obtained by a matrix-multiplication algorithm.

Suppose we are interested in the probability distribution of *the sum* of all particles in the system produced from the first through the $k$th generation. If we want the generating function for the probabilities of having a total of $n$ particles from the first through the $k$th generation, we proceed as follows.

The total of $n$ particles can be obtained in any one of the following mutually exclusive ways: We can have 1 in the first generation and $n - 1$ in the remaining $k - 1$, or 2 in the first generation and $n - 2$ in the remaining $k - 1$; in general, we can have $r$ in the first and $n - r$ in the remaining $k - 1$ generations. The required probability is therefore the sum

$$q(n) = \sum_{r=1}^{n} p_r p_{n-r}^{k-1}(n)$$

Here $p_{n-r}^{k-1}(n)$ denotes the probability that, starting from $r$ in the first generation, we shall attain from these $r$ a total of $n - r$ in $k - 1$ generations. But the $r$ particles are independent of each other. The probability of getting the total of $n - r$ from them is therefore the probability of $n - r$ in the sum of these $r$ variables. The generating function for the sum of the independent variables is the product of the generating functions corresponding to each of them. In our case it is the $r$th power of $f(x)$. We are looking for the coefficient of $x^{n-r}$ in $[f^{k-1}(x)]^r$. Our required probability $q_k$ equals, therefore, the sum with respect to $r$ of the coefficients of $x^{n-r}$ in $[f^{k-1}(x)]^r$, or the sum of the coefficients of $x^n$ in the sum

$$\sum_{r=1}^{n} p_n x^r [f^{k-1}(x)]^r \tag{11.2}$$

But the coefficient of $x^n$ in the expression (11.2) is the same as this coefficient in $f[xf^{k-1}(x)]$. This is true for all $n$. Therefore, the generating function for $q_n$ is $f[xf^{k-1}(x)]$. Since $n$ is arbitrary here, we get the generating function for the time sum:

$$u^k(x) = f[xu^{k-1}(x)]$$

If we count the original particle, this multiplies the generating function by $x$; expressing this slightly modified form recursively, we obtain the more convenient expression

$$u^k(x) = xf[u^{k-1}(x)]$$

We have, in general, a relation between *moments* of the $n$th order of a distribution function and the $n$th derivative of the generation function. We shall now show how one can compute the derivatives of $u^k(x)$ for any $k$ in an explicit manner.

Since, as was shown above,

$$u^k(x) = xf[u^{k-1}(x)]$$

we may obtain the desired results by repeated differentiations, and by solving the resulting finite-difference equations. But if $k$ is allowed to approach infinity, and if the system is subcritical, then

$$\lim_{k\to\infty} u^k(x) = \lim_{k\to\infty} u^{k-1}(x) = u(x)$$

Hence for the distribution of the total number produced, we have

$$u(x) = xf[u(x)]$$

By differentiating, we obtain

$$u'(1) = \frac{1}{1 - f'(1)}$$

$$u''(1) = \frac{f'' + f'(1 - f')}{(1 - f')^3}$$

Suppose that we do not start with a single particle but that there exists a source introducing particles in each generation according to probabilities given by a generating function $s(x)$. Then the generating functions for the zeroth, first, and second generations are given by $s(x)$, $s(x)s[f(x)]$, and $s(x)s[f(x)]s[f^2(x)]$, respectively.

In general, the generating function for the distribution of our multiplicative system in the $k$th generation is

$$g_k(x) = s(x)g_{k-1}[f(x)]$$

If the system is subcritical—that is to say, if $\nu < 1$— but sustained by the source, we shall have a limiting distribution for the generating functions. This limiting distribution $F(x)$ will be a solution of the functional equation

$$F(x) = s(x)F[s(x)]$$

Even without solving this equation, one can at once obtain useful statistical information about moments of $F(x)$ by differentiating it. Thus

$$F'(1) = \frac{s'}{1 - f'}$$

and

$$F''(1) = \frac{s''}{1 - (f')^2} + \frac{(2s'f' + f'')s'}{[1 - (f')^2](1 - f')}$$

## 11.4 Multidimensional Branching Processes

So far we have described a branching process starting from a particle producing other particles of the same type. If the particles differ in kind and have different characteristics, such as unequal spatial positions or momenta, we have a branching process of greater generality. By assuming that the positions or momenta, for example, are able to assume discrete values, we are led to the following $N$-dimensional branching process:

Suppose that a system of particles consists of $N$ different types, that a particle of type $i$ can transform into one or more particles of different types, and that the probability of a particle of type $i$ producing

$$j_1 + j_2 + \cdots + j_N$$

particles, $j_k$ of type $k$, is $p_1(i;j_1, \ldots ,j_N)$, $i = 1, 2, \ldots , N$. We assume that for every set of nonnegative integers $j_1, \ldots , j_N$ we have

$$p_1(i;j_1, \ldots ,j_N) \geq 0$$

and for each $i$ we have

$$\sum_j p_1(i; j_1, \ldots, j_N) = 1$$

As before, we imagine now that, starting with a particle, we have a continuing branching process and that, among other things, we are interested in the probability of having so many particles of each type in future generations. We again consider a generating function

$$g_i(x_1, \ldots, x_N) = \sum_j p_1(i; j_1, \ldots, j_N) x_1^{j_1} \cdots x_N^{j_N}$$

which defines the probabilities of progeny at the end of one generation from one particle of type $i$. Hence $x' = G(x)$; explicitly,

$$\begin{aligned} x_1' &= g_1(x_1, \ldots, x_N) \\ &\cdots\cdots\cdots \\ x_N' &= g_N(x_1, \ldots, x_N) \end{aligned}$$

with $0 \leq x_i \leq 1$, defines a *generating transformation* of the unit cube $I_N$ of the euclidean $N$ space into itself. Moreover, abbreviating $(1, \ldots, 1)$ as 1, we see that $G(1) = 1$, so that the point 1 is a fixed point of the transformation $G$.

If, in a given generation, the generating function is

$$f(x_1, \ldots, x_N) = \sum_k q(k_1, \ldots, k_N) x_1^{k_1} \cdots x_N^{k_N}$$

[i.e., the coefficient $q(k)$ is the probability in this generation of the state: $k_i$ particles of type $i$, $i = 1, \ldots, N$] then the generating function of the *next* generation is

$$f(g_1, \ldots, g_N) = \sum_k q(k) \left[\sum_j p_1(1;j) x_1^{j_1} \cdots x_N^{j_N}\right]^{k_1} \cdots \left[\sum_j p_1(N;j) x_1^{j_1} \cdots x_N^{j_N}\right]^{k_N}$$

If we begin with one particle of type $i$, then the generating functions are the following:

For the first generation, $g_i(x_1, \ldots, x_N)$
For the second generation, $g_i(g_1, \ldots, g_N)$
For the third generation, $g_i[g_1(g_1, \ldots, g_N), \ldots, g_N(g_1, \ldots, g_N)]$
. . . . . . . . . . . . . . . . . . . . . . . . . . . . . . . . . . . . . . .

Hence, by adopting the notation $G^k(x)$ for the $k$th iterate of $G(x)$:

$$x' = G^k(x)$$
$$x'_i = g_i^{(k)}(x_1, \ldots, x_N) = \sum_j p_k(i;j) x_1^{j_1} \cdots x_N^{j_N}$$

we see that $g_i^{(k)}(x)$ is the generating function for the $k$th generation of progeny from one particle of type $i$, for $i = 1, 2, \ldots, N$.

*First Moments; Jacobian.* Let

$$f(x_1, \ldots, x_N) = \sum_j q(j) x_1^{j_1} \cdots x_N^{j_N}$$

be the generating function for a particular generation. Then

$$\frac{\partial f}{\partial x_1} = \sum_j q(j) j_1 x_1^{j_1 - 1} x_2^{j_2} \cdots x_N^{j_N}$$

and

$$\frac{\partial f}{\partial x_1}\bigg]_{x=1} = \sum_j q(j) j_1 = \sum_{j_1=0}^{\infty} \bigg[ \sum_{j_2, \ldots, j_N} q(j_1, \ldots, j_N) \bigg] j_1 = \sum_{j_1=0}^{\infty} P(j_1) j_1$$

where $P(j_1)$ is the probability of $j_1$ particles of type 1 in this generation. Hence we define $\partial f/\partial x_1\big]_{x=1}$ as the first moment for particles of type 1; similarly, we define $\partial f/\partial x_j\big]_{x=1}$ as the first moment for particles of type $j$.

We adopt the notation

$$\frac{\partial g_i^{(k)}}{\partial x_j}\bigg]_{x=1} = m_{ij}^{(k)}$$

as the first moment of particles of type $j$ in the $k$th generation of progeny from one particle of type $i$.

Recall that, for a transformation $G(x)$, namely, $g_i(x_1, \ldots, x_N)$, the Jacobian matrix is

$$J(G) = \left[ \frac{\partial g_i}{\partial x_j} \right]$$

where we have exhibited the element in the $i$th row and $j$th column. If $H(x)$ is a second transformation, $h_i(x_1, \ldots, x_N)$, then the Jacobian of the composite transformation $H[G(x)]$, namely, $h_i(g_1, \ldots, g_N)$, is

$$J(H(G))\Big]_x = J(H)\Big]_{G(x)} J(G)\Big]_x$$

since

$$\frac{\partial h_i(g_1, \ldots, g_N)}{\partial x_k} = \sum_j \frac{\partial h_i}{\partial x_j}\bigg]_G \frac{\partial g_j}{\partial x_k}\bigg]_x$$

It follows, since $G^k = G(G^{k-1})$, that

$$J(G^k)_x = J(G)_{G^{k-1}} J(G)_{G^{k-2}} \cdots J(G)_G J(G)_x$$

and, for a fixed point,

$$\bar{x} = G(\bar{x}) \qquad J(G^k)_{\bar{x}} = (J(G)_{\bar{x}})^k$$

That is, the Jacobian of the $k$th iterate of $G$ at a fixed point is the $k$th power of the Jacobian matrix of $G$. In particular, since $1 = G(1)$, we have

$$J(G^k)_1 = (J(G)_1)^k$$

Now

$$J(G^k)_1 = \left[\frac{\partial g_i^{(k)}}{\partial x_j}\right]_1 = [m_{ij}^{(k)}]$$

so that the relationship

$$[m_{ij}^{(k)}] = [m_{ij}^{(1)}]^k$$

exists between the first moments of the $k$th generation and those of the first.

Since

$$(S^{-1}AS)^k = S^{-1}A^kS$$

we have $$[m_{ij}^{(k)}] = S(S^{-1}M^kS)S^{-1} = S(S^{-1}MS)^kS^{-1}$$

where $S^{-1}MS$ is the canonical form of

$$M = [m_{ij}^{(1)}]$$

This permits more rapid computation of $m_{ij}^{(k)}$.

Again the study of properties of a branching system of this general type can be reduced to an investigation of *iterates* of a transformation in $N$ dimensions, giving us a probability flow in time on the $N$-dimensional cube. The *expected values* of the number of the particles of each type are given by a linear transformation defined by the first derivatives of our functions. This matrix $M$ has all its elements nonnegative ($m_{ij}$ may be interpreted as the expected value of the number of particles of type $j$ produced by a particle of type $i$), and we recall a theorem of Frobenius-Perron on the properties of iterates of such a matrix. This theorem guarantees, among other things, that, if the elements are actually all positive and if we start with an arbitrary vector in the positive octant of space (a vector with all its coordinates nonnegative) and apply the matrix to this vector successively, then the iterated images of it will converge, *in direc-*

*tion*, to a unique vector $\bar{v}$ with nonnegative coefficients. Actually, it is sufficient for this purpose that some power of the matrix has all its elements positive. There is a considerable amount of literature on strengthened forms and ramifications of this theorem. The limiting vector $\bar{v}$ has the property that $M(\bar{v}) = \lambda\bar{v}$. That is to say, it is a characteristic vector (eigenvector); $\lambda$ is the characteristic root (eigenvalue) of highest absolute value for the matrix $M$.

The applications of this theorem are obviously important. As an illustration, we may consider a system of neutrons in a spatial assembly that has been divided into $N$ regions. If $m_{ij}$ is interpreted as the average number of neutrons produced, in unit time, in region $j$ by a neutron issuing from region $i$, the above theorem can be interpreted as asserting that a steady state—that is to say, an asymptotic distribution of neutrons —will obtain. In other words, the ratios of the numbers of neutrons in various zones will approach fixed values. These values can be obtained by computing the direction cosines of the vector $\bar{v}$; the number of neutrons in the system will increase geometrically in time, the ratios in successive generations being given by $\lambda$.

The foregoing discussion refers to the expected values or first moments of the numbers in successive generations. In reality, we are given a probability flow and need an analogue of the law of large numbers asserting that with very great probability the actual numbers of particles of each type will approach those given by the theorem. *If the system is supercritical*, this law can be demonstrated.[3,4]

We can see now that a Monte Carlo procedure may be justified, to some extent, by the foregoing theorem. Suppose a problem involves the determination of the asymptotic distribution of neutrons and their growth in time in a given medium. We may subdivide the medium into a large number $N$ of zones, compute the expected values of the $m_{ij}$, and iterate the matrix. An alternative procedure would be, instead of dealing with *all* the regions, to proceed in a stochastic manner (in analogy with our attempt to compute definite integrals). We start with a particle given in a zone and select its direction at random with a correct distribution. We then determine at random again, but with correct chances, the length of the path that the particle traverses before "producing" another particle (or several of them) in a fixed region $j$. This latter distribution is an exponential, the chances of the particle being absorbed or producing other particles varying from region to region that it traverses. In actual problems the situation is even more general. Not only do the different positions of the particle require separate indices, but also the energy of the particle assumes different values; thus our $N$ is actually the product of two numbers: the number of regions and the number of energy intervals for the particle. In many cases, a stochastic calculation of the properties of such a system is more practical than a computation of the eigenvalues and eigenvectors by iteration of very large matrices. A large number of examples illustrating the Monte Carlo approach can be found in the book by Cashwell and Everett.[2] The problems treated there involve the distribution of the number of particles starting from given sources in systems with a given geometry, undergoing various processes, and terminating in a number of different categories. The Monte Carlo problem involves following a large number of particles undergoing, with probabilities that are supposed to be known, events of different types. By a final census of the numbers produced, we obtain an idea of the terminal distribution.

In actual computations, it is often advantageous to use, instead of the literal imitation of the physical process as indicated above, certain special tricks. For example, instead of considering a neutron as producing an integral number of particles 0, 1, 2, . . . (with given probabilities $p_0$, $p_1$, $p_2$, . . .), one may use a *weight* representing an average of the numbers that it may reproduce and follow the change and growth of weights deposited by the particle in various regions of space as the branching process continues.

## 11.5 Statistical Sampling Methods

Even with fast electronic computing machines, there is a premium on the efficacy of statistical sampling methods, since some problems require, for moderate accuracy, enormous numbers of "particles" to be processed. This is especially true when one has a situation involving systems that are just about critical multiplicatively, or a situation requiring estimates of the fraction—very small—of particles that succeed in penetrating given shields involving many mean free paths. This is done by more sophisticated techniques, the so-called *importance sampling* being widely used.[5,6,8]

We might indicate, by an example, the kind of practical problems computed by the statistical sampling method. This is Problem 10 discussed in the book of Cashwell and Everett.[2] A cylinder of cadmium contains a gaseous boron compound and a vacuum space. This is surrounded by a hydrocarbon in a cylindrical outer container. There is a parallel beam of neutrons that are monoenergetic impinging on the base of the outer container. The neutrons can be captured in the hydrogen, in the cadmium, or in the boron; they can be scattered by the first nuclei according to a known scattering law. One is interested in the number of escapes from the base—escape out of the cylinder through the opposite face—the amount of capture, and the energy losses by collisions. The purpose is to estimate the efficiency of capture by boron as a function of the energy of the source and its distance from the axis. The system is to be used as a neutron counter. Each neutron is characterized by many parameters.

It is apparent that in a problem of this sort, in such a complicated geometry, an analytical study through an attempt to solve the integro-differential equations by ordinary numerical methods would be quite impractical.

For a description of another case, a calculation of resonance-capture probability of neutrons in the reactor lattice, we refer the reader to the article by R. D. Richtmyer.[10]

## 11.6 Reactions in a Heavy Nucleus

An instructive case of a branching process showing considerable combinatorial and analytical complexities is provided by the problem of reactions produced in a heavy nucleus by an incident particle of high energy. A nucleus of a heavy atom—say copper or silver—is composed of individual nucleons bound together. Imagine now that a cosmic-ray proton, or a neutron, or a high-energy particle produced in, e.g., the Brookhaven machine—a proton or a $\pi$ meson—strikes this assembly of nucleons. One of the nucleons constituting the target may be knocked off, with some of the energy of the incident particle lost, and the incident particle may continue inside the assembly of other nucleons, exciting them by elastic collisions or producing new particles on the way. The variety of the new particles produced is quite great. They may be $\pi$ mesons, $k$ mesons, many of the so-called strange particles, and $\gamma$ rays. Such production of "strange" particles will occur if the incident particle has enough energy—over a few billion electron volts. We have here a great variety of possible branching processes characterized by transmutation of indices denoting the various particles, their positions inside the nucleus, their equations, and their energies. For high energies, the mechanics of the processes has to be treated relativistically. For lower energies of the incident particle, the problem deals mainly with the question of "evaporation" of constituent nucleons. A 100-Mev proton may knock off several nucleons after exciting "thermally" the whole nucleus. M. L. Goldberger[7] treated this problem by a Monte Carlo method. A study of phenomena at higher energies involving pion production was made by N. Metropolis and A. Turkevitch.[9] The most recent work including the study of the strange-particle production is that of L. Sartori, A. Werbrouck, and J. Wooten, to appear in *Physical Review*.

## 11.7 The Petit Canonical Ensemble

An interesting study of a problem in statistical mechanics by a Monte Carlo method was undertaken in recent years to obtain statistical mechanical averages in the so-called *petit canonical ensemble*. The problem is to obtain thermodynamic properties of dense fluids. We consider $N$ spherical molecules occupying, say, a cubic volume and having a given temperature, and we try to obtain an insight into the spatial distributions of these molecules, which are close to a hard packing, by letting them move from one position in configuration space into another by performing, one at a time, allowed displacement of each molecule. We assume various forces between the molecules. One limiting case is that of hard spheres, but we may consider forces between pairs of these molecules such as, e.g., the Lennard-Jones potentials. Various thermodynamic properties, e.g., distribution functions, or pressures, may be obtained by averaging over the class of possible configurations. The equations of state and transition phases could be obtained from such an empirical study on computing machines with rather surprising indications of the real facts by using for $N$ such small numbers as 128 or even 32. In reality, $N$ is of the order of $10^{23}$ $cm^{-3}$.

The problem involved, of course, hundreds of thousands of configurations obtained from the starting one by an allowed elementary displacement and computations of the thermodynamic quantities given by configurations. It is impossible here to summarize the results and the procedure. The reader is referred to the articles by M. Rosenbluth, Metropolis, Teller, et al.,[9] and also the more recent work by Z. W. Salsburg, W. W. Wood, et al.[11,15,16]

In all these problems, the branching or multiplicative feature does not appear. The problems involve, rather, a Markovian process of transitions in a space of a high number of variables from one point to another, and instead of multiplication of the matrices, they employ a sampling process on individual chains of such transitions, sampled by many runs.

## 11.8 Iterates of Transformations, Ergodic Properties, and Time Averages

We have seen how the Monte Carlo method can be used to obtain information concerning the behavior of problems that involve iteration of operators—either for the study of a steady state, if it is established, or for an investigation of the history of these processes in time. (The random numbers used in the method are themselves also obtained by an iteration procedure!) The theorems regarding the existence of a steady state can often be obtained from generalizations of the Frobenius-Perron theorem concerning the iterates of linear transformations with nonnegative coefficients. The probabilistic branching process can be described by the iteration of the generating transformation so chosen that its first moments form this linear-transformation matrix.

In case the linear operator to be iterated is not positive in the sense of Frobenius and Perron, there will in general be no convergence of the vectors, under iteration, to a fixed direction. Nevertheless, the *ergodic* properties of the iterates of an arbitrary linear transformation allow us to obtain information concerning the properties of *time averages* of the sequence of iterates.[14] Suppose $A$ is an arbitrary linear transformation of the $n$-dimensional euclidean space. We consider $A^n(\mathbf{r})$, where $\mathbf{r}$ is any vector. If $A$ represents a rotation in the plane through an irrational angle, the sequence of vectors $A^n(\mathbf{r})$ will not converge. On the contrary, these vectors will become *uniformly dense* on the circumference of a circle; this is a theorem of Weyl. In a higher number of dimensions, the theorem of Kronecker asserts that in case $A$ is a rotation in $n$ space with the angles of rotation in each plane rationally independent, the points formed by the iteration of the vector will lie densely on the surface of the $n$-dimensional torus. Weyl's theorem permits us to assert that they will be *uniformly* dense, in the sense that if we take an arbitrary area on the torus and normalize the entire area of the torus to 1, then the fraction of the iterates of the vector $A^i(\mathbf{r})$ that fall into this area, divided by $N$, will tend to the area of $A$. The behavior of the iterates of a rotation is therefore in a sense diametrically opposite to that of the iterates of a matrix with positive coefficients. One can prove, however, for an arbitrary matrix $A$, the following property: Consider the iterates $A^i(\mathbf{r})$ and the points on the unit sphere defined, for $i = 1, 2, \ldots$, by

$$\mathbf{p}_i = \frac{A^i(\mathbf{r})}{|A^i(\mathbf{r})|}$$

Let $R$ be any region on the surface of the unit sphere. The sojourn time for the $\mathbf{p}_i$'s, under the iterates of the matrix, that fall into the region $R$ *exists* for almost every starting vector $\mathbf{r}$.

One might be interested in the properties of iterates of transformations that are more general than the linear ones, in particular, in case $A$ is a *quadratic* transformation—that is, a transformation such that the coordinates of the transformed point are quadratic functions of the original point. Such transformations occur in describing processes involving *binary* reactions between particles. In the problems discussed above, each particle behaved, in a given generation, independently of the other particles; i.e., the chances determining the fate of a neutron in a given generation were assumed to be independent of the fate of the coexisting neutrons. This is reasonable if the time can be chosen sufficiently short. In problems encountered in genetic studies and in biological processes, this is not so; quite to the contrary, the production of the new generation may be determined by pairs of particles. A mathematical study of the evolution of such processes could then necessitate a study of iterates of quadratic transformations.

As an illustration of such a problem, consider a system of a large number $N$ of particles, each of which is of one of $k$ possible "colors." Assume that in a given generation our particles combine in pairs at random and that each pair produces exactly two particles with colors that are given functions of the colors of the parents. The original particles, we may assume, disappear in the generation. In the next generation this process continues. We might be interested in the population ratios of particles of each kind—in particular, whether these will tend to a steady distribution. Assume, for example, that $k = 3$ and that we have the following rule determining the index of the offspring as a function of the indices of its parents. A pair of particles of type 1 and 1 produces particles of type 1. Particles of type 1 with particles of type 2 produce again particles of type 1. Particles of type 1 with particles of type 3 produce particles of type 2. Two particles of type 2 produce together particles of type 3. Particles of type 2 with particles of type 3 produce particles of type 3. Finally, two particles of type 3 produce particles of type 1. If $x_1$, $x_2$, $x_3$ denote the fractions of each kind in the total population originally present, among the $N$ particles, then the new proportions, $x_1'$, $x_2'$, $x_3'$, in the second generation will be given by the formulas

$$\begin{aligned} x_1' &= x_1^2 + x_3^2 + 2x_1x_2 \\ x_2' &= 2x_1x_3 \\ x_3' &= x_2^2 + 2x_2x_3 \end{aligned}$$

These formulas define a quadratic transformation of the three-dimensional space into itself. Actually, since $x_1 + x_2 + x_3 = 1$ and, in virtue of our special convention, $x_1' + x_2' + x_3' = 1$ also, the transformation is really one of a two-dimensional triangular area into itself. The proportions of the population of each kind will be given in subsequent generations by iteration of this transformation. The particular transformation corresponds to our particular rule. There are many different rules of this kind possible—actually for $k = 3$, 97 nonequivalent (by permutation of indices) ones—and a study of properties of iterates of all these was made by P. Stein and the author.[12] Under iteration of some of the transformations of the above type, any nondegenerate initial vector converges to a unique limiting value. In some other cases, the behavior approaches a periodic permutation among a finite number of limiting vectors. (In one case, for $k = 4$, there is a convergence of initial vectors to one of 12 fixed vectors permutating successively among themselves!)

For $k > 3$, the number of different cases increases rapidly. The ergodic properties of such transformations are not well known, and a study of them by sampling methods would have heuristic value: The transformation that we wrote above refers to the *expected* values of the fractions of particles of the different types. For a finite—even quite large—$N$, fluctuations would make the process deviate from the behavior predicted by means of the first moments.

It may be conjectured that, at least for transformations of the above type, the *time-average limits* exist for almost every initial condition. That is to say, if we denote by $x_i^n$ the fraction of the population of the $i$th type in the $n$th generation, then

$$\lim_{N\to\infty} \frac{1}{N} \sum_{n=1}^{N} x_i^n$$

exists for each $i$, $i = 1, \ldots, k$, and for almost every vector

$$\mathbf{x} = (x_1, x_2, \ldots, x_k)$$

with

$$\sum_{i=1}^{k} x_i = 1$$

The transformations like the above are, of course, still of a very special form. The assumption that each pair produces exactly two particles with identical indices may be generalized to allow a variable number of offspring with possibly different characteristics. The equations describing the expected values of the particles of each kind will still be quadratic, but with nonintegral coefficients. If the system of this type should be supercritical, that is to say, the expected value of the total number of particles should increase to infinity, an analogue of the theorem on branching processes mentioned may hold: In the space of all possible branching processes based on binary reactions as above, those processes that lead to population ratios approaching the ratios predicted by the "deterministic" method—that is to say, the limiting values given by iterates of the transformations of expected values—have measure equal to the measure of all processes. In other words, with great probability the probabilistic processes will converge to the states given by the iterates of the transformation concerning the first moments.

## REFERENCES

1. Brown, George W., Monte Carlo Methods, chap. 12 in "Modern Mathematics for the Engineer," First Series, edited by E. F. Beckenbach, McGraw-Hill Book Company, Inc., New York, 1956.
2. Cashwell, E. D., and C. J. Everett, "A Practical Manual on the Monte Carlo Method," Pergamon Press, New York, 1959.
3. Everett, C. J., and S. Ulam, "Multiplicative Systems," I, AECD-2164, II, AECD-2165, III, AECD-2532, Technical Information Branch, Oak Ridge, Tenn., 1948.
4. ——— and ———, Multiplicative Systems, I, *Proc. Nat. Acad. Sci. U.S.A.*, vol. 34, pp. 403–405, 1948.
5. Goad, W., and R. Johnston, A Monte Carlo Method for Criticality Problems, *Nuc. Sci. and Eng.*, vol. 5, pp. 371–375, 1959.

6. Goertzel, G., "Quota Sampling and Importance Functions in Stochastic Solution of Particle Problems," AECD-2793, Technical Information Branch, Oak Ridge, Tenn., 1949.
7. Goldberger, M. L., The Interaction of High Energy Neutrons and Heavy Nuclei, *Phys. Rev.*, vol. 74, pp. 1269–1277, 1948.
8. Kahn, H., Random Sampling (Monte Carlo) Techniques in Neutron Attenuation Problems. I, II, *Nucleonics*, vol. 6, pp. 27–33, 37, 60–65, 1950.
9. Metropolis, N., et al., Monte Carlo Calculations on Intranuclear Cascades I. Low Energy Studies, *Phys. Rev.*, vol. 110, pp. 185–203, 1958; II. High Energy Studies and Pion Processes, *Phys. Rev.*, vol. 110, pp. 204–219, 1958.
10. Richtmyer, R. D., Monte Carlo Methods, in "Symposium on Nuclear Reactor Theory," edited by Garrett Birkhoff, American Mathematical Society (to appear).
11. Salsburg, Z. W., et al., Application of the Monte Carlo Method to the Lattice Gas Model, *J. Chem. Phys.*, vol. 30, pp. 65–72, 1959.
12. Stein, P., and S. Ulam, "Quadratic Transformations, I," LA-2305, Office of Technical Services, U.S. Department of Commerce, 1959.
13. Taussky, Olga, and John Todd, Generation and Testing of Pseudo-random Numbers, in "Symposium on Monte Carlo Methods," John Wiley & Sons, Inc., New York, 1956.
14. Wiener, Norbert, The Theory of Prediction, chap. 8 in "Modern Mathematics for the Engineer," First Series, edited by E. F. Beckenbach, McGraw-Hill Book Company, Inc., New York, 1956.
15. Wood, W. W., and F. R. Parker, Monte Carlo Equation of State of Molecules Interacting with the Lennard-Jones Potential. I. A Supercritical Isotherm at about Twice the Critical Temperature, *J. Chem. Phys.*, vol. 27, pp. 720–733, 1957.
16. ——— et al., Recent Monte Carlo Calculations of the Equation of State of Lennard-Jones and Hard Sphere Molecules, *Nuovo Cimento*, ser. 10, vol. 9, pp. 133–143, 1958.

# 12

# Difference Equations and Functional Equations in Transmission-line Theory†

**RAYMOND REDHEFFER**
PROFESSOR OF MATHEMATICS
UNIVERSITY OF CALIFORNIA, LOS ANGELES

## 12.1 Introduction

Historically, the subject of this chapter starts in 1862 with a paper by G. G. Stokes, in which difference equations governing the optical behavior of $m + n$ identical glass plates are written down and solved by treating the discrete variable $n$ as continuous. It is probable that Stokes attached no physical significance to the continuous problem but regarded it merely as a mathematical artifice. If one applies Stokes' method to a dielectric sheet of thickness $m + n$, however, one gets the same equations as before, except that $m$ and $n$ are really continuous. The analysis can thus be thought of as solving the cascade problem by the construction of a suitable dielectric medium and looking at the behavior for integral values of the thickness.

Not much more appears to have been done along these lines until the advent of microwave technology during and after the Second World War. In this period, it was found that equations similar to Stokes' can be used to form an algebra of obstacles, which yields an interesting approach to certain aspects of transmission-line theory. Further impetus was given by the simultaneous development of the theory of the *scattering matrix*, and the subject has by now attained maturity.

The reader unfamiliar with engineering terminology may welcome a brief digression at this point. What we have in mind is a "disturbance"

† The preparation of this chapter was sponsored, in part, by the Office of Naval Research. 

or "wave" that propagates through an obstacle. The scattering matrix specifies the obstacle by its transmission and reflection coefficients, which are defined as follows. The left-hand transmission coefficient $t$ is the complex amplitude of the wave emerging from the right when a wave of amplitude 1 is incident at the left; thus, $|t|$ is the amplitude of the wave, while $\angle t$ is its phase. The left-hand reflection $r$ is the complex amplitude of the reflected wave when a wave of amplitude 1 is incident at the left, and similar definitions hold for the right-hand transmission $\tau$ and reflection $\rho$. The obstacle is supposed to be followed by a matched termination; that is, the terminating reflection $r_1$ is 0 when $r$ is measured, and similarly for $\rho$. Also, the propagation takes place in a uniform transmission line. To specify each transmission and reflection by a single number, it is assumed that only one mode is relevant at the given frequency, which remains fixed throughout the discussion.

Since the frequency is fixed, the main interest is in a space variable giving position along the line. Adding obstacles in succession leads to difference equations for the coefficients, whereas adding to the thickness of a continuous medium leads to functional equations. These equations, naturally, must be distinguished from those of network analysis and synthesis, in which the frequency is varied while the network is held constant. Their closest affinity is not to the methods of circuit analysis, but to a general class of methods for which Bellman has coined the descriptive phrase *invariant imbedding.* The latter have proved useful not only in circuit theory but also in probability, in neutron diffusion, and in other areas of mathematical physics.

The meaning of the term "invariant imbedding" is aptly illustrated by the functional equations for an inhomogeneous dielectric medium. The response of the part of the medium from $x$ to $y$, with $y > x$, is determined as if this part were in free space—in other words, without regard to the rest of the medium in which it is imbedded. The resulting equations, naturally, are *invariant* as far as this imbedding is concerned, and one speaks of the "method of invariant imbedding."

In the early part of this chapter, we present a résumé of certain topics in transmission-line theory using the combination of networks† with each other as leading idea. The discussion is then generalized, first to continuous variation, then to matrices, and finally to operators on a Hilbert space (see Chap. 4). Occasion is taken by the way to indicate an unexpected analogy between transmission-line theory and the theory of probability. The practical bearing of these methods is illustrated by examples in the form of problems, and the specific physical situation is kept firmly

† The colorless words *obstacle* and *object* are often used, since *network* conjures up a too vivid image of a resistance-inductance-capacitance (*RLC*) circuit. The obstacles need not be bilateral.

in mind throughout the discussion. There nevertheless emerges a unified mathematical structure, which has applications to several fields of technology.

## THE ALGEBRAIC FOUNDATIONS

### 12.2 An Instructive Special Case

Some interesting conclusions concerning the interaction of networks are suggested by the following example. Suppose a plane panel transmits 1 per cent of the power in a monochromatic plane wave incident on it and reflects 99 per cent, no power being absorbed. Converting from power to amplitude requires taking the square root, so that

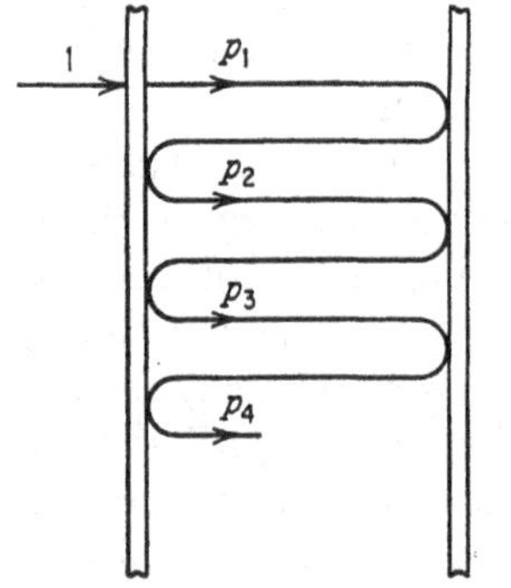


**Fig. 12.1** Two lossless plane sheets spaced for maximum transmission.

$$|t| = \sqrt{0.01} \qquad \text{and} \qquad |r| = \sqrt{0.99}$$

represent the absolute value of the amplitude transmission and reflection coefficients, respectively.

Let this panel be followed by another just like it, as shown in Fig. 12.1, at such a distance as to maximize the over-all transmission. For maximum transmission, the round-trip path from $p_1$ to $p_2$ in the figure (with due regard to the phase shift introduced by the two reflections) must be a whole number of wavelengths. In that case all the waves at $p_1$, $p_2$, $p_3$, . . . add in phase, so that the resulting amplitude of the wave moving from left to right between the panels is

$$\sqrt{0.01}\,[1 + 0.99 + (0.99)^2 + \cdots] = \frac{0.1}{1 - 0.99} = 10 \qquad (12.1)$$

Upon letting this wave traverse the second sheet, we get the value $(10)(0.1) = 1$.

The result has an air of paradox. It is as if we had a security guard who lets only 1 per cent of the visitors through; to make security doubly sure, that guard is followed by a second one with the same duty; and then we find, to our surprise, that everyone gets past both guards.

The reason for this odd behavior in the optical example is not hard to find. The two panels form a resonant cavity, so that the field between them is much stronger than the incident field. The attenuation introduced by the second panel is just what is needed to cut that stronger field down to the incident value, 1.

Since the over-all transmission is 100 per cent, the over-all reflection must be zero. The reflection depends, however, on the phase shifts associated with the complex reflection coefficient $r$ and transmission coefficient $t$. Only for certain values of $\angle t - \angle r$ does the over-all reflection in fact reduce to 0. Thus, we see that a relation involving phases can be deduced even though the underlying physical principle, conservation of energy, involves the magnitudes alone.

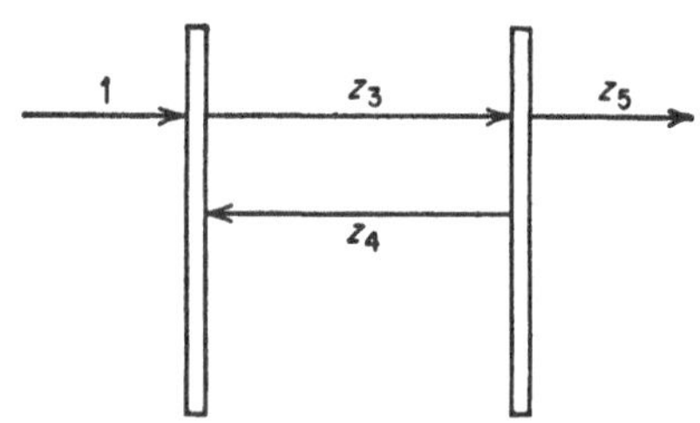

**Fig. 12.2** Resultant amplitudes of waves in Fig. 12.1.

The foregoing results can be obtained without introduction of infinite series. In Fig. 12.2, the quantities $z_i$ represent the absolute value of the amplitudes of the waves traveling in the indicated directions and regions. By superposition, $z_3$ is equal to the part of the incident wave, 1, that is transmitted, plus the part of $z_4$ that is reflected:

$$z_3 = \sqrt{0.01} + \sqrt{0.99}\, z_4 \tag{12.2}$$

Similarly, $z_4 = \sqrt{0.99}\, z_3$ and $z_5 = \sqrt{0.01}\, z_3$. These equations imply $z_5 = 1$, as before.

## 12.3 The Composition of Networks in General

Let the obstacle with coefficients $t$, $\tau$, $r$, and $\rho$ in Fig. 12.3 be placed adjacent to a second obstacle having coefficients $t_1$, $\tau_1$, $r_1$, and $\rho_1$ as shown

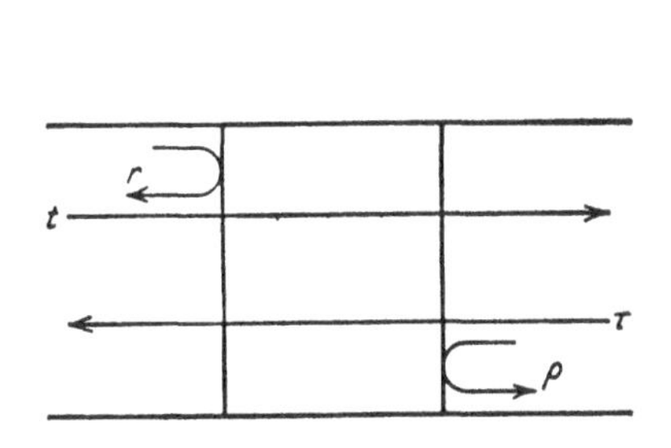

**Fig. 12.3** An obstacle with left-hand coefficients $(t,r)$ and right-hand coefficients $(\tau,\rho)$.

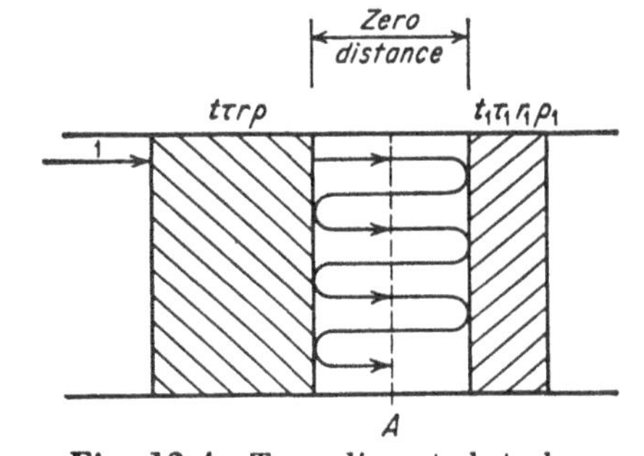

**Fig. 12.4** Two adjacent obstacles.

in Fig. 12.4. If the incident wave has amplitude 1, then the complex amplitude of the wave moving from left to right between the obstacles is

$$A = t(1 + r_1\rho + r_1^2\rho^2 + r_1^3\rho^3 + \cdots) = \frac{t}{1 - r_1\rho}$$

provided $|r_1\rho| < 1$, which is assumed without comment henceforth. Upon letting this resultant wave progress through the second obstacle, we get

$$\frac{tt_1}{1 - r_1\rho}$$

for the transmitted amplitude. Similarly, letting the wave $A$ be reflected at the second obstacle and then transmitted from right to left through the first gives

$$r + \frac{t\tau r_1}{1 - r_1\rho}$$

as the amplitude of the reflected wave.

Since the incident amplitude was 1, these expressions give the left-hand transmission and reflection coefficients, respectively. In the same way, we can determine the right-hand coefficients. Upon introducing the *scattering matrix*

$$S = \begin{pmatrix} t & \rho \\ r & \tau \end{pmatrix} \equiv (t\tau r\rho)$$

we can summarize the four equations as follows:

$$\begin{pmatrix} t & \rho \\ r & \tau \end{pmatrix} * \begin{pmatrix} t_1 & \rho_1 \\ r_1 & \tau_1 \end{pmatrix} = \begin{pmatrix} t_1(1 - \rho r_1)^{-1}t & \rho_1 + t_1\rho(1 - r_1\rho)^{-1}\tau_1 \\ r + \tau r_1(1 - \rho r_1)^{-1}t & \tau(1 - r_1\rho)^{-1}\tau_1 \end{pmatrix} \tag{12.3}$$

Here, the *star product* denotes the scattering matrix for the composite obstacle consisting of the first immediately followed by the second.

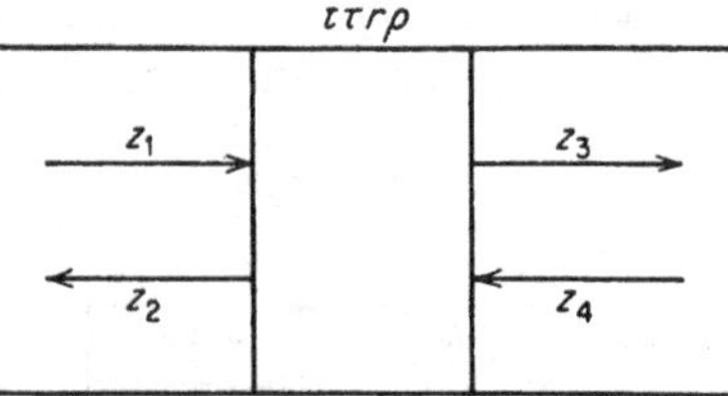

**Fig. 12.5** Resultant amplitudes of waves in the presence of an obstacle.

This star multiplication imposes a definite algebraic structure on our obstacles—a structure so detailed that it can be used as the definition of the obstacles themselves. From that standpoint a linear four-terminal network is a quadruple of complex numbers that combines with other quadruples in the manner prescribed by Eq. (12.3).

A somewhat different approach to Eq. (12.3) proves useful for the further mathematical development. In Fig. 12.5, the complex numbers $z_i$ represent amplitudes of waves propagating as shown, so that odd subscripts refer to propagation from left to right and even subscripts from right to left. Superposition combines with the definition of $t$, $\tau$, $r$, and $\rho$ to yield

$$\begin{pmatrix} z_3 \\ z_2 \end{pmatrix} = \begin{pmatrix} t & \rho \\ r & \tau \end{pmatrix} \begin{pmatrix} z_1 \\ z_4 \end{pmatrix} \tag{12.4}$$

as in the discussion accompanying Fig. 12.2. Thus, the scattering matrix is associated with a certain linear transformation.

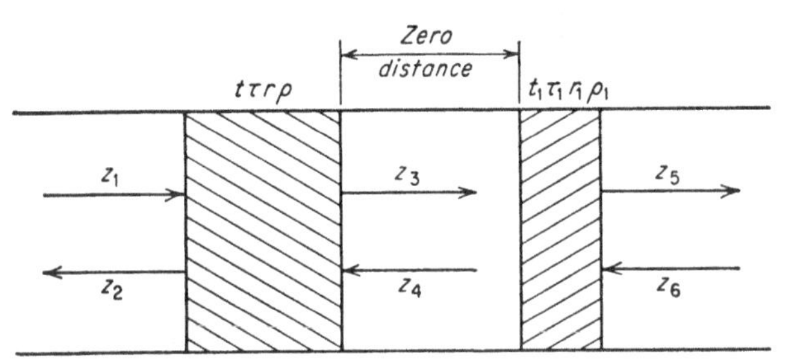


**Fig. 12.6** Amplitudes of waves in the presence of two adjacent obstacles.

Let us determine the transformation associated with the star product. In Fig. 12.6, we have the relation (12.4) and also

$$\begin{pmatrix} z_5 \\ z_4 \end{pmatrix} = \begin{pmatrix} t_1 & \rho_1 \\ r_1 & \tau_1 \end{pmatrix} \begin{pmatrix} z_3 \\ z_6 \end{pmatrix} \tag{12.5}$$

Upon solving for $z_5$ and $z_2$ in terms of $z_1$ and $z_6$, we get

$$\begin{pmatrix} z_5 \\ z_2 \end{pmatrix} = \begin{pmatrix} t_1(1 - \rho r_1)^{-1}t & \rho_1 + t_1\rho(1 - r_1\rho)^{-1}\tau_1 \\ r + \tau_1 r(1 - \rho r_1)^{-1}t & \tau(1 - r_1\rho)^{-1}\tau_1 \end{pmatrix} \begin{pmatrix} z_1 \\ z_6 \end{pmatrix} \tag{12.6}$$

Since the scattering matrix is wholly determined by the associated linear transformation, the matrix in the foregoing equation must be the scattering matrix for the composite obstacle. This yields a new proof of Eq. (12.3). Also, Eq. (12.3) has been shown consistent with Eqs. (12.4) and (12.5), just as the physics of the problem suggests.

## EXERCISES

**1.** Show that the scattering matrix for a length $x$ of transmission line with propagation constant $k$ is $(t\tau r\rho) = (e^{jkx}e^{jkx}00)$. The line is *lossless* if $|t| \equiv 1$, *passive* if $|t| \leq 1$ for $x > 0$. What can you say about $k$ in these two cases?

**2.** An obstacle is in *shunt* if the electrical fields ($E$ fields) on each side are always equal. Show that $(t\tau r\rho)$ is in shunt if and only if

$$t = 1 + r \qquad \text{and} \qquad \tau = 1 + \rho$$

(In the notation of Fig. 12.5, the desired condition is $z_1 + z_2 \equiv z_3 + z_4$.)

**3.** In the notation of Fig. 12.7, show that

$$\frac{|E + E'|^2}{|1 + p + t|^2} = \left| \frac{1 - pr + rt}{1 - pr - g[p + r(t + p)(t - p)]} \right|^2$$

where $g$ is the reflection coefficient looking toward the generator at point $A$ and $r$ is

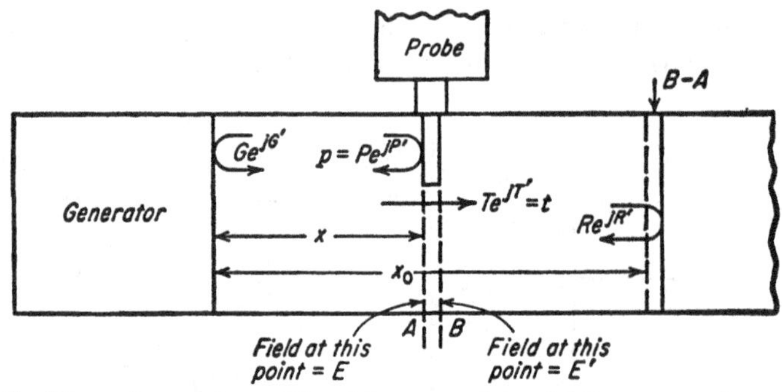


**Fig. 12.7** Measuring probe, with arbitrary probe reflection, generator mismatch, and load mismatch.

the reflection coefficient looking toward the load at point $B$. If the propagation constant is $k$, then

$$g = Ge^{j(G'+2kx)} \qquad r = Re^{j(R'+2kx_0-2kx)}$$

(The probe is symmetric, and the generator power to a *matched* line is unity. See Exercise 6.)

**4.** In Exercise 3 the probe is connected to a detector and voltmeter such that the reading $V$ is proportional to $|E + E'|^2$. The power standing-wave ratio $(\text{SWR})^2$ is defined by $(\text{SWR})^2 = V_{\max}/V_{\min}$ as the probe is moved along the line. Assuming a lossless line and neglecting the effect of the probe, show that the load reflection is

$$R = \frac{\text{SWR} - 1}{\text{SWR} + 1}$$

independently of the generator reflection $G$. (Take $p = 0$, $t = 1$ in the result of Exercise 3 and vary $x$.)

**5.** In Exercise 12.4 suppose that the probe reflection is not negligible but that $R = 0$. Show that the apparent standing-wave ratio is

$$(\text{SWR})_{\text{measured}} = \frac{1 + PG}{1 - PG}$$

and thus describe an experimental method of getting $G$.

**6.** A generator with complex reflection $g$ (looking toward the generator) would deliver a wave of amplitude $A_0$ to a matched line at its right. If the line is not matched but is terminated by the complex reflection $R_1$, as seen from the generator, show that the complex amplitudes of waves moving from left to right and from right to left at the generator are, respectively,

$$\frac{A_0}{1 - R_1 g} \qquad \text{and} \qquad \frac{A_0 R_1}{1 - R_1 g}$$

Obtain the field by adding these expressions.

## 12.4 Matrix Multiplication

The algebra of four-terminal networks is known to be equivalent to the algebra of 2 by 2 complex matrices, but the multiplication in Eq. (12.3) is certainly not matrix multiplication. To reconcile these two observations, note that Eq. (12.4) can be written as a relation between $(z_2,z_1)$ and $(z_4,z_3)$:

$$\begin{pmatrix} z_4 \\ z_3 \end{pmatrix} = \begin{pmatrix} \tau^{-1} & -\tau^{-1}r \\ \rho\tau^{-1} & t - \rho\tau^{-1}r \end{pmatrix} \begin{pmatrix} z_2 \\ z_1 \end{pmatrix} \tag{12.7}$$

When the first obstacle is followed by a second as in Fig. 12.6, we have a similar relation between $(z_4,z_3)$ and $(z_6,z_5)$. The matrix taking $(z_2,z_1)$ into $(z_6,z_5)$ is the ordinary matrix product, and this gives an alternative description of Eq. (12.6).

It follows that the correspondence

$$\begin{pmatrix} t & \rho \\ r & \tau \end{pmatrix} \leftrightarrow \begin{pmatrix} \tau^{-1} & -\tau^{-1}r \\ \rho\tau^{-1} & t - \rho\tau^{-1}r \end{pmatrix}' \tag{12.8}$$

where the prime denotes the transpose, is an isomorphism. The star product of two matrices on the left corresponds, under this relationship, to the matrix product of the two matrices on the right. This shows, incidentally, that star multiplication is associative, i.e., that

$$S_1 * (S_2 * S_3) = (S_1 * S_2) * S_3$$

for 2 by 2 complex matrices $S_i$. A physical interpretation is readily given.

For purposes of algebraic description, either the star product or the matrix product is equally appropriate. The star product applies if our four-terminal networks are specified by their reflection and transmission coefficients $t$, $\tau$, $r$, $\rho$, whereas the matrix product must be used if the network is specified by the four complex numbers

$$\frac{1}{\tau} \qquad \frac{\rho}{\tau} \qquad -\frac{r}{\tau} \qquad \frac{t\tau - r\rho}{\tau}$$

Sooner or later it must be stated that $\tau \neq 0$, if we wish to contemplate the foregoing expressions without uneasiness. A good many of the following results also require $\tau \neq 0$, or $t \neq 0$, or both, and these conditions are hereby assumed once and for all. By Eq. (12.3), $t\tau \neq 0$ in the star product if $t\tau \neq 0$ in each factor.

## EXERCISES

**1.** With $d = t\tau - r\rho$, let the obstacle be specified by

$$Q = \frac{1+d}{2t} - \frac{r+\rho}{2t}\mathbf{i} + \frac{r-\rho}{2jt}\mathbf{j} - \frac{1-d}{2jt}\mathbf{k}$$

where $j = \sqrt{-1}$ and $\mathbf{i}$, $\mathbf{j}$, $\mathbf{k}$ are the units of quaternion algebra. Show that the star-product rule for $(t\tau r\rho)$ makes these symbols combine by the rules of quaternion multiplication.

**2.** Show that the angles $\phi_0 = \angle t + \angle\tau - \angle r - \angle\rho$ and $\phi_1 = \angle t - \angle\tau$ are invariant under addition of lossless line lengths; i.e., they are unchanged if $(t\tau r\rho)$ is replaced by

$$(aa00) * (t\tau r\rho) * (bb00) \qquad \text{with } |a| = |b| = 1$$

**3.** Prove that the expressions $\phi_0$ and $\phi_1$ of Exercise 2 are the only phases invariant under addition of line lengths, in the following sense: If $f(\angle t, \angle\tau, \angle r, \angle\rho)$ is such an invariant, then

$$f(\angle t, \angle\tau, \angle r, \angle\rho) \equiv F(\phi_0, \phi_1)$$

for some function $F$.

## 12.5 Lossless Networks and the Reciprocity Theorem

Since power is proportional to $|\text{amplitude}|^2$, the condition for a network to be *lossless* is

$$|t|^2 + |r|^2 = 1 \qquad |\tau|^2 + |\rho|^2 = 1 \tag{12.9}$$

If we agree that the composition of two lossless networks is lossless, and that a network remains lossless when turned end for end,† we can deduce a reciprocity theorem, and also a relation among the phases.

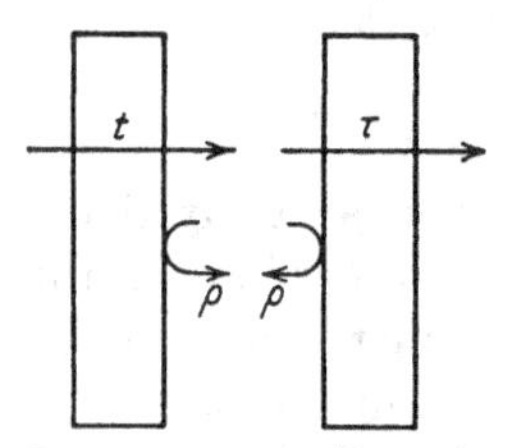


**Fig. 12.8** Obstacle followed by its mirror image at such a distance as to maximize the overall power transmission.

Let $(t\tau r\rho)$ and $(\tau t\rho r)$ be spaced for maximum transmission (Fig. 12.8). The maximum transmission is, in magnitude,

$$|t|(1 + |\rho|^2 + |\rho|^4 + \cdots)|\tau| = \frac{|t|\,|\tau|}{1 - |\rho|^2}$$

as in the discussion of Eq. (12.1). Since the transmission does not exceed 1, we get

$$|t|\,|\tau| \le 1 - |\rho|^2 = |\tau|^2$$

Thus, $|t| \le |\tau|$. Turning both networks end for end in Fig. 12.8 and repeating the argument gives $|\tau| \le |t|$. These two inequalities together show that $|t| = |\tau|$.

The full reciprocity theorem $t = \tau$ cannot be deduced by any argument involving power transfer at the terminals only; the equality $\angle t = \angle\tau$ requires information about the interior of the obstacle. If, however, the two obstacles in Fig. 12.8 are separated by an electrical distance $x$ of lossless line, the transmission

$$\left| \frac{t\tau e^{jx}}{1 - \rho^2 e^{2jx}} \right|$$

† The proof given here tacitly assumes the further hypothesis: "An arbitrary length of lossless line is lossless." The reader sufficiently steeped in pure mathematics to be bothered by this assumption will easily show that it is dispensable.

is maximum when $x = -\angle\rho$. (This choice of $x$ leads to the situation analyzed in the foregoing derivation.) Since

$$|t|^2 = |\tau|^2 = 1 - |\rho|^2$$

the maximum value is 1, so that the over-all reflection is

$$r + \frac{\rho t\tau e^{2jx}}{1 - \rho^2 e^{2jx}} = 0$$

The latter equation is equivalent to

$$\phi_0 \equiv \angle t + \angle\tau - \angle r - \angle\rho = (2n + 1)\pi \tag{12.10}$$

where $n$ is an integer. Here is a relationship among the phases that does follow from energy considerations at the terminals, even though the expected equality $\angle t = \angle\tau$ does not.

## EXERCISES

**1.** Prove that every lossless obstacle $(t\tau r\rho)$ with $r \neq 0$ satisfies the equation

$$t\tau - r\rho = -\exp[j(\angle r + \angle\rho)]$$

**2.** If the lossless obstacle $(t\tau r\rho)$ is followed by the complex reflection $r_1$, show that the over-all left-hand reflection is

$$e^{j\angle r}\frac{|r| - |r_1|e^{j\phi}}{1 - |r|\,|r_1|e^{j\phi}} \qquad \text{where } \phi = \angle\rho + \angle r_1$$

(Note that $|r| = |\rho|$ since $|t| = |\tau|$, and use Exercise 1.)

**3.** A lossless obstacle is followed by a second one at such a distance as to maximize the over-all left-hand reflection. Show that the phase shift associated with the left-hand reflection is the same as it was before the second obstacle was added. What if the spacing is such as to minimize the over-all reflection? (Use Exercise 2 with $\phi$ variable.)

**4.** The transformation from normalized impedance to reflection is

$$Z = \frac{1 - r}{1 + r}$$

If this function is denoted by $[\![r]\!]$, show that $[\![\tan\theta]\!] = \tan(\frac{1}{4}\pi - \theta)$ and

$$[\![1/r]\!] = -[\![r]\!] \qquad [\![-r]\!] = 1/[\![r]\!] \qquad [\![[\![r]\!]]\!] = r$$

Show also that $[\![(r + r_1)/(1 + rr_1)]\!] = [\![r]\!][\![r_1]\!]$.

**5.** The standing-wave ratio associated with a reflection $r$ is SWR $= [\![-|r|]\!] \equiv (1 + |r|)/(1 - |r|)$; cf. Sec. 12.3, Exercise 4. If two lossless obstacles are spaced for maximum reflection, show that the corresponding standing-wave ratio is the product

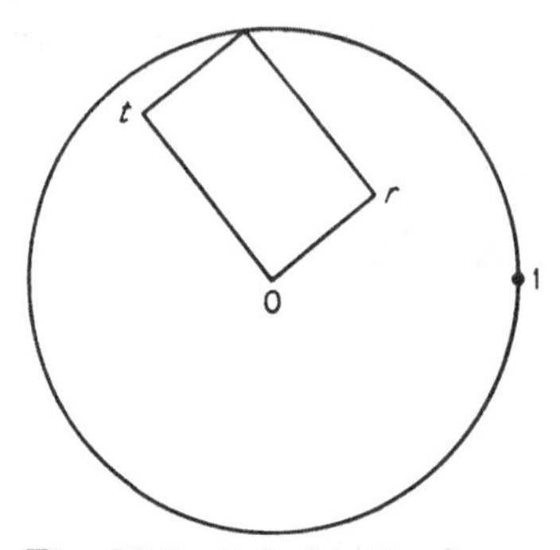


**Fig. 12.9** Relationship of complex transmission and complex reflection for symmetrical, lossless obstacles.

of the individual standing-wave ratios. What if they are spaced for minimum reflection? (Use Exercise 2 with $\phi$ variable, and also Exercise 4.)

**6.** Show that for a symmetrical lossless obstacle, $r$ and $t$ are related as in Fig. 12.9.

## 12.6 Passive Networks

A *dissipative* or *passive* network produces no energy, so that Eqs. (12.9) are replaced by

$$|t|^2 + |r|^2 \leq 1 \qquad |\tau|^2 + |\rho|^2 \leq 1 \tag{12.11}$$

Now, these inequalities are not preserved by star multiplication; hence they are not the only conditions needed to ensure passivity. The additional conditions can be obtained by noting that

$$|z_2|^2 + |z_3|^2 \leq |z_1|^2 + |z_4|^2 \tag{12.12}$$

in Fig. 12.5, since the power output does not exceed the power input. By Eq. (12.4) we get

$$|r\bar{\tau} + \bar{\rho}t|^2 \leq (1 - |t|^2 - |r|^2)(1 - |\rho|^2 - |\tau|^2) \tag{12.13}$$

together with the inequalities (12.11) as necessary and sufficient conditions for the inequality (12.12).

It is clear from the derivation that the conditions (12.11) and (12.13) are preserved by matrix multiplication, and we show that they are also preserved by star multiplication. Indeed, if the networks of Fig. 12.6 are dissipative, then we have, besides the inequality (12.12),

$$|z_4|^2 + |z_5|^2 \leq |z_3|^2 + |z_6|^2$$

Addition of this and (12.12) yields

$$|z_2|^2 + |z_5|^2 \leq |z_1|^2 + |z_6|^2$$

and hence the star product is dissipative, $(z_1,z_6)$ being arbitrary.

To compare these results with those for the lossless case, we introduce the power-absorption coefficients,

$$a^2 = 1 - |t|^2 - |r|^2 \qquad \alpha^2 = 1 - |\tau|^2 - |\rho|^2$$

Then the inequality (12.13) yields, after brief calculation,

$$\left| \left|\frac{r}{t}\right| - \left|\frac{\rho}{\tau}\right| \right| \leq \frac{a\alpha}{|t\tau|} \tag{12.14}$$

and, with $\phi_0 = \angle t + \angle \tau - \angle r - \angle \rho$,

$$|\cos \tfrac{1}{2}\phi_0| \leq \frac{1}{2} \frac{a\alpha}{\sqrt{|t\tau r\rho|}} \tag{12.15}$$

The former reduces to $|t| = |\tau|$ and the latter to Eq. (12.10) when $a\alpha = 0$. Since the inequality (12.14) compares the magnitudes of the coefficients from opposite sides, and the inequality (12.15) compares their phases, the two together constitute a reciprocity theorem.

## EXERCISES

**1.** A shunt obstacle $(t\tau r\rho)$ satisfies $t = 1 + r$, $\tau = 1 + \rho$ (Sec. 12.3, Exercise 2). Show that such an obstacle is passive if and only if $t = \tau$ and $|r| + \cos(\angle r) \leq 0$, so that $r$ lies in the shaded circle of Fig. 12.10. Thus show that a passive shunt obstacle is always symmetrical.

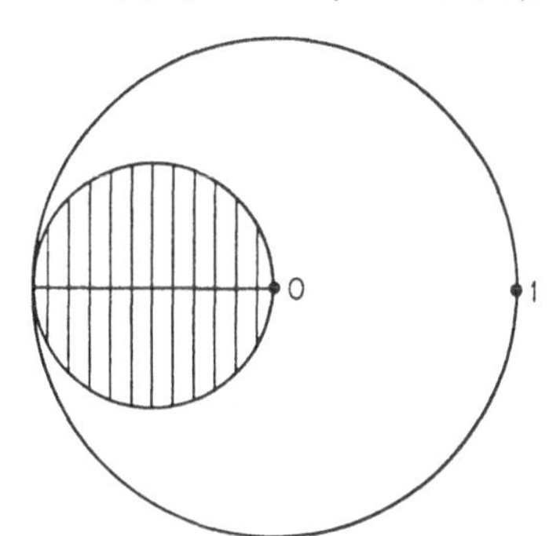


**Fig. 12.10** Possible locus of the complex reflection for a passive shunt obstacle.

**2.** A figure of merit for the performance of a measuring probe (Fig. 12.7) is $M = fa^2/(2|p|)$, where $p$ is the probe reflection coefficient, $a^2$ is the probe absorption coefficient, and $f$ is the fraction of the absorbed power that contributes to the voltmeter reading. Show that, for a passive shunt probe,

$$\frac{M}{f} = \cos(\pi - \angle p) - |p|$$

**3.** A passive obstacle $(t\tau r\rho)$ has small power-absorption coefficients $a^2$ and $\alpha^2$. With $d = t\tau - r\rho$ and $\phi_0$ as in the text, show that

$$|d|^2 = 1 - a^2 - \alpha^2 \qquad |\pi - \phi_0| \leq a\alpha/\sqrt{|t\tau r\rho|}$$

and

$$|\angle d - (\pi + \angle r + \angle\rho)| \leq a\alpha\sqrt{|t\tau r\rho|}$$

apart from higher-order terms in $a$ and $\alpha$.

**4.** Show that the over-all power absorption of $(ttrr) * (ttrr)$ is $2a^2/|t|^2$, apart from terms in $a^4$.

## 12.7 The Associated Linear Fractional Transformation

In Fig. 12.11, an obstacle is followed at the right by a variable reflection $z$. Symbolically, the situation is described by

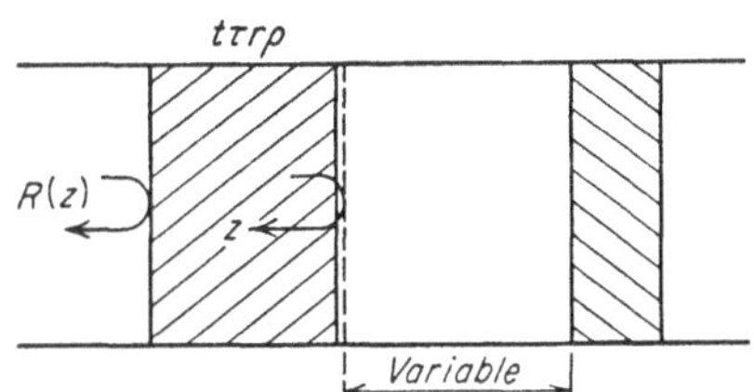


**Fig. 12.11** Obstacle followed by a reflection of constant magnitude but variable phase.

$$\begin{pmatrix} t & \rho \\ r & \tau \end{pmatrix} * \begin{pmatrix} - & - \\ z & - \end{pmatrix}$$

where the blank entries are irrelevant; hence the over-all left-hand reflection, given by Eq. (12.3), is

$$R(z) = r + \tau z(1 - \rho z)^{-1}t \qquad (12.16)$$

In just the same way, if the variable reflection $z$ is introduced on the left, then the expression

$$\mathrm{P}(z) = \rho + tz(1 - rz)^{-1}\tau \qquad (12.17)$$

gives the right-hand reflection of the composite obstacle.

The expressions (12.16) and (12.17) are linear fractional transformations of the complex variable $z$. To get a geometric interpretation, recall that such transformations take circles into circles. The circle $|z| = a < |\rho|^{-1}$ is mapped by $w = R(z)$ onto a circle with center and radius, respectively,

$$C_a = r + \frac{a^2 t\tau\bar{\rho}}{1 - a^2|\rho|^2} \qquad R_a = \frac{a|t\tau|}{1 - a^2|\rho|^2} \tag{12.18}$$

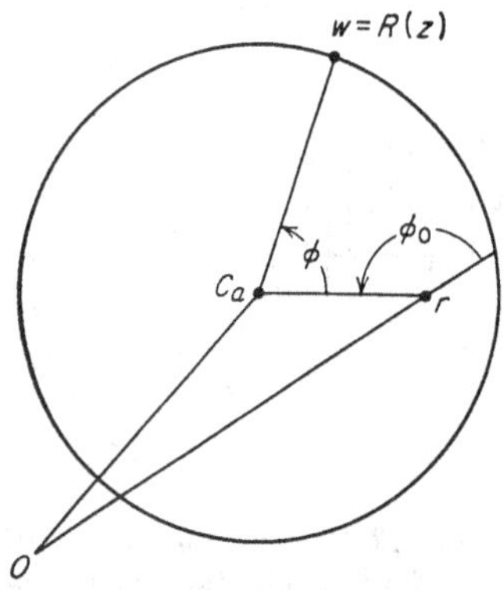


**Fig. 12.12** Image of the circle $|z|$ = constant after modification by an arbitrary obstacle $(t\tau\rho)$.

This gives the response to a terminating reflection of constant magnitude $a$ but variable phase. Since $C_a = R(a^2\bar{\rho})$, the center of the circle can be found as the response to the terminating reflection $z = a^2\bar{\rho}$.

The particular form of $C_a$ yields an interpretation of the angle

$$\phi_0 = \angle t + \angle\tau - \angle r - \angle\rho$$

that played a role in our energy considerations. Namely, $\phi_0$ is the positive angle from the line $Or$ produced to the line $rC_a$, no matter what value $a$ may have. See Fig. 12.12.

## EXERCISES

**1.** Prove that the origin lies inside the circle traced out by $R(z)$, for $|z| = a$, if and only if $|r/a| < |t\tau - r\rho|$.

**2.** Let

$$z = ae^{i(\theta - \angle\rho - \pi)}$$

where $a > 0$ and $\theta$ is a real variable. Let $\phi$ be the angle between the radius to $w = R(z)$ and the line joining $r$ and $C_a$ (Fig. 12.12). Show that

$$\frac{\tan \frac{1}{2}\phi}{\tan \frac{1}{2}\theta} = \frac{1 - a|\rho|}{1 + a|\rho|}$$

independently of the variables $t$, $\tau$, and $r$.

**3.** In Fig. 12.7, let the line be lossless, let $G = 0$, and let the probe be in shunt, so that $t = 1 + p$. Show that the measured standing-wave ratio gives

$$R\left(\frac{1 + 2P\cos P' + P^2}{1 + 2R^2P\cos P' + R^4P^2}\right)^{\frac{1}{2}}$$

for the apparent value of the load reflection $R$. Show also that the formula remains valid when the probe is not in shunt provided $P$, $R$, $P'$ are replaced respectively by

$$\frac{P}{D} \qquad RD \qquad P' - D' \qquad \text{with } t - p = De^{iD'}$$

(See Sec. 12.3, Exercises 3 and 4. The values $V_{max}$ and $V_{min}$ can be found by computing the radius and center of the image of the unit circle under a certain linear fractional transformation.)

**4.** In Exercise 3, let $t = 1 + p$ and let $\eta$ be the error in minimum position due to the probe reflection. Show that $\sin 2k\eta$ satisfies a certain quadratic equation, and thus obtain the approximation formulas

$$\sin 2k\eta \doteq -P \sin P' \qquad \sin 2k\eta \doteq -\frac{P \sin P'(1 - R)^2}{1 - 2P \cos P' + P^2}$$

for $R \doteq 0$ and $R \doteq 1$, respectively. Show also that the result for $t \neq 1 + p$ can be obtained by the substitution given in Exercise 3.

## 12.8 Another Characterization of Passive Networks

A discussion of energy can be based on $R(z)$ and $\mathrm{P}(z)$. We say that the network is *reflectively dissipative from the left*,† or *left-dissipative*, if $|R(z)| < 1$ whenever $|z| < 1$. Evidently, the condition implies $|r| < 1$ and $|\rho| < 1$. When that is the case, $R(z)$ is analytic for $|z| < 1$, so that the maximum of $|R(z)|$ occurs when $|z| = 1$. The desired condition is then equivalent to

$$|C_a| + R_a \leq 1 \qquad \text{for } a = 1$$

which becomes

$$1 + 2\Re(r\rho\bar{t}\bar{\tau}) \leq (1 - |t\tau|)^2 + (1 - |r|^2)(1 - |\rho|^2) \tag{12.19}$$

when we use Eqs. (12.18)

The inequality (12.19), together with $|r| < 1$ and $|\rho| < 1$, is necessary and sufficient to ensure $|R(z)| < 1$ for $|z| < 1$. Since the conditions are unaltered by interchanging $t$ with $\tau$, and $r$ with $\rho$, the same ensures $|\mathrm{P}(z)| < 1$ for $|z| < 1$. In other words, the network is left-dissipative if and only if it is right-dissipative.

To contrast this notion of passivity with that of Sec. 12.6, observe that the inequality (12.13) can be written in the form

$$1 + 2\Re(r\rho\bar{t}\bar{\tau}) \leq (1 - |t\tau|)^2 + (1 - |r|^2)(1 - |\rho|^2) - (|t| - |\tau|)^2 \tag{12.20}$$

Comparing the inequalities (12.19) and (12.20) shows that dissipative networks are also reflectively dissipative and that the converse is true provided $|t| = |\tau|$; the equivalence of the two notions, however, hinges on this reciprocity theorem.

It is always possible to choose a positive constant $h$ so that

$$|ht| = |h^{-1}\tau|$$

Since

$$(ht)(h^{-1}\tau) = t\tau$$

† The overworked term "stable" is sometimes used.

we deduce that the network specified by $t$, $\tau$, $r$, and $\rho$ is reflectively dissipative if and only if there is a positive constant $h$ for which

$$\begin{pmatrix} ht & \rho \\ r & h^{-1}\tau \end{pmatrix} \tag{12.21}$$

is dissipative in the sense of Sec. 12.6.

## EXERCISES

**1.** Prove that if a passive object $(t\tau r\rho)$ has $\rho = 0$, then $|t\tau| + |r| \leq 1$. (Let the object be backed by a perfect reflector, $|r_1| = 1$, with such a phase $\angle r_1$ as to maximize the over-all reflection.)

**2.** For $\rho = 0$ as in Exercise 1, use the inequality (12.13) to get the relationship

$$|r|^2 \leq (1 - |t|^2)(1 - |\tau|^2)$$

This inequality is stronger than that of Exercise 1 unless $|t| = |\tau|$.

**3.** A *tuner* is a lossless network $(t_1 t_1 r_1 \; -r_1)$ such that $(t\tau r\rho) * (t_1 t_1 r_1 \; -r_1)$ has zero right-hand reflection (Fig. 12.13). The network $(t\tau r\rho)$ with the tuner is then said to be *matched* from the right. Show that a match is obtained if and only if $r_1 = \bar{\rho}$. In this case, the over-all transmissions of the matched obstacle are, in magnitude,

$$|t|(1 - |\rho|^2)^{-1/2} \qquad \text{and} \qquad |\tau|(1 - |\rho|^2)^{-1/2}$$

(See Sec. 12.5, Exercise 2.)

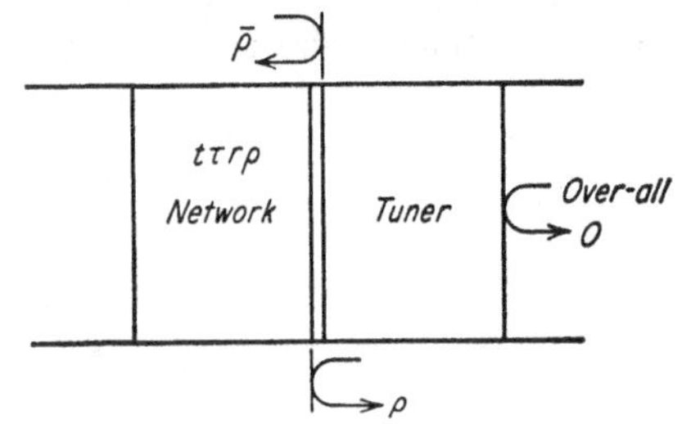

**Fig. 12.13** Obstacle matched by a symmetric lossless tuner.

**4.** Show that the left-hand reflection $R$ of the matched obstacle in Exercise 3 satisfies the inequality

$$|R| \leq \frac{a\alpha}{1 - |\rho|^2} \leq \frac{a}{|\tau|}\frac{\alpha}{|\tau|}$$

where $a^2$ and $\alpha^2$ are the power-absorption coefficients. (Assume that the obstacle is passive and use Exercise 2.)

**5.** A homogeneous, lossy dielectric sheet reflects 10 per cent of the incident power and transmits 80 per cent. Two identical sheets of this type are spaced for maximum transmission. Show that the over-all power reflection coefficient $|R|^2$ is at most 0.0124, although the over-all transmission $T$ satisfies $1 - |T|^2 = 0.210$. (Note that $R$ is unchanged if the second sheet is replaced by a lossless one having the same reflection coefficient, and use Exercise 4.)

**6.** Show that if $(t\tau r\rho)$ is lossless, then the image of $|z| = a$ under $R(z)$ or $\mathrm{P}(z)$ is a circle with radius and distance from the origin to its center respectively

$$R_a = a\frac{1 - |r|^2}{1 - a^2|r|^2} \qquad |C_a| = |r|\frac{1 - a^2}{1 - a^2|r|^2}$$

and thus that interchanging $a$ and $|r|$ has the effect of interchanging $|C_a|$ and $R_a$. [Use Eqs. (12.18) and Sec. 12.5.]

7. In Exercise 6 deduce $a|C_a|^2 = (a - R_a)(1 - aR_a)$; see Fig. 12.14. Show that the result for $P(z)$ remains valid even when the lossless tuner is accompanied by an arbitrary fixed network at its left as in Fig. 12.13.

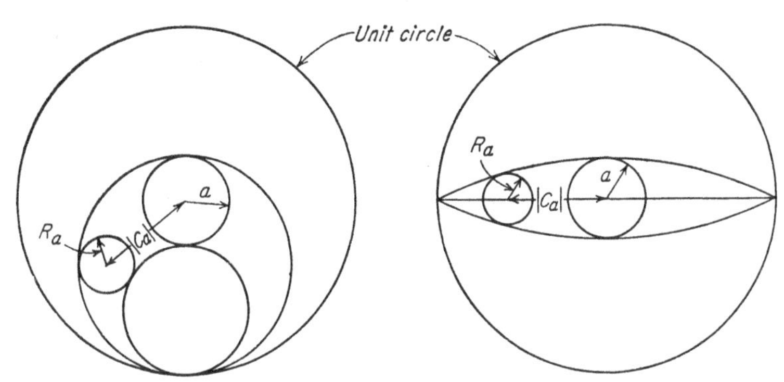


**Fig. 12.14** The Silver crescent and the Rieke banana.

## 12.9 Fixed Points and Commutativity

The condition $\xi = R(\xi)$ in Eq. (12.16) leads to

$$\rho\xi^2 + (t\tau - r\rho - 1)\xi + r = 0 \tag{12.22}$$

The two† roots $\xi_1$ and $\xi_2$ are fixed points of the transformation $w = R(z)$, and their reciprocals are fixed points of $w = P(z)$. The values $\xi_1$ and $\xi_2$ are the particular terminating reflectances that are not changed by the network.

If two networks both have the same fixed point $\xi$, then so also does their star product. Now the fixed points are determined by, and determine, the two ratios

$$A = \frac{r}{\rho} \qquad B = \frac{t\tau - r\rho - 1}{2\rho} \tag{12.23}$$

Hence if these equations hold for each factor, $A$ and $B$ being constant, then they hold for the star product.

A different interpretation of Eqs. (12.23) is obtained when we consider networks that permute—i.e., that give the same result independently of their order in the line. The equation

$$\begin{pmatrix} t & \rho \\ r & \tau \end{pmatrix} * \begin{pmatrix} t_1 & \rho_1 \\ r_1 & \tau_1 \end{pmatrix} = \begin{pmatrix} t_1 & \rho_1 \\ r_1 & \tau_1 \end{pmatrix} * \begin{pmatrix} t & \rho \\ r & \tau \end{pmatrix}$$

is equivalent to $r/\rho = r_1/\rho_1$ together with

$$\frac{t\tau - r\rho - 1}{\rho} = \frac{t_1\tau_1 - r_1\rho_1 - 1}{\rho_1}$$

† We refuse to worry about other possibilities.

Hence if each object of a series satisfies Eqs. (12.23), where $A$ and $B$ are constant, the objects permute with each other. It follows that each permutes with the whole series, and therefore the coefficients of the series satisfy Eqs. (12.23). This invariance of Eqs. (12.23) under star multiplication is an aid in the study of cascaded networks.

## EXERCISES

**1.** Prove that if a lossless network takes a match into a match, then it takes any reflection into a reflection of the same magnitude. [If $R(0) = 0$ in Eq. (12.16), then $r = 0$. Hence $|t| = 1$, hence $|\tau| = 1$, hence $\rho = 0$, and the result follows.]

**2.** Prove that if the linear fractional transformation $w = R(z)$ has the distinct fixed points $z_1$ and $z_2$, then it can be written in the form

$$\frac{w - z_1}{w - z_2} = H \frac{z - z_1}{z - z_2}$$

where $H$ is a complex constant.

**3.** Suppose two transformations $R_1(z)$ and $R_2(z)$ of the type considered in Exercise 2 have the same fixed points but different constants $H_1$ and $H_2$. Show that the iterated transformation $w = R_2\{R_1(z)\}$ has the same fixed points and the constant $H_1H_2$. Generalize to $n$ iterations.

## 12.10 Series of Obstacles; the Cascade Problem

The scattering matrix for a series of obstacles can be obtained by a continued star product, by a continued matrix product, or by compounding the transformations

$$z_{n+1} = \mathrm{P}_n(z_n)$$

(The latter can be made to lead again to a matrix product.) If the obstacles are not adjacent, one regards the line lengths separating them as being themselves obstacles and computes the product accordingly. The scattering matrix for a line of length $x$ is

$$\begin{pmatrix} e^{jkx} & 0 \\ 0 & e^{j\kappa x} \end{pmatrix} \tag{12.24}$$

where $k$ and $\kappa$ are the propagation constants in the two directions.

Dividing

$$\frac{tt_1}{1 - r_1\rho} \quad \text{by} \quad \frac{\tau\tau_1}{1 - r_1\rho}$$

shows that the quotient $t/\tau$ for two obstacles can be got by multiplying the individual quotients; hence, the analogous result is true for any number. [Since the determinant of the right-hand matrix (12.8) is $t/\tau$, this remark also follows from the fact that the determinant of the matrix product is the product of the determinants.]

No other relations of the foregoing kind can be stated for the cascade problem in full generality, though statements can be given in important special cases. For example, since the star product of two lossless networks is lossless, the relations

$$|t|^2 + |r|^2 = |\tau|^2 + |\rho|^2 = 1 \qquad |t| = |\tau| \qquad \angle t + \angle\tau - \angle r - \angle\rho = \pi$$

are valid for the series if they are for each member of the series.

Another example is the commutativity relations (12.23). Let the over-all coefficients for a series of $n$ obstacles be denoted by $t_n$, $\tau_n$, $r_n$, and $\rho_n$. If Eqs. (12.23) hold for each obstacle, $A$ and $B$ being constant, we can express $t_n$, $\tau_n$, and $r_n$ in terms of $\rho_n$, namely,

$$r_n = A\rho_n \qquad t_n\tau_n = 1 + 2B\rho_n + A\rho_n^2 \tag{12.25}$$

and $t_n/\tau_n$ is known, since it is equal to the product $\Pi(t_k/\tau_k)$.

Sometimes it is desirable to find the limit of the transmission and reflection as the number of obstacles increases indefinitely. If there is enough dissipation to make $t_n\tau_n \to 0$, then Eqs. (12.25) give

$$\lim r_n = \xi_1 \qquad \lim \rho_n^{-1} = \xi_2 \tag{12.26}$$

where $\xi_1\xi_2 = A$ and $\xi_1 + \xi_2 = -2B$. Thus, the $\xi_i$ are the roots of

$$\xi^2 + 2B\xi + A = 0$$

That is, they are the fixed points (12.22) that were discussed previously.

## EXERCISES

**1.** A network is *bilateral* (or *reciprocal*) if $t = \tau$. Prove by induction or otherwise: If each object in a series is bilateral, so is the series as a whole. Is the analogous statement for reflection also true?

**2.** Prove that the reflection from a series of lossless objects is maximal when the objects are so spaced that the reflection from each adjacent pair is maximal. (See Sec. 12.5, Exercise 3. In this and the following problems, the objects are in a lossless line.)

**3.** Show that the standing-wave ratio for a series of lossless objects cannot exceed the product of the individual standing-wave ratios and that this value is always attained for some spacing. (See Sec. 12.5, Exercise 5.)

**4.** Show that, if the amplitude reflection for any array of lossless objects is equal to $|R|$ with the objects in any given order, then for any other order there exists a spacing such that the reflection is again equal to $|R|$. (Consider the interchange of adjacent objects.)

**5.** Show that a series of lossless obstacles can be spaced for zero reflection if and only if the largest standing-wave ratio does not exceed the product of the others. (By Exercise 4, the largest can be put at the end. For zero reflection, the reflection of all but the last must be equal in magnitude to that of the last. Use Exercise 3, noting that the reflection is a continuous function of the spacing variables.)

## 12.11 Identical Networks in Cascade

Let the obstacle specified by $t$, $\tau$, $r$, and $\rho$ be repeated $n$ times, so that the over-all scattering matrix is

$$\begin{pmatrix} t_n & \rho_n \\ r_n & \tau_n \end{pmatrix} = \begin{pmatrix} t & \rho \\ r & \tau \end{pmatrix}^n \qquad \text{(star multiplication)} \tag{12.27}$$

Since identical obstacles permute, Eqs. (12.25) hold with $A$ and $B$ as in Eqs. (12.23), and also $t_n/\tau_n = (t/\tau)^n$. All that is lacking is the expression for $\rho_n$, which can be obtained by the difference equation

$$\rho_{n+1} = \rho + \frac{t\tau\rho_n}{1 - r\rho_n} \equiv \mathrm{P}(\rho_n) \tag{12.28}$$

or by a matrix product. Solution of Eq. (12.28) is facilitated by introducing the fixed points, and the $n$th power of a matrix can be obtained by writing it in diagonal form.†

The result of this calculation is

$$\begin{aligned} \rho_n &= \frac{\rho \sinh \delta n}{\sinh \delta n - \sqrt{t\tau} \sinh \delta(n-1)} \\ \tau_n &= \tau \left(\frac{\tau}{t}\right)^{(n-1)/2} \frac{\sinh \delta}{\sinh \delta n - \sqrt{t\tau} \sinh \delta(n-1)} \end{aligned} \tag{12.29}$$

where $\delta$ is defined by

$$2\sqrt{t\tau} \cosh \delta = 1 + t\tau - r\rho \tag{12.30}$$

Expressions for $r_n$ and $t_n$ are found by interchanging $r$ with $\rho$ and $t$ with $\tau$.

As we shall presently see, a less pedestrian proof of Eqs. (12.29) can be based on the identity

$$\begin{pmatrix} t_{m+n} & \rho_{m+n} \\ r_{m+n} & \tau_{m+n} \end{pmatrix} = \begin{pmatrix} t_m & \rho_m \\ r_m & \tau_m \end{pmatrix} * \begin{pmatrix} t_n & \rho_n \\ r_n & \tau_n \end{pmatrix} \tag{12.31}$$

[Equation (12.31) states that $m + n$ obstacles can be regarded as a group of $m$, adjacent to a group of $n$.] Written out in full, Eq. (12.31) yields the difference equations

$$r_{m+n} = r_m + \frac{t_m \tau_m r_n}{1 - \rho_m r_n} \qquad t_{m+n} = \frac{t_m t_n}{1 - \rho_m r_n}$$

and their analogues for $\rho_{m+n}$ and $\tau_{m+n}$. For $n = 1$, the equations become recurrence formulas, which show that the four coefficients are determined

† The two techniques are not unrelated. The characteristic vectors needed for the diagonalization are the fixed directions associated with the matrix transformation.

uniquely by Eq. (12.31) together with the initial values

$$t_1 = t \qquad \tau_1 = \tau \qquad r_1 = r \qquad \rho_1 = \rho \tag{12.32}$$

Thus, any method whatever of solving Eq. (12.31) subject to the conditions (12.32) must yield Eqs. (12.29), the desired result.

### EXERCISES

1. Show that for lossless media, Eq. (12.30) becomes $\cosh \delta = \Re(t\tau)^{-1/2}$.
2. Let $(t_n t_n r_n \rho_n)$ be a bilateral network for which the limits

$$\lim \delta_n n = jp \qquad \lim (t_n - 1)n = jq$$

exist as $n \to \infty$, with $\delta_n$ as in Eq. (12.30). This network is repeated $n$ times to give

$$(T_n T_n R_n \mathrm{P}_n) = (t_n t_n r_n \rho_n)^n$$

Show that the over-all transmission $T_n$ satisfies

$$\lim \frac{1}{T_n} = \cos p - j\frac{q}{p}\sin p \qquad \text{as } n \to \infty$$

## FUNCTIONAL EQUATIONS

### 12.12 Homogeneous Anisotropic Media

The dielectric medium shown in cross section in Fig. 12.15 is anisotropic, in the sense that it has different properties for waves traveling from right to left and from left to right. (This anisotropy should be distinguished from that in which the medium properties depend on polarization.) Since a medium of thickness $x + y$ can be thought of as a medium of thickness $x$ adjacent to a medium of thickness $y$, the transmission and reflection coefficients $t(x)$, $\tau(x)$, $r(x)$, and $\rho(x)$ satisfy the equation

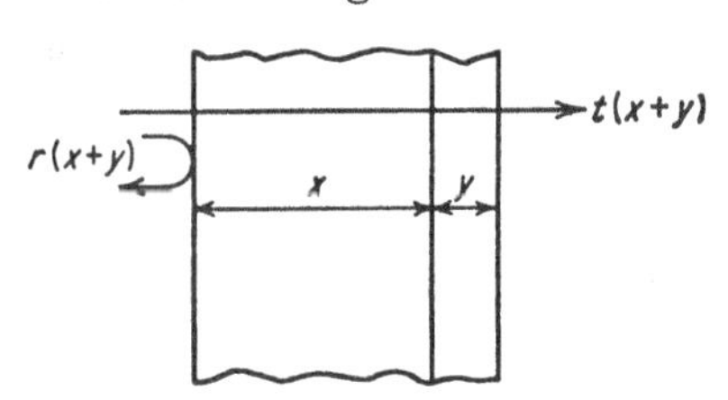


Fig. 12.15 Uniform dielectric medium of thickness $x + y$, in cross section.

$$\begin{pmatrix} t(x+y) & \rho(x+y) \\ r(x+y) & \tau(x+y) \end{pmatrix} = \begin{pmatrix} t(x) & \rho(x) \\ r(x) & \tau(x) \end{pmatrix} * \begin{pmatrix} t(y) & \rho(y) \\ r(y) & \tau(y) \end{pmatrix} \tag{12.33}$$

Writing out in full gives the functional equations

$$r(x+y) = r(x) + \frac{t(x)\tau(x)r(y)}{1 - \rho(x)r(y)} \qquad t(x+y) = \frac{t(x)t(y)}{1 - \rho(x)r(y)} \tag{12.34}$$

and similarly for $\rho(x + y)$ and $\tau(x + y)$. We assume the initial conditions

$$t(0) = \tau(0) = 1 \qquad r(0) = \rho(0) = 0$$

and also existence of the derivatives

$$a = \rho'(0) \qquad b_1 = t'(0) \qquad b_2 = \tau'(0) \qquad c = r'(0) \tag{12.35}$$

Subtracting $r(x)$ from the first equation (12.34) and dividing by $y$ yields

$$\frac{r(x+y) - r(x)}{y} = \frac{t(x)\tau(x)[r(y)/y]}{1 - \rho(x)r(y)}$$

Upon letting $y \to 0$ we get†

$$\frac{dr}{dx} = ct\tau \qquad [t = t(x) \qquad \tau = \tau(x)] \tag{12.36}$$

In just the same way the remaining equations lead to

$$\frac{dt}{dx} = t(c\rho + b_1) \tag{12.37}$$

$$\frac{d\tau}{dx} = \tau(c\rho + b_2) \tag{12.38}$$

$$\frac{d\rho}{dx} = a + 2b\rho + c\rho^2 \tag{12.39}$$

where $2b = b_1 + b_2$.

### 12.13 Solution of the Equations

Since the medium is homogeneous, the two matrices on the right of Eq. (12.33) can be permuted; hence, by Sec. 12.9,

$$\frac{r}{\rho} = \text{const} \qquad \frac{t\tau - r\rho - 1}{2\rho} = \text{const}$$

Letting $x \to 0$ shows that the constants are $c/a$ and $b/a$, respectively, so that

$$ar = c\rho \tag{12.40}$$

$$at\tau = a + 2b\rho + c\rho^2 \tag{12.41}$$

Equations (12.37) and (12.38) give $(\log t)'$ and $(\log \tau)'$, and by subtraction,

$$\frac{t}{\tau} = e^{(b_1 - b_2)x} \tag{12.42}$$

These three relationships among the four coefficients of our dielectric medium were obtained by quite general principles, which presuppose no knowledge of the solution. Of course one can actually solve Eq. (12.39),

† The result could also be found by differentiating formally with respect to $y$ and then putting $y = 0$. The method of the text has the merit that existence of $r'(x)$ is deduced rather than assumed, and this feature, irrelevant physically, is a great comfort to the mathematician.

since the coefficients are constant; the result is

$$\rho = \frac{a \sinh x\Delta}{\Delta \cosh x\Delta - b \sinh x\Delta} \qquad \text{where } \Delta^2 = b^2 - ac \tag{12.43}$$

Equation (12.41) now gives

$$\sqrt{t\tau} = \frac{\Delta}{\Delta \cosh x\Delta - b \sinh x\Delta} \tag{12.44}$$

We get $r$ from Eq. (12.40) and $t$ and $\tau$ by combining Eq. (12.44) with Eq. (12.42):

$$\tau = e^{\frac{1}{2}(b_1 - b_2)x} \frac{\Delta}{\Delta \cosh x\Delta - b \sinh x\Delta}$$

### EXERCISE

Show that a homogeneous isotropic dielectric sheet is in shunt (Sec. 12.3, Exercise 2) only if $t = 1$ and $r = 0$.

## 12.14 Application to the Cascade Problem

The expressions obtained in the foregoing discussion satisfy the functional equations (12.33). If $m$ and $n$ are integers and

$$t_n = t(n) \qquad \tau_n = \tau(n) \qquad r_n = r(n) \qquad \rho_n = \rho(n)$$

the choice $x = m$ and $y = n$ makes Eq. (12.33) identical with Eq. (12.31). In other words, the expressions satisfy not only the functional equations for the anisotropic medium but also the difference equations for the iterated network.

The physical basis of the relationship is clear from Fig. 12.16. If a dielectric medium of unit thickness has the prescribed coefficients $t$, $\tau$, $r$, and $\rho$, then the $n$th iterate $t_n$, $\tau_n$, $r_n$, $\rho_n$ corresponds to the same dielectric medium at a thickness $n$. This remark gives a new method of solving the cascade problem.

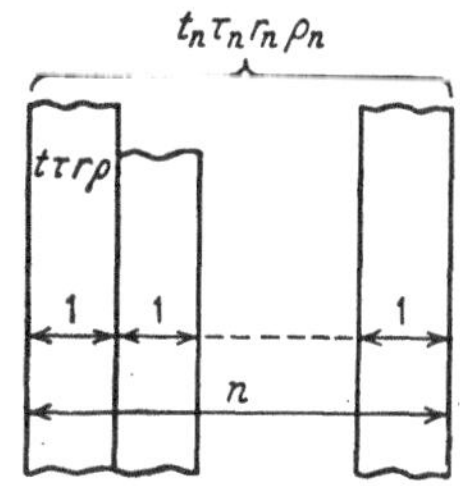


**Fig. 12.16** Cross section of $n$ identical dielectric sheets, stacked.

The desired initial conditions

$$t(1) = t \qquad \tau(1) = \tau \qquad r(1) = r \qquad \rho(1) = \rho$$

lead to four equations in the four unknowns $a$, $b_1$, $b_2$, and $c$. If a new unknown $X$ is defined by

$$\Delta \cosh \Delta = X \sinh \Delta$$

then Eqs. (12.40), (12.41), and (12.43) with $x = 1$ give

$$\frac{c}{a} = A \qquad \frac{b}{a} = B \qquad \frac{X}{a} = \frac{1}{\rho} + B$$

where $A$ and $B$ are as in Eqs. (12.23). The condition $b^2 - ac = \Delta^2$ reduces to

$$\cosh^2 \Delta = (t\tau)^{-1}(1 + \rho B)^2$$

The coefficients $r(1)$, $\rho(1)$, and $t(1)\ \tau(1)$ now have the correct values, and $t(1) = t$, $\tau(1) = \tau$ determine $b_i$ through

$$b_1 + b_2 = b \qquad b_1 - b_2 = \log \frac{t}{\tau}$$

Thus, the cascade problem is solved by Eq. (12.43) with these values of the parameters.

## EXERCISES

**1.** Given any solution $a$, $b_1$, $b_2$, and $c$, show that all others can be found by replacing $\Delta$, $b_1$, and $b_2$, respectively, by

$$\pm\Delta + n_1\pi j \qquad b_1 + n_2\pi j \qquad \text{and} \qquad b_2 - n_2\pi j$$

where $n_1$ and $n_2$ are integers.

**2.** Show that the particular solution for which $\Delta = \delta$ in Eq. (12.30) gives

$$\frac{b}{a} = \frac{\sqrt{t\tau}}{\rho}\cosh \Delta - \frac{1}{\rho} \qquad \frac{\Delta}{a} = \frac{\sqrt{t\tau}}{\rho}\sinh \Delta$$

Thus obtain Eqs. (12.29) from Eq. (12.43).

## 12.15 Interpretation of the Constants

Although Eqs. (12.35) give the mathematical meaning of the four parameters $a$, $b_1$, $b_2$, $c$, their physical significance requires clarification. Let the dielectric medium of Fig. 12.17 have the propagation constant $k$ for waves traveling from left to right and $\kappa$ for the opposite direction. The interface reflections at the left are $r_1$ and $\rho_1$, and those at the right are $r_2$ and $\rho_2$. Continuity of the $E$ field gives the corresponding transmissions,

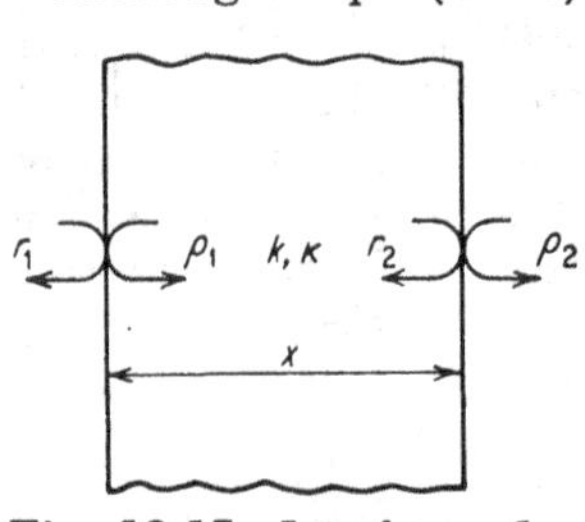

**Fig. 12.17** Interface reflections and propagation constants for homogeneous anisotropic dielectric sheet.

$$t_i = 1 + r_i \qquad \tau_i = 1 + \rho_i$$

The multiple-reflection argument of Sec. 12.3, or Eq. (12.3) with the matrix (12.24), leads to

$$\begin{aligned} t(x) &= \frac{(1 + r_1)(1 + r_2)e^{jkx}}{1 - \rho_1 r_2 e^{j(k+\kappa)x}} \\ r(x) &= r_1 + \frac{(1 + r_1)(1 + \rho_1) r_2 e^{j(k+\kappa)x}}{1 - \rho_1 r_2 e^{j(k+\kappa)x}} \end{aligned} \tag{12.45}$$

with corresponding expressions for $\tau(x)$ and $\rho(x)$.

Since $t(0) = \tau(0) = 1$ and $r(0) = \rho(0) = 0$, the constants $r_i$ and $\rho_i$ are not independent, but satisfy the equations

$$r_2(1 + r_1 + \rho_1) + r_1 = 0 \qquad \rho_2(1 + r_1 + \rho_1) + \rho_1 = 0 \tag{12.46}$$

Thus, as before, our problem is characterized by *four* complex constants: $k$, $\kappa$, $r_1$, and $\rho_2$.

Differentiating and setting $x = 0$ yields

$$a = \frac{j\rho_1(k + \kappa)}{(1 + r_1)(1 + \rho_1)} \qquad b_1 = \frac{j[k(1 + r_1 + \rho_1) - \kappa r_1 \rho_1]}{(1 + r_1)(1 + \rho_1)} \tag{12.47}$$

with similar expressions for $b_2$ and $c$. As a check, we note the relationships

$$\frac{r}{\rho} = \frac{r_1}{\rho_2} \qquad \frac{t\tau - r\rho - 1}{\rho} = -\left(r_1 + \frac{1}{\rho_2}\right) \tag{12.48}$$

which agree with the general principles set forth previously.

The fact that solutions of the functional equations always involve just four complex constants is not surprising. From the point of view of Maxwell's equations, the behavior is characterized by $\epsilon^+$, $\epsilon^-$, $\mu^+$, $\mu^-$, the dielectric constants and permeabilities in the two directions. In terms of the free-space wavelength $\lambda$, we have

$$k = \frac{2\pi}{\lambda}\sqrt{\frac{\epsilon^+\mu^+}{\epsilon_0\mu_0}} \qquad \kappa = \frac{2\pi}{\lambda}\sqrt{\frac{\epsilon^-\mu^-}{\epsilon_0\mu_0}}$$

The quantities

$$Z = \sqrt{\frac{\epsilon^+/\epsilon_0}{\mu^+/\mu_0}} \qquad \zeta = \sqrt{\frac{\epsilon^-/\epsilon_0}{\mu^-/\mu_0}}$$

determine $r_1 = (1 - Z)/(1 + Z)$ and $\rho_2$, with the result

$$\begin{aligned} a &= \frac{-j(k + \kappa)(1 + Z)(1 - \zeta)}{2(Z + \zeta)} \\ b_1 &= j\,\frac{k(1 + Z)(1 + \zeta) + \kappa(1 - Z)(1 - \zeta)}{2(Z + \zeta)} \end{aligned} \tag{12.49}$$

Also Eqs. (12.46) hold, as they should. For isotropic media, we have

$$a = c = \tfrac{1}{2}jk(Z - Z^{-1}) \qquad b_1 = b_2 = \tfrac{1}{2}jk(Z + Z^{-1}) \tag{12.50}$$

With these results it is easy to show (if we had not known it before) that the method of functional equations is consistent with the methods of field theory.

It should be mentioned, in conclusion, that $t = \tau$ leads to $k = \kappa$, but $r_1 = \rho_2$ only if $r = \rho$. Thus the representation even of bilateral lossless networks by the process of Sec. 12.14 requires an anisotropic medium.

## EXERCISES

**1.** In books on electromagnetic theory it is shown that the formulas governing behavior of a dielectric sheet at normal incidence can be used at arbitrary incidence $\theta$ provided the propagation constant is taken as

$$k = \frac{2\pi}{\lambda}\sqrt{\frac{\epsilon\mu}{\epsilon_0\mu_0} - \sin^2\theta}$$

and provided the quantity $Z$ determining the interface reflection is replaced by

$$Z_\perp = \frac{[(\epsilon/\epsilon_0)(\mu/\mu_0) - \sin^2\theta]^{1/2}}{(\mu/\mu_0)\cos\theta} \qquad Z_\parallel = \frac{(\epsilon/\epsilon_0)\cos\theta}{[(\epsilon/\epsilon_0)(\mu/\mu_0) - \sin^2\theta]^{1/2}}$$

at perpendicular and parallel polarization, respectively.

Show that the transmission $t$ of an isotropic, homogeneous dielectric sheet at incidence $\theta$ and thickness $x_0$ is given by

$$t^{-1} = \cos kx_0 - \frac{j}{2}(Z + Z^{-1})\sin kx_0$$

**2.** For any dielectric sheet, the derivatives $r'(0)$, $\rho'(0)$, . . . depend on the behavior as the thickness approaches zero, while for lossy sheets the interface reflections $r_1$, $\rho_1$, . . . depend on the behavior as the thickness approaches infinity. Prove that these quantities are related by the formulas

$$\frac{r'(0)}{\rho'(0)} = \frac{r_1}{\rho_2} \qquad \frac{t'(0) + \tau'(0)}{\rho'(0)} = -r_1 - \frac{1}{\rho_2}$$

[Apply L'Hospital's rule to Eqs. (12.48).]

**3.** A homogeneous isotropic dielectric sheet with thickness $x$ and propagation constant $k$ is preceded by a metal plate having right-hand reflection $K = -1$ (see Fig. 12.22). Show that the over-all right-hand reflection can be written in the form

$$\rho = -\frac{\cos(kx - j\xi)}{\cos(kx + j\xi)}$$

where $\xi$ is constant. [At normal incidence, $\coth^2\xi = (\epsilon/\epsilon_0)/(\mu/\mu_0)$.]

**4.** Show that if $\xi$ and $k$ are real in Exercise 3, then $|\rho| = 1$ and

$$\tan\angle\rho = 2\tan kx \tanh\xi$$

## 12.16 Nonuniform Dielectric Media

Let the quantities $\epsilon^+$, $\epsilon^-$, $\mu^+$, $\mu^-$ of Sec. 12.15 be arbitrary functions of the coordinate $s$ perpendicular to the interfaces. We shall regard these parameters as being given for $-\infty < s < \infty$. The dielectric slab is obtained by taking that part of the infinite medium that extends from† $x$ to $y > x$, as shown in Fig. 12.18. In other words, $\epsilon^+(s)$, $\mu^+(s)$, $\epsilon^-(s)$,

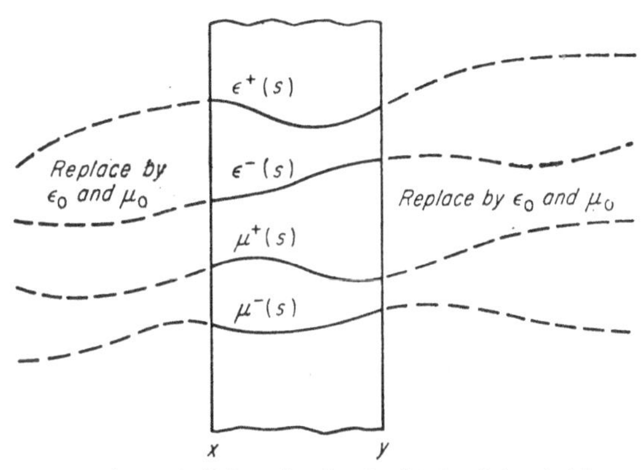


**Fig. 12.18** Cross section of dielectric sheet obtained by taking a slice from an infinite inhomogeneous anisotropic medium.

and $\mu^-(s)$ are used for $x < s < y$, but $\epsilon_0$ and $\mu_0$ elsewhere. If the transmission and reflection coefficients are denoted by

$$t(x,y) \qquad \tau(x,y) \qquad r(x,y) \qquad \rho(x,y)$$

then the same procedure as that employed previously yields

$$\begin{pmatrix} t(x,z) & \rho(x,z) \\ r(x,z) & \tau(x,z) \end{pmatrix} = \begin{pmatrix} t(x,y) & \rho(x,y) \\ r(x,y) & \tau(x,y) \end{pmatrix} * \begin{pmatrix} t(y,z) & \rho(y,z) \\ r(y,z) & \tau(y,z) \end{pmatrix} \tag{12.51}$$

for $z > y > x$. Writing out in full gives four functional equations, of which

$$r(x,z) = r(x,y) + \frac{t(x,y)\tau(x,y)r(y,z)}{1 - \rho(x,y)r(y,z)} \tag{12.52}$$

is typical. The initial conditions are

$$t(x,x) = \tau(x,x) = 1 \qquad r(x,x) = \rho(x,x) = 0$$

and it is assumed that the partial derivatives exist for $y = x$:

$$a(x) = \rho_y(x,x) \qquad b_1(x) = t_y(x,x) \qquad b_2(x) = \tau_y(x,x)$$
$$c(x) = r_y(x,x) \tag{12.53}$$

In terms of $\epsilon^+(x)$, $\epsilon^-(x)$, $\mu^+(x)$, and $\mu^-(x)$, these quantities have the values assigned in Sec. 12.15.

† The thickness of the slab is not $x$ as in the previous discussion, but $y - x$.

Subtracting $r(x,y)$ from both sides of Eq. (12.52), dividing by $z - y$, and letting $z \to y$, we obtain

$$r_y \equiv \frac{\partial r}{\partial y} = ct\tau \qquad r(x,x) = 0 \tag{12.54}$$

just as in the proof of Eqs. (12.36). Similarly,

$$t_y = t(c\rho + b_1) \qquad t(x,x) = 1 \tag{12.55}$$
$$\tau_y = \tau(c\rho + b_2) \qquad \tau(x,x) = 1 \tag{12.56}$$
$$\rho_y = a + 2b\rho + c\rho^2 \qquad \rho(x,x) = 0 \tag{12.57}$$

where $2b = b_1 + b_2$. In these equations, $t$, $\tau$, $r$, and $\rho$ are evaluated at $(x,y)$, and $a$, $b_1$, $b_2$, and $c$ at $y$.

The system has the same structure as Eqs. (12.36) to (12.39) except that the coefficients are variable. It might be expected that $t$, $\tau$, and $r$ can be expressed in terms of $\rho$, just as before, and we shall see that this is the case. Equations (12.55) and (12.56) give

$$\left(\frac{t}{\tau}\right)_y = \left(\frac{t}{\tau}\right)(b_1 - b_2) \qquad (t\tau)_y = 2t\tau(c\rho + b)$$

and these, together with Eq. (12.54), yield the desired formulas:

$$\frac{t}{\tau} = \exp \int_x^y (b_1 - b_2)\, dy$$
$$t\tau = \exp 2 \int_x^y (c\rho + b)\, dy$$
$$r = \int_x^y ct\tau\, dy$$

## EXERCISES

**1.** If the medium is isotropic, prove that the coefficients satisfy $t = \tau$ and

$$r^2 \left(\frac{t^2}{r}\right)_y + (1 - r^2)\rho_y = 0$$

**2.** Let $t_n$, $r_n$, and $\rho_n$ be the transmission and reflection coefficients obtained for an isotropic inhomogeneous dielectric sheet of thickness $x_0$ when the free-space wavelength $\lambda$ is replaced by $n\lambda$ (Fig. 12.19). If $a(x)$ and $b(x)$ are given in terms of $\epsilon(x)$ and $\mu(x)$ by Eqs. (12.50), show that, as $n \to \infty$,

$$\rho_n = \frac{1}{n}\int_0^{x_0} a(x)\, dx + \frac{2}{n^2}\int_0^{x_0} b(x) \int_0^x a(\xi)\, d\xi\, dx$$
$$t_n = 1 + \frac{1}{n}\int_0^{x_0} b(x)\, dx + \frac{1}{2n^2}\left\{\left[\int_0^{x_0} a(x)\, dx\right]^2 + 2\int_0^{x_0} b(x)\, dx\right\}$$

apart from terms of order $1/n^3$. [If $a(x)$ and $b(x)$ correspond to $\lambda$, note that $a(x)/n$

and $b(x)/n$ correspond to $n\lambda$. By the method of Sec. 12.15, Exercise 1, the result remains valid for arbitrary incidence $\theta$.]

**3.** Two lossless, homogeneous, isotropic media of dielectric constants $\epsilon_1$ and $\epsilon_3$ and permeabilities $\mu_0$ are separated by a similar medium of constants $\epsilon_2$ and $\mu_0$. Show that, for zero over-all reflection, $\epsilon_2^2 = \epsilon_1\epsilon_3$. Show further that the electrical thickness of the central medium must be a quarter wave.

**4.** In Exercise 3, let the media 1 and 3 have the constant separation $d$ and let the intervening medium consist of many matched quarter-wave steps of the type there described. Show that, as $\lambda \to 0$, the limiting behavior of the central medium is such that $(1/\epsilon)^{1/2}$ is a linear function of position $x$, agreeing with $(1/\epsilon_1)^{1/2}$ and $(1/\epsilon_3)^{1/2}$ at $x = 0$ and $x = d$, respectively. (Note that the layers have different thicknesses.)

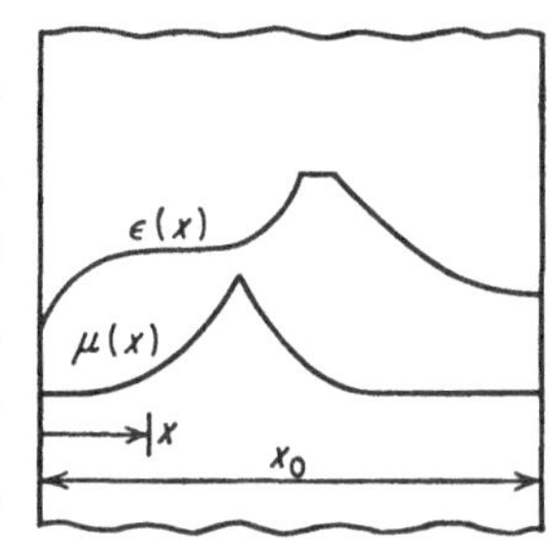


**Fig. 12.19** An inhomogeneous, lossless, isotropic sheet in cross section.

**5.** Express the reflection of the transition step of Exercise 4 in closed form as a function of $\lambda$.

## 12.17 Linearization

The nonlinear equation (12.57) can be linearized by the familiar substitution $\rho = -q_y/(cq)$. But a linearization more tightly knit with the underlying physics is determined by the matrix

$$M = \begin{pmatrix} \dfrac{1}{\tau} & \dfrac{\rho}{\tau} \\ -\dfrac{r}{\tau} & \dfrac{t\tau - r\rho}{\tau} \end{pmatrix}$$

of the correspondence (12.8). If

$$C = \begin{pmatrix} -b_2 & a \\ -c & b_1 \end{pmatrix}$$

then Eqs. (12.54) to (12.57) are found to be equivalent to

$$M_y(x,y) = M(x,y)C(y) \tag{12.58}$$

The initial condition is $M(x,x) = I$, the identity matrix.

If $M$ and $C$ were scalars rather than matrices, we should obtain

$$M(x,y) = \exp \int_x^y C(s)\, ds \tag{12.59}$$

by inspection of Eq. (12.58). The latter would give

$$M(x,z) = M(x,y)M(y,z) \tag{12.60}$$

which, by the isomorphism (12.8), is equivalent to Eq. (12.51).

Although the foregoing considerations can be justified even when $C(s)$ is a matrix, it is preferable to give a direct proof. We have

$$\frac{\partial}{\partial z} M(x,z)M^{-1}(y,z) = 0$$

upon noting that $(M^{-1})_z = -M^{-1}M_zM^{-1}$ and using Eq. (12.58). Since the expression has the value $M(x,y)$ when $z = y$, Eq. (12.60) holds. What this shows is that the functional equations follow from the differential system.

## EXERCISES

**1.** For an isotropic homogeneous dielectric sheet, the interface reflections satisfy

$$-r_1 = \rho_1 = r_2 = -\rho_2 = \frac{Z-1}{Z+1}$$

(Sec. 12.15). If $\chi$ is defined by $r_1 = \tan \frac{1}{2}\chi$, show that

$$Z^2 \equiv \frac{\epsilon/\epsilon_0}{\mu/\mu_0} = \tan^2 (\tfrac{1}{2}\chi + \tfrac{1}{4}\pi)$$

**2.** In Exercise 1, let $X = kx$ be the electrical thickness of the sheet. Show that

$$\begin{aligned} t^{-1} &= \cos X - j \sin X \sec \chi \\ rt^{-1} &= j \sin X \tan \chi \\ dt^{-1} &= \cos X + j \sin X \sec \chi \end{aligned}$$

where $d = t^2 - r^2$.

**3.** A medium, lossless but anisotropic with respect to polarization, is specified in the notation of Exercise 2 by $X_1$ and $\chi_1$ at a certain polarization and by $X_2$ and $\chi_2$ when the polarization is rotated through 90°. To construct a quarter-wave plate giving circular polarization, it is desired to make

$$\angle t_1 - \angle t_2 = \frac{\pi}{2}$$

Show that this condition reduces to

$$\cos \chi_1 \cos \chi_2 + \tan X_1 \tan X_2 = 0$$

Show also that a measure of the transmitting efficiency is

$$|t_1|^{-2} + |t_2|^{-2} = 2 + \sin^2 X_1 \tan^2 \chi_1 + \sin^2 X_2 \tan^2 \chi_2$$

## 12.18 Conditions for a Passive Solution

We inquire next: What conditions on $a$, $b_1$, $b_2$, and $c$ ensure that the dielectric medium represents a passive network for $y \geq x$? The answer to this question can be found by physical considerations, as follows. Regard the medium as composed of many thin layers. If each layer is dissipative, Sec. 12.6 shows that the star product is also dissipative, and the latter gives the scattering matrix for the whole medium.

When $y - x = \Delta y$ is small, the differential system (12.54) to (12.57) together with the initial conditions yields

$$\begin{pmatrix} t(x,y) & \rho(x,y) \\ r(x,y) & \tau(x,y) \end{pmatrix} = \begin{pmatrix} 1 & 0 \\ 0 & 1 \end{pmatrix} + \begin{pmatrix} b_1 & a \\ c & b_2 \end{pmatrix} \Delta y + o(\Delta y) \qquad (12.61)$$

where $o(\Delta y)$ denotes an expression such that $o(\Delta y)/\Delta y \to 0$ as $\Delta y \to 0$. For the matrix (12.61) to satisfy the criterion (12.12) as $\Delta y \to 0$ through positive values, it is necessary that

$$\Re\{\bar{z}_1(b_1 z_1 + a z_4) + \bar{z}_4(c z_1 + b_2 z_4)\} \leq 0$$

Since $z_1$ and $z_4$ are unrestricted, the condition reduces to

$$\Re b_1 \leq 0 \qquad \Re b_2 \leq 0 \qquad |a + \bar{c}|^2 \leq 4(\Re b_1)(\Re b_2) \qquad (12.62)$$

For the lossless case, the three inequalities are replaced by equalities.

When $a$, $b_1$, $b_2$, and $c$ are piecewise continuous on a given interval $J$, it is not difficult to put the foregoing considerations into proper mathematical form.† The inequalities (12.62) are necessary and sufficient conditions that the scattering matrix be dissipative for $x$ in $J$, $y$ in $J$, and $y > x$. This is an example, *par excellence*, of a theorem in which the mathematician's task is facilitated by a modicum of physical insight.

## EXERCISES

**1.** Prove that an isotropic dielectric medium is lossless if and only if $\epsilon(s)$ and $\mu(s)$ are real, $-\infty < s < \infty$.

**2.** Let $r_1$, $r_0$, $t_1$, $t_0$, $\rho_1$, and $\rho_0$ be six real functions of the real variable $y$ for $y \geq 0$, satisfying the following conditions:

*a.* $t_0'$ and $r_1' \sin r_0 - r_1\rho_0' \cos r_0 \equiv H$ are piecewise continuous.
*b.* $1 - r_1^2 = t_1^2 = 1 - \rho_1^2 \neq 0$, and the common value is continuous.
*c.* $2t_0 - r_0 - \rho_0 = (2n + 1)\pi$ for an integer $n$.
*d.* $r_1' \cos r_0 + r_1\rho_0' \sin r_0 = 0$, $t_1(0) = 1$.

Show that these functions can be realized as coefficients

$$r_1 e^{ir_0} = r(0,y) \qquad t_1 e^{it_0} = t(0,y) \qquad \rho_1 e^{i\rho_0} = \rho(0,y)$$

of a lossless dielectric medium having piecewise continuous $\epsilon(y)$ and $\mu(y)$. [With

$$\frac{2\pi}{\lambda}\frac{\epsilon}{\epsilon_0} = t_0' - \left(\frac{H}{t_1^2}\right)(r_1 \cos r_0 - 1)$$

$$\frac{2\pi}{\lambda}\frac{\mu}{\mu_0} = t_0' - \left(\frac{H}{t_1^2}\right)(r_1 \cos r_0 + 1)$$

it will be found that the desired differential equations hold. The medium is lossless by Exercise 1.]

† In each interval of continuity, approximate the coefficients by *continuously differentiable* ones that satisfy the given conditions (12.62) with *strict inequality*.

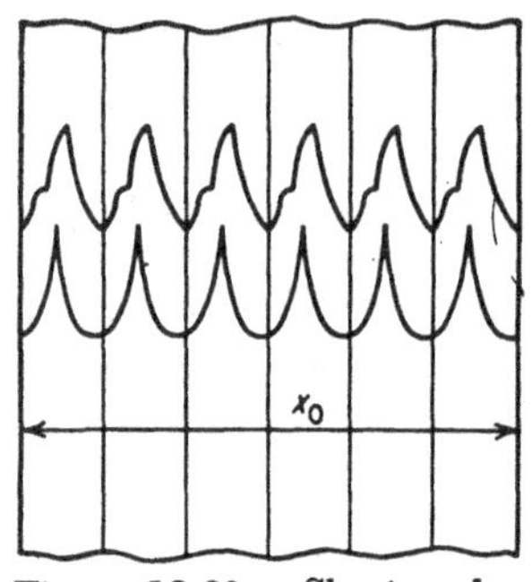


**Fig. 12.20** Sheet obtained by contracting and stacking the sheets shown in Fig. 12.19.

**3.** Prove conversely that conditions $a$ to $d$ of Exercise 2 must be satisfied if $r_1, r_0, \ldots$ can be realized as coefficients of a dielectric medium of the type described. (Use the results of Sec. 12.5. It is necessary to give a suitable interpretation to the phases $r_0$ and $\rho_0$ when $r_1$ or $\rho_1$ is zero.)

**4.** A lossless, isotropic, dielectric medium (Fig. 12.19) is contracted by a factor $n$; then $n$ of these contracted media are stacked together as shown in Fig. 12.20. If $T_n$ is the transmission coefficient in Fig. 12.20, find $t = \lim T_n$ as $n \to \infty$. (The limits needed in Sec. 12.11, Exercise 2, are given by the result of Sec. 12.16, Exercise 2. See also Sec. 12.11, Exercise 1.)

**5.** Let the symbol [ ] denote "mean value of," so that

$$[f(x)] = \frac{1}{x_0} \int_0^{x_0} f(x)\, dx$$

Show that the limiting medium considered in Exercise 4 is equivalent† to a homogeneous medium with dielectric constant, permeability, and thickness

$$[\epsilon][\mu]^{-1/2} \left[\frac{1}{\mu}\right]^{-1/2} \qquad [\mu]^{1/2} \left[\frac{1}{\mu}\right]^{-1/2} \qquad x_0[\mu]^{1/2} \left[\frac{1}{\mu}\right]^{1/2}$$

at perpendicular polarization, while for parallel polarization the respective values are

$$[\epsilon]^{1/2} \left[\frac{1}{\epsilon}\right]^{-1/2} \qquad [\mu][\epsilon]^{-1/2} \left[\frac{1}{\epsilon}\right]^{-1/2} \qquad x_0[\epsilon]^{1/2} \left[\frac{1}{\epsilon}\right]^{1/2}$$

(Extend the result of Exercise 4 to arbitrary $\theta$ as in Sec. 12.15, Exercise 1, and compare your expression for $t^{-1}$ with the one given there. Equality of the reflections follows from Fig. 12.9.)

**6.** In Exercise 5 let $\epsilon(x)$ be continuous and not constant, and let $\mu \equiv \mu_0$. Show that the equivalent permeability $\mu_e$ for the limit medium satisfies $\mu_e < \mu_0$ at parallel polarization. (Use the Schwarz inequality.)

## 12.19 Probability: A Reinterpretation

A particle moves at random on a line $L$ in such a way that it can be transmitted, reflected, or absorbed when it encounters a given segment $J$ of $L$. If $x$ and $y$, with $y > x$, give the coordinates of the ends of $J$, the probabilities of transmission and reflection are denoted by $t(x,y)$ and $r(x,y)$ when the particle is incident on $J$ from the left, and by $\tau(x,y)$ and $\rho(x,y)$ when the particle is incident from the right. (For purposes of this definition, the segment $J$ is thought to be withdrawn from the rest of the line.) When the line is homogeneous, e.g., has a uniform mass distribution on it, then the probabilities are functions of the length $y - x$ only.

† That is, the complex transmission and reflection coefficients for the one medium equal those for the other, as functions of both $\theta$ and $\lambda$.

Use of two variables allows for inhomogeneity, as in the foregoing discussion of dielectric media.

The probabilities associated with a given segment are assumed independent of all information about adjacent segments. With this understanding, one can get functional equations as follows: Consider the interval $(x,y)$ and an adjacent interval $(y,z)$, $z > y > x$. The particle can pass through the interval $(x,z)$ in the mutually exclusive ways suggested by Fig. 12.21; namely, it can pass with no internal reflections, with two, with four, and so on. The assumed independence yields

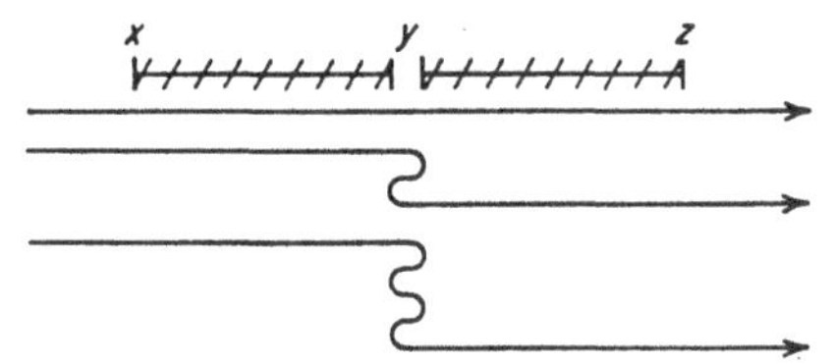


**Fig. 12.21** Modes of passage through two adjacent segments.

$$t(x,y)t(y,z) \qquad t(x,y)r(y,z)\rho(x,y)t(y,z) \qquad t(x,y)r^2(y,z)\rho^2(x,y)t(y,z) \quad \cdots$$

for the probabilities of these various events, and the sum represents the probability of getting through the segment $(x,z)$. Thus we are led to the functional equation

$$t(x,z) = \frac{t(x,y)t(y,z)}{1 - \rho(x,y)r(y,z)}$$

Analogous considerations give $\tau(x,z)$, $r(x,z)$, and $\rho(x,z)$. The resulting functional equations are identical with those of transmission-line theory; in other words, Eq. (12.51) holds.

Because of this equivalence, much of the foregoing analysis applies without change to the probability problem. The differential system (12.54) to (12.57) is valid, as are the results of Sec. 12.13 for uniform media. Interpreting the constants as in Sec. 12.15 enables us even to assign a "dielectric constant" and "permeability." It is only in conditions for the *dissipative* or *passive* character of the process that the difference in the two physical models becomes mathematically significant. Here these conditions result from the fact that our probabilities cannot exceed 1, whereas for the transmission line they result from conservation of energy. Analysis of the inequalities is facilitated by introducing the scattering matrix, which we proceed to do.

## 12.20 The Scattering Matrix

Let the segment $(x,y) = J$ be withdrawn from the line $L$, and let the particle be presented to $J$ in a random manner. Specifically, $z_1$ represents the probability that the particle is tossed at $J$ from the left, and $z_4$ that it is tossed from the right (see Fig. 12.5). In these circumstances, the molecule can emerge from the right in two mutually exclusive ways:

It can be presented from the left and then transmitted, or from the right and then reflected. If $z_3$ is the corresponding probability, we have

$$z_3 = tz_1 + \rho z_4$$

upon writing $t$ for $t(x,y)$ and $\rho$ for $\rho(x,y)$. A similar equation gives $z_2$, the probability of emerging at the left. Together the equations lead to Eq. (12.4), which is repeated for convenience:

$$\begin{pmatrix} z_3 \\ z_2 \end{pmatrix} = \begin{pmatrix} t & \rho \\ r & \tau \end{pmatrix} \begin{pmatrix} z_1 \\ z_4 \end{pmatrix} \tag{12.63}$$

The role of the probabilities $z_1$ and $z_4$ is now clear: They give a suitable vector for our scattering matrix to scatter.

The conditions of physical realizability can be presented from two viewpoints. On the one hand, we have

$$t \geq 0 \qquad r \geq 0 \qquad \tau \geq 0 \qquad \rho \geq 0 \qquad t + r \leq 1 \qquad \tau + \rho \leq 1 \tag{12.64}$$

because of the probability interpretation. On the other hand, if $z_1 \geq 0$ and $z_4 \geq 0$ in Eq. (12.63) then we expect

$$z_2 \geq 0 \qquad z_3 \geq 0 \qquad z_2 + z_3 \leq z_1 + z_4 \tag{12.65}$$

It is readily seen that the inequalities (12.65) and (12.64) are equivalent. When they hold, we say that the matrix is *p dissipative* (*p* for probability). The matrix is *p lossless* if the two inequalities on the right of (12.64), and the one on the right of (12.65), are replaced by equalities. There is then zero probability of absorption.

The criterion (12.65) shows at once that the star product of two matrices is *p* dissipative if each factor is; we simply add the inequalities

$$z_2 + z_3 \leq z_1 + z_4 \qquad \text{and} \qquad z_5 + z_4 \leq z_3 + z_6$$

referring to Fig. 12.6. This fact makes it possible to pass from relationships *in the small* to those *in the large*, just as in the transmission-line case. Upon combining Eq. (12.61) with the inequalities (12.64), we get

$$a \geq 0 \qquad c \geq 0 \qquad b_1 + c \leq 0 \qquad b_2 + a \leq 0 \tag{12.66}$$

as necessary and sufficient conditions on the coefficients of the differential system for the solution matrix to be *p* dissipative in the region $x$ in $L$, $y$ in $L$, $y > x$. Replacing the last two inequalities in (12.66) by equalities makes the matrix be *p* lossless.†

It appears, then, that the sole difference between one-dimensional wave propagation and one-dimensional diffusion lies in the constraints that the coefficients must satisfy. In the former case, they are complex and

† The mathematical details cause no trouble when $a$, $b_1$, $b_2$, and $c$ are piecewise continuous.

satisfy the inequalities (12.62), whereas in the latter they are real and satisfy the inequalities (12.66).

### EXERCISE

Show that, in the lossless isotropic case,

$$1 - r = t = 1 - \rho = \left[1 + \int_x^y a(y)\,dy\right]^{-1}$$

## 12.21 The Underlying Closure Principle

Let $\rho(x,y,K)$ be the solution of

$$\rho_y = a + 2b\rho + c\rho^2$$

that equals $K$ when $y = x$. Figure 12.22 and Eq. (12.17) suggest that

$$\rho(x,y,K) = \rho(x,y) + t(x,y)K[1 - r(x,y)K]^{-1}\tau(x,y) \qquad (12.67)$$

where $\rho$, $t$, $\tau$, and $r$ are given by Eqs. (12.54) to (12.57); the result is verified by direct substitution. Here we have a concrete physical interpretation of the familiar fact that all solutions of Riccati's equation can be expressed in terms of a single solution by quadrature.

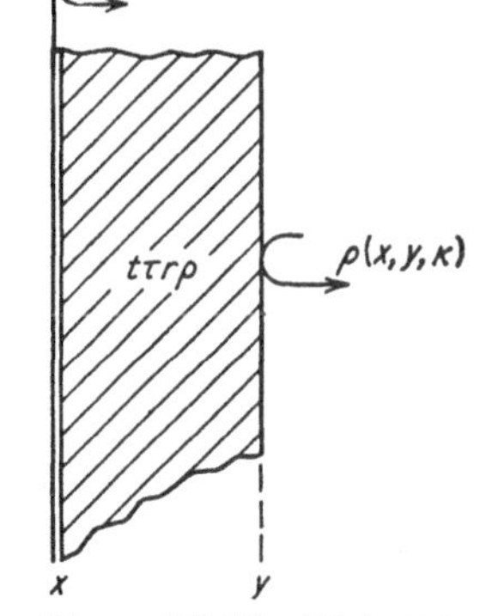


Fig. 12.22 Dielectric sheet backed by a termination of arbitrary reflection $K$.

Let $\rho(x,y,K)$ have the value $K_1$ at $(x_0,y_1)$, so that

$$\rho(x_0,y,K) = K_1 \qquad \text{at } y = y_1$$

Since $\rho(x,x,K) \equiv K$ we have also

$$\rho(y_1,y,K_1) = K_1 \qquad \text{at } y = y_1 \quad (12.68)$$

as shown in Fig. 12.23; hence, by uniqueness,

$$\rho(x_0,y,K) \equiv \rho(y_1,y,K_1)$$

The result becomes

$$\rho(x,z,K) \equiv \rho[y,\, z,\, \rho(x,y,K)] \qquad (12.69)$$

when Eq. (12.68) is used and the variables are renamed.

Depending on one's preference, Eq. (12.69) is a uniqueness theorem, a semigroup property (see Chap. 4), or "the major premise of Huygens' principle according to Hadamard." It has to do with the consistency of our differential equation as a description of the underlying process, and it forms the starting point for the Lie-group theory of that equation. On the other hand, the result is also meaningful without reference to differential equations. By the foregoing proof, Eq. (12.69) holds for *any* function $\rho(x,y,K)$, provided $\rho(x,x,K) \equiv K$ and provided the curves

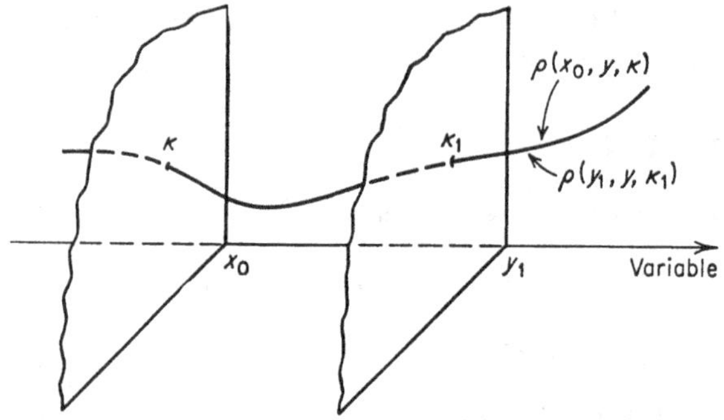


**Fig. 12.23** Trajectory traced out by the complex solution of a first-order differential equation.

traced out by $\rho(x,y,K)$, as $y$ varies, form a field in a suitable region. (A family of curves forms a *field* in a given region if there is one and only one curve of the family through each point of the region.)

If Eq. (12.69) is written in terms of Eq. (12.67), the choice $K = 0$ yields

$$\rho(x,z) = \rho(y,z) + \frac{t(y,z)\tau(y,z)\rho(x,y)}{1 - \rho(x,y)r(y,z)} \tag{12.70}$$

while $K = 1/r(x,z)$ gives

$$r(x,z) = r(x,y) + \frac{t(x,y)\tau(x,y)r(y,z)}{1 - \rho(x,y)r(y,z)} \tag{12.71}$$

The choice $K = 1/r(x,y)$, together with the two preceding equations, leads to

$$T(x,z) = \frac{T(x,y)T(y,z)}{1 - \rho(x,y)r(y,z)} \tag{12.72}$$

where $T(x,y) = [t(x,y)\tau(x,y)]^{1/2}$. Conversely, these three relationships ensure the validity of Eq. (12.69).

If $t = \tau$, the foregoing equations are identical with the functional equations (12.51). Since these have formed the basis of our whole discussion, a substantial part of transmission-line theory appears to be contained in the general closure property (12.69). Here we have a unifying principle behind the diversity of special methods.

To see what becomes of this principle in the nonbilateral case, let a function $h = h(x,y)$ be defined by

$$T(x,y) = h(x,y)t(x,y)$$

Then $ht = T$ and $h^{-1}\tau = T$ both satisfy Eq. (12.72), and hence

$$\begin{pmatrix} ht & \rho \\ r & h^{-1}\tau \end{pmatrix}$$

satisfies Eq. (12.51). In other words, there exists a function $h$ such that the desired equations hold for $ht$, $h^{-1}\tau$, $r$, and $\rho$, though they may not hold for $t$, $\tau$, $r$, and $\rho$.

Nothing more can be deduced from the Riccati equation alone. But if, in addition, Eqs. (12.54) to (12.57) hold, a brief calculation shows that $h$ is constant and hence, by the initial conditions, $h = 1$. Thus the functional equations express the closure property of Eq. (12.57) in the bilateral case, and the closure property of the system (12.54) to (12.57) in the general case.

## REVIEW EXERCISES

**1.** An isotropic inhomogeneous dielectric slab specified by $\epsilon(x)/\epsilon_0 \equiv \epsilon$ and $\mu(x)/\mu_0 \equiv m$ extends from the plane $x = 0$ to the plane $x = x$. If the reflection at incidence $\theta$ is $\rho = \rho(x)$, define

$$w = (1 + \rho)(1 - \rho)^{-1} \qquad Y_\perp = w \sec \theta \qquad Y_\parallel = w \cos \theta$$

where the subscript specifies polarization perpendicular or parallel to the plane of incidence, respectively. Prove that

$$\frac{dY_\perp}{dx} = \frac{2\pi j}{\lambda}\left(\frac{m\epsilon - \sin^2\theta}{m} Y_\perp^2 - m\right)$$

$$\frac{dY_\parallel}{dx} = \frac{2\pi j}{\lambda}\left(\epsilon Y_\parallel^2 - \frac{m\epsilon - \sin^2\theta}{\epsilon}\right)$$

where $\lambda$ is the free-space wavelength. (See Sec. 12.15, Exercise 1.)

**2.** The reflection at near-grazing incidence is important in the study of radio-wave propagation over distances that are small compared with the earth's radius. In Exercise 1, let $\alpha$ and $\beta$ be those solutions of the Riccati equations

$$\frac{d\alpha}{dx} = \frac{2\pi j}{\lambda}\left(\frac{m\epsilon - 1}{m}\alpha^2 - m\right) \qquad \frac{d\beta}{dx} = \frac{2\pi j}{\lambda}\left(\epsilon\beta^2 - \frac{m\epsilon - 1}{\epsilon}\right)$$

that satisfy $\alpha(0) = \beta(0) = 0$. Prove that

$$\lim_{\theta\to\pi/2} \frac{1 + r_\perp}{\cos\theta} = 2\alpha \qquad \frac{d}{d\theta}|r_\perp|^2 = 4\Re(\alpha) \qquad \text{at } \theta = \frac{\pi}{2}$$

$$\lim_{\theta\to\pi/2} \frac{1 - r_\parallel}{\cos\theta} = \frac{2}{\beta} \qquad \frac{d}{d\theta}|r_\parallel|^2 = \Re\left(\frac{4}{\beta}\right) \qquad \text{at } \theta = \frac{\pi}{2}$$

**3.** A microwave absorber consists of a thin layer of material having parameters $\epsilon(x)$ and $\mu(x)$, backed by a metal plate at $x = 0$ (see Fig. 12.22). Suppose $|\epsilon(x)\mu(x)|/(\epsilon_0\mu_0)$ is so large that the quantity $m\epsilon - \sin^2\theta$ in Exercise 1 can be replaced, with negligible error, by $m\epsilon - \frac{1}{2}$ as $\theta$ varies. Show that the absorber satisfies

$$w_\perp \sec\theta = \text{const} \qquad w_\parallel \cos\theta = \text{const}$$

as $\theta$ varies, the thickness $x$ being fixed. (The result follows at once from the uniqueness theorem for first-order equations.)

**4.** In Exercise 3, the performance of the absorber, reflection vs. $\theta$, is given by the explicit formula

$$|\rho_\perp(\theta)|^2 = \frac{T^2 - 2AT\cos B + A^2}{1 - 2AT\cos B + A^2T^2}$$

where $T = \tan^2(\theta/2)$ and where $\rho_\perp = Ae^{iB}$ at $\theta = 0$. The result for $\rho_\parallel$ is obtained by adding $\pi$ to $B$. Prove these results.

**5.** Show that if an absorber of the type considered in Exercise 3 has zero reflection at a given angle $\theta$ and polarization, then the reflection at the angle $\theta$ for the other polarization is

$$|\rho(\theta)| = \frac{1 - \cos^2\theta}{1 + \cos^2\theta}$$

**6.** An absorber of the type considered in Exercise 3 is designed for use at both polarizations simultaneously. Show that the performance is optimum when $r(0)$ is pure imaginary, in which case the reflection is independent of polarization. Show that the optimum performance possible is

$$|\rho_\parallel(\theta)| = |\rho_\perp(\theta)| = \frac{1 - \cos\theta}{1 + \cos\theta}$$

## TRANSMISSION AND REFLECTION OPERATORS

### 12.22 Transmission, Reflection, and Scattering Matrices

Let the obstacle be supplied with $2n$ terminal pairs, of which $n$ are thought to be at the left and $n$ more at the right. At the high frequencies

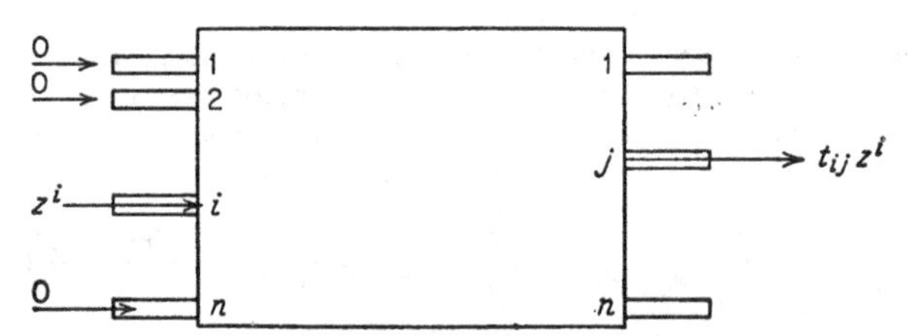


**Fig. 12.24** Obstacle with $2n$ terminal pairs.

for which the analysis is especially suited, these terminal pairs are presumably waveguides, but only one mode is considered relevant.† Also, without real loss of generality we can, and do, assume that the waveguides all have the same characteristic impedance.

If a wave of complex amplitude $z^i$ is incident on the $i$th guide at the left, as in Fig. 12.24, let

$$t_{ij}z^i \qquad \text{and} \qquad r_{ij}z^i$$

represent the amplitudes of the transmitted waves and reflected waves at the $j$th guide on the right or left, respectively. When the amplitudes

† Finitely many modes can be accounted for by imagining extra (fictitious) terminal pairs, one for each mode.

$z^1, z^2, \ldots, z^n$ are simultaneously present, the input is represented by a column vector $z$ having components $z^i$. Superposition gives

$$\sum_{i=1}^{n} t_{ij}z^i \qquad \text{and} \qquad \sum_{i=1}^{n} r_{ij}z^i$$

for the transmitted and reflected vectors, respectively.

In terms of the transposed matrices $t = (t_{ji})$ and $r = (r_{ji})$, the vectors are $tz$ and $rz$, respectively. Similarly, $\tau z$ and $\rho z$ represent the vectors transmitted or reflected when the vector $z$ is incident from the right.

If the vector $z_1$ is incident from the left and simultaneously $z_4$ from the right, as in Fig. 12.5, then the vector $z_3$ emerging from the right can be computed by taking the part of $z_1$ that is transmitted together with the part of $z_4$ that is reflected. The calculation is done at first for the $j$th terminal, but the single matrix equation

$$z_3 = tz_1 + \rho z_4 \tag{12.73}$$

describes the result for each value of $j$. Upon computing $z_2$, the wave reflected at the left, we get

$$\begin{pmatrix} z_3 \\ z_2 \end{pmatrix} = \begin{pmatrix} t & \rho \\ r & \tau \end{pmatrix} \begin{pmatrix} z_1 \\ z_4 \end{pmatrix} \equiv S \begin{pmatrix} z_1 \\ z_4 \end{pmatrix} \tag{12.74}$$

just as in Eq. (12.4).

The $n$ by $n$ matrices $t$, $\tau$, $r$, and $\rho$ play the same algebraic role as the complex numbers $t$, $\tau$, $r$, and $\rho$ used previously. Hence, $t$ and $\tau$ are called the *transmission coefficients* and $r$ and $\rho$ the *reflection coefficients*. Here, however, these quantities are matrices, i.e., *operators* on the space of $n$-dimensional complex vectors. The $2n$ by $2n$ matrix $S$ in Eq. (12.74) is the *scattering matrix*, as in the case $n = 1$. In order to use the previous formulas, we denote the $n$ by $n$ identity matrix by 1.

## 12.23 The Star Product and Closure

Distinguishing $n$ terminal pairs as input and $n$ as output enables us to combine obstacles (and that is the reason for making the distinction). If the obstacle of the foregoing discussion is adjacent to a second one, the over-all coefficients can be computed in principle by multiple reflection. But it is much simpler to use the alternative derivation, the one leading to Eq. (12.6) by way of Fig. 12.6. We get the same formula (12.3) in the matrix case; in fact, the order of factors in Eq. (12.3) was chosen to obviate rewriting the result for $n > 1$. The isomorphism (12.8) is also valid, and proved in the same way. Hence, star multiplication is associative.

So much for the algebra of star products. In the case $n = 1$, we found that there was also an associated calculus; namely, the functional equation (12.51) turned out to be the closure or semigroup property for a certain differential system. The analogous result is true for arbitrary $n$, as will be seen next. For precision of statement, we say that a region $R$ of the $xy$ plane is admissible if it is open, connected, and such that the closed line segment joining $(x,x)$ to $(x,y)$ is in $R$ whenever $(x,y)$ is in $R$.

THEOREM 12.1. *Let $t$, $\tau$, $r$, and $\rho$ be complex $n$ by $n$ matrix functions of the two real variables $x$ and $y$, with $t\tau$ nonsingular, and set*

$$S(x,y) = \begin{pmatrix} t(x,y) & \rho(x,y) \\ r(x,y) & \tau(x,y) \end{pmatrix}$$

*If $S(x,x)$ equals the identity matrix, if $S_y(x,x)$ exists, and if*

$$S(x,z) = S(x,y) * S(y,z) \tag{12.75}$$

*in an admissible region $R$, then there are complex $n$ by $n$ matrix functions $a(y)$, $b_1(y)$, $b_2(y)$, and $c(y)$ such that the entries of $S$ satisfy*

$$\begin{aligned} \rho_y &= a + b_1\rho + \rho b_2 + \rho c \rho & \rho(x,x) &= 0 \\ t_y &= (b_1 + \rho c)t & t(x,x) &= 1 \\ \tau_y &= \tau(b_2 + c\rho) & \tau(x,x) &= 1 \\ r_y &= \tau c t & r(x,x) &= 0 \end{aligned} \tag{12.76}$$

*throughout $R$. Conversely, if $t$, $\tau$, $r$, $\rho$ is any solution of this differential system, with $t\tau$ nonsingular, then Eq. (12.75) holds.*

It should be noticed that no continuity hypothesis on the coefficients is needed. In Theorems 12.3 and 12.4, below, which assume continuity, the fact that $t\tau$ is nonsingular can be deduced from the differential system.

Derivation of the differential system is similar to that in the case $n = 1$. To proceed in the opposite direction, consider the equivalent linear system $M_y = MC$, where

$$M = \begin{pmatrix} \tau^{-1} & -\tau^{-1}r \\ \rho\tau^{-1} & t - \rho\tau^{-1}r \end{pmatrix}' \qquad C = \begin{pmatrix} -b_2 & a \\ -c & b_1 \end{pmatrix}$$

The matrix $M$ is nonsingular since

$$M^{-1} = \begin{pmatrix} \tau - rt^{-1}\rho & rt^{-1} \\ -t^{-1}\rho & t^{-1} \end{pmatrix}'$$

Hence, $M(x,z) = M(x,y)M(y,z)$ follows from

$$\frac{\partial}{\partial z} M(x,z)M^{-1}(y,z) = 0$$

as in Sec. 12.17, and Eq. (12.75) follows from this.

If $K$ is a complex $n$ by $n$ matrix such that the expression $\rho(x,y,K)$ of Eq. (12.67) is well defined, then the latter satisfies the equations

$$\rho_y = a + b_1\rho + \rho b_2 + \rho c\rho \qquad \rho(x,x,K) = K \qquad (12.77)$$

It is possible to discuss functional equations by means of the closure principle (12.69), which is proved for matrices just as for scalars. The result of an elementary but long calculation is that Eq. (12.69) holds if and only if there is a scalar function $h$ such that

$$S_h(x,y) = \begin{pmatrix} ht(x,y) & \rho(x,y) \\ r(x,y) & h^{-1}\tau(x,y) \end{pmatrix}$$

satisfies the relationship $S_h(x,z) = S_h(x,y) * S_h(y,z)$. The system (12.76) implies $h = 1$, giving another proof of Eq. (12.75). At the same time, we see to what extent the closure principle for the whole system is more demanding than that for Eqs. (12.77) alone.

## EXERCISES

**1.** The identity $I$ is a matrix such that $S * I = I * S = S$ for all $2n$ by $2n$ scattering matrices $S$. The inverse $S^{-1}$ satisfies $S^{-1} * S = S * S^{-1} = I$. Show that the identity and inverse under star multiplication are the same as those under matrix multiplication. [Set $z_5 = z_1$ and $z_6 = z_2$ in Eqs. (12.4) to (12.6). In this and the following problems the results are to be established for arbitrary $n$.]

**2.** The network $(t\tau r\rho)$ has left-hand reflection $R$ when followed by an open circuit $r_1 = 1$. The network $(t\tau r\rho) * (\tau t\rho r)$ has left-hand reflection $R_0$ and transmission $T_0$. Show that

$$R = R_0 + T_0$$

and interpret by the method of images (Fig. 12.25). How would the result change if $(t\tau r\rho)$ were backed by a short circuit, $r_1 = -1$?

**3.** Show that the solutions of Eqs. (12.76) satisfy the equations

$$\begin{aligned} -\rho_x &= ta\tau \\ -\tau_x &= (b_2 + ra)\tau \\ -t_x &= t(b_1 + ar) \\ -r_x &= c + rb_1 + b_2r + rar \end{aligned}$$

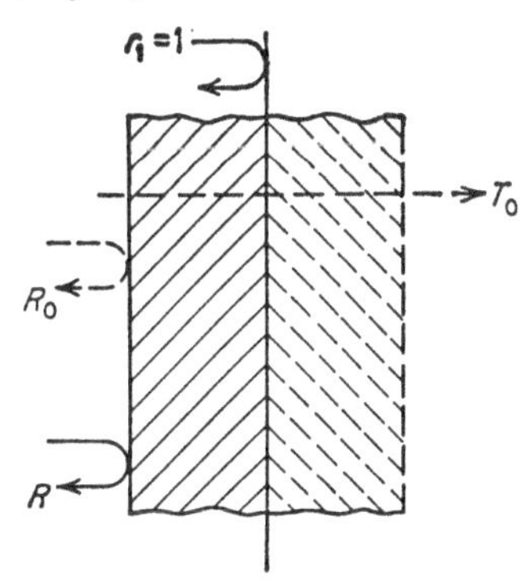


**Fig. 12.25** Obstacle backed by a perfect reflector, and its mirror image.

and interpret physically. (Refer the proof to the linear system. The minus sign is needed because the "slab thickness" increases with increasing $y$ but decreases with increasing $x$.)

**4.** Prove that in Sec. 12.7 we have $R(z_1) = z_2^{-1}$ if and only if $P(z_2) = z_1^{-1}$. (Assume that the needed inverses exist.)

**5.** Show that the star product of two symmetric matrices is symmetric.

## 12.24 The Norm and Energy Transfer

If $A = (A_1, A_2, \ldots, A_n)$ is a complex row vector, its length is defined by

$$|A| = (\Sigma A_i \bar{A}_i)^{1/2}$$

By the Schwarz inequality, we have

$$|A| = \max |Az| \qquad \text{for } |z| = 1 \tag{12.78}$$

where $z$ ranges over all complex column vectors $(z^1, z^2, \ldots, z^n)$ of unit length.

The merit of this alternative description is that the latter applies to any matrix $A$, provided $A$ has $n$ columns. The resulting measure is called the *norm* of $A$. Since $|A|$ reduces to the ordinary absolute value when $A$ has a single element, or to the length of $A$ when $A$ is a vector, there is no need to introduce a new symbol.

If $A^*$ denotes the adjoint of $A$, then

$$|Az|^2 = (Az)^*Az = z^*A^*Az \tag{12.79}$$

Hence $|A|^2$ is given by the Rayleigh quotient,

$$|A|^2 = \max \frac{z^*A^*Az}{z^*z} \qquad \text{for } z \neq 0 \tag{12.80}$$

If $A$ and $B$ are self-adjoint matrices, the condition $A \leq B$ means

$$z^*Az \leq z^*Bz \qquad \text{for all } z$$

It can be shown that this relationship has many of the properties of ordinary inequality. By Eq. (12.80), the condition $|A| \leq 1$ is equivalent to $A^*A \leq 1$. If $A^*A = 1$, that is, if $A$ is unitary, then Eq. (12.79) gives $|Az| = |z|$ for all $z$.

A network specified by the four matrices $t$, $\tau$, $r$, and $\rho$ is *dissipative* if and only if the scattering matrix $S$ of (12.74) is such that

$$|z_3|^2 + |z_2|^2 \leq |z_1|^2 + |z_4|^2$$

for all vectors $z_1$ and $z_4$. In terms of the foregoing notation, the condition is simply $|S| \leq 1$. When equality holds—i.e., when $S$ is unitary—the network is lossless.

As in Sec. 12.8, the network is reflectively dissipative from the left if the matrix

$$R(K) = r + \tau K(1 - \rho K)^{-1}t \tag{12.81}$$

satisfies $|R(K)| < 1$ whenever $|K| < 1$. (Here $K$ is an $n$ by $n$ complex matrix, not a vector.) The corresponding condition from the right involves

$$P(K) = \rho + tK(1 - rK)^{-1}\tau \tag{12.82}$$

If, in addition, $R(K)$ is unitary for all unitary $K$, the network is reflectively lossless from the left.

### EXERCISES

1. Prove that if a unitary scattering matrix $(t\tau r\rho)$ has $|r| < 1$, then also $|\rho| < 1$.
2. Show that when $|S_1| \leq 1$ and $|S_2| \leq 1$, then $|S_1 * S_2| \leq \max(|S_1|, |S_2|)$. (In this and the following exercises assume that the relevant star products exist.)
3. If two matrices are unitary, show that their star product also is unitary.
4. If $S$ is unitary, show that the corresponding bilinear function $R(K)$ is unitary for all unitary $K$. [Consider $(t\tau r\rho) * (00KK)$ and use the result of Exercise 3.]

## 12.25 The Matching Problem

As the reader will recall, for $n = 1$ a lossless network with right-hand reflection $\bar{r}$ gives zero left-hand reflection when followed by a reflection $r$. The lossless network is called a *tuner*, and the new network, tuner plus load, is said to be *matched* from the left.

Construction of a lossless tuner for arbitrary $n$ hinges on the following fact of matrix theory: Given any $n$ by $n$ complex matrix $K$, with $|K| \leq 1$, there is an $n$ by $n$ matrix $X$ such that

$$\begin{pmatrix} X & K \\ -K & X \end{pmatrix} \qquad \text{is unitary}$$

If $|K| < 1$, then $X$ is nonsingular. The proper choice of $K$ for match is $K = r^*$, as is shown by a brief calculation. In other words, if $|r| < 1$ then

$$\begin{pmatrix} X & r^* \\ -r^* & X \end{pmatrix} * \begin{pmatrix} t & \rho \\ r & \tau \end{pmatrix} = \begin{pmatrix} t_1 & \rho_1 \\ 0 & \tau_1 \end{pmatrix} \tag{12.83}$$

The problem of matching a network simultaneously from both sides is more subtle. We try to choose reflections $r_0$ and $\rho_0$ with corresponding unitary scattering matrices such that

$$\begin{pmatrix} X & r_0 \\ -r_0 & X \end{pmatrix} * \begin{pmatrix} t & \rho \\ r & \tau \end{pmatrix} * \begin{pmatrix} Y & -\rho_0 \\ \rho_0 & Y \end{pmatrix} = \begin{pmatrix} t_1 & 0 \\ 0 & \tau_1 \end{pmatrix} \tag{12.84}$$

By Eq. (12.83), the attempt is successful if simultaneously

$$\rho_0^* = P(r_0) \qquad r_0^* = R(\rho_0) \tag{12.85}$$

with $R$ and P as in Eqs. (12.81) and (12.82). Now, the Brouwer fixed-point theorem guarantees existence of a solution $(r_0,\rho_0)$ of Eq. (12.85), provided P and $R$ map the unit operator sphere into itself, i.e., provided

$$|R(K)| < 1 \qquad \text{and} \qquad |\mathrm{P}(K)| < 1$$

for all $|K| < 1$. The latter, in turn, is precisely the condition for the network to be reflectively dissipative.†

## EXERCISES

**1.** A scattering matrix $S = (t\tau r\rho)$ with $|\rho| < 1$ is matched from the right by each of two lossless networks, $T_1$ and $T_2$. Show that there are unitary matrices $t_0$ and $\tau_0$ such that

$$T_1 = T_2 * (t_0\tau_0 00)$$

For $n = 1$, interpret the result as stating that a tuner is unique, except for a length of anisotropic lossless line. (Let $S * T_1 = S_1$ and $S * T_2 = S_2$ both have right-hand reflections $\rho = 0$. The matrix $T_0 = T_2^{-1} * T_1$ is unitary by Exercise 3 of Sec. 12.24, and it satisfies $S_2 * T_0 = S_1$.)

**2.** A network $(t\tau r\rho)$ is *lossless from the left* provided $R(K)$ is unitary for all unitary $K$. Assuming $t\tau$ nonsingular, show that $(t\tau r\rho)$ is lossless from the left if and only if $(hth^{-1}\tau r\rho)$ is lossless for some positive scalar $h$. [To investigate the converse of Exercise 4, Sec. 12.24, match $(t\tau r\rho)$ from the right. A matrix that commutes with every unitary matrix is necessarily scalar.]

**3.** Show that when $t\tau$ is nonsingular, the network $(t\tau r\rho)$ is lossless from the left if and only if it is from the right. (Use Exercise 2.)

**4.** Let $(t\tau r\rho)$ be a lossless network with $t\tau$ nonsingular. If $K$ runs through the set of all unitary matrices, show that $R(K)$ does also. [Let $K_0$ be unitary. With $h$ as given by Exercise 2, solve the equation

$$(hth^{-1}\tau r\rho) * (00KK_1) = (00K_0K_0)$$

for $(00KK_1)$. The fact that $K$ is unitary follows from Exercise 3, Sec. 12.24.]

**5.** Let $S = (t\tau r\rho)$ be a scattering matrix with $|\rho| < 1$. Show that the operators

$$I_1 = t^*(1 - \rho\rho^*)^{-1}t \qquad I_2 = \tau(1 - \rho^*\rho)^{-1}\tau^* \qquad I_3 = r + \tau\rho^*(1 - \rho\rho^*)^{-1}t$$

are *energy invariants;* i.e., they are unchanged if $S$ is replaced by $S * U$, where $U$ is any lossless network with nonsingular transmissions. (Establish the invariance when $U$ is a tuner $T$, and also when $\rho = 0$. The desired result then follows from $S * U = S * T * T^{-1} * U = S_1 * U_1$.)

**6.** The operators $I$ in Exercise 5 are the sole energy invariants in the following sense: If $f(t,\tau,r,\rho)$ is such an invariant, then

$$f(t,\tau,r,\rho) = f_1(I_1,I_2,I_3)$$

for some function $f_1$. [Let $S * T = (t_1\tau_1 r_1 0)$, where $T$ is a tuner. If the polar decompositions for $t_1$ and $\tau_1$ are $t_1 = u_1h_1$, $\tau_1 = h_2u_2$, consider $S * T * (u_1^*u_2^*00)$.]

† To ensure that the product (12.84) is well defined, it is necessary that we have $|\rho_0| < 1$ and $|r_0| < 1$. These inequalities can be attained by arbitrarily small changes in $t$, $\tau$, $r$, and $\rho$; hence, their failure to hold does not invalidate the conclusions we shall draw later (Theorem 12.2).

7. For $n = 1$ obtain invariants by inspection of Eqs. (12.18). Show that there is the additional invariant $\angle t - \angle\tau$ if the lossless network $U$ is required to be bilateral. (See Sec. 12.10. The geometric interpretation can be extended to arbitrary $n$ by use of Exercise 4.)

## 12.26 Further Discussion of Passive Networks

If a dissipative obstacle has $t\tau$ nonsingular, it is easily shown that $|\rho| < 1$ and $|r| < 1$, so that the star product of two such obstacles is well defined. Indeed,

$$|\rho r_1| \leq |\rho|\,|r_1| < 1$$

so that

$$(1 - \rho r_1)z \neq 0 \qquad \text{for } z \neq 0$$

Therefore, $1 - \rho r_1$ is nonsingular. Proceeding as in the case $n = 1$ (Sec. 12.6), we deduce that *if two networks are dissipative and have nonsingular transmissions, the same is true of their star product.* The analogous statement for reflective dissipativity from the right or left is also true. These considerations lead to the following result:

THEOREM 12.2. *Let $t$, $\tau$, $r$, and $\rho$ be complex $n$ by $n$ matrices with $t\tau$ nonsingular. Then the following statements a, b, and c are equivalent:*

*a.* $|r + \tau K(1 - \rho K)^{-1}t| < 1$ *whenever* $|K| < 1$
*b.* $|\rho + tK(1 - rK)^{-1}\tau| < 1$ *whenever* $|K| < 1$
*c. There is a positive scalar $h$ such that*

$$\left|\begin{pmatrix} ht & \rho \\ r & h^{-1}\tau \end{pmatrix}\right| \leq 1$$

*Also, if the latter matrix is unitary, then the former are unitary for all unitary $K$, and conversely.*

In other words, if the network is reflectively dissipative (or lossless) from the left, then it is also from the right, but it need not be dissipative or lossless in the general sense, unless $h$ in statement $c$ happens to be 1.

A proof of Theorem 12.2 can be based on the theory of quadratic forms.† Here, however, we sketch an argument more closely related to the engineering situation.

To show that statement $c$ implies statements $a$ and $b$, choose $X$ so that the matrix containing it is unitary and consider

$$\begin{pmatrix} X & K \\ -K & X \end{pmatrix} * \begin{pmatrix} ht & \rho \\ r & h^{-1}\tau \end{pmatrix} = \begin{pmatrix} t_1 & \mathrm{P}(K) \\ r_1 & \tau_1 \end{pmatrix}$$

[One gets $\mathrm{P}(K)$ by Eq. (12.3), since the scalar $h$ commutes.] Being the product of two dissipative matrices, the matrix containing $\mathrm{P}(K)$ is dis-

† See the problems at the end of this section.

sipative; hence, $|\mathrm{P}(K)| < 1$ by consideration of

$$\begin{pmatrix} t_1 & \mathrm{P}(K) \\ r_1 & \tau_1 \end{pmatrix} \begin{pmatrix} 0 \\ z \end{pmatrix} \qquad \text{with } |z| = 1$$

Thus statement *c* implies *b*; similarly, *c* implies *a*.

We show next that statement *b* implies *a*. The proof depends on the fact that a condition like *b* gives not only the obvious inequality $|\rho| < 1$ got by setting $K = 0$, but also (provided $t\tau$ is nonsingular) $|r| < 1$. With this in mind, observe that the product

$$\begin{pmatrix} t & \rho \\ r & \tau \end{pmatrix} * \begin{pmatrix} X & -K \\ K & X \end{pmatrix}$$

is reflectively dissipative from the right, since each factor is. Hence, by the foregoing remark, the left-hand reflection for the product cannot have norm as large as 1. This gives the desired condition, $|R(K)| < 1$.

It remains to show that *b* implies *c*. Since *a* holds when *b* does, we can use the bilateral matching of Eq. (12.84). The resulting matrix is reflectively dissipative both from the left and from the right, since each factor is. By computing the associated functions $\mathrm{P}_1(K)$ and $R_1(K)$, we see that

$$|t_1 K \tau_1| < 1 \qquad \text{whenever } |K| < 1$$

Because $K$ is arbitrary, the latter is equivalent to $|t_1|\,|\tau_1| \leq 1$.

Since $t_1$ and $\tau_1$ are nonsingular, their norms are not zero. If

$$h = \left(\frac{|\tau_1|}{|t_1|}\right)^{1/2}$$

we have

$$h|t_1| = h^{-1}|\tau_1|$$

Each quantity is $\leq 1$, since the product is; i.e., we have

$$|ht_1| \leq 1 \qquad |h^{-1}\tau_1| \leq 1 \tag{12.86}$$

Now, it is easily shown that the inverse under star multiplication is the same as that under matrix multiplication. In particular, the inverse of a unitary matrix is its adjoint. By solving Eq. (12.84) for the $(t\tau r\rho)$ matrix and writing $ht$ for $t$, and $h^{-1}\tau$ for $\tau$, we get

$$\begin{pmatrix} ht & \rho \\ r & h^{-1}\tau \end{pmatrix} = \begin{pmatrix} X & r_0 \\ -r_0 & X \end{pmatrix}^* * \begin{pmatrix} ht_1 & 0 \\ 0 & h^{-1}\tau_1 \end{pmatrix} * \begin{pmatrix} Y & -\rho_0 \\ \rho_0 & Y \end{pmatrix}^*$$

By the inequalities (12.86), the matrix involving $t_1$ and $\tau_1$ is dissipative; therefore, the whole product is dissipative. This completes the proof.

## EXERCISES

**1.** Suppose we say that $(t\tau r\rho)$ is reflectively dissipative from the left if $|R(K)| \leq 1$ whenever $|K| \leq 1$ and $1 - \rho K$ is nonsingular. Show that this definition is equivalent to that of the text provided $t\tau$ is nonsingular.

**2.** (In this and the following problems, "reflectively dissipative" has the meaning given in Exercise 1.) Show that $(t\tau r\rho)$ is reflectively dissipative from the left if

$$|z_2| \leq |\rho z_2 + tz_1| \qquad \text{implies} \qquad |z_1| \leq |\tau z_2 + rz_1|$$

for all complex $n$ vectors $z_i$. Show further that the converse is true provided $|\rho| \leq 1$ and $t \neq 0$.

**3.** Let $t \neq 0$, $\tau \neq 0$, $|r| \leq 1$, $|\rho| \leq 1$. Show that $(t\tau r\rho)$ is reflectively dissipative from the left if and only if it is from the right. (Use Exercise 2.)

**4.** If there is a positive scalar $h$ such that

$$\left| \begin{pmatrix} ht & \rho \\ r & h^{-1}\tau \end{pmatrix} \right| \leq 1$$

show that $(t\tau r\rho)$ is reflectively dissipative from the left. (Use Exercise 2. Note that $t\tau$ is not assumed nonsingular.)

**5.** In Exercise 4, prove that the converse holds provided $t \neq 0$, $\tau \neq 0$, and $|\rho| \leq 1$. (Use Exercise 2 together with the theory of pencils of quadratic forms.)

## 12.27 Inequalities for the Differential System

If $\Delta y = y - x$ is small, the system (12.76) gives the estimate (12.61) for the scattering matrix $S(x,y)$. The condition $|S| \leq 1$ is equivalent to $SS^* \leq 1$. Since $\Delta y > 0$, and since $(\Delta y)^2$ is negligible compared with $\Delta y$, we get

$$\begin{pmatrix} b_1 & a \\ c & b_2 \end{pmatrix} + \begin{pmatrix} b_1 & a \\ c & b_2 \end{pmatrix}^* \leq 0 \tag{12.87}$$

as the desired inequality. Such is the condition for the matrix-propagating medium to be passive in the small.

Since the algebra of the star product for arbitrary $n$ is the same as that for $n = 1$, passing to relationships in the large causes no more trouble now than it did in Sec. 12.18. Without further ado we state the following result:

THEOREM 12.3. *Let the hypothesis of Theorem* 12.1 *hold for $x$ in $J$ and $y$ in $J$, where $J$ is a given interval. Suppose also that the matrix coefficients $a(y)$, $b_1(y)$, $b_2(y)$, and $c(y)$ are piecewise continuous. Then*

$$|S(x,y)| \leq 1 \qquad \text{for } x \text{ in } J,\ y \text{ in } J, \text{ and } y > x$$

*if and only if the inequality* (12.87) *holds for $y$ in $J$. In case equality holds in* (12.87) *for all $y$, then $S(x,y)$ is unitary, and conversely.*

The analogous criterion for reflectively dissipative networks is based on part $c$ of Theorem 12.2. It leads to the condition

$$\begin{pmatrix} b_1 & a \\ c & b_2 \end{pmatrix} + \begin{pmatrix} b_1 & a \\ c & b_2 \end{pmatrix}^* \leq \begin{pmatrix} H & 0 \\ 0 & -H \end{pmatrix} \qquad y \text{ in } J \tag{12.88}$$

where $H$ is an arbitrary real scalar matrix. The inequality (12.88) is necessary and sufficient for the function $\rho(x,y,K)$ of Eq. (12.67) to satisfy

$$|\rho(x,y,K)| < 1$$

whenever $|K| < 1$, $x$ is in $J$, $y$ is in $J$, and $y > x$.

## REVIEW EXERCISES

**1.** If a one-to-one correspondence between $n$ by $n$ matrices is denoted by $L = f(S)$, we can define a star product by

$$S_1 * S_2 = f^{-1}[f(S_1)f(S_2)]$$

where the product on the right is ordinary matrix multiplication. Prove that all such star products are associative.

**2.** Show that the particular star product of this chapter is generated by

$$f(S) = [(1 - S) + (1 + S)J]^{-1}[(S - 1) + (S + 1)J]$$

where
$$J = \begin{pmatrix} -1 & 0 \\ 0 & 1 \end{pmatrix}$$

**3.** Denoting the $n$ by $n$ zero matrix by 0, we define

$$A = \begin{pmatrix} a & 0 \\ 0 & 0 \end{pmatrix} \qquad W = \begin{pmatrix} \rho & t \\ \tau & r \end{pmatrix} \qquad J = \begin{pmatrix} 0 & 1 \\ 1 & 0 \end{pmatrix}$$

and $B_1$, $B_2$, $C$ by analogy with $A$. Show that the entire system (12.76) is equivalent to the single Riccati equation

$$W_y = A + B_1W + WB_2 + WCW$$

together with the initial condition $W(x,x) = J$.

**4.** A *two-sided invariant* is a function $(t\tau r\rho)$ that is unchanged when $S$ is replaced by $U_1 * S * U_2$, where $U_1$ and $U_2$ are any lossless networks with nonsingular transmissions. Prove that any two-sided invariant must be of the form

$$f[\sigma(t_1^*t_1),\sigma(\tau_1^*\tau_1)]$$

where $t_1$ and $\tau_1$ are given in Eq. (12.84) and where $\sigma$ means "the spectrum of." (Proceed as in Exercise 6 of Sec. 12.25.)

## 12.28 The Probability Scattering Matrix

The analogy between transmission-line theory and diffusion that was noted for $n = 1$ is also valid in the general case. Let a particle move at random on $n$ parallel lines in such a way that it can be transmitted, reflected, or absorbed when it encounters a given segment. The particle can also jump from one line to another, but only in a direction perpendicular to the lines. If $J$ is a system of aligned parallel segments (Fig. 12.26), let $t_{ij}$ denote the probability that the particle emerges from the $j$th segment at the right when the particle is introduced into the $i$th segment at the left. Under the same hypothesis, $r_{ij}$ represents the probability of emerging from the $j$th segment at the left. Similarly, one can define the right-hand coefficients $\tau_{ij}$ and $\rho_{ij}$.

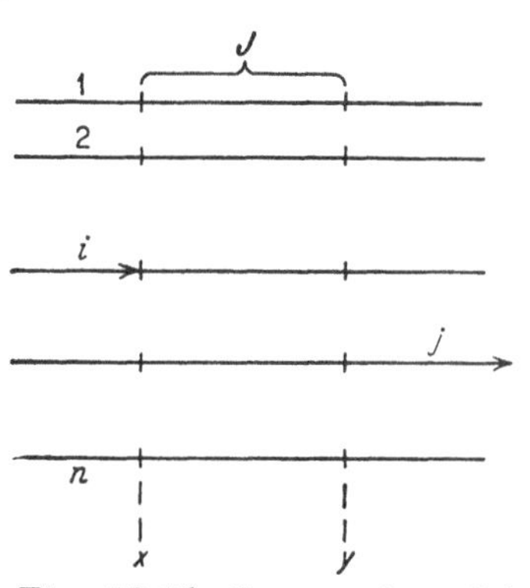

**Fig. 12.26** System of parallel segments, permitting formulation of diffusion problem.

Let the particle be introduced at random from the left in such a way that $z^i$, the $i$th component of a column vector $z$, gives the probability that we toss the particle at the $i$th segment. The laws of total and compound probability lead to

$$\sum_{i=1}^{n} t_{ij}z^i \qquad \text{and} \qquad \sum_{i=1}^{n} r_{ij}z^i$$

for the probabilities of transmission and reflection, respectively, at the $j$th segment. In terms of the transposed matrices $t = (t_{ji})$ and $r = (r_{ji})$, the results become $tz$ and $rz$, just as in the electromagnetic case.

If the column vector $z_1$ describes the probabilities of presentation at the left, and $z_4$ at the right, then the vectors $z_2$ and $z_3$ giving probabilities of egress at the left and right, respectively, are obtained from Eq. (12.74). Hence the algebraic structure of this problem is identical with that of the transmission-line problem considered previously. The role taken by the complex amplitude vectors in the former case is now taken by the real probability vectors $z_i$.

So far the set of intervals $J$ was regarded as fixed. If the left-hand ends are at $x$ and the right-hand ends at $y$, with $y > x$, then the coefficients $t$, $\tau$, $r$, and $\rho$ are functions of $x$ and $y$, and so is the scattering matrix $S$. Assuming independence of the phenomena for adjacent sets of intervals, we get the same functional equations as in the electromagnetic case. Theorem 12.1 now shows that the probabilities are characterized by the differential system (12.76). Thus, as for $n = 1$, here too the sole distinction between the two models lies in the constraints.

The probability interpretation gives

$$t_{ij} \geq 0 \qquad \tau_{ij} \geq 0 \qquad r_{ij} \geq 0 \qquad \rho_{ij} \geq 0 \tag{12.89}$$

$$\sum_{j=1}^{n} (t_{ij} + r_{ij}) \leq 1 \qquad \sum_{j=1}^{n} (\tau_{ij} + \rho_{ij}) \leq 1 \tag{12.90}$$

with equality in (12.90) when there is no possibility of absorption. On the other hand, if $z_1$ and $z_4$ have nonnegative components then the same must be true of $z_3$ and $z_2$ in Eq. (12.74). Also, we have

$$\sum_{i=1}^{n} (z_3^i + z_2^i) \leq \sum_{i=1}^{n} (z_1^i + z_4^i) \tag{12.91}$$

with equality when there is zero probability of absorption.

It is easily seen that the latter conditions are equivalent to the former. When they hold, the system is called $p$ dissipative or $p$ lossless, just as in the case $n = 1$. The formulation (12.91) shows that the conditions are preserved by the star product, and hence that the foregoing methods can be used. The result is the following:

THEOREM 12.4. *Let the hypothesis of Theorem* 12.1 *hold for $x$ in $J$ and $y$ in $J$, where $J$ is an interval of real values. Suppose also that the matrix coefficients $a(y)$, $b_1(y)$, $b_2(y)$, and $c(y)$ are piecewise continuous. Then the matrix $S(x,y)$ is $p$ dissipative for $x$ in $J$, $y$ in $J$, $y > x$, if and only if the matrix*

$$\begin{pmatrix} b_1 & a \\ c & b_2 \end{pmatrix}$$

*is such that its off-diagonal elements are nonnegative and its column sums nonpositive for $y$ in $J$. The condition for $S(x,y)$ to be $p$ lossless is that the column sums be zero.*

## 12.29 A More General Interpretation

It is small comfort to the engineer who asks a mathematician for help when the latter fails to solve the problem and then proceeds to imbed the problem in a broad class of similar ones that he also cannot solve. A penchant for facile generalization must not be used as a cloak for perplexity.

The problems posed in the foregoing discussion were actually solved; therefore, it is only fair to claim a certain latitude in this matter of abstraction. We wish to regard $t$, $\tau$, $r$, and $\rho$ as operators on a Hilbert space—i.e., on a vector space of infinitely many dimensions in which the notions of "length" and "angle" are defined. The resulting mathematical structure is related to potential theory, to the theory of heat flow, and to other areas of mathematical physics; see Chap. 4.

The particular Hilbert space to be used consists of functions in $L^2$, that is, functions $z$ such that $z$ is Lebesgue integrable and

$$\int_{-\infty}^{\infty} |z(\xi)|^2 \, d\xi$$

is finite. To see how this space enters the situation, one can imagine a continuous rather than a discrete variation across the transmission line or across the diffusion medium. In that case, sums such as

$$tz = \sum_{i=1}^{n} t_{ij} z^i$$

are replaced by integrals,

$$tz = \int_{-\infty}^{\infty} t_0(\xi, \xi_1) z(\xi) \, d\xi$$

Much of the foregoing analysis goes through without change,† but the $n$-dimensional vector space is now the Hilbert space of complex $L^2$ functions $z$.

If the cross section of the generalized transmission line is uniform, the result depends on $\xi - \xi_1$ only; hence it can be written as a convolution,

$$tz = \frac{1}{\sqrt{2\pi}} \int_{-\infty}^{\infty} t_0(\xi - \xi_1) z(\xi) \, d\xi \tag{12.92}$$

Assuming that the functions are in $L^2$, we introduce the Fourier transform $\mathbf{T}$ and inverse transform $\mathbf{T}^{-1}$ defined by

$$\mathbf{T}z = \frac{1}{\sqrt{2\pi}} \operatorname*{l.i.m.}_{a \to \infty} \int_{-a}^{a} e^{-ju\xi} z(\xi) \, d\xi \qquad \mathbf{T}^{-1}w = \frac{1}{\sqrt{2\pi}} \operatorname*{l.i.m.}_{a \to \infty} \int_{-a}^{a} e^{ju\nu} w(u) \, du$$

If $\Lambda(u) = \mathbf{T}t_0$, then the convolution theorem, applied to Eq. (12.92), gives $\mathbf{T}(tz) = \Lambda \mathbf{T}z$, or

$$tz = \mathbf{T}^{-1} \Lambda(u) \mathbf{T}z \tag{12.93}$$

Thus, the operator $t$ admits the representation $t = \mathbf{T}^{-1}\Lambda\mathbf{T}$.

## 12.30 A Special Case and Examples

Having generalized the transmission and reflection operators, let us specialize the rest of the model. If there is no multiple reflection, the functional equation gives

$$t(x,z) = t(x,y)t(y,z) \tag{12.94}$$

where, as before, $t(x,y)$ is the operator $t$ that effects the transmission from the cross section at $x$ to the cross section at $y$, with $y > x$. When the

† One new feature is that the fixed-point theorem invoked for the double-matching problem of Sec. 12.25 is no longer valid.

transmission line is longitudinally uniform, we have

$$t(x,y) = t(y - x)$$

On writing $x$ for $y - x$ and $y$ for $z - y$ as in Sec. 12.12, we see that Eq. (12.94) leads to a one-parameter semigroup:

$$t(x + y) = t(x)t(y) \tag{12.95}$$

Let the function $\Lambda(u)$ corresponding to $t(x)$ be denoted by $\Lambda_x(u)$. Substituting Eq. (12.95) into Eq. (12.93) gives

$$\Lambda_{x+y} = \Lambda_x \Lambda_y \tag{12.96}$$

The advantage of this formulation over that of Eq. (12.95) is that here we have ordinary multiplication, while in Eq. (12.95) there is a more elusive, operator multiplication. Subject to mild continuity restrictions, Eq. (12.96) gives $\Lambda_x = (\Lambda_1)^x$, so that

$$tz = \mathbf{T}^{-1}[\Lambda(u)]^x \mathbf{T}z \tag{12.97}$$

The propagation process is wholly characterized by $\Lambda(u) \equiv \Lambda_1(u)$, a single complex function of the real variable $u$.

Formula (12.97) represents a *generalization* of transmission-line theory in that the propagating quantity $z$ is an element of Hilbert space and the transmission coefficient $t$ is an operator on that space. But the formula is a *specialization* of transmission-line theory in that the transmission line is uniform both longitudinally and transversely, and multiple internal reflection is absent.

The function $\Lambda(u)$ in Eq. (12.97) is determined by the partial differential equation underlying the propagation process. For instance, the Laplace equation†

$$z_{xx} + z_{yy} = 0$$

gives

$$\Lambda(u) = e^{-|u|}$$

when suitable behavior at $\infty$ is required. The resulting formula

$$z(x,y) = \mathbf{T}^{-1} e^{-|u|x} \mathbf{T} z(0,\xi)$$

represents a harmonic function for $x > 0$ that reduces to $z(0,\xi)$ for $x = 0^+$, provided the latter is in $L^2$. Returning to Eq. (12.92) via the convolution theorem gives the Poisson formula for a half plane:

$$z(x,y) = \frac{y}{\pi} \int_{-\infty}^{\infty} \frac{z(0,\xi)}{(x - \xi)^2 + y^2} \, d\xi$$

† The $y$ here is measured along the $y$ axis while that in $t(x,y)$ is measured along the $x$ axis.

The reader may experience a certain dismay at seeing the Dirichlet problem treated as a problem concerning a propagation or transmission-line phenomenon. On the other hand, the equation

$$z_{xx} + z_{yy} + k^2 z = 0$$

seems eminently suited for such treatment; it describes the radiation field of an antenna when the initial illumination $z(0,\xi)$ is a known function in $L^2$. In this case, the differential equation gives

$$\Lambda(u) = e^{j\sqrt{k^2-u^2}} \tag{12.98}$$

when suitable regularity conditions are postulated, and the resulting formula (12.97) is the plane-wave expansion of the antenna field.† Since we get Dirichlet's problem upon letting $k \to 0$, it is no wonder that the two topics can both be treated in the same conceptual framework.

Sometimes the propagation occurs in time rather than space. The absence of multiple reflection then means that the future does not influence the past. As an illustration, the equation

$$z_x = kz_{yy} \qquad x = \text{time}$$

for heat flow yields

$$\Lambda(u) = e^{-ku^2}$$

and Eq. (12.97) gives the temperature at time $x > 0$ in terms of that at time 0. The familiar form

$$z(x,y) = (4\pi kx)^{-1/2} \int_{-\infty}^{\infty} z(0,\xi) e^{-(y-\xi)^2/(4kx)}\, d\xi$$

is obtained from Eq. (12.97) by the convolution theorem.

The implication to be drawn from these examples is that the development presents many points of contact with other disciplines, as soon as one takes a sufficiently broad viewpoint. Transmission-line theory not only is of interest in its own right but serves to illuminate adjacent areas of mathematical physics.

## EXERCISES

1. With $u = k \sin \theta$, show that Huygens' wavelets give $F(u) = \mathbf{T}z$ for the secondary pattern of a cylindrical antenna with primary illumination density equal to $z(y)/\sqrt{2\pi}$. Deduce the formula of the text,

$$z(x,y) = \mathbf{T}^{-1}\Lambda^x(u)\mathbf{T}z(0,y) \qquad \Lambda(u) = e^{j\sqrt{k^2-u^2}}$$

from the fact that $z(x,y)$ must give the same secondary pattern as does $z(0,y)$, except

† The need for infinitely many modes to satisfy the boundary conditions gives the problem an infinite-dimensional character, even though the free-space region outside the antenna is a "single" transmission line.

for a phase shift due to the change in origin. (In Fig. 12.27, the change of origin introduces a phase shift

$$kx \cos \theta \equiv x\sqrt{k^2 - u^2}$$

for the plane waves traveling in direction $\theta$. Convergence of the infinite integrals is assumed in this and the following problems.)

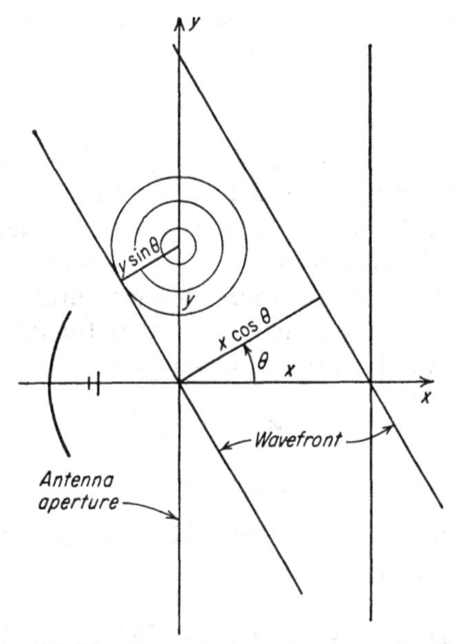


**Fig. 12.27** Computation of antenna far field.

**2.** The effect of a thin, plane obstacle at $x = x_0$ is approximated by applying a complex transmission function $T_0(y)$ to the near field $z(x,y)$ and neglecting all other effects. Show that the secondary pattern with obstacle can be obtained by applying the operator

$$\Lambda^{-x}(u)\mathbf{T}T_0(y)\mathbf{T}^{-1}\Lambda^x(u)$$

to the original unperturbed secondary pattern $F(u)$. [The pattern with obstacle is

$$F_0(u) = \Lambda^{-x}(u)\mathbf{T}T_0(y)z(x,y)$$

Use Exercise 1.]

**3.** In Exercise 2, let $t_0(u) = \mathbf{T}T_0(y)$. Show that the far pattern with obstacle is given in terms of that without obstacle by

$$F_0(u) = \Lambda^{-x}(u) \int_{-\infty}^{\infty} t_0(u - u_1)\Lambda^x(u_1)F(u_1)\, du_1$$

(Change the order of integration indiscriminately.)

**4.** Let the function $t_0(u,u_1)$ characterize an obstacle in the following sense: Given an incident plane wave of amplitude 1 in direction $\sin^{-1}(u/k)$, the amplitude of the transmitted wave in direction $\sin^{-1}(u_1/k)$ is $t_0(u,u_1)\, du_1$. Show that the pattern with obstacle in terms of that without obstacle is

$$F_0(u) = \Lambda^{-x}(u) \int_{-\infty}^{\infty} t_0(u,u_1)\Lambda^x(u_1)F(u_1)\, du_1$$

if all multiple reflection between obstacle and antenna can be neglected.

**5.** Prove that the obstacle of Exercise 4 can be specified by a single transmission function $T_0(y)$, after the manner of Exercise 2, if and only if $t_0(u,u_1)$ is a function of the single variable $u - u_1$. (Prove a converse to Exercise 3 by use of the convolution theorem.)

**6.** If the propagation process of the text is inhomogeneous, so that Eq. (12.94) holds instead of Eq. (12.95), then Eq. (12.97) becomes

$$t(x,y)z = \mathbf{T}^{-1}\left[\exp \int_x^y \Lambda(\xi,u)\, d\xi\right] \mathbf{T}z$$

when suitable auxiliary conditions are postulated. Show that this form satisfies Eq. (12.94).

## REFERENCES


1. Ambarzumian, V., Diffuse Reflection of Light by a Foggy Medium, *Compt. rendus Acad. Sci. U.R.S.S.*, vol. 38, pp. 229–232, 1943.
2. Barrar, R. B., and R. M. Redheffer, On Nonuniform Dielectric Media, *I.R.E. Trans. on Antennas and Propagation*, vol. AP-3, pp. 101–107, 1955.
3. Belevitch, V., Synthèse des réseaux électriques passifs à *n* paires de bornes de matrice de répartition prédéterminée, *Ann. Télécommun.*, vol. 6, pp. 302–312, 1951.
4. Bellman, R., and R. Kalaba, On the Principle of Invariant Imbedding and Propagation through Inhomogeneous Media, *Proc. Nat. Acad. Sci. U.S.A.*, vol. 42, pp. 629–632, 1956.
5. ——— and ———, On the Principle of Invariant Imbedding and Diffuse Reflection from Cylindrical Regions, *Proc. Nat. Acad. Sci. U.S.A.*, vol. 43, pp. 514–517, 1957.
6. ——— and ———, Random Walk, Scattering and Invariant Imbedding—I: One-dimensional Case, *Proc. Nat. Acad. Sci. U.S.A.*, vol. 43, pp. 930–933, 1957.
7. ——— and ———, Invariant Imbedding, Wave Propagation and the WKB Approximation, *Proc. Nat. Acad. Sci. U.S.A.*, vol. 44, pp. 317–319, 1958.
8. ——— and ———, Invariant Imbedding, Random Walk and Scattering, *J. Math. Mech.*, vol. 8, pp. 683–704, 1959.
9. ———, ———, and G. M. Wing, On the Principle of Invariant Imbedding and One-dimensional Neutron Multiplication, *Proc. Nat. Acad. Sci. U.S.A.*, vol. 43, pp. 517–520, 1957.
10. ———, ———, and ———, On the Principle of Invariant Imbedding and Neutron Transport Theory—I: One-dimensional Case, *J. Math. Mech.*, vol. 7, pp. 149–162, 1958.
11. ———, ———, and ———, Invariant Imbedding and Neutron Transport Theory—II: Functional Equations, *J. Math. Mech.*, vol. 7, pp. 741–756, 1958.
12. Carlin, H. J., The Scattering Matrix in Network Theory, *I.R.E. Trans. on Circuit Theory*, vol. CT-3, no. 2, pp. 88–97, 1956.
13. Cazenave, R., Représentation géométrique de la transformation de Gudermann, *Ann. Télécommun.*, vol. 7, pp. 330–333, 1954.

14. Chandrasekhar, S., "Radiative Transfer," Oxford University Press, New York, 1950.
15. Desoer, C. A., On the Characteristic Frequencies of Lossless Nonreciprocal Networks, *I.R.E. Trans. on Circuit Theory*, vol. CT-5, pp. 374–375, 1958.
16. Dicke, R. H., A Computational Method Applicable to Microwave Networks, *J. Appl. Phys.*, vol. 18, pp. 873–878, 1947.
17. Dowker, Y. N., and R. M. Redheffer, "An Investigation of RF Probes," Report No. 483–14, Massachusetts Institute of Technology Radiation Laboratory, Cambridge, Mass., 1946.
18. Guillemin, E. A., "Synthesis of Passive Networks," John Wiley & Sons, Inc., New York, 1957.
19. ———, "Communication Networks," 2 vols., John Wiley & Sons, Inc., New York, 1946, 1947.
20. Hurewicz, W., and H. Wallman, "Dimension Theory," p. 40, Princeton University Press, Princeton, N.J., 1941.
21. Levin, J. J., On the Matrix Riccati Equation, *Proc. Amer. Math. Soc.*, vol. 10, pp. 519–524, 1959.
22. Llewellyn, F. B., Some Fundamental Properties of Transmission Systems, *Proc. I.R.E.*, vol. 40, pp. 271–283, 1952.
23. Luneberg, R. K., "The Propagation of Electromagnetic Plane Waves in Plane Parallel Layers," Research Report No. 172-3, New York University, 1947.
24. Mason, S. J., Power Gain in Feedback Amplifiers, *I.R.E. Trans. on Circuit Theory*, vol. CT-1, pp. 20–25, 1954.
25. McMillan, E. M., Violation of the Reciprocity Theorem in Linear Passive Electromechanical Systems, *J. Acoust. Soc. Amer.*, vol. 18, pp. 344–347, 1946.
26. Montgomery, C. G., R. H. Dicke, and E. M. Purcell (eds.), "Principles of Microwave Circuits," chap. 5, McGraw-Hill Book Company, Inc., New York, 1948.
27. Mycielski, J., and S. Paszkowski, Sur un problème du calcul de probabilité (I), *Studia Math.*, vol. 15, pp. 188–200, 1956.
28. Oono, Y., and K. Yasuura, Synthèse des réseaux passifs à $n$ paires de bornes donnés par leurs matrices de répartition, *Ann. Télécommun.*, vol. 9, pp. 73–80, 109–115, 133–140, 1954.
29. Paszkowski, S., Sur un problème du calcul de probabilité (II), *Studia Math.*, vol. 15, pp. 273–298, 1956.
30. Preisendorfer, R., Invariant Imbedding Relation for the Principles of Invariance, *Proc. Nat. Acad. Sci. U.S.A.*, vol. 44, pp. 320–323, 1958.
31. ———, Functional Relations for the $R$ and $T$ Operators on Plane-Parallel Media, *Proc. Nat. Acad. Sci. U.S.A.*, vol. 44, pp. 323–328, 1958.
32. ———, Time-dependent Principles of Invariance, *Proc. Nat. Acad. Sci. U.S.A.*, vol. 44, pp. 328–332, 1958.
33. Raisbeck, G., A Definition of Passive Linear Networks in Terms of Time and Energy, *J. Appl. Phys.*, vol. 25, pp. 1510–1514, 1954.
34. Redheffer, R. M., Microwave Antennas and Dielectric Surfaces, *J. Appl. Phys.*, vol. 20, pp. 397–411, 1949.
35. ———, Remarks on the Basis of Network Theory, *J. Math. Phys.*, vol. 28, pp. 237–258, 1950.
36. ———, Operators and Initial-value Problems, *Proc. Amer. Math. Soc.*, vol. 4, pp. 617–629, 1953.
37. ———, Novel Uses of Functional Equations, *J. Rational Mech. Anal.*, vol. 3, pp. 271–279, 1954.
38. ———, Limit-periodic Dielectric Media, *J. Appl. Phys.*, vol. 27, pp. 1136–1140, 1956.

39. ———, Limit-periodic Dielectric Media, *J. Appl. Phys.*, vol. 28, pp. 820–821, 1957.
40. ———, The Riccati Equation: Initial Values and Inequalities, *Math. Ann.*, vol. 133, pp. 235–250, 1957.
41. ———, Inequalities for a Matrix Riccati Equation, *J. Math. Mech.*, vol. 8, pp. 349–367, 1959.
42. ———, The Mycielski-Paszkowski Diffusion Problem, *J. Math. Mech.*, vol. 9, pp. 607–621, 1960.
43. ———, The Dependence of Reflection on Incidence Angle, *I.R.E. Trans. on Microwave Theory and Technique*, vol. 7, pp. 423–429, 1959.
44. ———, Supplementary Note on Matrix Riccati Equations, *J. Math. Mech.*, vol. 9, pp. 745–748, 1960.
45. ———, On a Certain Linear Fractional Transformation, *J. Math. Phys.* (to appear).
46. Reid, W. T., Solutions of a Riccati Matrix Differential Equation as Functions of Initial Values, *J. Math. Mech.*, vol. 8, pp. 221–230, 1959.
47. ———, Properties of Solutions of a Riccati Matrix Differential Equation, *J. Math. Mech.*, vol. 9, pp. 749–770, 1960.
48. Roberts, S., Conjugate-image Impedances, *Proc. I.R.E.*, vol. 34, pp. 198–204, 1946.
49. Siegel, C. L., Symplectic Geometry, *Amer. J. Math.*, vol. 65, pp. 1–86, 1943.
50. Stokes, G. G., On the Intensity of the Light Reflected from or Transmitted through a Pile of Plates, *Proc. Roy. Soc.*, vol. 11, pp. 545–556, 1862.
51. Titchmarsh, E. C., "Theory of Functions," chap. VI, Oxford University Press, New York, 1939.
52. Twersky, V., Scattering Theorems for Bounded Periodic Structures, *J. Appl. Phys.*, vol. 27, pp. 1118–1122, 1956.
53. ———, On the Scattering of Waves by an Infinite Grating, *I.R.E. Trans. on Antennas and Propagation*, vol. AP-4, pp. 330–345, 1956.
54. ———, On Scattering and Reflection of Sound by Rough Surfaces, *J. Acoust. Soc. Amer.*, vol. 29, pp. 209–225, 1957.
55. ———, On Scattering and Reflection of Electromagnetic Waves by Rough Surfaces, *I.R.E. Trans. on Antennas and Propagation*, vol. AP-5, pp. 81–90, 1957.
56. Walther, K., Polarisations und Winkelabhängigkeit des Reflexionsfaktors von Absorben für elektromagnetische Zentimeterwellen, *Z. angew. Phys.*, vol. 10, pp. 285–295, 1958.
57. Youla, D. C., Direct Single-frequency Synthesis from a Prescribed Scattering Matrix, *I.R.E. Trans. on Circuit Theory*, vol. CT-6, pp. 340–344, 1959.
58. Youla, D. C., L. J. Castriota, and H. J. Carlin, Bounded Real Scattering Matrices and the Foundations of Linear Passive Network Theory, *I.R.E. Trans. on Circuit Theory*, vol. CT-6, pp. 102–124, 1959.

# 13

# Characteristic-value Problems in Hydrodynamic and Hydromagnetic Theory

S. CHANDRASEKHAR
MORTON D. HULL DISTINGUISHED SERVICE PROFESSOR
DEPARTMENTS OF PHYSICS AND ASTRONOMY, UNIVERSITY OF CHICAGO

## 13.1 Introduction

The general theme of both this book and its predecessor is modern mathematics for the engineer. "Modern mathematics" might be interpreted in this connection as new mathematical methods or techniques for solving problems in applied mathematics. Very often the discovery of new methods and techniques consists in reducing to an elementary level problems once considered difficult or complicated. Whether or not this point of view is justified under all circumstances, it is our purpose in this chapter to illustrate it by considering a very old and a very classical problem; it is a problem to which some of the great applied mathematicians have contributed. The problem concerns the stability of purely rotational flow between two concentric cylinders. The mathematical problems that arise are rather different when the fluid is considered nonviscous and when it is considered viscous. In the absence of viscosity, the subject was investigated by Lord Rayleigh in a famous paper;[5] in the presence of viscosity, the subject was investigated in an equally famous paper by G. I. Taylor.[6] Taylor's investigation in fact provided the first example of a problem in hydrodynamical stability some aspects of which were successfully solved both analytically and experimentally. A complete solution of the basic mathematical problem was not available, however, until recently.[1,2,3] We shall consider the two cases separately.

## 13.2 The Rayleigh Criterion for the Stability of Inviscid Rotational Flow

If we consider the fluid as incompressible and inviscid, then the equations of motion allow the rotational velocity $u_\theta$ to be an arbitrary function $V(r)$ of the distance $r$ from the axis:

$$u_\theta = V(r) \tag{13.1}$$

Since the equations of motion do not restrict $V(r)$, the question arises: Are there any restrictions on $V(r)$ that result from requirements of stability? Rayleigh has stated the following criterion:

*In the absence of viscosity, a necessary and sufficient condition for a distribution of angular velocity $\Omega(r)$ to be stable is*

$$\Phi(r) = \frac{d}{dr}(r^2\Omega)^2 > 0 \qquad \Omega = \frac{V}{r} \tag{13.2}$$

*everywhere in the interval; further, the distribution is unstable if $\Phi(r)$ should decrease anywhere inside the interval.*

In establishing this criterion, Lord Rayleigh argued as follows:

If we restrict our attention to axisymmetric motions, then it is a direct consequence of the equations of motion that the angular momentum

$$L = r^2\Omega$$

of a fluid element, per unit mass, remains constant as we follow it with its motion. Suppose now that we interchange the fluid contained in two elementary rings, of equal heights and masses, at

$$r = r_1 \qquad \text{and} \qquad r = r_2 > r_1$$

If $dr_1$ and $dr_2$ are the radial extents of the rings, the equality of their masses requires

$$\pi r_1\, dr_1 = \pi r_2\, dr_2 = dS$$

(say). In view of the constancy of $L$ with the motion, the fluid at $r_2$, after the interchange, will have the same angular momentum (namely, $L_1$) that it had at $r_1$ before the interchange; similarly, the fluid at $r_1$, after the interchange, will have the same angular momentum (namely, $L_2$) that it had at $r_2$. As a result, the change in the kinetic energy (or, what is the same thing, the change in the centrifugal potential energy) is proportional to

$$\left[\left(\frac{L_2^2}{r_1^2} + \frac{L_1^2}{r_1^2}\right) - \left(\frac{L_1^2}{r_1^2} + \frac{L_2^2}{r_2^2}\right)\right] dS = (L_2^2 - L_1^2)\left(\frac{1}{r_1^2} - \frac{1}{r_2^2}\right) dS \tag{13.3}$$

Remembering that $r_2 > r_1$, we observe that this is positive or negative according as $L_2^2$ is greater than $L_1^2$ or less than $L_1^2$. Consequently, if $L^2$

is a monotonic increasing function of $r$, no interchange of fluid rings such as we have imagined can occur without a source of energy, and this means stability. On the other hand, if $L^2$ should decrease anywhere, then an interchange of fluid rings in this region will result in a liberation of energy, and this means instability.

### 13.3 Analytical Discussion of the Rayleigh Criterion

While the foregoing arguments of Rayleigh make his criterion a very likely one by drawing attention to its physical origin, one should like to establish it directly from the relevant perturbation equations. These equations are

$$\frac{\partial u_r}{\partial t} + \frac{V}{r}\frac{\partial u_r}{\partial \theta} - 2\frac{V}{r}u_\theta = -\frac{\partial \omega}{\partial r} \tag{13.4}$$

$$\frac{\partial u_\theta}{\partial t} + \frac{V}{r}\frac{\partial u_\theta}{\partial \theta} + \left(\frac{V}{r} + \frac{dV}{dr}\right)u_r = -\frac{1}{r}\frac{\partial \omega}{\partial \theta} \tag{13.5}$$

$$\frac{\partial u_z}{\partial t} + \frac{V}{r}\frac{\partial u_z}{\partial \theta} = -\frac{\partial \omega}{\partial z} \tag{13.6}$$

and the equation of continuity,

$$\frac{\partial u_r}{\partial r} + \frac{u_r}{r} + \frac{1}{r}\frac{\partial u_\theta}{\partial \theta} + \frac{\partial u_z}{\partial z} = 0 \tag{13.7}$$

In accordance with the general procedure of treating these problems, we analyze the disturbance into normal modes. In the present instance, it is natural to suppose that the various quantities describing the perturbation have a $(t,\theta,z)$ dependence given by

$$e^{j(pt+m\theta+kz)} \tag{13.8}$$

where $p$ is a constant (which can be complex), $m$ is an integer (which can be positive, zero, or negative), and $k$ is the wave number of the disturbance in the $z$ direction. Let $u_r(r)$, $u_\theta(r)$, $u_z(r)$, and $\omega(r)$ now denote the amplitudes of the respective perturbations for which the $(t,\theta,z)$ dependence is given by the expression (13.8). Equations (13.4) to (13.7) then give

$$j\sigma u_r - 2\Omega u_\theta = -\frac{d\omega}{dr} \tag{13.9}$$

$$j\sigma u_\theta + \left(\Omega + \frac{d}{dr}r\Omega\right)u_r = -\frac{jm}{r}\omega \tag{13.10}$$

$$j\sigma u_z = -jk\omega \tag{13.11}$$

and

$$\frac{du_r}{dr} + \frac{u_r}{r} + \frac{jm}{r}u_\theta + jku_z = 0 \tag{13.12}$$

where

$$\sigma = p + m\Omega \tag{13.13}$$

For our present discussion we shall restrict our attention to the axisymmetric case, for which none of the quantities describing the perturbation depend on $\theta$. Then

$$m = 0 \qquad \sigma = p$$

and the equations become

$$jpu_r - 2\Omega u_\theta = -D\omega \tag{13.14}$$

$$jpu_\theta + (\Omega + Dr\Omega)\, u_r = 0 \tag{13.15}$$

and

$$jpD_* u_r = -k^2\omega \tag{13.16}$$

where

$$D = \frac{d}{dr} \qquad \text{and} \qquad D_* = D + \frac{1}{r}$$

By eliminating $\omega$ from Eqs. (13.14) and (13.16), we find that

$$jpu_r - 2\Omega u_\theta = \frac{jp}{k^2}\, DD_* u_r \tag{13.17}$$

Now by making use of Eq. (13.15), we find, after some further reductions, that

$$DD_* u_r - k^2 u_r = -\frac{k^2}{p^2}\,\Phi(r) u_r \tag{13.18}$$

We must seek solutions of this equation that satisfy the boundary condition

$$u_r = 0 \qquad \text{for } r = R_1 \text{ and } R_2 \tag{13.19}$$

The problem presented by Eq. (13.18) and the boundary conditions (13.19) constitutes a characteristic-value problem. This problem is in fact of the classical Sturm-Liouville type,[4] and by appealing to the standard theorems of the subject, we can reach the following conclusion: *The characteristic values of $k^2/p^2$ are all positive if $\Phi(r)$ is everywhere positive, and they are all negative if $\Phi(r)$ is everywhere negative. If $\Phi(r)$ should change sign anywhere in the interval $(R_1, R_2)$, then there are two sets of real characteristic values that have the limit points $+\infty$ and $-\infty$.*

Since a negative $p^2$ means instability, it is clear that the foregoing statements regarding the sign of the characteristic values of $k^2/p^2$ are equivalent to a restatement of Rayleigh's criterion.

An alternative way of establishing the same results is instructive. Multiply Eq. (13.18) by $ru_r$ and integrate over the range of $r$. The integral on the left-hand side can be transformed by an integration by parts. We obtain

$$\int_{R_1}^{R_2} \left\{ \frac{1}{r}\left[\frac{d}{dr}(ru_r)\right]^2 + k^2 r u_r^2 \right\} dr = \frac{k^2}{p^2} \int_{R_1}^{R_2} \Phi(r) r u_r^2\, dr \tag{13.20}$$

or

$$\frac{p^2}{k^2} = \frac{\int_{R_1}^{R_2} \Phi(r) r u_r^2 \, dr}{\int_{R_1}^{R_2} \left\{ \frac{1}{r} \left[ \frac{d}{dr} (r u_r) \right]^2 + k^2 r u_r^2 \right\} dr} = \frac{I_1}{I_2} \tag{13.21}$$

(say). From Eq. (13.21), it is immediately apparent that $p^2/k^2$ is positive if $\Phi(r)$ is everywhere positive, and is negative if $\Phi(r)$ is everywhere negative. The further result, that there exist unstable modes if $\Phi(r)$ is anywhere negative, can be deduced from the fact that we may regard Eq. (13.18) as the condition that

$$I_1 = \int_{R_1}^{R_2} \Phi(r) r u_r^2 \, dr \tag{13.22}$$

is a maximum, or a minimum, for a given

$$I_2 = \int_{R_1}^{R_2} \left\{ \frac{1}{r} \left[ \frac{d}{dr} (r u_r) \right]^2 + k^2 r u_r^2 \right\} dr \tag{13.23}$$

The ratio $p^2/k^2$ is then the value of $I_1/I_2$. If $\Phi(r)$ is negative anywhere, then $I_1$ admits a negative value and therefore a negative minimum, so that one value at least of $p^2$ is negative, and one mode of disturbance is unstable.

In deriving Rayleigh's criterion for axisymmetric perturbations, we have not, of course, solved the mathematical problems completely: The complete solution will require a detailed consideration of the general nonaxisymmetric modes distinguished by $m$. It may be surprising, but it is true, that up to the present this more general problem has not been adequately solved. The reason is that in the general case the characteristic-value parameter occurs nonlinearly in the problem. It nevertheless appears possible that, by inverting the problem in the sense of regarding $k^2$ as the characteristic-value parameter, some progress can be made. For example, one can show that, considered from this latter point of view, the general nonaxisymmetric problem allows a variational formulation exactly as in the axisymmetric case. In fact, if $p$ is real then one finds that $k^2$, given by the formula

$$k^2 = \frac{\int_{R_1}^{R_2} r \left[ (\Phi - \sigma^2) \xi_r^2 - m^2 \omega^2 / (r^2 \sigma^2) \right] dr}{\int_{R_1}^{R_2} r \omega^2 / \sigma^2 \, dr} \tag{13.24}$$

together with the constraint

$$(\sigma^2 - \Phi) \xi_r \equiv \frac{d\omega}{dr} + \frac{2m\Omega}{r\sigma} \omega \tag{13.25}$$

and the boundary condition

$$\xi_r = 0 \qquad \text{for } r = R_1 \text{ and } R_2 \tag{13.25a}$$

has an extremal character when the proper solutions of the problem are inserted in it. The problem of deriving Rayleigh's criterion from Eq. (13.24) seems, however, to be a very delicate matter. But one conclusion is obvious: If $\Phi(r)$ is everywhere negative, then $k^2$ cannot admit a positive characteristic value. For real $p$'s, the characteristic values of $k$ are necessarily imaginary. Therefore for real $k$'s, $p$ must necessarily be complex, and this means instability.

## 13.4 The Stability of Viscous Rotational Flow

We now turn to the question of how the inclusion of viscosity modifies the physical and the mathematical aspects of the problem. It is remarkable that, no matter how small viscosity may be, an arbitrary distribution of the rotational velocity (permitted when viscosity is exactly zero) is not consistent with the Navier-Stokes equations of hydrodynamics. Instead of $\Omega(r)$ being an arbitrary function, the allowed solution degenerates into a two-parameter family. The two parameters may, in fact, be taken as the angular velocities of the two cylinders between which the fluid is confined; that this should be so is physically understandable, since an experimenter can choose to rotate the two cylinders with different angular velocities at his discretion. It is, however, not under his discretion to distribute the angular velocities inside the fluid once he has decided to rotate the two cylinders at some assigned speeds; in other words, given the angular velocities of rotation of the two cylinders, the distribution of the angular velocity inside the fluid is uniquely determined. This is the content of the formula

$$\Omega(r) = A + \frac{B}{r^2} \tag{13.26}$$

which follows from the Navier-Stokes equations. In Eq. (13.26), $A$ and $B$ are two arbitrary constants; we may relate them to the angular velocities $\Omega_1$ and $\Omega_2$ with which the inner and the outer cylinders are rotated.

The problem of the stability of viscous flow between rotating cylinders reduces to the following: As the initial stationary flow, we consider one in which $A$ and $B$ have some preassigned values; we then ask whether, other things being equal, there are restrictions on $A$ and $B$ that follow from requirements of stability.

One might formulate the problem of stability in a somewhat different way. A distribution of angular velocity for arbitrarily assigned $A$ and $B$ will not, of course, satisfy Rayleigh's criterion derived for nonviscous fluids, and one does not expect that in the presence of viscosity a violation of Rayleigh's criterion would be followed instantly by instability. On general grounds, viscosity will have an inhibiting effect, and we expect to be able to transgress Rayleigh's criterion to some finite extent before viscosity is unable to prevent the onset of instability. The question is: What is the extent of the transgression we are allowed?

When $\Omega$ has the form given by Eq. (13.26), Rayleigh's discriminant becomes

$$\Phi = 4A\left(A + \frac{B}{r^2}\right) \tag{13.27}$$

On expressing $A$ and $B$ in terms of the angular velocities of the two cylinders and measuring $r$ in units of the radius $R_2$ of the outer cylinder, we find

$$\Phi = -4\Omega_1^2\eta^4 \frac{(1-\mu)(1-\mu/\eta^2)}{(1-\eta^2)^2}\left(\frac{1}{r^2} - \kappa\right) \tag{13.28}$$

where
$$\eta = \frac{R_1}{R_2} \qquad \mu = \frac{\Omega_2}{\Omega_1} \qquad \kappa = \frac{1-\mu/\eta^2}{1-\mu} \tag{13.29}$$

Clearly,
$$\Phi > 0 \qquad \text{for } \eta \leq r \leq 1 \tag{13.30}$$

so long as
$$\mu = \frac{\Omega_2}{\Omega_1} > \eta^2 \tag{13.31}$$

Therefore, Rayleigh's criterion applied to the distribution (13.26) requires that, *for stability, the outer cylinder must rotate with an angular speed greater than $\eta^2$ times that of the inner cylinder and in the same sense.* In the $(\Omega_2,\Omega_1)$ plane, the regions of stability are delimited by the positive $\Omega_2$ axis and the "Rayleigh line,"

$$\Omega_2 = \Omega_1\eta^2$$

If the effects of viscosity are ignored, we must have instability to the left of the Rayleigh line; we should like to know how viscosity extends the domain of stability.

Formulated in an entirely general way, the mathematical problem that we must solve in order to establish a criterion for stability applicable under all circumstances is a difficult and complicated one, and it may be said that the complete solution has not yet been found. There is, however, one case that has been solved with relative completeness. This is the case when the gap

$$d = R_2 - R_1$$

between the two cylinders is small compared with the mean radius

$$\tfrac{1}{2}(R_2 + R_1)$$

With suitable approximations appropriate for this case, we eventually find that the characteristic-value problem we have to solve is the following:

$$(D^2 - a^2)^2u = (1 + \alpha z)v \tag{13.32}$$

and
$$(D^2 - a^2)v = -Ta^2u \tag{13.33}$$

where
$$T = -\frac{4A\Omega_1}{\nu^2}d^4 \tag{13.34}$$

$$\alpha = -1 + \mu \tag{13.35}$$

and $\nu$ denotes the kinetic viscosity of the fluid.

Equations (13.32) and (13.33) must be considered together with the boundary conditions

$$u = Du = v = 0 \qquad \text{for } z = 0 \text{ and } 1 \tag{13.36}$$

The characteristic-value problem presented by Eqs. (13.32), (13.33), and (13.36) is typical of a large class of such problems that arise in the theory of hydrodynamic and hydromagnetic stability. For this reason it may be useful to describe how we might solve Eqs. (13.32) and (13.33) by a method that converges rapidly. The method of solution we shall adopt is the following: Since $v$ is required to vanish at $z = 0$ and 1, we expand it in a sine series of the form

$$v = \sum_{m=1}^{\infty} C_m \sin m\pi z \tag{13.37}$$

Having chosen $v$ in this manner, we next *solve* the equation

$$(D^2 - a^2)^2 u = (1 + \alpha z) \sum_{m=1}^{\infty} C_m \sin m\pi z \tag{13.38}$$

obtained by inserting the series (13.37) in Eq. (13.32), and arrange that the solution satisfies the four remaining boundary conditions on $u$. With $u$ determined in this fashion and $v$ given by Eq. (13.37), Eq. (13.33) will lead to a secular equation for $T$.

When the details of the method described in the preceding paragraphs are carried out, we find that the process of solving the infinite-order characteristic equation for $T$, by setting the determinant formed by the first $n$ rows and columns equal to zero and letting $n$ take increasingly larger values, converges very rapidly indeed. Thus for $\alpha = -2.5$ and $a = 5.00$, 5.05, and 5.10, the values of $T$ obtained in the third and the fourth approximations (the "order" of the approximation being the order of the determinant that is set equal to zero in the determination of $T$) are

$$a = \begin{cases} 5.00 \\ 5.05 \\ 5.10 \end{cases} \qquad T \text{ (3d approx)} = \begin{cases} 4.607 \times 10^4 \\ 4.600 \times 10^4 \\ 4.604 \times 10^4 \end{cases}$$

$$T \text{ (4th approx)} = \begin{cases} 4.626 \times 10^4 \\ 4.619 \times 10^4 \\ 4.623 \times 10^4 \end{cases}$$

It is seen that the values of $T$ given in the third and the fourth approximations differ by only four parts in a thousand. The origin of this rapid convergence clearly lies in the splitting of the original equation of order six into a pair of order two and four, respectively, and satisfying the equation of order four exactly. This basic idea underlying the method is capable of extension and application to a wide class of problems.

## 13.5 On Methods of Solving Characteristic-value Problems in High-order Differential Equations

The methods described in the preceding section, while they are elementary, are adapted to the solution of a large class of problems that arise in problems of hydrodynamic and hydromagnetic stability. The basic idea is the following: When we wish to solve a characteristic-value problem in differential equations of high order, we try to separate the original differential equation and boundary conditions into two systems and seek to solve one of the systems exactly, and it is important that the order of the system we solve exactly is as high as can be managed. This manner of separation is useful even when a variational formulation is possible. In the latter connection, one carries out the variations with subsidiary constraints in the form of equations and boundary conditions, again of as high order as one can manage.

### REFERENCES

1. Chandrasekhar, S., On Characteristic Value Problems in High Order Differential Equations Which Arise in Studies on Hydrodynamic and Hydromagnetic Stability, *Amer. Math. Monthly*, vol. 61 supplement, pp. 32–45, 1954.
2. ———, The Stability of Viscous Flow between Rotating Cylinders, *Proc. Roy. Soc. London, Ser. A*, vol. 246, pp. 301–311, 1958.
3. ——— and W. H. Reid, On the Expansion of Functions Which Satisfy Four Boundary Conditions, *Proc. Nat. Acad. Sci. U.S.A.*, vol. 43, pp. 521–527, 1957.
4. Ince, E. L., "Ordinary Differential Equations," chap. 10, Longmans, Green & Co., Ltd., London, 1927.
5. Rayleigh, Lord, On the Dynamics of Revolving Fluids, *Proc. Roy. Soc. London, Ser. A*, vol. 96, pp. 148–154, 1917; also "Scientific Papers," vol. 6, pp. 447–453, Cambridge University Press, Cambridge, England, 1920.
6. Taylor, G. I., Stability of a Viscous Liquid Contained between Two Rotating Cylinders, *Phil. Trans. Roy. Soc. London, Ser. A*, vol. 223, pp. 289–343, 1923.

# 14

# Applications of the Theory of Partial Differential Equations to Problems of Fluid Mechanics

PAUL R. GARABEDIAN

PROFESSOR OF MATHEMATICS
NEW YORK UNIVERSITY

## 14.1 Introduction

In this chapter we shall be concerned with Cauchy's problem for a linear or quasi-linear partial differential equation in two independent variables. We shall be interested in applications of this subject to fundamental, and also to less familiar, problems in mathematical physics.

The usual physical context in which one encounters Cauchy's problem is that of a mechanical configuration, with known initial position and velocity, for which the subsequent motion is governed by a partial differential equation of the second order. The classical example is the case of the vibrating string, with behavior described by the wave equation[2,4]

$$u_{xx} - u_{tt} = 0$$

Another example is the motion of a perfect inviscid gas in a pipe, which leads to a nonlinear partial differential equation. In these familiar cases, one of the two independent variables is the time $t$, and the Cauchy data, or initial data, are simply the values of the displacements and velocities of a mechanical system at the time $t = 0$. Thus it is truly an initial-value problem to which such physical situations lead, and, furthermore, the partial differential equations involved are of the hyperbolic type.[2] This is as it should be, since we can then deduce that the Cauchy problem, or initial-value problem, is properly set in the sense that the solution depends in a continuous manner on the given physical data, so that a small error in measurement of the latter would not lead to significant errors in the results.

Although the great bulk of the theory of Cauchy's problem has been developed for hyperbolic partial differential equations, and although the direct physical applications of the theory involve hyperbolic equations, it is nevertheless true that Cauchy's problem is of significance in the elliptic case as well. Hadamard has shown, of course, as stated in Chap. 4, that the problem is not properly set for an elliptic equation, in the sense that the solution does not in general depend continuously on the data. To illustrate this, one usually cites (cf. pages 111 to 112 of Ref. 10) the case of the Laplace equation

$$u_{xx} + u_{yy} = 0$$

with Cauchy data

$$u(x,0) = 0 \qquad u_y(x,0) = n^{-1} \sin nx$$

The expression

$$u(x,y) = n^{-2} \sin nx \operatorname{sh} ny$$

is easily verified to be the solution of this particular Cauchy problem. The solution, however, does not approach zero as $n \to \infty$, despite the fact that the Cauchy data do tend toward zero in this limiting process.

We have thus established without difficulty the incorrectness, in Hadamard's sense, of the Cauchy problem for an elliptic partial differential equation, but we can also present physical applications in which it would be quite useful to find the solution even in this case. Such situations do not usually arise from the direct physical formulation of a problem, but they can occur in the construction, by inverse methods, of solutions possessing prescribed properties that would not ordinarily be part of a given set of physical data. Thus, with suitable idealizations, if we set out to describe an electromagnetic field guiding an electron beam in a prescribed channel, then we must discuss a Cauchy problem for Laplace's equation. The direct, as opposed to the inverse, physical problem here would have been to locate the paths of the electrons when the field is given. Another question that results in Cauchy's problem for an elliptic equation concerns the determination of flows of an incompressible, inviscid fluid possessing prescribed free streamlines. Finally, the inverse construction of steady gas flows through a given detached shock wave in front of a blunt body can be effected by solving Cauchy's problem for a nonlinear partial differential equation of mixed elliptic-hyperbolic type. A common feature of all these examples is the appearance of an inverse solution to what is essentially a free-boundary problem.

From the list of applications we have mentioned, it becomes evident that a successful analytical treatment of the Cauchy problem is called for even in the case of an elliptic equation. The Cauchy-Kowalewski solution by means of power-series expansions clearly is not satisfactory here any more than it is for hyperbolic problems. What is actually required is a method that can be adapted for numerical integration based on a finite-difference scheme. We shall present such an approach in this chapter. We emphasize some of the numerical aspects of this topic because of the frequent occurrence in practice of situations that are so complicated that only such an analysis proves to be adequate. A treatment of Cauchy's problem that lends itself to numerical computation is, of course, as important for the hyperbolic as for the elliptic case, and we shall bear this in mind in presenting our material.

We begin with a discussion of the method of characteristics for a hyperbolic partial differential equation in two independent variables. This is followed by the description of a finite-difference scheme for the numerical solution of Cauchy's problem in a characteristic coordinate system. Next we take up elliptic equations and study them by means of analytic extension into the complex domain. This material is followed by an elementary application to a free-streamline flow problem concerned with a bubble that rises under the influence of gravity. We close the chapter with a brief treatment of an inverse solution of the detached-shock problem for supersonic flow past a blunt body.

## 14.2 Cauchy's Problem for a Hyperbolic Partial Differential Equation in Two Independent Variables

We consider the quasi-linear partial differential equation

$$au_{xx} + 2bu_{xy} + cu_{yy} + d = 0 \tag{14.1}$$

where the coefficients $a$, $b$, $c$, and $d$ are given functions of the two independent variables $x$ and $y$ and of the three quantities $u$, $u_x$, and $u_y$. For the moment, we are interested primarily in the hyperbolic case of a solution $u$ in a region where

$$b^2 - ac > 0 \tag{14.2}$$

Cauchy's problem for Eq. (14.1) consists in finding a solution $u$ that assumes on a given curve $\Gamma$ prescribed values of $u$, $u_x$, and $u_y$. Of course, in order that $u_x$ and $u_y$ should be the first partial derivatives of the function $u$, the Cauchy data $u$, $u_x$, and $u_y$ must fulfill along $\Gamma$ the familiar condition

$$du = u_x\,dx + u_y\,dy \tag{14.3}$$

so that we actually prescribe there only two independent functions of a single variable.

A natural first step in any attempt to solve the Cauchy problem just described would be to try to calculate the second derivatives $u_{xx}$, $u_{xy}$, and $u_{yy}$ of $u$ along $\Gamma$. The partial differential equation (14.1) itself yields one linear relationship among these three unknowns, and two further linear equations of the form

$$du_x = u_{xx}\,dx + u_{xy}\,dy \tag{14.4}$$
$$du_y = u_{yx}\,dx + u_{yy}\,dy \tag{14.5}$$

can be found for them along Γ by differentiation there of the Cauchy data. We are assured of the existence of a unique solution of this system of three simultaneous linear equations in three unknowns if and only if the determinant

$$\begin{vmatrix} a & 2b & c \\ dx & dy & 0 \\ 0 & dx & dy \end{vmatrix} = a\,dy^2 - 2b\,dx\,dy + c\,dx^2 \tag{14.6}$$

of the coefficients does not vanish along Γ. Curves Γ along which this determinant vanishes identically must be treated as inappropriate for the solution of the Cauchy problem, and they are known as the *characteristics* of the partial differential equation (14.1). Along a characteristic curve Γ, we can form a linear combination of Eqs. (14.4) and (14.5) that must reduce to Eq. (14.1), so that an ordinary differential equation is obtained there for the Cauchy data themselves. Even when the Cauchy data fulfill this additional condition along a characteristic, they do not determine uniquely the second or higher derivatives of $u$, so that in this case more than one solution of Cauchy's problem can exist. A consequence of this remark is that a curve across which the higher derivatives of a solution $u$ of Eq. (14.1) exhibit discontinuities must be a characteristic. Thus, for example, the characteristics of the equation of one-dimensional gas dynamics can represent sound waves.

The relationship

$$a\,dy^2 - 2b\,dx\,dy + c\,dx^2 = 0 \tag{14.7}$$

which defines the characteristics of Eq. (14.1), can be interpreted as an ordinary differential equation for a family of curves in the $xy$ plane. Of course, when $a$, $b$, and $c$ depend on $u$, $u_x$, or $u_y$, the same can be true of the characteristics. Under our assumption (14.2) of hyperbolicity, the quadratic equation (14.7) for the slope $dy/dx$ has two distinct roots $\sigma$ and $\tau$, so that actually two separate one-parameter families of characteristics are defined. Incidentally, there is no real loss of generality here if we assume for the sake of simplicity that $a \neq 0$, since when this is not true, it can easily be achieved by rotating the $xy$ plane. Now it would seem natural to try to introduce our two families of characteristic curves as a new coordinate network. This can be done by defining two functions $\alpha(x,y)$ and $\beta(x,y)$ such that one family of characteristics is represented in implicit form by the equation $\alpha(x,y) = \text{const}$, while the other family is represented by the equation $\beta(x,y) = \text{const}$. Of course, $\alpha$ becomes a natural parameter to use along each curve of the latter family, whereas $\beta$ is the parameter to be used along the characteristics of the first family.

Let us rewrite the quadratic ordinary differential equation (14.7) for the characteristics in the form

$$(dy - \sigma\, dx)(dy - \tau\, dx) = 0 \tag{14.8}$$

where $\sigma$ and $\tau$ depend in an obvious fashion on $a$, $b$, and $c$, and let us suppose that it is the first factor in Eq. (14.8) that vanishes on the characteristics along which $\alpha$ is the parameter. If we think of $\alpha$ and $\beta$ as new independent variables, it follows that

$$y_\alpha - \sigma x_\alpha = 0 \tag{14.9}$$

whereas by symmetry

$$y_\beta - \tau x_\beta = 0 \tag{14.10}$$

If our partial differential equation (14.1) is linear, so that $a$, $b$, and $c$, and hence $\sigma$ and $\tau$, do not depend on $u$, $u_x$, or $u_y$, then the pair of partial differential equations (14.9) and (14.10) can be solved for $x$ and $y$ independently of Eq. (14.1). We are interested, however, in the quasi-linear case when $\sigma$ and $\tau$ do involve $u$, $u_x$, and $u_y$, and therefore we shall reformulate the whole problem in terms of the characteristic coordinates $\alpha$ and $\beta$, and we shall study Eq. (14.1) simultaneously with Eqs. (14.9) and (14.10).

Our next step is to express the original partial differential equation (14.1) in terms of the new independent variables $\alpha$ and $\beta$. This is achieved by following through on our earlier remark that along a characteristic we can combine Eqs. (14.4) and (14.5) with Eq. (14.1) to derive an ordinary differential equation there for the Cauchy data. Let the characteristic in question be one along which $\alpha$ is the parameter, and introduce the familiar notation $p = u_x$, $q = u_y$. We notice that along our characteristic the determinant (14.6) vanishes, while Eqs. (14.1), (14.4), and (14.5) express the fact that a linear combination of the columns of this determinant with coefficients chosen to be $u_{xx}$, $u_{xy}$, and $u_{yy}$ is equal to the triple of quantities $-d$, $p_\alpha$, and $q_\alpha$. Hence any three-by-three determinant from the matrix

$$\begin{pmatrix} a & 2b & c & -d \\ x_\alpha & y_\alpha & 0 & p_\alpha \\ 0 & x_\alpha & y_\alpha & q_\alpha \end{pmatrix}$$

must vanish. We consider, in particular, the determinant composed of the first, third, and fourth columns, and we divide it by $x_\alpha$ and use Eq. (14.9) to obtain

$$a\sigma p_\alpha + cq_\alpha + dy_\alpha = 0 \tag{14.11}$$

This is the ordinary differential equation to which we have been referring for the Cauchy data $p$ and $q$ along a characteristic with the parameter

$\alpha$ variable. In a symmetric fashion we can derive along a characteristic of the opposite family, with $\beta$ variable, the analogous equation

$$a\tau p_\beta + cq_\beta + dy_\beta = 0 \tag{14.12}$$

If $u$ itself does not appear as an argument of the coefficients $a$, $b$, $c$, and $d$, then Eqs. (14.9) to (14.12) comprise an elegant canonical system of four first-order quasi-linear partial differential equations for the new unknowns $x$, $y$, $p$, and $q$ as functions of the new independent variables $\alpha$ and $\beta$. The system is equivalent to Eq. (14.1), but it has the advantage that in each equation partial derivatives with respect to only one of the two independent variables appear. In the new formulation, the characteristics are simply the parallels to the coordinate axes. Thus, the characteristics and the solution of Eq. (14.1) are found simultaneously in the $xy$ plane when we solve the canonical system (14.9) to (14.12) and then invert to introduce $x$ and $y$ once more as the independent variables.

If $u$ occurs as an argument of the given functions $a$, $b$, $c$, and $d$, then an additional artifice is required in our derivation of the canonical system. This consists in choosing either of the two equations

$$u_\alpha = px_\alpha + qy_\alpha \tag{14.13}$$

or

$$u_\beta = px_\beta + qy_\beta \tag{14.14}$$

as a fifth equation of the system. These latter equations are readily obtained from the requirement that $p$ and $q$ are the first derivatives of $u$ with respect to $x$ and $y$, and they are, indeed, equivalent to this requirement. It is an instructive exercise for the reader to verify that either of the two relationships (14.13) and (14.14) implies, in conjunction with the remainder of the system (14.9) to (14.12), that the other must hold for a solution of a Cauchy problem with data satisfying Eq. (14.3) along a noncharacteristic initial curve $\Gamma$. Thus it is not difficult to obtain a canonical system of five first-order partial differential equations in characteristic coordinates for the five unknowns $x$, $y$, $p$, $q$, and $u$ that can be used in place of the original Eq. (14.1) in the general quasi-linear case, but the final results are somewhat unattractive because of the lack of symmetry made necessary by our choice of either Eq. (14.13) or Eq. (14.14) as the final equation.

A considerable amount of freedom remains in our choice of a characteristic coordinate system, for we can always replace $\alpha$ by an arbitrary function of $\alpha$, and $\beta$ by an arbitrary function of $\beta$, in any of our canonical equations (14.9) to (14.14) without changing their form. For a given initial curve that is not characteristic, we can always pick the characteristic coordinates so that the initial curve corresponds to the line $\alpha + \beta = 0$ in the $\alpha\beta$ plane and so that on this line $x = \alpha - \beta$, whereupon no further transformations or normalizations are possible. It is also instructive to use $s = \alpha - \beta$ and $t = \alpha + \beta$ as the independent variables and to reformulate our canonical system in terms of matrix notation. To do this, we set

$$U = \begin{pmatrix} x \\ y \\ p \\ q \\ u \end{pmatrix}$$

and

$$B = \frac{Q}{(b^2 - ac)^{1/2}} \tag{14.15}$$

where

$$Q = \begin{pmatrix} b & -a & 0 & 0 & 0 \\ c & -b & 0 & 0 & 0 \\ 0 & d & b & c & 0 \\ -d & 0 & -a & -b & 0 \\ p(b^2 - ac)^{1/2} + pb + qc & q(b^2 - ac)^{1/2} - pa - qb & 0 & 0 & -(b^2 - ac)^{1/2} \end{pmatrix}$$

It then turns out that the system (14.9) to (14.13) assumes the simple matrix form

$$U_t = BU_s \tag{14.16}$$

The reader can easily verify as an exercise that the square $B^2$ of the matrix $B$ is the identity matrix $I$. Thus also

$$U_s = BU_t \tag{14.17}$$

If we differentiate Eq. (14.16) with respect to $t$ and Eq. (14.17) with respect to $s$ and subtract the latter from the former, we find

$$U_{tt} - U_{ss} = B_t U_s - B_s U_t \tag{14.18}$$

This second-order system of partial differential equations helps to motivate our reduction to a canonical form, since the wave-equation operator now appears on the left. One application that this formulation suggests, for example, is a solution of the Cauchy problem based on a sequence of successive approximations found by inserting an established approximation into the terms on the right in Eq. (14.18) and calculating the next approximation by substituting it into the operator on the left. The Cauchy data here would be given at $t = 0$.

We are now in possession of the more formal aspects of the theory of characteristics for the hyperbolic equation (14.1). We turn our attention to the matter of examples and applications. Perhaps the most instructive case that we could treat here is that of the isentropic one-dimensional flow of a perfect inviscid gas. Such a flow can be visualized as the motion of a gas in a very narrow rectilinear pipe. Let $x$ denote a space coordinate measured along the pipe, let $t$ denote the time, let $\rho$ denote the density, and let $v$ denote the velocity of the gas; $v$ is actually a scalar, since the flow is only one-dimensional. The pressure $P$ is given explicitly as a function of the density $\rho$ alone by an equation of state

$$P = P(\rho) \tag{14.19}$$

Conservation of mass provides us with the first-order partial differential equation

$$\rho_t + (\rho v)_x = 0 \tag{14.20}$$

connecting the quantities $\rho$ and $v$, while conservation of momentum yields the further equation

$$\rho v_t + \rho v v_x + P_x = 0 \tag{14.21}$$

When the initial distributions of density $\rho$ and velocity $v$ are given in the pipe, the subsequent motion of the gas can be determined by solving the system of equations (14.19) to (14.21).

We can easily reformulate the initial-value problem just described as a Cauchy problem for a second-order partial differential equation of the type (14.1). Indeed, the first-order equation (14.20) asserts the existence of a function $u$ of $x$ and $t$ such that

$$u_x = \rho \qquad u_t = -\rho v \tag{14.22}$$

and according to Eqs. (14.19) and (14.21) this function must satisfy the second-order hyperbolic partial differential equation

$$u_x^2 u_{tt} - 2u_t u_x u_{tx} + [u_t^2 - u_x^2 P'(u_x)]u_{xx} = 0$$

The coefficients in this equation do not depend on $u$ itself, whence our equivalent canonical system in characteristic coordinates $\alpha$ and $\beta$ reduces to the four first-order equations

$$x_\alpha - [v - P'(\rho)^{1/2}]t_\alpha = 0 \tag{14.23}$$
$$x_\beta - [v + P'(\rho)^{1/2}]t_\beta = 0 \tag{14.24}$$
$$p_\alpha + [v + P'(\rho)^{1/2}]q_\alpha = 0 \tag{14.25}$$
$$p_\beta + [v - P'(\rho)^{1/2}]q_\beta = 0 \tag{14.26}$$

where $p = u_t$ and $q = u_x$ and where for clarity we have replaced $u_x$ and $u_t$ in the coefficients by their values from Eqs. (14.22). The last two Eqs. (14.25) and (14.26) can also, of course, be written in the form

$$\rho v_\alpha = P'(\rho)^{1/2}\rho_\alpha \tag{14.27}$$
$$\rho v_\beta = -P'(\rho)^{1/2}\rho_\beta \tag{14.28}$$

From Eqs. (14.23) and (14.24) we see that at each point in the $tx$ plane the two slopes of the characteristics through that point are given by

$$\frac{dx}{dt} = v \pm P'(\rho)^{1/2} \tag{14.29}$$

We recall that the characteristics can be interpreted in our time-space diagram as the possible trajectories of infinitesimal disturbances or sound waves. Equation (14.29) shows that, relative to the motion of the gas with speed $v$, these trajectories advance along the pipe with the velocity $\pm P'(\rho)^{1/2}$. Thus we recognize that, at each point, the quantity $P'(\rho)^{1/2}$ represents the local speed of sound.

If we introduce the expression

$$R(\rho) = \int \rho^{-1} P'(\rho)^{1/2} \, d\rho$$

we can put Eqs. (14.27) and (14.28) in the form

$$(v - R)_\alpha = 0 \tag{14.30}$$

$$(v + R)_\beta = 0 \tag{14.31}$$

Consider a situation in which the initial distributions of velocity $v$ and density $\rho$ in the pipe are so arranged that $v - R$ has a constant value there at the time $t = 0$. By integrating Eq. (14.30), we conclude that $v - R$ must then be constant throughout the subsequent flow. But if we now advance along any specific characteristic on which $\beta$ varies, Eq. (14.31) tells us that $v + R$ also remains constant, whence both $v$ and $\rho$ must have constant values there. Hence from Eq. (14.24) the slope of this characteristic does not vary, and it must actually be a straight line. A flow of this special type, for which one family of the characteristics in the $tx$ plane consists exclusively of straight lines, is called a simple wave. Consider now the case in which the slopes $dt/dx$ of these lines increase as we move along our pipe in the direction of increasing $x$. As time goes on, the lines must reach an envelope, and this envelope must form a singularity in the flow. Thus by a consideration of the geometry of the characteristics alone we have been able to establish that discontinuities must eventually appear in the flow; see Ref. 2, pages 101 to 105. These discontinuities are *shock waves;* as is well known, their theory[3] is extremely important in fluid mechanics.

A useful exercise for the reader is to show that, if $\rho$ and $v$ are introduced as the independent variables, the resulting partial differential equations obtained for $t$ and $x$ are linear. Riemann[9] has used this remark to derive an elegant explicit solution of Cauchy's problem for the equation of one-dimensional gas dynamics in the case of a polytropic gas with the equation of state

$$P = A\rho^\gamma$$

where $A$ is a positive constant and $\gamma$, the ratio of specific heats, is a constant exceeding 1. His results provide one of the few examples where an exact solution of a nonlinear problem in the mechanics of continua can be achieved. The details of his method, however, would carry us too far from our present discussion.

## 14.3 The Method of Finite Differences

Cauchy's problem for the hyperbolic partial differential equation (14.1) reduces, through our transformation to the characteristic coordinates $\alpha$ and $\beta$, to an initial-value problem for the canonical system (14.9) to (14.13) in which the values of $x$, $y$, $p$, $q$, and $u$ are prescribed, for example, along the line $\alpha + \beta = 0$. We are interested in setting up an approximate solution of this problem by means of the method of finite differences. This method consists in laying down a grid of mesh points over the $\alpha\beta$ plane, in replacing the derivatives in the canonical system by suitable differences of function values at the mesh points in order to obtain a discrete set of equations, and in solving these equations step by step for the function values involved. There are many possible formulations of such a technique that differ from one another in the type of grid that is used and in the type of differences that are selected to approximate the derivatives. We shall describe here some of the simpler difference schemes that are in common use for the solution of initial-value problems.

First, let us consider a square grid over the $\alpha\beta$ plane, having mesh-point coordinates of the form

$$\alpha = mh \qquad \beta = nh$$

where $m$ and $n$ are arbitrary integers and where $h$ is a fixed positive constant called the *mesh size* of the grid. We have been given the values of the unknowns $x$, $y$, $p$, $q$, and $u$ at the grid points such that $m + n = 0$, and our objective is to calculate step by step, from the values of the unknowns at grid points with a fixed choice of $m + n$, the corresponding values at grid points with $m + n$ one unit greater. We use the subscripts $m$, $n$ to indicate quantities evaluated at the mesh point $(mh,nh)$; thus, for example, we write

$$x_{m,n} = x(mh,nh)$$

With this notation, a useful difference approximation to the canonical system (14.9) to (14.13) is given by the equations

$$y_{m,n} - \sigma_{m-1,n}x_{m,n} = y_{m-1,n} - \sigma_{m-1,n}x_{m-1,n} \tag{14.32}$$

$$y_{m,n} - \tau_{m,n-1}x_{m,n} = y_{m,n-1} - \tau_{m,n-1}x_{m,n-1} \tag{14.33}$$

$$\begin{aligned} a_{m-1,n}\sigma_{m-1,n}p_{m,n} &+ c_{m-1,n}q_{m,n} + d_{m-1,n}y_{m,n} \\ &= a_{m-1,n}\sigma_{m-1,n}p_{m-1,n} + c_{m-1,n}q_{m-1,n} + d_{m-1,n}y_{m-1,n} \end{aligned} \tag{14.34}$$

$$\begin{aligned} a_{m,n-1}\tau_{m,n-1}p_{m,n} &+ c_{m,n-1}q_{m,n} + d_{m,n-1}y_{m,n} \\ &= a_{m,n-1}\tau_{m,n-1}p_{m,n-1} + c_{m,n-1}q_{m,n-1} + d_{m,n-1}y_{m,n-1} \end{aligned} \tag{14.35}$$

$$\begin{aligned} u_{m,n} &- p_{m-1,n}x_{m,n} - q_{m-1,n}y_{m,n} \\ &= u_{m-1,n} - p_{m-1,n}x_{m-1,n} - q_{m-1,n}y_{m-1,n} \end{aligned} \tag{14.36}$$

This set of equations is so arranged that it is possible to solve for the unknowns $x$, $y$, $p$, $q$, $u$ at the grid point $(mh,nh)$ directly in terms of their values at the previous pair of grid points $((m - 1)h, nh)$ and $(mh,$

$(n - 1)h$). Such a formulation enables us to march forward, calculating the values of the unknowns at successive levels of mesh points in the desired fashion. The rules of the procedure are altogether elementary and straightforward, but in principle one should also justify the convergence of the answer to the solution of the initial-value problem for the actual differential equations (14.9) to (14.13) as the mesh size $h$ tends to the limit zero. We shall not go into this more difficult question in the present short chapter, where we must content ourselves with a heuristic discussion of the order of magnitude of the error in the difference scheme when the convergence is granted.

There are two kinds of errors that should be distinguished; see page 486 of Ref. 7. The first of these is the *truncation error* caused by approximating derivatives by finite differences, and the second is the *rounding error* that occurs at each step in the computation when we retain only a fixed number of decimal places in our arithmetic. As to the truncation error, we can estimate it crudely by noting that, as $h \to 0$, a forward difference such as $(x_{m,n} - x_{m-1,n})/h$ approaches the corresponding derivative $x_\alpha[(m - 1)h, nh]$ with a remainder of the order of magnitude $h$, provided we assume that the second derivative $x_{\alpha\alpha}$ is uniformly bounded. Thus, substitution of such finite differences into our system of partial differential equations in place of the actual derivatives can be viewed as equivalent to the introduction of additional terms of the order of magnitude $h$. Such new terms might be expected to alter the solution by an equivalent amount, and therefore we can assume that the truncation error itself is of the order of magnitude $h$. It is significant in this connection that, if we replace the coefficients in the difference equations (14.32) to (14.36), such as $\sigma_{m-1,n}$, by the appropriate averages, such as $(\sigma_{m,n} + \sigma_{m-1,n})/2$, we achieve a symmetric arrangement more or less equivalent to the use of a central-difference approximation to the derivatives. An expansion in Taylor series, which the reader can easily carry out as an exercise, shows that this symmetric approximation is equivalent to introducing into the differential equations extra terms only of the order of magnitude $h^2$, whence the truncation error of the modified scheme should be proportional only to $h^2$ instead of $h$; since, however, the coefficients now have the unknowns at the advanced grid point

$$\alpha = mh \qquad \beta = nh$$

among their arguments, the new system of difference equations is in implicit form and can be solved only by a sequence of iterations; this fact, of course, is a disadvantage of the refinement.

The magnitude of the rounding error can be estimated by a similar technique. Let $\epsilon$ denote the maximum contribution we can omit in any single arithmetic operation by retaining only a certain fixed number of significant figures or decimal places in the computation. It is then clear that an error of the order of magnitude $\epsilon$ could appear in each of the difference equations (14.32) to (14.36) because of rounding. Notice, however, that these equations must be divided by $h$ before the differences appearing in them occur in a form that approximates the derivatives in our system of differential equations. The rounding error should therefore be reviewed as arising from insertion in the differential equations of additional terms proportional to $\epsilon/h$. We conclude that the rounding error itself is of the order of magnitude $\epsilon/h$. It follows that, to obtain a significant numerical solution of the Cauchy problem, we must always take $\epsilon$ much smaller than $h$. Therefore, as the mesh size $h$ is taken smaller and smaller, it is necessary to retain more and more decimal places in the calculations.

In our numerical treatment of the Cauchy problem by finite differences, we notice that a certain segment of Cauchy data along the initial line $\alpha + \beta = 0$ determines the solution in a triangular region bounded by this segment and by two intersecting characteristics $\alpha = \text{const}$ and $\beta = \text{const}$ through its end points; see Fig. 14.1. By letting the mesh size $h \to 0$, we see that the same result is valid for our canonical system (14.9) to (14.13) or for the original hyperbolic partial differential equation (14.1) itself. This analysis of the precise amount of data upon which the solution of the Cauchy problem at any specific point must depend is one of the most important features of our theory. It is interesting and instructive now to manipulate with difference schemes that do not take this phenomenon into account but are designed instead to yield the solution in too large a triangle. We shall indicate how such a procedure leads to unstable results as the mesh size $h$ diminishes.

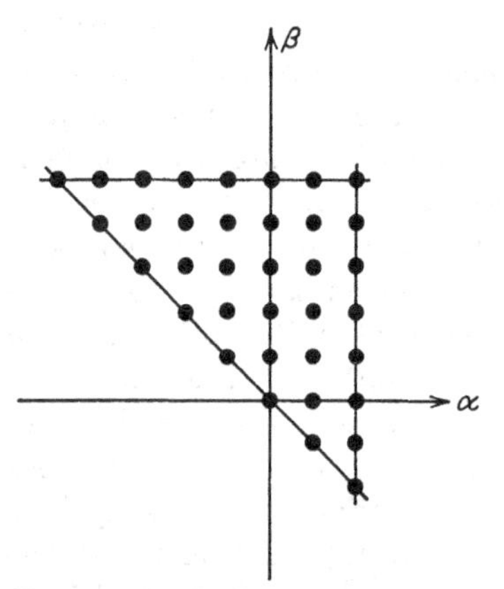


**Fig. 14.1** Finite-difference grid for Cauchy's problem.

For this purpose we refer to the canonical system of partial differential equations (14.18). In the $ts$ plane, we introduce the grid of mesh points

$$t = mh \qquad s = nk$$

with $h$ and $k$ not necessarily equal, and the corresponding difference equation is seen to be

$$\begin{aligned} h^{-2}(U_{m+1,n} - 2U_{m,n} + U_{m-1,n}) - k^{-2}(U_{m,n+1} - 2U_{m,n} + U_{m,n-1}) \\ = 2^{-1}h^{-1}k^{-1}[(B_{m,n} - B_{m-1,n})(U_{m,n+1} - U_{m,n-1}) \\ - (B_{m,n+1} - B_{m,n-1})(U_{m,n} - U_{m-1,n})] \end{aligned} \tag{14.37}$$

in an obvious notation. For our analysis of the stability of this difference scheme, we shall confine our attention to the simplest case $B \equiv 0$ of the wave equation. Here we can find explicit solutions of Eq. (14.37) of the form

$$U_{m,n} = T^m S^n \tag{14.38}$$

where $T$ and $S$ are appropriately selected constants. Now if $|T| \leq 1$ whenever $|S| = 1$, we shall say that the difference scheme (14.37) is stable, since then a bounded initial choice of $U_{m,n}$ at $m = 0$ corresponds to a bounded solution as $m$ increases. On the other hand, if there exists a $T$ with $|T| > 1$ for $|S| = 1$, we shall say that the difference equation (14.37) is unstable, since we can find a solution with bounded initial values that grows exponentially as $m$ increases.

If we set $S = e^{j\theta}$, we see that Eq. (14.38) is a solution of Eq. (14.37) if and only if

$$T = 1 - 2\left(\frac{h}{k}\right)^2 \sin^2\left(\frac{\theta}{2}\right) \pm 2j\left(\frac{h}{k}\right)\left[1 - \left(\frac{h}{k}\right)^2 \sin^2\left(\frac{\theta}{2}\right)\right]^{1/2} \sin\left(\frac{\theta}{2}\right) \tag{14.39}$$

a result that the reader can check for himself as an exercise. When $h \leq k$, this shows that $|T| = 1$ for all real choices of $\theta$, and Eq. (14.37) is therefore stable. The restriction $h \leq k$ on our grid in the $ts$ plane means that a triangle of mesh points used in a step-by-step solution of Cauchy's problem for the difference equation (14.37) lies within the corresponding triangle bounded by a pair of characteristics emanating from the ends of the segment of Cauchy data involved. Thus all is as it should be and we have not attempted to compute the solution of Eq. (14.18) from data insufficient to determine it. On the other hand, when $h > k$, there is a real choice of $\theta$ such that one of the two values of $T$ given by Eq. (14.39) has a modulus larger than unity, $|T| > 1$. Hence the difference scheme (14.37) is unstable; indeed, in this case the triangle of mesh points where the difference equation has a solution depending on given Cauchy data reaches beyond the characteristic triangle in which the solution of the original partial differential equation (14.18) is determined by these data. Thus we must never use an $h > k$ in numerical applications; actually, the best choice would be $h = k$, which corresponds more or less to the formulation of the finite-difference method that we presented relative to the $\alpha\beta$ plane.

Of course, the difficulty with the difference scheme (14.37) in the unstable case $h/k > 1$ is that, when we pass to the limit $h \to 0$, $k \to 0$ with the mesh sizes in such a ratio, our solution can exhibit large oscillations in spite of any restrictions on its initial data. It is interesting at this point to remark that this phenomenon occurs for any ratio $h/k$ of mesh sizes, no matter how small, if we attempt to treat by these methods Laplace's equation

$$U_{tt} + U_{ss} = 0$$

instead of the hyperbolic equation (14.18). For this elliptic case, indeed, our whole discussion could proceed along the same lines except for a difference in sign in one place, but this change in sign would bring Eq. (14.39) into the new form

$$T = 1 + 2\left(\frac{h}{k}\right)^2 \sin^2\left(\frac{\theta}{2}\right) \pm 2\left(\frac{h}{k}\right)\left[1 + \left(\frac{h}{k}\right)^2 \sin^2\left(\frac{\theta}{2}\right)\right]^{1/2} \sin\left(\frac{\theta}{2}\right)$$

Here for any positive pair of mesh sizes $h$ and $k$ there is a real choice of $\theta$ such that $|T| > 1$, so that the difference scheme for the solution of Cauchy's problem is always unstable. This result is consistent with the comments in Sec. 14.1 to the effect that Cauchy's problem is incorrectly set for Laplace's equation. Despite all this, however, we shall have occasion later on to develop a procedure that yields a valid numerical solution of Cauchy's problem for elliptic equations in those cases that are of practical significance for physical applications.

### 14.4 Cauchy's Problem in the Elliptic Case

We shall present here a theory of Cauchy's problem for an elliptic partial differential equation in two independent variables with analytic coefficients and analytic data. Our treatment will be suitable for numerical analysis by the method of finite differences when the Cauchy data have an analytic continuation into the complex domain in the large. In Sec. 14.1, we have described applications of this technique to problems of plasma physics, stationary shock waves, and flow with free streamlines, but we shall begin by discussing a much simpler special example.

Consider the linear elliptic equation

$$u_{xx} + u_{yy} = u \tag{14.40}$$

with Cauchy data

$$u(0,y) = f(y) \qquad u_x(0,y) = g(y) \tag{14.41}$$

where $f$ and $g$ are assumed to be real analytic functions of the real argument $y$. This problem has a solution $u$ in the form of a convergent power series, and thus $u$ can be extended analytically to a complex-valued function of the complex arguments

$$x = x_1 + jx_2 \qquad y = y_1 + jy_2$$

by direct substitution of these variables into its series representation. By holding $x_2$ and $y_1$ fixed, we find from Eq. (14.40) and the Cauchy-Riemann equations that $u = u_1 + ju_2$ satisfies the telegraph equation

$$u_{x_1x_1} - u_{y_2y_2} = u \tag{14.42}$$

which is of hyperbolic type. In particular, we set $x_2 = 0$ and find that $u = u(x_1, y_1 + jy_2)$ has as a function of $x_1$ and $y_2$ the Cauchy data

$$u(0, y_1 + jy_2) = f(y_1 + jy_2) \qquad u_{x_1}(0, y_1 + jy_2) = g(y_1 + jy_2) \tag{14.43}$$

along the initial line $x_1 = 0$, where $f(y_1 + jy_2)$ and $g(y_1 + jy_2)$ are the analytic continuations into the complex $y$ plane of the two functions $f$ and $g$ appearing in Eq. (14.41). Thus $u$ satisfies the properly set Cauchy problem defined by Eqs. (14.42) and (14.43) when it is considered as a complex-valued function of the pair of real variables $x_1$ and $y_2$. This Cauchy problem can be treated either numerically by the method of finite differences or explicitly by means of Bessel functions. The original elliptic Cauchy problem (14.40) and (14.41) can therefore be solved in two stages consisting of the analytic continuation of the data into the complex $y$ plane and of the solution by any standard procedure of the hyperbolic problem (14.42) and (14.43). To obtain the final real answer, it suffices to set $y_2 = 0$ once more and to take $u = u_1$; but, of course, to calculate this result in a full region of the real $xy$ plane, we must repeat the solution of the hyperbolic problem for a complete interval of values of the parameter $y_1$.

In the foregoing analysis, the unstable dependence of the solution of the elliptic equation (14.40) on the real Cauchy data (14.41) appears exclusively in the step where we extend the data to complex values of the independent variable $y$. Therefore, in cases when this analytic continuation of $f$ and $g$ can be performed in an elementary way—e.g., by direct substitution of complex values of $y$ into explicit expressions for $f$ and $g$—we encounter no difficulty in solving the elliptic Cauchy problem in the manner described. Furthermore, it is evident from inspection of our construction (see Fig. 14.2) that the region of existence of the solution in the real $xy$ plane is identical with the region of regularity of the analytic functions $f(y)$ and $g(y)$ in the $y_2y_1$ plane.

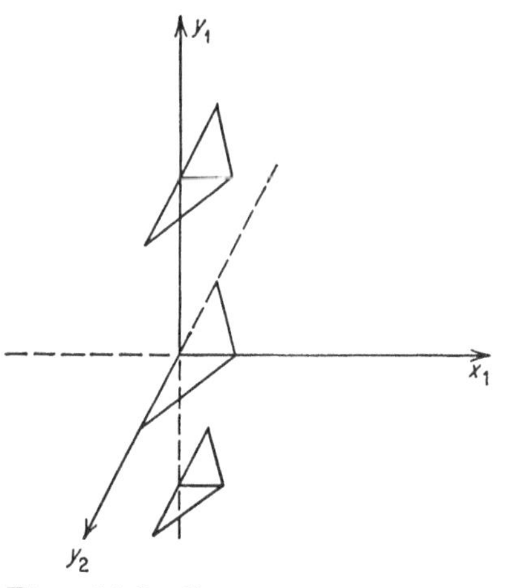


**Fig. 14.2** Geometry of Cauchy's problem in the complex domain.

Our next concern will be the more complicated situation of Cauchy's problem for a quasi-linear elliptic equation (14.1) with analytic coefficients $a$, $b$, $c$, and $d$ satisfying

$$b^2 - ac < 0 \tag{14.44}$$

For analytic data, a convergent power-series solution still exists, and we can again extend it into the complex domain by direct substitution of complex arguments into the series expansion. In this fashion we have at hand an analytic function $u$ of two complex variables $x$ and $y$, and there is no reason why we cannot perform with it all the calculations that led us earlier from the partial differential equation (14.1) to the canonical systems (14.9) to (14.13) or (14.16). All the manipulations required here are just as valid for analytic functions of several complex variables as they are for differentiable real functions of real arguments. This reduction to canonical form will prove to be an essential feature of our treatment of Cauchy's problem in the complex domain, where it no longer makes sense to speak of the sign of the discriminant $b^2 - ac$ and where, in general, the roots $\sigma$ and $\tau$ of the quadratic equation (14.7) are complex numbers.

If we think of $u$ as known, we can determine a general solution of the ordinary differential equation (14.7) for the characteristics in the implicit form

$$\alpha(x,y) = \text{const} \tag{14.45}$$

solved for the arbitrary constant of integration. This can be done, for example, by working out the power-series solution of Eq. (14.7) that assumes the value $y = y_0$ at $x = x_0$ and then expressing $y_0$ as a function of the remaining variables in the result. Since

$$\frac{dy}{dx} = -\frac{\alpha_x}{\alpha_y}$$

according to Eq. (14.45), the function $\alpha = \alpha_1 + j\alpha_2$ satisfies the first-order partial differential equation

$$a\alpha_x^2 + 2b\alpha_x\alpha_y + c\alpha_y^2 = 0 \tag{14.46}$$

and this is true in particular for strictly real values of $x$ and $y$. In this real domain, $a$, $b$, and $c$ are real, whence $\bar{\alpha} = \alpha_1 - j\alpha_2$ also satisfies Eq. (14.46) there. Thus if we define

$$\beta(x,y) = \alpha_1(x,y) - j\alpha_2(x,y)$$

for real values of $x$ and $y$ and extend it afterward to complex values by analytic continuation, we obtain a second solution of Eq. (14.46) valid in the entire complex domain. Evidently, the implicit equation

$$\beta(x,y) = \text{const}$$

defines another family of solutions of the quadratic ordinary differential equation (14.7). The pair of functions $\alpha$ and $\beta$ that we have introduced in this manner form a possible choice of the characteristic coordinates in the canonical system (14.9) to (14.13), and they have the special property that for real $x$ and $y$ the expression $t = \alpha + \beta$ is real, while $s = \alpha - \beta$ is pure imaginary.

We set $s = \xi + j\eta$ and introduce $t$ and $\eta$ as new real independent variables, whereupon the canonical system (14.16), which is equivalent to Eq. (14.1), takes the form

$$U_t = -jBU_\eta \tag{14.47}$$

The point here is that for $\xi = 0$, and for real choices of $t$ and $\eta$, the components $x$, $y$, $p$, $q$, and $u$ of the solution $U$ of Eq. (14.47) are all real. Since the radical $(b^2 - ac)^{1/2}$ appearing in our definition (14.15) of $B$ is pure imaginary in the real domain, it even turns out that all the elements of the matrix $jB$ except those in the last row are real there. If we had elected to take the average of Eqs. (14.13) and (14.14) instead of Eq. (14.13) alone as our fifth equation in the canonical system (14.16), we would actually have been led to a matrix equation of the type (14.47) that would have been entirely real in the real domain. The choice we have made, however, has the advantage that by cross differentiation we obtain the result

$$U_{tt} + U_{\eta\eta} = jB_\eta U_t - jB_t U_\eta \tag{14.48}$$

analogous to Eq. (14.18), in which the highest derivatives appear only as a Laplacian. The reader should verify as an exercise that in this form the canonical system is invariant under conformal mappings[1] of the $t\eta$ plane.

Cauchy's problem for the elliptic system (14.48) can be handled by the same technique that we used for studying the special case (14.40), and thus the general quasi-linear elliptic equation (14.1) from which Eq. (14.48) arose can also be treated. For this purpose, we have only to exploit the analytic continuation that has led us to Eq. (14.48). First, however, we perform a preliminary conformal transformation of the $t\eta$ plane, mapping the initial curve onto the $\eta$ axis. In this connection, it should be emphasized again that in our change of coordinates the real $t\eta$ plane corresponds to the real $xy$ plane. Now our method is to allow $s = \xi + j\eta$ to have arbitrary complex values once more and to pick $t$ and $\xi$ as the independent variables, while $\eta$ is held fixed as a parameter. The system (14.47) then reverts to its hyperbolic form

$$U_t = BU_\xi \tag{14.49}$$

for which the Cauchy problem can be solved by the classical procedures that we described earlier. The Cauchy data for Eq. (14.49) at $t = 0$ now appear, to be sure, as the analytic extensions into the full complex $s$ plane of the Cauchy data for the elliptic system (14.47), which were originally defined only as real analytic functions of the real argument $\eta$. Thus

Cauchy's problem for the general quasi-linear elliptic partial differential equation (14.1) is solved in three steps consisting of a reduction to the canonical form (14.47), an analytic continuation of the data, and, finally, a solution of the hyperbolic problem (14.49) by, for example, the method of finite differences.

We should point out at this stage that further conformal transformations of the $t\eta$ plane are still at our disposal, provided that they preserve the relevant segment of the $\eta$ axis where the Cauchy data are prescribed. Locally, such a conformal transformation can be obtained by analytic continuation to complex arguments $t + j\eta$ of an arbitrary purely imaginary analytic function of the purely imaginary variable $j\eta$. Thus, at the beginning, any analytic change of scale along the $\eta$ axis can be introduced. This can be arranged, in fact, so that the initial values of any one of the unknowns $x$, $y$, $p$, $q$, and $u$, say of $x$, are given by a quite arbitrary analytic expression in $\eta$. Consider what becomes of this freedom in our choice of $x$ as a function of $\eta$ after we have continued the Cauchy data analytically into the complex $s$ plane. Our remark then implies that for a fixed value of $\eta$ we can pick the initial values of $x$ to be any complex-valued analytic expression in the real variable $\xi$ that assumes at $-\xi$ the conjugate of its value at $\xi$. But these data are to be substituted into a stable Cauchy problem for the hyperbolic system (14.49) in the $t\xi$ plane, and therefore any convergent limiting process performed on them still leaves us with valid results. In particular, we can approximate any continuously differentiable function $x = x(\xi)$ arbitrarily well by analytic expressions in $\xi$, and hence such a function is admissible as our initial choice of $x$ in terms of $\xi$, even when it is not analytic. The remaining unknowns $y$, $p$, $q$, and $u$, however, must still be related initially to $x$ by the analytic conditions of our original Cauchy problem in the $xy$ plane.

That our initial choice of $x = x(\xi)$ need not be analytic is a somewhat surprising feature of our procedure for solving Cauchy's problem for an elliptic equation by analytic extension into the complex domain; the underlying explanation of this phenomenon, however, is that the function $x = x(\xi)$ essentially determines what might be thought of as a path of integration in the real $xy$ plane. Indeed, if we solve the hyperbolic Cauchy problem (14.49) in the $t\xi$ plane for a fixed value of $\eta$, the real portion of the answer is found along the line $\xi = 0$, which corresponds to a curve in the real $xy$ plane along which the values of $p$, $q$, and $u$ are determined. This curve would feature as a path of integration if we were to solve the hyperbolic problem (14.49) by the classical method of successive approximations. Furthermore, with each initial value of $x$ is associated a specific initial value of $y$, and through the corresponding point $(x,y)$ in the complex domain we can pass a complex characteristic of the partial differential equation (14.1) that intersects the real $xy$ plane in a unique point lying on this same path. Therefore, for each fixed $\eta$ our choice of the initial function $x = x(\xi)$ serves to determine in a specific manner the path in the real domain along which the results of the solution of the Cauchy problem will appear. The path need not be analytic because $x(\xi)$ need not be analytic; it could, for example, have several corners. It is of interest that, in the actual execution of a finite-difference scheme, the characteristic coordinates do not occur explicitly, so that the freedom in the selection of $x(\xi)$ allows in practice for a situation in which we can assign an arbitrary initial column of complex values of $x$ and subsequently can march forward with a valid solution by fixed elementary rules.

We have used the method of characteristics and of analytic continuation into the complex domain to define a stable finite-difference scheme for the solution of Cauchy's problem in the elliptic case. There are many other significant applications of this theory. For example, Lewy[8] has used it to establish that a sufficiently differentiable solution of an analytic elliptic partial differential equation in two independent variables is necessarily itself analytic. It is also possible to show by this procedure that a free streamline in the steady flow of an incompressible, inviscid fluid must always be an analytic curve.[6] Such results concerning the analyticity of solutions of elliptic boundary-value problems are of interest for our discussion of inverse methods because they show that we start in the right class of functions when we assume analyticity of our data. We shall not devote space to a more detailed presentation of such theoretical matters here, however, since our principal concern is with the applications.

## 14.5 Flow around a Bubble Rising under the Influence of Gravity

We shall consider the plane, steady, irrotational motion of an incompressible, inviscid liquid that falls over a free streamline bounding a bubble of gas held at constant pressure. The flow is described (see pages 375, 384 of Ref. 1) by a velocity potential $\phi$ and a stream function $\psi$ that can be combined to form a complex potential $\zeta = \phi + j\psi$ that is analytic as a function of the complex variable $z = x + jy$ in the physical plane. The velocity of the motion is the gradient of $\phi$, and the pressure $P$ is given by Bernoulli's equation

$$\tfrac{1}{2}(\phi_x^2 + \phi_y^2) + gy + P = \text{const} \tag{14.50}$$

in which we have taken the density $\rho$ to be identically 1. We assume that the fluid has significant weight, and $g$ denotes the acceleration of gravity.

Along a free streamline bounding a bubble of the type described above, two boundary conditions are required, stating that $\psi$ and $P$ are constant. There is no loss of generality if we assume that the constant value of $\psi$ along the free streamline is

$$\psi = 0 \tag{14.51}$$

and by translating the coordinate system up or down appropriately we can write the constant-pressure boundary condition in the form

$$\psi_x^2 + \psi_y^2 + 2gy = 0 \tag{14.52}$$

according to Bernoulli's equation (14.50). We shall study the inverse problem in which the free streamline is prescribed to be the principal arc of the cycloid

$$z = x + jy = \theta + je^{-j\theta} - j \tag{14.53}$$

corresponding to the interval $-\pi \leq \theta \leq \pi$. This is a Cauchy problem for the Laplace equation

$$\psi_{xx} + \psi_{yy} = 0 \tag{14.54}$$

with Cauchy data defined by Eqs. (14.51) and (14.52) along the initial curve (14.53). We shall solve it in closed form by using the fundamental procedures we have already outlined.

A first remark is that Eq. (14.54) is already in canonical form. We can make a conformal transformation that brings the initial curve into a segment of the $\eta$ axis in the $t\eta$ plane by interpreting Eq. (14.53) as a conformal mapping of the $z$ plane onto the complex $\theta$ plane and then setting

$$\theta = j(t + j\eta) \tag{14.55}$$

Of course, $\psi$ is still harmonic as a function of $t$ and $\eta$,

$$\psi_{tt} + \psi_{\eta\eta} = 0 \tag{14.56}$$

From Eqs. (14.51) and (14.52), we easily find for $\psi$ the initial conditions

$$\psi = 0 \qquad \psi_t = -2g^{1/2} \sin \eta \tag{14.57}$$

at $t = 0$. Now we can extend $\psi$ analytically to complex values $\eta - j\xi$ of its second argument and consider it as a function of $t$ and $\xi$ with $\eta$ held fixed as a parameter. Under this complex substitution of variables, the Cauchy problem (14.56) and (14.57) assumes the hyperbolic form

$$\psi_{tt} - \psi_{\xi\xi} = 0 \tag{14.58}$$

with

$$\psi = 0 \qquad \psi_t = -2g^{1/2} \sin (\eta - j\xi) \tag{14.59}$$

for $t = 0$.

We can use the general solution

$$\psi = \alpha(\xi + t) + \beta(\xi - t)$$

of the wave equation (14.58) in terms of two arbitrary functions $\alpha$ and $\beta$ in order to solve this problem explicitly. We find from Eqs. (14.59) that

$$\alpha(\xi) + \beta(\xi) = 0$$

and

$$\alpha'(\xi) - \beta'(\xi) = -2g^{1/2} \sin (\eta - j\xi)$$

whence

$$\psi = jg^{1/2} \cos (\eta - j\xi - jt) - jg^{1/2} \cos (\eta - j\xi + jt)$$

Specializing to the case $\xi = 0$, we obtain the real solution

$$\psi = -2g^{1/2} \sin \eta \sinh t \tag{14.60}$$

of the original elliptic Cauchy problem (14.56) and (14.57).

In view of Eq. (14.55), it is an easy matter to derive, from Eq. (14.60), the relationship

$$\zeta = -2g^{1/2} \cos \theta$$

for the complex potential $\zeta = \phi + j\psi$; therefore, according to our interpretation of Eq. (14.53) as a conformal transformation of the complex $\theta$ plane onto the complex $z$ plane, we have

$$z = \cos^{-1}\left(-\frac{\zeta}{2g^{1/2}}\right) + \frac{(4g - \zeta^2)^{1/2}}{2g^{1/2}} - j\frac{\zeta + 2g^{1/2}}{2g^{1/2}}$$

Although it is not possible to invert this formula in order to express $\zeta$ explicitly in terms of $z$, it is easy to read off the essential properties of the resulting flow directly. The liquid moves downward with symmetry about the $y$ axis in the upper half plane. The central streamline divides at the origin into two forks along the principal arc of the cycloid (14.53), which forms the free boundary of a gas bubble underneath. At the ends of this arc of the cycloid, the two forks of the central streamline change over into vertical segments that descend indefinitely. Far to the right and to the left the velocity approaches a constant downward value. Thus the flow represents the motion of a liquid falling around a semi-infinite channel of gas bounded at the top by a smoothly forked free streamline in the shape of a cycloid. See Fig. 14.3. If we think of our coordinate system in the $xy$ plane as moving upward uniformly at a speed just sufficient to cancel out the velocity of the liquid at infinity, the plane flow we have described can be interpreted as the motion of a gas bubble rising in a liquid of infinite extent. The part of the gas under the bubble, which is bounded by vertical walls, can be understood in this connection to represent a wake.

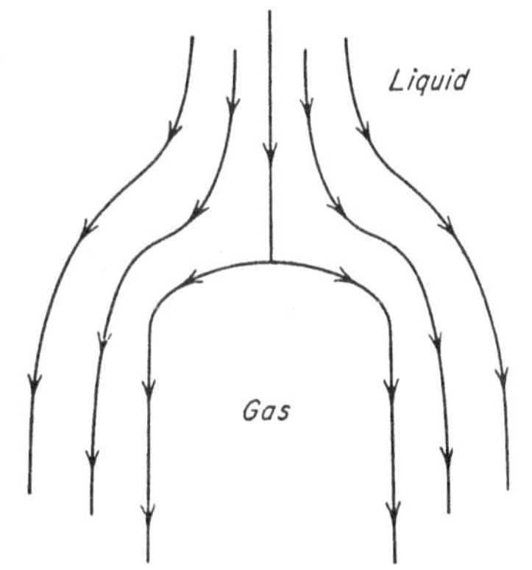


**Fig. 14.3** Flow over a gas bubble.

We have gone out of our way to solve the elementary Cauchy problem defined by Eqs. (14.51), (14.52), and (14.54) by using general procedures applicable to a wider variety of cases. This particular example, however, could be treated more directly by means of analytic continuation alone. It suffices for this purpose to derive a relationship between $\zeta$ and $z$ along the cycloid (14.53) and then continue it analytically over the whole $z$ plane. As an exercise, the reader should carry out the details of such a solution based exclusively on the theory of analytic functions of a complex variable.

### 14.6 The Detached-shock Problem

We are finally in a position to develop with reasonable facility an inverse method for calculating steady plane or axially symmetric flows of a compressible inviscid fluid behind a prescribed shock wave.[5] In our presentation, we shall confine our attention to the axially symmetric case, which we can study in a meridian $xy$ plane. Here $y$ plays the role of the radius in a cylindrical coordinate system. We denote by $u$ and $v$ the components of velocity parallel, respectively, to the $x$ axis and the $y$ axis. Although the flow we shall consider is rotational, the law of conservation of mass permits us to describe it by means of a stream function $\psi = \psi(x,y)$ such that

$$\psi_x = -y\rho v \qquad \psi_y = y\rho u$$

where $\rho$ is the density. The fundamental laws of conservation of momentum and energy in fluid mechanics can be used to show that $\psi$ satisfies the second-order partial differential equation

$$(C^2 - u^2)\psi_{xx} - 2uv\psi_{xy} + (C^2 - v^2)\psi_{yy} + C^2y^2\rho^2 \frac{2H + u^2 + v^2}{2\gamma} \frac{A'(\psi)}{A(\psi)} - C^2\rho u = 0 \qquad (14.61)$$

of the quasi-linear type (14.1). Here $A$ is an arbitrary function of $\psi$, $C$ is the local speed of sound defined by

$$C^2 = \gamma A \rho^{\gamma-1}$$

where $\gamma$ is the ratio of specific heats, and

$$H = \frac{u^2 + v^2}{2} + \frac{C^2}{\gamma - 1}$$

is a constant called the *stagnation enthalpy*.

We are interested in the flow behind a shock wave that occurs along a prescribed curve

$$x = F(y^2) \qquad (14.62)$$

We assume the flow in front of the shock wave to be uniform and in the direction of the $x$ axis, with a Mach number $M > 1$. Across the shock wave, of course, the various flow quantities have jumps that are determined by the laws of conservation of mass, momentum, and energy, and that can be expressed at each point of the wave in terms of the slope

$$\frac{dx}{dy} = 2yF'(y^2)$$

there alone. In particular, conservation of energy implies that the stagnation enthalpy $H$ is continuous across the shock, and this fact explains why it is to be constant throughout the flow. Similarly, the law of conservation of mass shows that the stream function $\psi$ has no jump. Conservation of momentum in the directions of the $x$ axis and the $y$ axis can be used to calculate the arbitrary function $A$ explicitly in terms of the given function $F$ and to determine the normal derivative of $\psi$ at each point of the shock wave. If we normalize so that the pressure $P$ and density $\rho$ in front of the shock wave are identically 1, and if we call the speed there $Q$, then the formulas derived from the above shock conditions for flow quantities evaluated just behind the shock are

$$\psi = \frac{Qy^2}{2} \tag{14.63}$$

$$\psi_x = \frac{2Qy^2F'(y^2)}{L}\left(1 - \frac{1}{V}\right) \tag{14.64}$$

and

$$A = \frac{V^\gamma}{(\gamma + 1)L}\,[2Q^2 - (\gamma - 1)L]$$

where $L = 1 + \dfrac{8\psi}{Q}F'\left(\dfrac{2\psi}{Q}\right)^2$ and $V = \dfrac{\gamma - 1}{\gamma + 1} + \dfrac{2\gamma}{\gamma + 1}\dfrac{L}{Q^2}$

Somewhat more complicated expressions would feature in these relationships if we were to consider a more involved, but occasionally useful, formulation in which the constant $\gamma$ merely plays the role of an effective ratio of specific heats behind the shock and differs from the actual ratio of specific heats $\gamma_1$ in front.

The conditions (14.63) and (14.64) along the prescribed curve (14.62) present us with a Cauchy problem for the partial differential equation (14.61). The coefficients of this equation depend in an essential way on the quantities $y$, $\psi_x$, $\psi_y$, and $\psi$, and therefore we must use our full theory to obtain a solution. Furthermore, the discriminant of Eq. (14.61) has the form

$$b^2 - ac = C^2(u^2 + v^2 - C^2)$$

so that this partial differential equation is of the hyperbolic type in a region of supersonic flow with $u^2 + v^2 > C^2$ and is of the elliptic type in a region of subsonic flow with $u^2 + v^2 < C^2$. As we have seen, Cauchy's problem is straightforward for a hyperbolic equation, but can be handled in the elliptic case only after an extension into the complex domain. We can treat both situations by means of our reduction to canonical form, but the details of the analysis are really quite different for the two types of flow. In particular, we must assume that any arc of the shock wave (14.62) bounding a subsonic region is analytic.

We have given a complete description of the method of characteristics and of finite differences for the solution of a Cauchy problem such as the one with which we are confronted in the present situation. In this special application of our procedure, all the steps to be taken are quite routine except for the analytic continuation, into the complex domain, of the Cauchy data along a subsonic arc of the shock wave (14.62). Even here, the only difficulty is with the extension of the function $F(y^2)$ to complex values of the variable $y$, since the remaining expressions that appear in the data (14.63) and (14.64) or in the canonical system (14.49) are elementary in the sense that they involve only additions, multiplications, divisions, or possibly square roots, and hence their extension is merely a matter of direct substitution of complex arguments. Thus, in case $F(y^2)$ is also defined by an explicit formula that can be continued analytically over the $y$ plane by straightforward substitution of complex values of $y$, the Cauchy problem (14.61), (14.63), and (14.64) is completely solved in a satisfactory fashion. Since a wide variety of shock waves (14.62) of an elementary form like this can be found, we can obtain a large class of flows with detached shock waves by applying our inverse method based on prescribing $F$. The body that generates the shock wave (14.62) appears, of course, as the level curve $\psi = 0$ of the stream function $\psi$. It is not too hard in practice to determine a choice of the shock wave that corresponds to a blunt body within reasonable approximation to a desired shape. This can be achieved with moderate success by studying the behavior of the analytic function $F(y^2)$ in the complex $y$ plane.

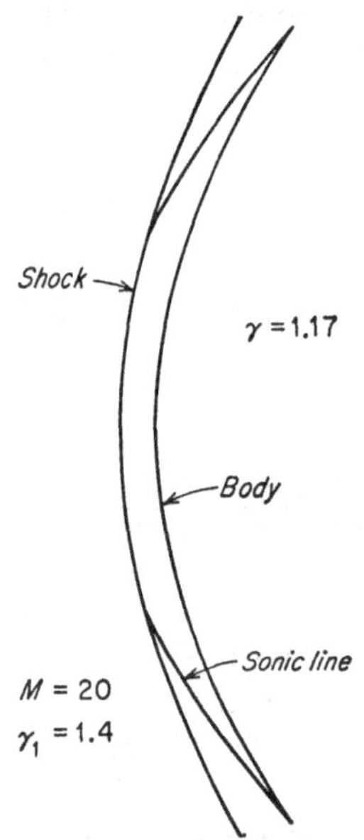


**Fig. 14.4** Detached shock in front of a blunt body in hypersonic flow.

In Fig. 14.4 we have sketched an axially symmetric flow of the type under consideration here, for which the shock wave was selected to be a hyperbola. In this calculation, the Mach number in front of the shock wave was taken to be $M = 20$, and the ratio of specific heats there was $\gamma_1 = 1.4$. Behind the shock, however, an effective value $\gamma = 1.17$ of the ratio of specific heats was assumed in order to approximate in some degree the effects of dissociation. It is this small choice of $\gamma$ that brings the shock wave so close to the body, which turned out to be very nearly spherical.

Certain qualitative properties of flows with a detached shock wave can be deduced from our analysis based on analytic continuation. We observe, for example, that, in order for the flow constructed by our inverse method to exist in the large and to reach as far as the body without exhibiting a singularity, it is necessary that the portion of the Cauchy data prescribed along the subsonic section of the shock wave should have analytic continuations that are regular in the large in the complex domain. It follows that the subsonic arc of the shock wave lies in the interior of the region of regularity of the stream function $\psi$ in the four-dimensional domain of the two complex variables $x$ and $y$. For this reason, the values of $\psi$ or of its derivatives on this arc should not change very much when the boundary values of $\psi$ are perturbed along the body, since in the closed interior of its region of regularity an analytic function is not greatly affected by alterations of its boundary values. In other words, the shape of the shock wave in front of a subsonic portion of flow should not be very sensitive to changes in the shape of the body that generates it. Those readers familiar with experimental work on detached shock waves will recognize that the phenomenon that we can predict in this manner does actually occur in nature.

A final comment about our construction of flows through prescribed shock waves is that it applies to the case of attached shocks, too. The difficulty here is to formulate the correct asymptotic form for the equation of the shock wave near its point of attachment. This is easy if the flow is entirely supersonic, but if it is subsonic in the neighborhood of the point of attachment, we are led to a problem concerning the behavior of the solution of an elliptic partial differential equation near the confluence of two analytic boundary conditions. The method of analytic extension into the complex domain indicates that the appropriate asymptotic expansion (see Chap. 5) here should proceed in fractional powers of $z$ and of $z \log z$ when we normalize so that the point of attachment lies at the origin. However, we must close with the admission that this interesting question is far too involved for further discussion in the present chapter.

## REFERENCES

1. Beckenbach, Edwin F., Conformal Mapping Methods, chap. 15 in "Modern Mathematics for the Engineer," First Series, edited by E. F. Beckenbach, McGraw-Hill Book Company, Inc., New York, 1956.
2. Courant, Richard, Hyperbolic Partial Differential Equations and Applications, chap. 5 in "Modern Mathematics for the Engineer," First Series, edited by E. F. Beckenbach, McGraw-Hill Book Company, Inc., New York, 1956.
3. ——— and K. O. Friedrichs, "Supersonic Flow and Shock Waves," Interscience Publishers, Inc., New York, 1948.
4. ——— and D. Hilbert, "Methoden der mathematischen Physik," vol. 2, Springer-Verlag, Berlin, Vienna, 1937.
5. Garabedian, P. R., and H. M. Lieberstein, On the Numerical Calculation of Detached Bow Shock Waves in Hypersonic Flow, *J. Aero. Sci.*, vol. 25, pp. 109–118, 1958.

6. ———, H. Lewy, and M. Schiffer, Axially Symmetric Cavitational Flow, *Ann. Math.*, vol. 56, pp. 560–602, 1952.
7. Lehmer, Derrick H., High-speed Computing Devices and Their Applications, chap. 19 in "Modern Mathematics for the Engineer," First Series, edited by E. F. Beckenbach, McGraw-Hill Book Company, Inc., New York, 1956.
8. Lewy, H., Neuer Beweis des analytischen Charakters der Lösungen elliptischer Differentialgleichungen, *Math. Ann.*, vol. 101, pp. 609–619, 1929.
9. Riemann, B., "Gesammelte mathematische Werke," Leipzig, 1876.
10. Schiffer, Menahem M., Boundary-value Problems in Elliptic Partial Differential Equations, chap. 6 in "Modern Mathematics for the Engineer," First Series, edited by E. F. Beckenbach, McGraw-Hill Book Company, Inc., New York, 1956.

# 15

# The Numerical Solution of Elliptic and Parabolic Partial Differential Equations

DAVID YOUNG

PROFESSOR OF MATHEMATICS AND DIRECTOR OF
THE COMPUTATION CENTER
THE UNIVERSITY OF TEXAS

## 15.1 Introduction

Many important problems in science and engineering can be formulated in terms of linear second-order partial differential equations. For instance, a number of problems in elasticity,[44] steady-state heat conduction,[8,39] and fluid mechanics[8,39] involve the solution of Laplace's equation in a two-dimensional region subject to certain auxiliary conditions at the boundary of the region. Sometimes the function itself is prescribed on the boundary; in other cases, the normal derivative or even a linear combination of the function and its normal derivative may be prescribed. Some problems—for instance, those involving steady-state heat conduction with variable conductivity—involve linear second-order elliptic partial differential equations other than Laplace's. Other problems, such as those arising in the study of the neutron-flux distribution in a nuclear reactor, may involve the simultaneous solution of several equations.

Except in very special cases, it is not practical to obtain numerical solutions to such problems by analytic methods. Instead, one is nearly always forced to resort to approximate numerical methods. Of such methods, probably the most popular and the one that is most frequently used is the method of finite differences.[8] A network of regularly spaced straight lines, each parallel to one of the coordinate axes, is superimposed over the region. Instead of seeking the value of the solution of the differential equation at every point in the continuous region, one seeks to determine approximate values of the solution at the *mesh points*, the intersections of the network lines with each other and with the boundary of the region. The partial differential equation is replaced by a partial difference equation. Although more elaborate techniques may be used, one method of obtaining the difference equation at a given interior mesh point is to replace the partial derivatives in the differential equation by partial difference quotients in the standard way, thus obtaining a linear algebraic equation corresponding to each interior mesh point. In the case of a boundary-value problem, since the values at the boundary mesh points are prescribed, the solution of the difference equation involves the solution of a system of $N$ linear algebraic equations with $N$ unknowns, $N$ being the number of interior mesh points.

While it is easy to show that under rather general conditions the system of equations has a unique solution, the actual numerical determination of the solution presents a serious computational problem. Since very little is known about the accuracy of the solution of the difference equation as a solution of the differential equation, it is desirable to make the distance between adjacent mesh points, here denoted as the *mesh size*, small in order to be reasonably sure of obtaining satisfactory accuracy. Thus the number of interior mesh points and hence the number of equations will generally be large; in some cases it may be in the thousands. Direct methods such as the use of determinants, as in Cramer's rule, and the Gauss elimination method appear out of the question. Since the matrix of the system is very sparse, each equation involving the unknown values only at the corresponding mesh point and at a few neighboring points, the use of iterative methods appears appropriate. One selects an arbitrary initial approximation to each of the unknowns and successively modifies the approximate values according to a given rule of procedure. If the iterative method is properly chosen, the approximate values for any given point will converge to the value of the exact solution of the difference equation at that point. Of course, what is important from the computational standpoint is not only whether the method converges but also how fast it converges.

The Gauss-Seidel method is probably one of the simplest iterative methods and is very easy to use on a high-speed computer; however, although convergence is guaranteed for a wide class of problems, in many cases the convergence is exceedingly slow. For instance, for Laplace's equation the number of complete iterations—i.e., the number of times that the value of the function at each point is modified—necessary to achieve a specified degree of convergence is approximately proportional to $h^{-2}$, where $h$ is the mesh size. As $h$ decreases, the number of iterations increases very rapidly. By a simple modification of the Gauss-Seidel method, the number of iterations can be substantially reduced.

The modified method, which was called the "extrapolated Liebmann method" by Frankel[17] and the "successive-overrelaxation method" by Young,[57] involves a parameter known as the *relaxation factor*. If this relaxation factor is properly chosen, then in the case involving Laplace's equation discussed above the number of iterations varies approximately as $h^{-1}$, thus affording a very substantial improvement over the Gauss-Seidel method.

Another iterative method that is designed to give rapid convergence is the alternating-direction implicit method of Peaceman and Rachford.[35] Whereas both the Gauss-Seidel method and the successive-overrelaxation method are "point" iterative procedures in the sense that at each step the value at one single point is modified, in the Peaceman-Rachford method the values for all points of a line of mesh points, either a row or a column, are modified simultaneously. The modifications are carried out alternately first for all rows and then for all columns. Although in order to perform a row or a column iteration it is necessary to solve a linear system with as many equations and unknowns as there are mesh points in the row or column, nevertheless since the matrix of the system is usually tridiagonal, the solution of the system presents very little practical difficulty. If a set of iteration parameters is properly chosen, then for the case of Laplace's equation on a rectangle the number of complete iterations required to achieve convergence varies approximately as $|\log h|^{-1}$. The theoretical improvement over the successive-overrelaxation method in this case is considerable. Moreover, although the theoretical advantage in convergence has so far been proved only in the special case mentioned above, nevertheless actual numerical experiences indicate a much greater generality. At the present time, the successive-overrelaxation method and the Peaceman-Rachford method are each finding considerable usage for problems involving elliptic partial differential equations while other methods are being used only occasionally. Quite recently, successive-line overrelaxation and even two-line over-relaxation has been used by Varga and others.[11]

Parabolic partial differential equations arising in scientific and engineering problems are often of the form

$$U_t = L(U)$$

where $L(U)$ is a second-order elliptic partial differential operator with either one or two independent "space" variables. Typical examples are the equation

$$U_t = U_{xx}$$

involving one space variable, and the equation

$$U_t = U_{xx} + U_{yy}$$

involving two space variables. Let $R$ denote the region of the space variable or variables and let $S$ denote the boundary of $R$. One frequently seeks to solve an initial-value problem in which the values of $U$ in $R$ are specified for $t = t_0$ and the values of $U$ on $S$ are specified for all $t > t_0$. As in the case of problems involving elliptic equations, the method of finite differences is frequently used to obtain approximate numerical solutions. In addition to the network that is superimposed over the region $R$, a time increment $k$ is introduced. One seeks approximate numerical values of $U$ for all mesh points in $R$ and for $t = t_0 + k, t_0 + 2k, \ldots$, etc., the values of $U$ for $t = t_0$ being already given. Here, in contrast to the situation for elliptic equations, it is possible to develop an explicit procedure for determining the values of $U$ for any given value of $t$ in terms of those values that have already been obtained for previous values of $t$. Unfortunately, however, this procedure, which is sometimes known as the "forward-difference method," may be numerically unstable unless $k$ is chosen so small relative to the space-mesh size that the computational labor involved in proceeding from $t_0$ to another value of $t$, say $t_1$, in steps of length $k$ may be prohibitive even with a fast computer. A larger value of $k$ could not be used because the excessive growth of rounding errors that occurs with an unstable process would destroy all accuracy within a few time steps.

For problems involving one space dimension, this difficulty may be avoided by use of the Crank-Nicolson method.[10] The procedure is implicit in that the process of proceeding from one time step to another involves the solution of a system of linear algebraic equations, there being as many equations as there are interior space-mesh points. Since the matrix of the system is tridiagonal, however, there is little difficulty involved. The extension of the Crank-Nicolson method to problems involving two space dimensions would involve solving an elliptic boundary-value problem for each time step. Since this procedure would be very slow indeed, the alternating-direction method of Peaceman, Rachford, and Douglas is recommended. This method involves computing approximate values of $U$ successively at $t_0 + k/2$, $t_0 + k$, $t_0 + 3k/2$, etc., using a procedure that is alternately implicit on rows and explicit on columns and vice versa. Since each step involves the solution of a linear system with a tridiagonal matrix, the computing time is much less than would be required if the Crank-Nicolson method were used. The local accuracy of the method is of the same order of magnitude as for the Crank-Nicolson method. The alternating-direction method is also numerically stable.

Prior to the advent of high-speed computing machines, numerical solutions were obtained by manual methods. For boundary-value problems involving equations of elliptic type, the difference equations were solved by "relaxation methods"; see, for instance, the books of Southwell[45] and Allen.[2] The relaxation method is basically an iterative method in which, instead of following a fixed procedure, the computer is allowed considerable freedom in modifying the approximate values. A skilled and experienced computer could often obtain a numerical solution to a complicated problem in a remarkably short time. Nevertheless, the amount of drudgery involved with manual methods places a limit on the accuracy, on the speed with which a solution can be obtained, and on the number of related cases that it is practical to solve in a given study.

The availability of modern high-speed machines and the development of efficient and effective numerical procedures have made practical the solution of a number of problems. In Sec. 15.9, the solutions of three such problems are described; these include a flow problem, a nuclear-reactor problem, and a thermal-ignition problem. It would be interesting to try to estimate how many millions of hours of tedious hand computing have been avoided by the use of these programs.

One difficulty in the use of large computers for solving partial differential equations is the amount of effort required in the preparation of a program for the machine. This is of less relative importance when a great many cases are to be solved, but if only a few cases are required, then it may not always be easy to decide whether or not the problem should be done on a computer. In any case, there is a considerable time lag between the formulation of the problem and its solution on the computer.

Attempting to remedy this situation, a group of computing installations, primarily on the West Coast, established a collaborative project known as the SPADE project. The objective of the SPADE project was the development of a set of programs for the IBM 704 and 709 computers for solving a class of problems involving systems of elliptic and parabolic partial differential equations. It would be necessary for the program user to specify the boundary (for instance, by giving the equations of a finite number of boundary segments), the mesh size, formulas for the determination of the coefficients of the differential equations, and the iterative method to be used. The program of SPADE would then determine the coordinates of the mesh points, determine the coefficients of the difference equations, and solve the difference equations by using the prescribed iterative procedure. Since the programs were to be available as closed subroutines that might be used in various combinations, a considerable degree of flexibility would be afforded. By means of the SPADE programs, the amount of labor and the time involved in preparing a problem for solution on a high-speed computer would be greatly reduced. Actually, in fact, the SPADE project has recently been abandoned. Its failure can be attributed to lack of sufficient effort and does not imply that the program could not be successfully carried out.

### 15.2 Boundary-value Problems and the Method of Finite Differences

Let $R$ be a bounded plane region with boundary $S$ and let $f$ be a function defined and continuous on $S$. We consider the problem of finding a function $U$ that is continuous in $R + S$, is twice differentiable in $R$, satisfies in $R$ the linear second-order partial differential equation

$$AU_{xx} + CU_{yy} + DU_x + EU_y + FU = G \tag{15.1}$$

and satisfies on $S$ the condition

$$U = f \tag{15.2}$$

Here $A$, $C$, $D$, $E$, $F$, and $G$ are analytic functions of the independent variables $x$ and $y$ in $R + S$ and satisfy the conditions $A > 0$, $C > 0$, and $F \leq 0$. Because of the conditions on $A$ and $C$, Eq. (15.1) is said to be *elliptic.*[8,39] More generally, an equation of the form

$$AU_{xx} + 2BU_{xy} + CU_{yy} = D(x,y,U,U_x,U_y) \tag{15.3}$$

is said to be elliptic in $R + S$ if $B^2 - AC < 0$ in $R + S$. We remark that by introducing new variables $\xi$ and $\eta$ satisfying the conditions

$$\frac{\xi_y}{\xi_x} = \sqrt{\frac{A}{C}} \qquad \frac{\eta_y}{\eta_x} = -\sqrt{\frac{A}{C}}$$

one obtains an equation of the form

$$\alpha U_{\xi\xi} + \gamma U_{\eta\eta} = \delta(\xi,\eta,U,U_\xi,U_\eta) \tag{15.4}$$

where $\alpha > 0$, $\gamma > 0$ in $R' + S'$, the region in the $\xi\eta$ plane corresponding to $R + S$. Equation (15.4) is similar to Eq. (15.1) in so far as the terms involving the second-order derivatives are concerned. Much of our subsequent discussion will apply to Eq. (15.4) as well as to Eq. (15.1) and will also apply to certain cases in which, instead of prescribing $U$ on $S$, one prescribes on $S$ the normal derivative $\partial U/\partial n$ or a linear combination of $U$ and $\partial U/\partial n$.

A special case of the above is the solution of Laplace's equation

$$U_{xx} + U_{yy} = 0 \tag{15.5}$$

in the circle $x^2 + y^2 < 1$ with values prescribed on the boundary of the circle. Such a problem corresponds to the solution of a steady-state heat-conduction problem for an infinite right-circular cylinder with axis normal to the $xy$ plane and with external temperature held constant along lines parallel to the axis of the cylinder. The variation of the temperature on the circumference of any cross section of the cylinder is prescribed. If, instead of prescribing the temperature everywhere, one assumes that part of the surface is insulated, then the corresponding condition for the appropriate part of the circle $x^2 + y^2 = 1$ is $\partial U/\partial n = 0$.

In order to apply the method of finite differences to obtain an approximate numerical solution of the problem defined by Eqs. (15.1) and (15.2), the first step is to construct a network. We choose an arbitrary but fixed point $(\bar{x},\bar{y})$ and a positive number $h$ that we shall call the *mesh size.* We choose one mesh size for simplicity, although we could have different mesh sizes in the two coordinate directions. We let $\Sigma_h$ denote the set of points $(x,y)$ such that both $h^{-1}(x - \bar{x})$ and $h^{-1}(y - \bar{y})$ are integers. Such points are called *mesh points.* Two mesh points $(x,y)$ and $(x',y')$ are *adjacent* if

$$(x - x')^2 + (y - y')^2 = h^2$$

We let $R_h$ denote the set of all points that are in both $R$ and $\Sigma_h$. Points of $R_h$ are known as *interior mesh points.* We determine the set $S_h$ of *boundary mesh points* as follows: For each point $(x,y)$ of $R_h$, we consider the four adjacent mesh points. Let $(x',y')$ denote one such point. If no point of $S$ lies on the line segment joining $(x,y)$ to $(x',y')$, then no point of $S_h$ lies on that segment. Otherwise, if at least one point of $S$ lies on the line segment, then the nearest such point to $(x,y)$, which may be $(x',y')$, is said to belong to $S_h$. The set $S_h$ consists of all points found by considering each point of $R_h$ and the corresponding four line segments to the four adjacent points.

As an example, consider the ellipse

$$\frac{x^2}{9} + \frac{y^2}{4} = 1$$

with $\bar{x} = \bar{y} = 0$ and $h = 1$; see Fig. 15.1. Points of $R_h$ and $S_h$ are designated by solid black circles and by open circles, respectively. Other mesh points are designated by open squares.

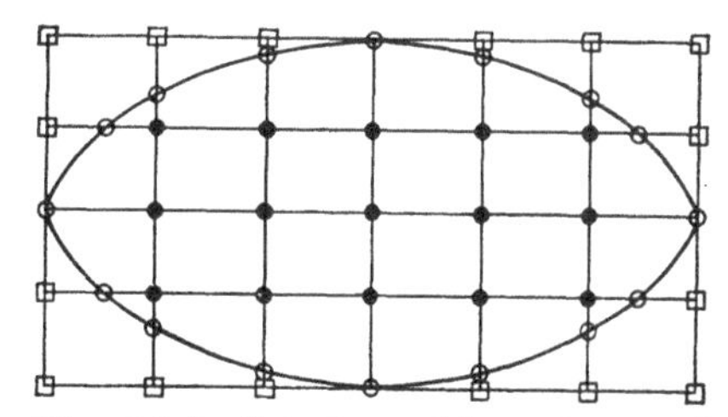

**Fig. 15.1** Interior and boundary mesh points.

Once we have determined a set of mesh points, our objective will be to determine approximate values of $U$ at these points. To this end, we next construct a difference-equation representation of Eq. (15.1). First we consider a typical *regular* point $(x_0,y_0)$ of $R_h$, that is, a point of $R_h$ such that the four adjacent points of $\Sigma_h$ belong either to $R_h$ or to $S_h$.

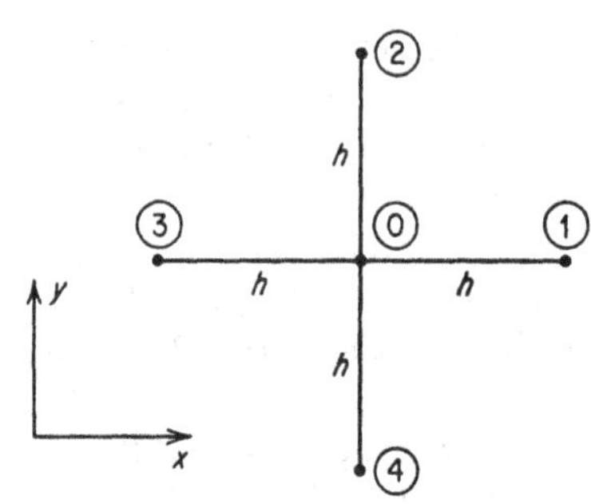


**Fig. 15.2** Typical neighborhood configuration for a regular interior mesh point.

We seek to determine a difference equation corresponding to Eq. (15.1) for the point $(x_0,y_0)$.

Let us consider the configuration shown in Fig. 15.2; this consists of $(x_0,y_0)$ and the four adjacent mesh points, which for convenience are labeled as shown. We first seek an expression for $U_{xx}$. We adopt the notation that $U_i = U(x_i,y_i)$, $i = 0, 1, 2, 3, 4$, and $(U_x)_i = U_x(x_i,y_i)$, etc. Assuming that $U(x,y)$ has partial derivatives of the fourth order in a sufficiently large neighborhood about $(x_0,y_0)$, we have, by Taylor's theorem,

$$U_1 = U_0 + h(U_x)_0 + \frac{h^2}{2!}(U_{xx})_0 + \frac{h^3}{3!}(U_{xxx})_0 + \frac{h^4}{4!}(\bar{U}_{xxxx})_{0,1} \quad (15.6)$$

with

$$(\bar{U}_{xxxx})_{0,1} = U_{xxxx}(\xi,y_0)$$

where $\xi$ is some value of $x$ such that

$$x_0 < \xi < x_1$$

Similarly,

$$U_3 = U_0 - h(U_x)_0 + \frac{h^2}{2!}(U_{xx})_0 - \frac{h^3}{3!}(U_{xxx})_0 + \frac{h^4}{4!}(\bar{U}_{xxxx})_{0,3} \quad (15.7)$$

By combining Eqs. (15.6) and (15.7), we obtain

$$(U_{xx})_0 = \frac{U_1 + U_3 - 2U_0}{h^2} - \frac{h^2}{24}[(\bar{U}_{xxxx})_{0,1} + (\bar{U}_{xxxx})_{0,3}] \quad (15.8)$$

By using the same method, we can also obtain

$$(U_{yy})_0 = \frac{U_2 + U_4 - 2U_0}{h^2} - \frac{h^2}{24}[(\bar{U}_{yyyy})_{0,2} + (\bar{U}_{yyyy})_{0,4}] \quad (15.9)$$

$$(U_x)_0 = \frac{U_1 - U_3}{2h} - \frac{h^3}{12}[(\bar{U}_{xxx})_{0,1} + (\bar{U}_{xxx})_{0,3}] \quad (15.10)$$

$$(U_y)_0 = \frac{U_2 - U_4}{2h} - \frac{h^3}{12}[(\bar{U}_{yyy})_{0,2} + (\bar{U}_{yyy})_{0,4}] \quad (15.11)$$

By neglecting the remainder terms, i.e., the terms involving brackets, in Eqs. (15.8) to (15.11) and by substituting in Eq. (15.1) for each regular point $(x,y)$ in $R_h$, we obtain the difference equation

$$\begin{aligned}(A + \tfrac{1}{2}Dh)U(x + h, y) &+ (A - \tfrac{1}{2}Dh)U(x - h, y) \\ &+ (C + \tfrac{1}{2}Eh)U(x, y + h) + (C - \tfrac{1}{2}Eh)U(x, y - h) \\ &- 2(A + C - \tfrac{1}{2}Fh^2)U(x,y) = Gh^2 \quad (15.12)\end{aligned}$$

where $A$, $C$, $D$, $E$, $F$, and $G$ are evaluated at $(x,y)$. If we write Eq. (15.12) in the form

$$\alpha_1 U(x+h,\, y) + \alpha_3 U(x-h,\, y) + \alpha_2(x,\, y+h) + \alpha_4 U(x,\, y-h) - \alpha_0 U(x,y) = t(x,y) \qquad (15.13)$$

where the $\alpha_i$, $i = 0, 1, 2, 3, 4$, are functions of $x$ and $y$, we observe that, because $A > 0$, $C > 0$, and $F \leq 0$, if the $\alpha_i$ are positive then

$$\alpha_0 \geq \alpha_1 + \alpha_2 + \alpha_3 + \alpha_4 \qquad (15.14)$$

Even for *irregular* interior mesh points, i.e., points of $R_h$ that are not regular interior mesh points, we can write the difference equation in the form (15.13). Thus if, as in Fig. 15.3, $x_1 - x_0 = s_1 h$ and $x_3 - x_0 = s_3 h$, where $0 < s_1 \leq 1$, $0 < s_3 \leq 1$, with the aid of Taylor's theorem we can easily derive the following formulas:

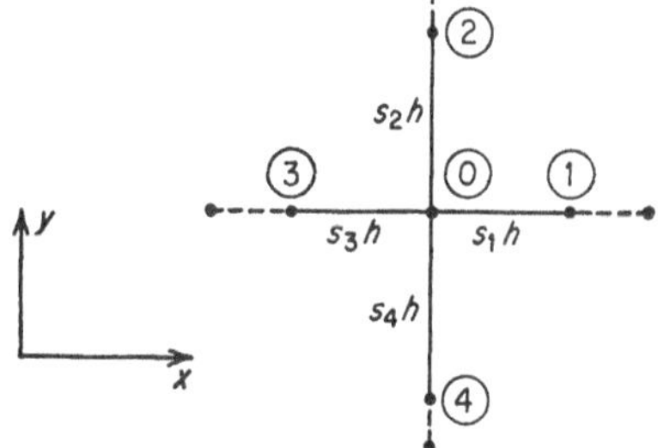


**Fig. 15.3** Typical neighborhood configuration for an irregular interior mesh point.

$$(U_x)_0 \doteq h^{-1}\left[\frac{s_3}{s_1(s_1+s_3)}\,U_1 - \frac{s_1}{s_3(s_1+s_3)}\,U_3 - \frac{s_3 - s_1}{s_1 s_3}\,U_0\right] \qquad (15.15)$$

$$(U_{xx})_0 \doteq 2h^{-2}\left[\frac{1}{s_1(s_1+s_3)}\,U_1 + \frac{1}{s_3(s_1+s_3)}\,U_3 - \frac{1}{s_1 s_3}\,U_0\right] \qquad (15.16)$$

Similar expressions can be obtained for $(U_y)_0$ and $(U_{yy})_0$. If the resulting expressions are substituted into Eq. (15.1), then one obtains

$$\alpha_1 U(x+s_1h,\, y) + \alpha_3 U(x-s_3h,\, y) + \alpha_2 U(x,\, y+s_2h) + \alpha_4 U(x,\, y-s_4h) - \alpha_0 U(x,y) = t(x,y) \qquad (15.17)$$

where

$$\begin{aligned}
\alpha_1 &= \frac{2A}{s_1(s_1+s_3)} + \frac{hs_3D}{s_1(s_1+s_3)} & \alpha_3 &= \frac{2A}{s_3(s_1+s_3)} - \frac{hs_1D}{s_3(s_1+s_3)} \\
\alpha_2 &= \frac{2C}{s_2(s_2+s_4)} + \frac{hs_4E}{s_2(s_2+s_4)} & \alpha_4 &= \frac{2C}{s_4(s_2+s_4)} - \frac{hs_2E}{s_4(s_2+s_4)} \qquad (15.18)\\
\alpha_0 &= \alpha_1 + \alpha_2 + \alpha_3 + \alpha_4 - Fh^2 & t &= Gh^2
\end{aligned}$$

It is evident that every $\alpha_i$ appearing in Eqs. (15.13) and (15.17) will be positive provided $h$ is chosen so small that

$$h < \min\,(M_1, M_2)$$

where
$$M_1 = \min \frac{2A}{|D|} \qquad M_2 = \min \frac{2C}{|E|}$$

For $M_1$ and $M_2$, the minima are taken over all points of $R + S$. That $M_1$ and $M_2$ are positive is easily seen by noting that the functions $A$, $C$,

$D$, and $E$ are continuous in the closed region $R + S$ and that $A > 0$, $C > 0$. It is evident that $D$ and $E$ are bounded and that $A$ and $C$ have positive minima.

If Eq. (15.1) is self-adjoint, i.e., if it can be written in the form

$$(AU_x)_x + (CU_y)_y + FU = G \tag{15.19}$$

then one can obtain a difference equation with the following properties: The local accuracy is as good as that for Eqs. (15.13) and (15.17); the coefficients are positive; and if $R_h$ contains only regular points, then the difference equation is *symmetric*. That is to say, the coefficient of $U(x',y')$ in the difference equation for the point $(x,y)$ is the same as the coefficient of $U(x,y)$ in the equation for the point $(x',y')$. In such a case, the matrix of the linear system corresponding to the difference equation would be symmetric, a situation that has certain theoretical and, perhaps, practical advantages.

In order for Eq. (15.1) to be self-adjoint, it is necessary and sufficient that we have

$$D = A_x \qquad E = C_y \tag{15.20}$$

If Eqs. (15.20) hold, then the differential equation can readily be put in the self-adjoint form. In deriving the difference equation for the neighborhood configuration shown in Fig. 15.3, we use finite-difference expressions of the form

$$(AU_x)_x \doteq \frac{A(x + \frac{1}{2}s_1h, y)[(U_1 - U_0)/s_1h] - A(x - \frac{1}{2}s_3h, y)[(U_0 - U_3)/s_3h]}{h(s_1 + s_3)/2}$$

and derive a difference equation of the form (15.17) with

$$\begin{aligned} \alpha_1 &= \frac{A(x + \frac{1}{2}s_1h, y)}{s_1(s_1 + s_3)} & \alpha_3 &= \frac{A(x - \frac{1}{2}s_3h, y)}{s_3(s_1 + s_3)} \\ \alpha_2 &= \frac{C(x, y + \frac{1}{2}s_2h)}{s_2(s_2 + s_4)} & \alpha_4 &= \frac{C(x, y - \frac{1}{2}s_4h)}{s_4(s_2 + s_4)} \\ \alpha_0 &= \alpha_1 + \alpha_2 + \alpha_3 + \alpha_4 - \tfrac{1}{2}h^2F & t &= \tfrac{1}{2}h^2G \end{aligned} \tag{15.21}$$

We remark that, if there are irregular interior mesh points, then the matrix of the linear system that one obtains will not in general be symmetric. So far, experience indicates that this is probably not a serious disadvantage. We further remark that there are many other possible difference equations as well as other methods for deriving them, but space does not permit us to discuss them here.

Here and subsequently we assume that our difference equation can be written in the form (15.17), where the $\alpha_i$ are positive and where

$$\alpha_0 \geq \alpha_1 + \alpha_2 + \alpha_3 + \alpha_4 \tag{15.22}$$

We further assume that the region $R_h$ is *connected;* i.e., any two points of

$R_h$ can be joined by a broken line consisting of line segments connecting pairs of adjacent points of $R_h$.

Under the foregoing assumptions, it is not difficult to show that Eq. (15.17) has a unique solution. We shall give here a well-known proof. It is clearly sufficient to show that the determinant of the matrix of the related system of linear equations does not vanish. This will be true provided that the homogeneous system, which we obtain by letting the function $G$ and the values of $U$ on the boundary of the region vanish, has only the trivial solution that vanishes throughout $R_h$. Suppose that the homogeneous system had a nontrivial solution $V$. Then for some point of $R_h$ we would have $V \neq 0$. We can assume that $V > 0$ at this point; otherwise, we could consider the function $-V$, which would also satisfy the homogeneous system. Let $M > 0$ denote the maximum value of $V$ in $R_h$, and let $(x_0,y_0)$ be the interior mesh point for which $V(x_0,y_0) = M$. By Eqs. (15.17) and (15.22), and the fact that the $\alpha_i$ are positive, it follows that $V$ must assume the value $M$ at each of the four adjacent points. By repeating this argument and using the connectedness of $R_h$, one can show that $V(x,y) = M$ for all $(x,y)$ in $R_h$ and in $S_h$. This contradicts the assumption that $V$ vanishes on $S_h$ and thus proves that $V$ vanishes identically in $R_h + S_h$. Consequently, the homogeneous system admits only the trivial solution. It then follows that Eq. (15.17) has a unique solution.

Before considering, as we shall in the next three sections, the formidable computational problem of finding the unique solution that we now know exists, we shall discuss the accuracy of the exact solution of the difference equation as a solution of the differential equation. It is shown in the well-known paper of Courant, Friedrichs, and Lewy[9] that, as the mesh size $h$ approaches zero, the solutions of the corresponding difference equations approach the solution of the differential equation; nevertheless, very little is known about the accuracy of the solution of the difference equation for a given value of $h$.

Besides the doubtful methods of intuition and guessing, one frequently used procedure is to solve the difference equation for some mesh size $h$ and then again for the mesh size $h/2$. The difference between the two solutions so obtained is taken as a measure of the error. Indeed, if one assumes that the error is precisely proportional to $h^2$, which assumption is fairly well satisfied in certain favorable cases (see, for instance, Ref. 59), then one can do even better by using a process that was introduced by L. F. Richardson[36] and that is referred to as "extrapolation to zero grid size." If $u_h$ denotes the solution of the difference equation with mesh size $h$ and if $U$ denotes the exact solution of the differential equation, then under the given assumptions we have

$$U(x,y) \doteq \tfrac{4}{3}u_{h/2}(x,y) - \tfrac{1}{3}u_h(x,y)$$

Although the theoretical basis for using this procedure is not very substantial, in some cases one obtains a remarkable improvement in accuracy by using it.

Some empirical studies on the accuracy of difference-equation solutions for Laplace's equation indicate that extrapolation to zero grid size might work rather well provided that the boundary has no corners with interior angle greater than 180° and provided that the boundary values are sufficiently smooth. The effect of curved boundaries, which would involve irregular mesh configurations, could be very unfavorable.

Gerschgorin[21] obtained a bound for the maximum error in terms of the local error. In the case of Laplace's equation for a region containing only regular interior points, Gerschgorin showed that, for any point $(x,y)$ in $R_h$,

$$|u_h(x,y) - U(x,y)| \leq \tfrac{1}{24} M_4 r^2 h^2$$

where $r$ is the radius of any circle containing the given region and where

$$M_4 = \max\,(X_4, Y_4)$$

with $X_4 = \max |U_{xxxx}| \qquad Y_4 = \max |U_{yyyy}|$

the maxima for $X_4$ and $Y_4$ being taken in $R + S$. Since the solution $U$ is not known, much less its fourth partial derivatives, this estimate has limited usefulness, particularly since in some important cases the fourth partial derivatives are not even bounded. In some cases, however, one can estimate $U_{xxyy}$ by using the difference quotient

$$\begin{aligned} U_{xxyy} \doteq h^{-4}\{ & u_h(x + h,\, y + h) - 2u_h(x,\, y + h) + u_h(x - h,\, y + h) \\ & - 2[u_h(x + h,\, y) - 2u_h(x,y) + u_h(x - h,\, y)] \\ & + u_h(x + h,\, y - h) - 2u_h(x,\, y - h) + u_h(x - h,\, y - h)\} \end{aligned}$$

and then estimate $U_{xxxx}$ and $U_{yyyy}$ by the relations

$$U_{xxyy} = -U_{xxxx} = -U_{yyyy}$$

which are derived from Laplace's equation. When the largest derivative thus found is substituted in Gerschgorin's error formula, we obtain an estimate of the accuracy.

Although a great deal of effort has been and is being spent in an attempt to find a satisfactory method of error estimation (see, for instance, papers by Rosenbloom,[38] Wasow,[55,56] Walsh and Young,[53,54] and Laasonen[27,28]), a great deal of research remains to be done in this direction.

## 15.3 Point Iterative Methods

Although the existence of a unique solution of the difference equation (15.17) under rather general conditions has already been established, nevertheless the problem of actually computing the solution presents a serious practical difficulty. In view of the uncertainty about the accuracy of the finite-difference solution, one is inclined to wish to use a small value of the mesh size $h$. This, of course, implies that the number of mesh points will be large. In many practical cases, several thousand mesh points are considered. While the use of direct methods—such as Cramer's rule, which involves determinants, and the Gauss elimination method—is almost out of the question, the fact that the matrix of the system is sparse suggests the use of iterative methods. We note from Eq. (15.17) that at most five elements in any row of the matrix are non-zero. The sparseness of the matrix is of considerable advantage for the use of iterative methods but does not help appreciably for direct methods.

The use of an iterative method involves first the selection of a function $u^{(0)}(x,y)$, which is arbitrary except that the closer it agrees with $u(x,y)$ the less computing effort will be required. One then modifies the values of $u^{(0)}(x,y)$ at the interior mesh points according to a prescribed procedure in such a way that the approximate values so obtained converge to $u(x,y)$. In this section, we consider "point" iterative methods for which at each stage the approximate value at one single point is modified. In the next section we shall consider iteration procedures such that at each step several values are modified simultaneously.

For convenience, we let $u(x,y)$ denote the exact solution of Eq. (15.17). If we solve Eq. (15.17) for $u(x,y)$, we get

$$u(x,y) = \beta_1 u(x + h_1, y) + \beta_3 u(x - h_3, y) + \beta_2 u(x,y + h_2) + \beta_4 u(x,y - h_4) + \tau(x,y) \qquad (15.23)$$

where
$$\tau = -\frac{t(x,y)}{\alpha_0}$$

and where, for $i = 1, 2, 3, 4$,

$$\beta_i = \frac{\alpha_i}{\alpha_0} \qquad h_i = hs_i$$

We note that every $\beta_i$ is positive and that

$$\sum_{i=1}^{4} \beta_i \leq 1 \qquad (15.24)$$

We remark that, in solving a problem on a high-speed computer, one would normally evaluate all of the coefficients $\beta_i$ and then store them in an auxiliary memory such as a magnetic drum or magnetic tape memory. Large blocks of $\beta_i$ could then be transferred to the high-speed memory when needed. Of course, if many of the coefficients are identical, as for instance in the case of Laplace's equation where for every regular point each $\beta_i$ equals $\frac{1}{4}$, then some other procedure would be more efficient.

Perhaps the simplest point iterative method is the Gauss-Seidel method, which is also known as the method of "successive displacements."[20] The approximate values of $u$ are improved at points of $R_h$ according to an arbitrary but fixed ordering, using Eq. (15.23) and using improved values as soon as available. Thus, if we consider the ordering in which $(x,y)$ follows $(x',y')$ provided that either $y > y'$ or that $y = y'$ and $x > x'$, then for $n = 1, 2, 3, \ldots$ the successive approximations are determined by

$$u^{(n+1)}(x,y) = \beta_1 u^{(n)}(x + h_1, y) + \beta_3 u^{(n+1)}(x - h_3, y) + \beta_2 u^{(n)}(x, y + h_2) + \beta_4 u^{(n+1)}(x, y - h_4) + \tau(x,y) \quad (15.25)$$

A complete iteration consists in improving the approximate values once at all points of $R_h$. Having traversed the points of $R_h$, one starts over again at the "first" point and repeats the process until $d_n < \epsilon$, where $\epsilon$ is the prescribed tolerance and where

$$d_n = \max |u^{(n)}(x,y) - u^{(n-1)}(x,y)|$$

the maximum being taken over all points of $R_h$.

We remark that in a similar method known as the "Jacobi method," or the "method of simultaneous displacements,"[20] one does not use improved values until after a complete iteration. The improvement formula for this method is

$$u^{(n+1)}(x,y) = \beta_1 u^{(n)}(x + h_1, y) + \beta_3 u^{(n)}(x - h_3, y) + \beta_2 u^{(n)}(x, y + h_2) + \beta_4 u^{(n)}(x, y - h_4) + \tau(x,y)$$

This method requires approximately twice as many iterations to attain a specified convergence as does the Gauss-Seidel method.

As an example of the use of the Gauss-Seidel method, let us consider the problem of solving Laplace's equation

$$U_{xx} + U_{yy} = 0$$

for the unit square $0 \leq x \leq 1$, $0 \leq y \leq 1$. Let us choose $M = h^{-1} = 3$. For boundary conditions, we let $U = 1{,}000$ on the side of the square where $y = 1$, and we let $U = 0$ elsewhere on the boundary. Evidently, each of the four points of $R_h$ is a regular interior mesh point. By Eqs. (15.17), (15.18), and (15.23), the difference equation for each point of $R_h$ becomes

$$u(x,y) = \tfrac{1}{4}u(x + h, y) + \tfrac{1}{4}u(x - h, y) + \tfrac{1}{4}u(x, y + h) + \tfrac{1}{4}u(x, y - h) \quad (15.26)$$

Table 15.1 illustrates the computational procedure. First, the initial approximation $u^{(0)}(x,y)$ to the solution of the difference equation was set equal to zero at the four interior mesh points. The ordering of the points was as follows: $(\tfrac{1}{3},\tfrac{1}{3})$, $(\tfrac{2}{3},\tfrac{1}{3})$, $(\tfrac{1}{3},\tfrac{2}{3})$, $(\tfrac{2}{3},\tfrac{2}{3})$.

Since the values of $u^{(6)}$ are identical with those of $u^{(7)}$, the iteration process was terminated after seven iterations. The final values represent the exact solution of the difference equation, as can be verified by substituting in Eq. (15.26). For convenience in studying the behavior of the method, the values of $d^{(n)} = u^{(n)} - u^{(n-1)}$ are given in Table 15.1. The ratios $d^{(n)}/d^{(n-1)}$ give an indication of how rapidly the errors are decreasing. In this case it can be seen that the ratios are approximately $\frac{1}{4}$; hence, the convergence is quite rapid. The general formula for the limiting ratio $\lambda$, however, is

$$\lambda = \cos^2 \frac{\pi}{M} \sim 1 - \left(\frac{\pi}{M}\right)^2 \tag{15.27}$$

See, for instance, the paper by Young.[57] We shall refer to $\lambda$ as the *spectral radius*† of the linear transformation defined by the Gauss-Seidel method.

**Table 15.1** An Example of the Use of the Gauss-Seidel Method

| $(x,y) = (\frac{1}{3},\frac{1}{3})$ | | | $(x,y) = (\frac{2}{3},\frac{1}{3})$ | | | $(x,y) = (\frac{1}{3},\frac{2}{3})$ | | | $(x,y) = (\frac{2}{3},\frac{2}{3})$ | | |
|---|---|---|---|---|---|---|---|---|---|---|---|
| $n$ | $u^{(n)}$ | $d^{(n)}$ | $n$ | $u^{(n)}$ | $d^{(n)}$ | $n$ | $u^{(n)}$ | $d^{(n)}$ | $n$ | $u^{(n)}$ | $d^{(n)}$ |
| 0 | 0 | ... | 0 | 0 | ... | 0 | 0 | ... | 0 | 0 | ... |
| 1 | 0 | 0 | 1 | 0 | 0 | 1 | 250 | 250 | 1 | 312 | 312 |
| 2 | 62 | 62 | 2 | 94 | 94 | 2 | 344 | 94 | 2 | 360 | 48 |
| 3 | 110 | 48 | 3 | 118 | 24 | 3 | 368 | 24 | 3 | 372 | 12 |
| 4 | 122 | 12 | 4 | 124 | 6 | 4 | 374 | 6 | 4 | 375 | 3 |
| 5 | 124 | 2 | 5 | 125 | 1 | 5 | 375 | 1 | 5 | 375 | 0 |
| 6 | 125 | 1 | 6 | 125 | 0 | 6 | 375 | 0 | 6 | 375 | 0 |
| 7 | 125 | 0 | 7 | 125 | 0 | 7 | 375 | 0 | 7 | 375 | 0 |

That the convergence of the Gauss-Seidel method is very slow can be seen from considering the case $M = 20$. Since $\lambda = 0.975$ in this case, the error in the approximate solution is reduced by only about 2.5 per cent in each iteration.

The number $N$ of iterations needed to reduce the error to a specified factor $\rho$ of its original value is given approximately by

$$\lambda^N = \rho$$

Hence,

$$N \sim \frac{-\log \rho}{-\log \lambda} \sim \frac{M^2}{\pi^2}(-\log \rho)$$

By this argument, it can be seen that the number of iterations varies as $M^2 = h^{-2}$. For a more rigorous discussion, the reader might consult Ref. 57.

† This quantity is sometimes referred to as the *spectral norm*, for instance in Ref. 57. The spectral radius is a norm in the mathematical sense for linear transformations associated with symmetric matrices, but not for general linear transformations.

By a simple modification of Eq. (15.25), a substantial improvement can be made in the rate of convergence. The modified formula is

$$u^{(n+1)}(x,y) = \omega[\beta_1 u^{(n)}(x + h_1, y) + \beta_3 u^{(n+1)}(x - h_3, y) + \beta_2 u^{(n)}(x, y + h_2) + \beta_4 u^{(n+1)}(x, y - h_4) + \tau(x,y)] - (\omega - 1)u^{(n)}(x,y) \quad (15.28)$$

Here $\omega$ is a parameter that we shall refer to as the *relaxation factor*. If $\omega = 1$, then the procedure reduces to the Gauss-Seidel method. The proper choice of the relaxation factor determines the rate of convergence of the method.

The method defined by Eq. (15.28) is referred to as the "successive overrelaxation method" by Young;[57] when applied to Laplace's equation, it is called the "extrapolated Liebmann method" by Frankel.[17] We remark that the ordinary "Liebmann method" is just the Gauss-Seidel method as applied to the difference-equation analogue of Laplace's equation; see the papers by Liebmann[33] and Shortley and Weller.[42]

The use of the successive-overrelaxation method will be illustrated by applying it to the same numerical example as was used in connection with the Gauss-Seidel method. Table 15.2 shows the results obtained using

**Table 15.2** An Example of the Use of the Successive-overrelaxation Method

| $(x,y) = (\frac{1}{3},\frac{1}{3})$ | | | $(x,y) = (\frac{2}{3},\frac{1}{3})$ | | | $(x,y) = (\frac{1}{3},\frac{2}{3})$ | | | $(x,y) = (\frac{2}{3},\frac{2}{3})$ | | |
|---|---|---|---|---|---|---|---|---|---|---|---|
| $n$ | $u^{(n)}$ | $d^{(n)}$ | $n$ | $u^{(n)}$ | $d^{(n)}$ | $n$ | $u^{(n)}$ | $d^{(n)}$ | $n$ | $u^{(n)}$ | $d^{(n)}$ |
| 0 | 0 | ... | 0 | 0 | ... | 0 | 0 | ... | 0 | 0 | ... |
| 1 | 0 | 0 | 1 | 0 | 0 | 1 | 275 | 275 | 1 | 351 | 351 |
| 2 | 76 | 76 | 2 | 118 | 118 | 2 | 365 | 90 | 2 | 373 | 22 |
| 3 | 126 | 50 | 3 | 126 | 8 | 3 | 376 | 11 | 3 | 376 | 3 |
| 4 | 126 | 0 | 4 | 126 | 0 | 4 | 376 | 0 | 4 | 376 | 0 |

1.1 as relaxation factor. It should be noted that, since the values of $u^{(3)}$ and $u^{(4)}$ are identical, the process is terminated after four iterations. The final values differ from the exact solution of Eq. (15.28) by one unit. This discrepancy could have been avoided if more significant figures had been retained throughout the iteration process. The erratic behavior of the ratios $d^{(n)}/d^{(n-1)}$ should be noted and contrasted to the relatively smooth behavior for the Gauss-Seidel method.

As shown in Ref. 57, the optimum value of $\omega$ is given by

$$\omega_b = 1 + \frac{\lambda}{(1 + \sqrt{1 - \lambda})^2} \quad (15.29)$$

where $\lambda$ is the spectral radius of the Gauss-Seidel method. In the example treated above, we have $\lambda = \frac{1}{4}$, and upon substituting this value in Eq. (15.29), we get $\omega_b = 1.072$. It is not difficult to show that the convergence of the successive-overrelaxation method with the best value of $\omega$ is approximately as rapid as it would be if the ratio $d_{n+1}/d_n$ tended to $\lambda^*$, where

$$\lambda^* = \omega_b - 1 \sim 1 - \frac{2\pi}{M}$$

Here $\lambda^*$ is the spectral radius of the successive-overrelaxation method. In the case $M = 20$, we get $\omega_b = 1.73$ and hence $\lambda^* = 0.73$, as compared with $\lambda = 0.975$ for the Gauss-Seidel method. Thus, on the average, one complete iteration of the successive-overrelaxation method reduces the error by approximately 27 per cent. The number $N$ of iterations required to reduce the error to a specified fraction $\rho$ of its initial value is given approximately by

$$(\lambda^*)^N = \rho$$

Hence
$$N \sim \frac{-\log \rho}{-\log \lambda^*} \sim \frac{M}{2\pi}(-\log \rho)$$

from which it follows that the number of iterations varies approximately as $M = h^{-1}$. This represents a very substantial improvement over the Gauss-Seidel method and one that has been well verified in actual practice; see, for instance, Ref. 59.

In order to achieve the maximum possible improvement in the speed of convergence, it is necessary to select the proper value of $\omega$. By Eq. (15.29), this choice depends on an estimate of $\lambda$, the spectral radius of the Gauss-Seidel method. For the case of Laplace's equation and for a rectangle with sides $a = Rh$ and $b = Sh$, where $R$ and $S$ are integers, $\lambda$ is given by

$$\lambda = \left[\frac{1}{2}\left(\cos\frac{\pi}{R} + \cos\frac{\pi}{S}\right)\right]^2$$

For other problems involving Laplace's equation but for regions other than rectangles, one may compute $\lambda$ for a rectangle that contains the given region and then use the value of $\omega$ computed from Eq. (15.29). The value of $\omega$ so obtained will be somewhat larger than the true optimum value. It has been shown theoretically and verified experimentally by Young[57,59] that an overestimation of the optimum value of $\omega$ does not result in a serious reduction in the rate of convergence. A slightly better estimate of the optimum value could probably be obtained by the use of a rectangle of approximately the same area and proportions.

The problem of choosing a suitable relaxation factor is in general more difficult for problems involving differential equations other than Laplace's. Sometimes special properties of the equation can be used. It may be possible to show that the appropriate value of $\lambda$ will be close to that for Laplace's equation. Alternatively, we can estimate $\lambda$ by performing a number of iterations with $\omega = 1$ and then estimating the average of the ratios $d^{(n)}(x,y)/d^{(n-1)}(x,y)$. Finally, and this is certainly a reasonable approach if a large number of cases are to be treated, one can simply try different values of $\omega$ for different cases and determine the best value by experiment.

The improvement in convergence rate that is afforded by the successive-overrelaxation method has been shown by Young[57] to hold under certain conditions. These conditions are satisfied by symmetric five-point difference equations of the form (15.17), provided that the ordering of the points is properly chosen. Recent theoretical work of Garabedian[19] and Kahan,[26] however, indicates that the improvement in convergence rate can be realized under much more general conditions. For instance, it was observed by Young[59] in certain numerical cases that improvement in convergence rate occurs when a nine-point difference equation is used to represent Laplace's equation. Garabedian[19] provided theoretical justification for this. Kahan's results indicate that the improvement in convergence rate will hold provided that the coefficients $\alpha_i$ of the difference equation when written in the form (15.17), though possibly with more terms, are positive, and provided that the matrix of the corresponding linear system is symmetric and positive definite. The ordering according to which the approximate values are modified can be arbitrary.

## 15.4 Peaceman-Rachford Iterative Method

Whereas the Gauss-Seidel and the successive-overrelaxation iterative methods involve at each step the modification of the approximate solution at one single point of $R_h$, the method that we shall consider in this section involves the simultaneous modification of the values at all points of a line of points. Such a line of points may be either a row of points or a column of points. The method was developed by Peaceman and Rachford[35] and is frequently known as the Peaceman-Rachford method. Each iteration consists of two half iterations: first, all the values on all the rows are modified; second, all the values on all the columns are modified. To illustrate, let us consider the five-point difference-equation representation of Laplace's equation for a region containing only regular interior mesh points. Our difference equation (15.13) becomes

$$u(x+h,\, y) + u(x-h,\, y) + u(x,\, y+h) + u(x,\, y-h) - 4u(x,y) = 0 \qquad (15.30)$$

As before, one selects an arbitrary function $u^{(0)}(x,y)$ defined at all points of $R_h$. In general, having determined $u^{(n)}(x,y)$, we then determine $u^{(n+\frac{1}{2})}(x,y)$ by a row iteration and $u^{(n+1)}(x,y)$ by a column iteration according to the following formulas:

$$\begin{aligned} u^{(n+1/2)}(x,y) &= u^{(n)}(x,y) \\ &\quad + r[u^{(n+1/2)}(x+h,\,y) + u^{(n+1/2)}(x-h,\,y) - 2u^{(n+1/2)}(x,y)] \\ &\quad + r[u^{(n)}(x,\,y+h) + u^{(n)}(x,\,y-h) - 2u^{(n)}(x,y)] \\ u^{(n+1)}(x,y) &= u^{(n+1/2)}(x,y) \\ &\quad + r[u^{(n+1/2)}(x+h,\,y) + u^{(n+1/2)}(x-h,\,y) - 2u^{(n+1/2)}(x,y)] \\ &\quad + r[u^{(n+1)}(x,\,y+h) + u^{(n+1)}(x,\,y-h) - 2u^{(n+1)}(x,y)] \end{aligned} \tag{15.31}$$

Here $r$ is a parameter that may depend on $n$ but must be the same for both halves of any iteration.

In order to determine $u^{(n+1/2)}(x,y)$ from the first of the above equations, it is necessary to solve, for each value of $y$, a system of $M$ simultaneous linear equations, where $M$ is the number of interior mesh points on the row. It is easy to see that the problem of doing this is the same as that of solving a system of $M$ linear algebraic equations of the form

$$\begin{aligned} B_1T_1 + C_1T_2 &= D_1 \\ A_iT_{i-1} + B_iT_i + C_iT_{i+1} &= D_i \qquad i = 2, 3, \ldots, M-1 \\ A_MT_{M-1} + B_MT_M &= D_M \end{aligned} \tag{15.32}$$

where the $T_i$ are unknown and where the $A_i$, $B_i$, $C_i$, and $D_i$ are given coefficients. Since the matrix of such a system is tridiagonal, i.e., has no nonzero elements except on the main diagonal and on the diagonals adjacent to the main diagonal, there is a very simple algorithm for solving the equations. In this algorithm, which is actually just the Gauss elimination method, the number of arithmetic operations is proportional to $M$, as opposed to approximately $\frac{1}{3}M^3$ operations as required with a general matrix. The algorithm appears to have been first used for solving elliptic partial differential equations by L. H. Thomas.[47] It was used first in connection with parabolic partial differential equations by Bruce, Peaceman, Rachford, and Rice.[6]

The following formulas are used in the determination of the $T_i$:

$$\begin{aligned} &b_1 = \frac{C_1}{B_1} \qquad b_i = \frac{C_i}{B_i - A_ib_{i-1}} \qquad i = 2, 3, \ldots, M-1 \\ &q_1 = \frac{D_1}{B_1} \qquad q_i = \frac{D_i - A_iq_{i-1}}{B_i - A_ib_{i-1}} \qquad i = 2, 3, \ldots, M \\ &T_M = q_M \qquad T_i = q_i - b_iT_{i+1} \qquad i = M-1, M-2, \ldots, 1 \end{aligned} \tag{15.33}$$

The numerical procedure defined by Eqs. (15.33) has been observed to be very satisfactory in a large number of cases, no large growth of rounding errors having been noted.

Peaceman and Rachford[35] and Wachspress and Habetler[52] have given methods for choosing iteration parameters. We first describe a procedure that is very similar to that given in Ref. 35. Let $M - 1$ denote the number of interior mesh points in the row or column of greatest length and let

$$\alpha = \sin^2 \frac{\pi}{2M} \tag{15.34}$$

We choose a set of $m$ different values of $r$ that will be used in a cyclic order: $r_1, r_2, \ldots, r_m, r_1, r_2, \ldots$. First, $m$ is chosen as the smallest positive integer such that

$$(\sqrt{2} - 1)^{2m} \leq \alpha \tag{15.35}$$

The $r_k$ are then determined by

$$r_k = \alpha^{(1-2k)/(2m)} \tag{15.36}$$

for $k = 1, 2, \ldots, m$. After each set of $m$ iterations, the error will be reduced by a factor

$$\theta = \left(\frac{1 - \alpha^{1/(2m)}}{1 + \alpha^{1/(2m)}}\right)^2 \tag{15.37}$$

As an example, let us consider the case of the unit square with $h = 1/20$. Since $M = 20$ we have, by Eq. (15.34),

$$\alpha = \sin^2 \frac{\pi}{40} = 0.00616$$

From inequality (15.35), we find that $m = 3$; hence by Eq. (15.36) we have

$$r_1 = 2.34 \qquad r_2 = 12.74 \qquad r_3 = 69.54$$

and, by Eq. (15.37),

$$\theta = 0.16$$

Thus with three double sweeps, the error will theoretically be reduced by a factor of 0.16. This may be compared with a factor of reduction of 0.73 for the same problem after one iteration of the successive-overrelaxation method and with 0.39 after three iterations. Because of the added complication involved in using the Peaceman-Rachford method and the fact that each iteration involves a double sweep, it would appear that the successive-overrelaxation method is slightly better for the particular problem under discussion. Actually, however, the reduction factor for the Peaceman-Rachford method is somewhat less than that indicated above. Moreover, for smaller values of $h$ the Peaceman-Rachford method is relatively better, the number of iterations being approximately proportional to $|\log h|^{-1}$ instead of to $h^{-1}$ as for the successive-overrelaxation method; see the discussion given in Ref. 35.

An alternative procedure for choosing the $r_k$, based on more detailed analysis, has been given by Wachspress and Habetler.[52] The method described below represents a slight variant of their procedure. Again letting $M - 1$ denote the largest number of interior mesh points in any row or column, one first computes a number $z$ from the relation

$$z = \sigma^{1/m-1}$$

where $\sigma = \frac{a}{b} \qquad a = 4 \sin^2 \frac{\pi}{2M} \qquad b = 4 \cos^2 \frac{\pi}{2M}$

The $r_k$ are then determined by

$$r_k = \frac{1}{bz^{k-1}} \qquad k = 1, 2, \ldots, m$$

According to the analysis given in Ref. 51, the factor of reduction of the error that is achieved after $m$ double sweeps is approximately

$$P_m = \left( \frac{1 - z^{1/2}}{1 + z^{1/2}} e^{-z^{3/2}/(1-z)} \right)^4$$

Since the *average* factor of reduction per double sweep is $S_m = (P_m)^{1/m}$, it seems reasonable to choose $m$ so as to minimize $S_m$. Because of the complicated formula for $S_m$ as a function of $m$, it is probably simplest to determine the optimum value of $m$ by computing $S_m$ for a number of different values of $m$.

For example, in the example considered above we have

$$a = 4 \sin^2 \frac{\pi}{40} = 0.024623 \qquad b = 4 \cos^2 \frac{\pi}{40} = 3.97538$$

$$\sigma = \frac{a}{b} = 0.0061940$$

By computing $S_m$ for various values of $m$, we obtain

$$S_6 = 0.3152 \qquad S_8 = 0.3060 \qquad S_9 = 0.3056 \qquad S_{10} = 0.3064$$

Other values of $m$ give larger values of $S_m$. Thus the optimum value for $S_m$ is taken as 0.3056, which is assumed for $m = 9$. The corresponding values of $r_k$ are

$$\begin{array}{lll} r_1 = 0.25155 & r_4 = 1.69291 & r_7 = 11.39320 \\ r_2 = 0.47492 & r_5 = 3.19623 & r_8 = 21.51045 \\ r_3 = 0.89666 & r_6 = 6.03450 & r_9 = 40.61191 \end{array}$$

The average factor of reduction per double sweep thus obtained, namely, 0.3056, compares favorably with 0.73 for the successive-over-relaxation method even though each iteration of the latter method involves only a single sweep.

As mentioned above, the number of iterations required to achieve a specified accuracy by using the Peaceman-Rachford method is approximately proportional to $|\log h|^{-1}$ as compared with $h^{-1}$ for the successive-overrelaxation method. Before one is tempted to discard the latter method entirely, however, there are several factors to be considered. In the first place, the result stated above has been proved only for the five-point difference-equation representation of Laplace's equation and for a rectangle. As a matter of fact, Birkhoff and Varga[5] showed that in the case of Laplace's equation the methods used in the theoretical analysis of the Peaceman-Rachford method are not applicable for regions other than rectangles. On the other hand, numerical experiences to date[52,60] indicate that the method works well for considerably more general problems. In the second place, for problems involving a moderately small mesh size the actual machine time required for using the successive-overrelaxation method may be less, either because the total number of sweeps is less or because the time required per sweep is less, even though the total number of sweeps might be slightly greater. Finally, the successive-overrelaxation method is much simpler both because of the arithmetic operations involved and because it is not necessary with the successive-overrelaxation method to obtain the approximate values first by rows and then by columns. This may be a considerable advantage if magnetic-tape storage must be used.

The theoretical analysis underlying the Peaceman-Rachford method has not yet been extended to include differential equations that are essentially different from Laplace's. Nevertheless, the method can be formally applied. One approach would be first to estimate $a$ and $b$ and then determine the $r_k$ as described above. After some experience with several cases involving a certain differential equation, improved estimates can probably be obtained.

## 15.5 Other Iterative Methods for Solving Elliptic Equations

It is probable that at the present time one would choose either the successive-overrelaxation method or the Peaceman-Rachford method when solving an elliptic partial differential equation by finite-difference methods; nevertheless, it is well to be aware of the existence of other methods. One of these other methods may actually be best suited for a given problem. Moreover, it may well happen that such a method, or a slight modification thereof, may turn out to be generally superior to either of the two methods that are now more frequently used.

A method of L. F. Richardson,[36] which dates back to 1910, can probably best be described in terms of its application to a linear system

$$A\mathbf{u} + \mathbf{d} = 0 \tag{15.38}$$

where $\mathbf{u}$ is an unknown vector, or column matrix, $\mathbf{d}$ is a given column matrix, and $A$ is a given square matrix that is positive definite and symmetric. The difference-equation representation described in Sec. 15.2 of a self-adjoint elliptic equation will have a positive-definite symmetric matrix if every interior mesh point is regular. Starting with an arbitrary initial approximation $\mathbf{u}^{(0)}$ to the exact solution, one computes $\mathbf{u}^{(1)}$, $\mathbf{u}^{(2)}$, . . . according to the formula

$$\mathbf{u}^{(n+1)} = \mathbf{u}^{(n)} + \beta_{n+1}(A\mathbf{u}^{(n)} + \mathbf{d}) \qquad n = 0, 1, 2, \ldots \tag{15.39}$$

Here the $\beta_n$ are constants, and their selection determines the rapidity of convergence. Normally, one chooses a set of $m$ different values of $\beta$ and employs them in a cyclic order, $\beta_1, \beta_2, \ldots, \beta_m, \beta_1, \beta_2, \ldots$. If we let $\mathbf{e}^{(n)}$ denote the error after $n$ iterations as defined by

$$\mathbf{e}^{(n)} = \mathbf{u}^{(n)} - \mathbf{u}$$

then, since $A\mathbf{u} + \mathbf{d} = 0$, we have, by Eq. (15.39),

$$\mathbf{e}^{(n+1)} = (I + \beta_{n+1}A)\mathbf{e}^{(n)} \qquad \text{and} \qquad \mathbf{e}^{(m)} = \left[\prod_{n=1}^{m} (I + \beta_n A)\right] \mathbf{e}^{(0)}$$

As described by Shortley[41] and by Young,[58] one chooses the coefficients $\beta_n$ so as to minimize the maximum absolute value of the polynomial

$$P_m(\lambda) = \prod_{n=1}^{m} (1 + \beta_n \lambda)$$

over the range $a \leqq \lambda \leqq b$, where $a$ and $b$ are positive lower and upper bounds, respectively, for the eigenvalues of $A$. The polynomial with the smallest maximum absolute value is found by using methods involving Chebyshev polynomials. It can be shown that, when Richardson's method is applied to Laplace's equation, the number of iterations varies as $h^{-1}$, as the mesh size $h$ tends to zero. The number of iterations is thus of the same order of magnitude as for the successive-overrelaxation method; however, since the latter method converges at least twice as fast as Richardson's method and since it is simpler and requires less storage, its use is recommended in preference to Richardson's method. Moreover, it has been found by Young and Warlick[63] that with Richardson's method there is serious difficulty in controlling rounding errors even when great care is taken in selecting the order in which the $\beta_n$ are to be used. This difficulty may be at least partially overcome by the use, in place of Eq. (15.39), of a formula involving $\mathbf{u}^{(n-1)}$ as well as $\mathbf{u}^{(n)}$ and $\mathbf{u}^{(n+1)}$, as described in Ref. 46. This formula is based on the use of a three-term recurrence relationship for Chebyshev polynomials. Even with this improvement, however, Richardson's method appears inferior to the successive-overrelaxation method for solving elliptic difference equations, particularly since the convergence is slower and since the method is more complicated and requires considerably more storage. [Quite recently, though, Golub and Varga (in "Chebyshev Semi-iterative Methods, Successive Overrelaxation Iterative Methods, and Second Order Richardson Iterative Methods," Report 1028, Case Institute of Technology, Cleveland, Ohio, February, 1960) have modified the method so that it converges more rapidly than the successive-overrelaxation method. The modified method closely resembles the successive-overrelaxation method.]

Sheldon[40] developed a method that combined the ideas used in Aitken's "to-and-fro" method,[1] the successive-overrelaxation method, and Richardson's method. The to-and-fro method as applied to the successive-overrelaxation procedure involves improving the approximate values according to a given ordering on the first half of an iteration and then proceeding in exactly the reverse order for the second half of the iteration. This procedure has the feature of making the eigenvalues of the associated linear transformation real, whereas in the ordinary successive-overrelaxation method the eigenvalues are complex. By having real eigenvalues, we can then combine successive iterants according to a procedure based on the use of Chebyshev polynomials, as in Richardson's method, in order to increase the rate of convergence still further. Unfortunately, however, since no definite results are available concerning the spectral radius of the successive-overrelaxation method under the to-and-fro ordering, it is difficult to estimate the effectiveness of the method. Until this is possible with some degree of confidence, it appears doubtful that Sheldon's method will receive general usage, particularly in view of the complications and extra storage requirements involved.

Arms and Gates,[3] with Zondek, and Friedman[18] considered the use of "line relaxation," by which at each step, instead of modifying the approximate value of $u$ at a single point, one modifies simultaneously the values on an entire line of points. The procedure for modifying the values in a line of points is the same as that described in the preceding section. With line relaxation, however, one treats the lines—which are usually either rows or columns—successively, new values being used as soon as available. By the introduction of overrelaxation, it is possible to obtain for the Dirichlet problem a rate of convergence that is faster by a factor of approximately $\sqrt{2}$ than the (point) successive-overrelaxation method. Moreover, as shown recently by Cuthill and Varga,[11] in many cases it is possible to perform the line relaxation in the *same* number of arithmetic operations per iteration as required by the successive-overrelaxation method. While the line-relaxation method is somewhat more complicated, nevertheless, if a number of cases are to be solved for the same problem, then line relaxation should be considered as a means of reducing machine time by nearly 30 per cent.

The method of Douglas and Rachford[14] is rather similar to the Peaceman-Rachford method described in the previous section. It has the advantage of applying to three-dimensional problems as well as to two-dimensional problems. On the other hand, for two-dimensional problems in which a direct comparison is possible, the Peaceman-Rachford method converges considerably more rapidly. It is thus not surprising that the Douglas-Rachford method has so far not been used as much. This situation may change, however, with the development of larger and faster computers capable of solving three-dimensional problems.

The discussions of the various iterative methods considered in this section have necessarily been brief. The objective has been to provide sufficient information concerning each method so that the reader can decide whether the method might be useful to his particular problem. More of the details involved in using and in estimating the effectiveness of the methods can be found in the appropriate references.

## 15.6 Parabolic Equations—Forward-difference Method

The parabolic partial differential equations that we shall consider in this section and in the next two sections are of the form

$$\frac{\partial U}{\partial t} = L(U) \tag{15.40}$$

where $L(U)$ is a second-order elliptic partial differential expression with either one or two "space" variables. With one space variable, $L(U)$ would have the form

$$L(U) = AU_{xx} + \phi(x,t,U,U_x) \tag{15.41}$$

whereas with two space variables it would have the form

$$L(U) = AU_{xx} + CU_{yy} + \psi(x,y,t,U,U_x,U_y) \tag{15.42}$$

Here $A$ and $C$ are functions of the space variables such that $A$ and $C$ are positive in the domain of the space variables under consideration. The functions $\phi$ and $\psi$ are assumed to be analytic in their respective arguments.

A typical problem involving Eq. (15.40) would be the following: Let $R$ be a region in the domain of the space variables and let $S$ be the boundary of $R$. We wish to find a function $U(\mathbf{x},t)$ that satisfies Eq. (15.40) for $t > t_0$ and for $\mathbf{x}$ in $R$ such that $U$ equals a prescribed function $g$ on $S$ for $t > t_0$. Here $\mathbf{x} = x$ in the case of one space variable and $\mathbf{x} = (x,y)$ in the case of two space variables.

The problem just formulated is known as an *initial-value problem.* Since, as in the case of boundary-value problems for elliptic equations, analytic methods are seldom effective as a means for obtaining numerical solutions, approximate methods are usually employed. As before, the method of finite differences is probably the best such method. In order to apply the method of finite differences, one superimposes a network over the region $R + S$ in the same manner as described in Sec. 15.2. In addition, a time-mesh size $k$ is selected. For each discrete value of $t$, one desires to determine approximate values of $U$ for all mesh points in $R$.

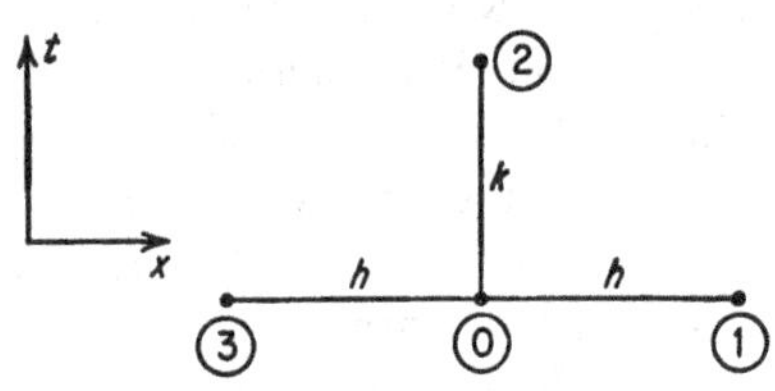

**Fig. 15.4** Forward-difference method: typical neighborhood configuration.

The *forward-difference method* (see Fig. 15.4) for determining approximate values of $U$ involves the replacement of the time derivative in Eq. (15.40) by the forward-difference quotient

$$\frac{\partial U}{\partial t} \doteq \frac{U(\mathbf{x}, t + k) - U(\mathbf{x},t)}{k}$$

The differential expression $L(U)$, on the right-hand side of Eq. (15.40), is replaced by a difference expression $L^*(U)$, which is derived by means of the methods of Sec. 15.2. By letting $u(\mathbf{x},t)$ denote the solution of the difference equation and solving for $u(\mathbf{x}, t + k)$, we obtain

$$u(\mathbf{x}, t + k) = u(\mathbf{x}, t) + kL^*(u) \tag{15.43}$$

In the case of the differential equation

$$U_t = U_{xx} + \phi(x,t,U) \tag{15.44}$$

for example, the difference equation would be

$$u(x, t + k) = u(x,t) + r[u(x + h, t) + u(x - h, t) - 2u(x,t)] + k\phi[x,t,u(x,t)] \tag{15.45}$$

where $h$ is the mesh size for the variable $x$ and where

$$r = \frac{k}{h^2} \tag{15.46}$$

Whereas for elliptic equations the value of the solution at any one point depends on all the boundary values, in the case of the parabolic equations considered here the value of $U(\mathbf{x},t_1)$ depends only on the initial values and on the values of $g$ for $t_0 \leq t \leq t_1$. This means that it should be possible to solve the difference equation by a kind of "marching" process rather than being required, as in the case of elliptic equations, to determine the solution at all points in order to be able to evaluate it at any one point. In the case of the forward-difference method, the marching process is particularly easy to carry out. By knowing $U(\mathbf{x},t_0)$, one can determine $L^*(U)$ for $t_0$ and thus compute $U(\mathbf{x}, t_0 + k)$. In the same way one can compute $U(\mathbf{x}, t_0 + 2k)$, $U(\mathbf{x}, t_0 + 3k)$, etc. Unfortunately, as we shall presently see by considering a specific example, the usefulness of this very simple procedure is very limited because of the problem of *stability*. Without attempting to formulate a precise mathematical definition of the term, we shall simply state that a method that is not stable is unsuitable for numerical computation because of the rapid growth of rounding errors that may, after a few steps, cause all accuracy to be lost.

In order to obtain stability with the forward-difference method, one is frequently forced to use a value of $k$ that is extremely small relative to the space-mesh size. In such a case, the number of time steps necessary to reach a desired value $t_1$ starting from $t = t_0$ may be prohibitively large.

To illustrate, let us consider a special case of Eq. (15.44), namely, the diffusion equation

$$U_t = U_{xx} \tag{15.47}$$

in the region $0 < x < 1$, $t > 0$, subject to the boundary conditions

$$\begin{aligned} U(0,t) &= g_1(t) & t &\geq 0 \\ U(1,t) &= g_2(t) & t &\geq 0 \\ U(x,0) &= f(x) & 0 &< x < 1 \end{aligned} \tag{15.48}$$

The interval $0 \leq x \leq 1$ is divided into subintervals of length $h = M^{-1}$, where $M$ is an integer. The difference equation (15.45) becomes

$$u(x, t + k) = u(x,t) + r[u(x + h, t) + u(x - h, t) - 2u(x,t)]$$

By letting $u_{i,j} = u(ih,jk)$, $f_i = f(ih)$, $g_{1,j} = g_1(jk)$, etc., we have

$$u_{i,j+1} = r(u_{i+1,j} + u_{i-1,j}) + (1 - 2r)u_{i,j}$$

for $i = 1, 2, \ldots, M - 1$; $j = 0, 1, 2, \ldots$. The boundary conditions are

$$\begin{aligned} u_{i,0} &= f_i & i &= 1, 2, \ldots, M - 1 \\ u_{0,j} &= g_{1,j} & j &= 0, 1, 2, \ldots \\ u_{M,j} &= g_{2,j} & j &= 0, 1, 2, \ldots \end{aligned}$$

Now, letting $f(x) \equiv 1$, $g_1(t) \equiv g_2(t) \equiv 0$, and $M = 4$, for the case $r = \frac{1}{2}$ we get

$$\begin{aligned} u_{i,j+1} &= \tfrac{1}{2}(u_{i+1,j} + u_{i-1,j}) & i &= 1, 2, 3; j = 0, 1, 2, \ldots \\ u_{i,0} &= 1 & i &= 1, 2, 3 \\ u_{0,j} &= u_{4,j} = 0 & j &= 0, 1, 2, \ldots \end{aligned}$$

The numerical solution for $j = 0, 1, 2, \ldots, 5$ is given in Table 15.3(I).

**Table 15.3** Examples of the Use of the Forward-difference Method

I. $f(x) \equiv 1; r = \frac{1}{2}$

| $j \backslash i$ | 0 | 1 | 2 | 3 | 4 |
|---|---|---|---|---|---|
| 0 | 0 | 1.000 | 1.000 | 1.000 | 0 |
| 1 | 0 | 0.500 | 1.000 | 0.500 | 0 |
| 2 | 0 | 0.500 | $0.500(+\epsilon)$ | 0.500 | 0 |
| 3 | 0 | $0.250\left(+\frac{\epsilon}{2}\right)$ | 0.500 | $0.250\left(+\frac{\epsilon}{2}\right)$ | 0 |
| 4 | 0 | 0.250 | $0.250\left(+\frac{\epsilon}{2}\right)$ | 0.250 | 0 |
| 5 | 0 | $0.125\left(+\frac{\epsilon}{4}\right)$ | 0.250 | $0.125\left(+\frac{\epsilon}{4}\right)$ | 0 |

II. $f(x) \equiv 1; r = 1$

| $j \backslash i$ | 0 | 1 | 2 | 3 | 4 |
|---|---|---|---|---|---|
| 0 | 0 | 1 | 1 | 1 | 0 |
| 1 | 0 | 0 | $1(+\epsilon)$ | 0 | 0 |
| 2 | 0 | $1(+\epsilon)$ | $-1(-\epsilon)$ | $1(+\epsilon)$ | 0 |
| 3 | 0 | $-2(-2\epsilon)$ | $3(+3\epsilon)$ | $-2(-2\epsilon)$ | 0 |
| 4 | 0 | $5(+5\epsilon)$ | $-7(-7\epsilon)$ | $5(+5\epsilon)$ | 0 |
| 5 | 0 | $-12(-12\epsilon)$ | $17(+17\epsilon)$ | $-12(-12\epsilon)$ | 0 |

III. $f(x) \equiv \sin x; r = 1$

| $j \backslash i$ | 0 | 1 | 2 | 3 | 4 |
|---|---|---|---|---|---|
| 0 | 0 | 0.707 | 1.000 | 0.707 | 0 |
| 1 | 0 | 0.293 | 0.414 | 0.293 | 0 |
| 2 | 0 | 0.121 | 0.172 | 0.121 | 0 |
| 3 | 0 | 0.051 | 0.070 | 0.051 | 0 |
| 4 | 0 | 0.019 | 0.032 | 0.019 | 0 |
| 5 | 0 | 0.013 | 0.006 | 0.013 | 0 |
| 6 | 0 | −0.007 | 0.020 | −0.007 | 0 |

Ignoring for the moment the quantities in parentheses, one sees that even though the values obtained are not smooth, they at least appear to be reasonable. It is not difficult to believe that, as $h$ and $k$ approach zero, with the ratio $r = k/h^2$ being held fixed, the numbers obtained by this process would approach the exact solution of the differential equation. To investigate the stability of the process, we introduce an error of amount $\epsilon$ in $u_{2,2}$. Because of linearity, we can study the effect of the error separately as indicated in the table. We note that the numbers in the parentheses decrease in magnitude as $j$ increases. This is characteristic of a stable process.

Let us now consider the case where $r = 1$. The difference equation becomes

$$u_{i,j+1} = u_{i+1,j} + u_{i-1,j} - u_{i,j}$$

The results obtained by using this formula are given in Table 15.3(II). Since it is easy to show that the exact solution of the differential equation is always bounded between zero and unity, it is clear that the computed values are very much in error and that the error increases rapidly with $j$. We are not surprised to learn that the method is not convergent. Moreover, a stability analysis similar to that performed previously for the case $r = \frac{1}{2}$ indicates that the process is highly unstable.

If the initial values were defined by $f(x) \equiv \sin \pi x$, then for $r = 1$ the forward-difference method would be convergent though not stable. One would obtain the values indicated in Table 15.3(III). Here the values are at least respectable until one reaches $j = 5$, when, because of rounding errors, they begin to oscillate. If more decimal places had been retained in the calculations, then the occurrence of the oscillations would have been delayed. Finally, if exact values had been used throughout—i.e., if, for instance, $\sqrt{2}/2$ had been used in place of 0.707—then the oscillations would never have occurred. Of course, this latter consideration is largely academic except that it does illustrate the distinction between the concepts of stability and convergence.

Detailed discussions of convergence and stability for the forward-difference method and other methods of solving parabolic equations have been given by John,[23] Douglas,[13] O'Brien, Hyman, and Kaplan,[34] Lax and Richtmyer,[30] Leutert,[32] Todd,[48] Juncosa and Young,[24] and others. The forward-difference method as applied to the problem specified by Eqs. (15.47) and (15.48), with $g_1(t) \equiv g_2(t) \equiv 0$ and with $f(x)$ piecewise continuous, is convergent and stable provided that

$$r = \frac{k}{h^2} \leq \frac{1}{2}$$

For the corresponding differential equation with two space variables,

$$U_t = U_{xx} + U_{yy}$$

the condition on $r$ for stability is

$$r = \frac{k}{h^2} \leq \frac{1}{4}$$

where the mesh size in each space-coordinate direction is taken equal to $h$. In both cases, the use of a larger value of $r$ would make the process unstable. Consequently, a time-mesh size that is of the order of the *square* of the space-mesh size must be used. Thus the total computational effort involved in proceeding from $t_0$ to $t_1$, with space-mesh size $h$ and with time-mesh size $k = rh^2$, is proportional to $h^{-3}$.

Thus, while the forward-difference method is very simple and straightforward for use on a high-speed computer, stability considerations usually require the use of such a small time increment $k$ that the method is generally not practical. In the next two sections, we shall consider methods that, though less simple and though requiring considerably more arithmetic operations per time step, nevertheless more than compensate for this by making possible the use of a much larger time-step size.

## 15.7 The Crank-Nicolson Method

The difference equation used in the Crank-Nicolson method (see Fig. 15.5) for solving Eq. (15.40) is the same as the forward-difference method except that the differential expression $L(U)$ is replaced by

$$\tfrac{1}{2}[L^*(U) + L^{**}(U)]$$

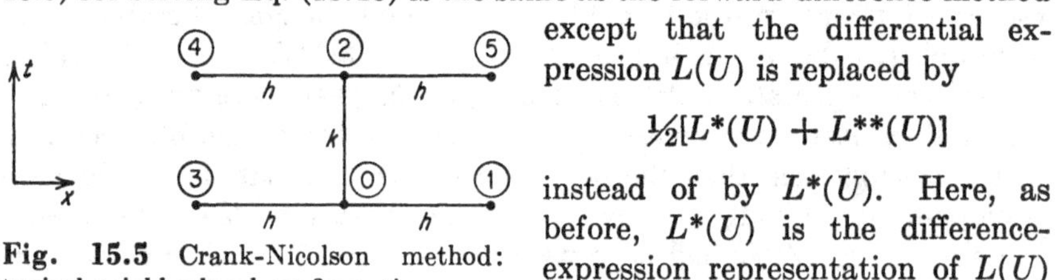


**Fig. 15.5** Crank-Nicolson method: typical neighborhood configuration.

instead of by $L^*(U)$. Here, as before, $L^*(U)$ is the difference-expression representation of $L(U)$ for the current value of $t$. The difference-expression representation of $L(U)$ for $t + k$ is represented by $L^{**}(U)$. The difference equation thus obtained can be written in the form

$$u(\mathbf{x}, t + k) = u(\mathbf{x},t) + \frac{r}{2}[L^*(u) + L^{**}(u)] \tag{15.49}$$

In the case of the differential equation (15.44), this becomes

$$\begin{aligned} u(x, t + k) = u(x,t) &+ \frac{r}{2}[u(x + h, t) + u(x - h, t) - 2u(x,t) \\ &+ u(x + h, t + k) + u(x - h, t + k) - 2u(x, t + k)] \\ &+ \frac{k}{2}\{\phi[x, t, u(x,t)] + \phi[x, t + k, u(x, t + k)]\} \end{aligned} \tag{15.50}$$

Evidently both Eq. (15.49) and Eq. (15.50) are *implicit* equations in the sense that values of $u$ for $t + k$ appear on both sides of the equations. Moreover, since in Eq. (15.50) the quantities $u(x, t + k)$, $u(x - h, t + k)$, and $u(x + h, t + k)$ appear, it is not possible to solve for any one of them explicitly. In order to proceed from one time step to another, it is thus necessary to solve a system of equations. If the function $\phi$ appearing in Eq. (15.50) is nonlinear in $u$, then the equations are nonlinear. For the problem defined by Eqs. (15.47) and (15.48), however, by using the notation $u_{i,j} = u(ih,jk)$, etc., we get

$$u_{i,j+1} = \frac{r}{2}(u_{i+1,j} + u_{i-1,j} + u_{i+1,j+1} + u_{i-1,j+1} - 2u_{i,j+1}) + (1 - r)u_{i,j}$$

$$i = 1, 2, \ldots, M - 1 \quad j = 0, 1, 2, \ldots \qquad (15.51)$$

One method of determining, for each $j$, the quantities $u_{1,j+1}$, $u_{2,j+1}$, $u_{3,j+1}$, . . . would be to use the successive-overrelaxation method. It is not difficult to show that the spectral radius of the Gauss-Seidel method for the system satisfies the inequality

$$\lambda \leq \left(\frac{r}{1 + r}\right)^2$$

from which the optimum value of $\omega$ can be determined by

$$\omega_b = 1 + \frac{\lambda}{(1 + \sqrt{1 - \lambda})^2}$$

This procedure was used by Young and Ehrlich.[61] Since each iteration of the successive-overrelaxation method on the average multiplies the magnitude of the error by $\omega_b - 1$, it is clear that, as $r$ increases, the rapidity of the convergence decreases. If $k$ is chosen proportional to $h$, so that $r$ increases as $h^{-1}$, then the number of iterations required varies as $h^{-1/2}$. Since the number of mesh points involved in proceeding from $t_0$ to $t_1$, with space-mesh size $h$ and with time-mesh size proportional to $h$, is proportional to $h^{-2}$, it follows that the total computational effort involved is proportional to $h^{-5/2}$, as compared with $h^{-3}$ for the forward-difference method.

Another method for solving Eq. (15.51) is to use the special procedure described in Sec. 15.4 for solving a linear system with a tridiagonal matrix. The specific formulas are

$$d_{i,j} = \frac{r}{2(1 + r)}(u_{i-1,j} + u_{i+1,j}) + \frac{1 - r}{1 + r}u_{i,j} \quad i = 1, 2, \ldots, M - 1$$

$$b_{0,j} = 0$$

$$b_{i,j} = -\frac{\dfrac{r}{2(1 + r)}}{1 + \dfrac{r}{2(1 + r)}b_{i-1,j}} \qquad i = 1, 2, \ldots, M - 2$$

$$q_{1,j} = d_{1,j}$$

$$q_{i,j} = \frac{d_{i,j} + \dfrac{r}{2(1 + r)}q_{i-1,j}}{1 + \dfrac{r}{2(1 + r)}b_{i-1,j}} \qquad i = 2, 3, \ldots, M - 1$$

$$u_{M-1,j+1} = q_{M-1,j}$$

$$u_{i,j+1} = q_{i,j} - b_{i,j}u_{i+1,j+1}$$

As an example, let us consider the numerical example given in the preceding section with $r = 1$. Table 15.4 shows the calculations involved in

**Table 15.4** Example of the Computations Involved in Using the Crank-Nicolson Method for $f(x) \equiv 1$, $r = 1$

| Item \ $i$ | 0 | 1 | 2 | 3 | 4 |
|---|---|---|---|---|---|
| $u_{i,0}$ | 0 | 1.000 | 1.000 | 1.000 | 0 |
| $d_{i,0}$ | ... | 0.250 | 0.500 | 0.250 | ... |
| $b_{i,0}$ | 0 | −0.250 | −0.266 | ..... | ... |
| $q_{i,0}$ | ... | 0.250 | 0.600 | 0.428 | ... |
| $u_{i,1}$ | 0 | 0.428 | 0.714 | 0.428 | 0 |

proceeding from $j = 0$ to $j = 1$. One can easily verify that the values obtained satisfy the Crank-Nicolson difference equation (15.51). Furthermore, as mentioned earlier, the process is stable and is otherwise satisfactory for use on a high-speed computer. Since the number of calculations per point is clearly independent of the number of points involved, it follows that the total computational effort involved in proceeding from $t_0$ to $t_1$ is proportional to $h^{-2}$, as compared with $h^{-5/2}$ with the successive-overrelaxation method and as compared with $h^{-3}$ with the forward-difference method.

In the above analysis we have assumed that it would be possible to let $k$ be proportional to $h$, rather than to $h^2$ as with the forward-difference method. For the problem specified by Eqs. (15.47) and (15.42), it is not difficult to show that the Crank-Nicolson method is stable for any positive $h$ and $k$; see for instance Ref. 34. For the case where $g_1(t) \equiv g_2(t) \equiv 0$ and where $f(x)$ is piecewise continuous, Juncosa and Young[25] have shown that, as $h$ approaches zero, the solutions obtained by the Crank-Nicolson method converge to the true solution of the differential equation provided that $k = O(h/|\log h|)$. The latter condition implies that $k$ approaches zero slightly faster than $h$. Actually, it is probable that the result could be proved for $k = O(h)$; some numerical experiments of Young and Ehrlich[61] suggest that this is the case. These experiments, together with theoretical results of Juncosa and Young,[24,25] suggest that the order of accuracy of the solution of the Crank-Nicolson difference equation is about the same as the order of the accuracy of the solution of the forward-difference method.

Let us now consider the differential equation (15.44), for which the Crank-Nicolson difference equation is given by Eq. (15.50). If $\phi$ is nonlinear in $u$, then one is faced with the problem of solving a system of nonlinear equations. One possibility would be to use the successive-overrelaxation method, wherein on any iteration the value of $u(x, t+k)$ from the previous iteration would be used to evaluate $\phi[x, t+k, u(x, t+k)]$. For the cases $\phi \equiv e^u$ and $\phi \equiv e^{-1/u}$, the experience of Young and Ehrlich[61] indicated no significant loss in convergence rate because of the nonlinear term. Probably a better procedure would be to iterate a fixed number of times by using the special procedure for solving a linear system with a tridiagonal matrix. The term $\phi[x, t+k, u(x, t+k)]$ would be evaluated from the previous iteration and treated as a known quantity. If, as seems reasonable, the use of a *fixed* number of iterations would suffice for all mesh sizes, then the total computational labor would still be proportional to $h^{-2}$ as for the linear case.

In applying the Crank-Nicolson method to problems involving two space variables, one would have to solve, at each time step, an elliptic boundary-value problem. For the case of the differential equation

$$U_t = U_{xx} + U_{yy}$$

however, we can get a fairly accurate estimate of the optimum relaxation factor to be used with the successive-overrelaxation method. It can be shown that the spectral norm $\lambda$ of the Gauss-Seidel method satisfies the inequality

$$\lambda \leq \left(\frac{2r}{1+2r}\right)^2$$

Consequently, if one chooses $k$ proportional to $h$, then the number of iterations would be proportional to $h^{-1/2}$, as compared with $h^{-1}$, as would be required if one were solving the equation

$$U_{xx} + U_{yy} = 0$$

In spite of this saving, it would certainly seem desirable to have a method that would permit one to proceed from $t$ to $t+k$ without the necessity of solving a system of equations and without the limitations of the forward-difference method.

## 15.8 The Alternating-direction Method for Parabolic Equations Involving Two Space Variables

Peaceman and Rachford[35] and Douglas[12] presented a method for solving the parabolic equation

$$U_t = U_{xx} + U_{yy} \tag{15.52}$$

The process of passing from one time step to the next is done in two parts, one implicit on rows and the other implicit on columns. The specific formulas are

$$u\left(x, y, t + \frac{k}{2}\right) = u(x,y,t) + r\left[u\left(x + h, y, t + \frac{k}{2}\right) + u\left(x - h, y, t + \frac{k}{2}\right) - 2u\left(x, y, t + \frac{k}{2}\right) + u(x, y + h, t) + u(x, y - h, t) - 2u(x,y,t)\right]$$

$$u(x, y, t + k) = u\left(x, y, t + \frac{k}{2}\right) + r\left[u\left(x + h, y, t + \frac{k}{2}\right) + u\left(x - h, y, t + \frac{k}{2}\right) - 2u\left(x, y, t + \frac{k}{2}\right) + u(x, y + h, t + k) + u(x, y - h, t + k) - 2u(x, y, t + k)\right] \quad (15.53)$$

where $r = k/2h^2$.

The formula (15.41) is very similar to those involved in performing an iteration by using the Peaceman-Rachford method for elliptic equations. As a matter of fact, the alternating-direction method for parabolic equations and the Peaceman-Rachford method for elliptic equations are very closely related, the latter method being suggested by the former.[12,35] The computations for both methods involve at each stage the solution of linear systems with tridiagonal matrices. After writing each system in the form (15.32), one uses Eqs. (15.33), first to obtain $u(x, y, t + k/2)$ for all $(x,y)$ and then to obtain $u(x, y, t + k)$.

It is shown in Ref. 12 that the local approximation of the difference equations (15.53) to the differential equation (15.52) is of the same order in $h$ and $k$ as that of the Crank-Nicolson difference equation. It is further shown that, for the case of the rectangle, the method is stable for any positive $h$ and $k$. On the other hand, either part of Eqs. (15.53) when used alone would not, in general, be stable. Although stability has so far been proved only for the case of the differential equation (15.52) for the rectangle, it may well hold for more general differential equations and for more general regions. Certainly, the method can be formally applied in more general situations.

It is clear that the method, where applicable, affords a substantial improvement over the Crank-Nicolson method. The amount of computation required per point is independent of $h$ and $k$, while for the Crank-Nicolson method, using the successive-overrelaxation method, the effort per point is proportional to $h^{-1/2}$, as stated in the previous section. Whenever the alternating-direction method is stable and whenever the time-mesh size $k$ can be taken proportional to $h$ without loss of accuracy, the method would appear to be superior. It may be very difficult, however, to estimate stability and accuracy in advance.

The method of Douglas and Rachford[14] is similar to the alternating-direction method; moreover, it can be extended to apply to three-dimensional problems. As previously stated in Sec. 15.5, the Douglas-Rachford method also leads to an effective iterative method for solving elliptic equations.

## 15.9 Illustrative Examples

Prior to the advent of modern high-speed computing machines, manual methods were used for solving elliptic and parabolic partial differential equations. Southwell[45] and his colleagues solved a number of problems by using relaxation methods, wherein the human computer was free to use ingenuity and intuition in order to speed up the convergence. Problems were solved in the following areas: torsion and shear stress for shafts with various cross sections; electrostatics; magnetism; conformal mapping; steady-state heat conduction; film lubrication; gas flow; and hydrodynamics, including flows where the boundaries are not initially known. Allen[2] also described the solution of a number of problems by relaxation methods. He considered membrane-vibration problems as well as plate-bending, compression, and stretching problems involving the biharmonic equation. Shortley, Weller, Darby, and Gamble[43] described the solution of problems in electrostatics and torsion. Instead of relaxation methods, systematic iteration methods were used, including the Gauss-Seidel method, block iteration, and an extrapolation procedure.

Many problems that involve the three space variables $x$, $y$, and $z$ were treated as two-dimensional problems by assuming that the functions involved are independent of $z$. The following examples are given of the mathematical formulations of typical problems:

*a.* The torsion problem for a shaft of uniform cross section involves the solution of the Poisson equation

$$\psi_{xx} + \psi_{yy} = -2$$

in a plane region subject to the requirement that $\psi = 0$ on the boundary of the region.

*b.* The steady-state temperature distribution in a homogeneous material of uniform cross section involves the solution of Laplace's equation

$$T_{xx} + T_{yy} = 0$$

in a region $R$. Frequently, the temperature distribution on the boundary is prescribed.

*c.* The problem of determining the current in a plane conducting sheet with resistivity $\rho$ involves the solution of the equation

$$(\rho\psi_x)_x + (\rho\psi_y)_y = 0$$

where the *current function* $\psi$ assumes prescribed values on the boundary. The currents in the $x$ and $y$ directions are $\psi_y$ and $-\psi_x$, respectively.

With the availability of high-speed computers, problems of the type just described could be handled very easily, once the necessary computer program had been prepared. A program for eliminating or reducing the effort required to prepare such programs is discussed in the next section. Our discussion for the remainder of this section will be devoted to describing work on three problems for which approximate numerical solutions have been obtained by use of large computers.

*a. Cavity-flow Problem.* An attempt to determine the axially symmetric flow past a disk was described by Young, Gates, Arms, and Eliezer.[62] In order to obtain a cavity of finite length behind the plate, a second artificial plate was introduced downstream, as shown in Fig. 15.6.

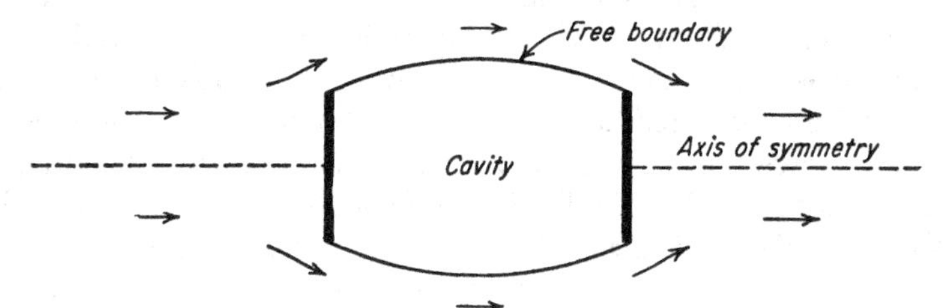


**Fig. 15.6** Cavity-flow problem.

This is the so-called *Riabouchinsky model.*[37] In the interior of the flow, the Stokes' stream function $\psi$ satisfies the equation

$$\psi_{xx} - \frac{1}{r}\,\psi_r + \psi_{rr} = 0 \tag{15.54}$$

where $x$ is the distance along the axis of symmetry and $r$ is the distance normal to the axis of symmetry. The stream function vanishes on the axis of symmetry, on the plates, and on the cavity boundary, the last of these being unknown. In addition, the stream function satisfies the conditions

$$\lim_{\substack{x^2+r^2\to\infty \\ r\neq 0}} \psi(x,r) = \tfrac{1}{2}r^2 \tag{15.55}$$

and, on the cavity boundary,

$$\frac{1}{r}\,\psi_n = \lambda$$

where $\psi_n$ denotes the inward normal derivative of $\psi$ and where $\lambda$ is a constant to be determined. The last condition, which states that the velocity along the cavity boundary must be constant, is derived from the

Bernoulli equation and from the requirement that the pressure in the cavity be constant.

In order to apply the method of finite differences, artificial boundaries $DC$ and $DE$ for the flow were introduced as shown in Fig. 15.7. The limiting values of $\psi$, as $x^2 + r^2 \to \infty$, were assumed on these artificial boundaries. By symmetry, only one-quarter of the flow was considered, the symmetry condition $\psi_x = 0$ being imposed on $BC$. A network was superimposed over the region. The spacing of the mesh points was made to vary so that the spacing became finer as one approached the separation point $A$. The basic procedure was as follows: The curve $AB$ and initial values for $\psi$ at all interior mesh points were assumed. By means of the successive-overrelaxation method, values of $\psi$ were determined at all

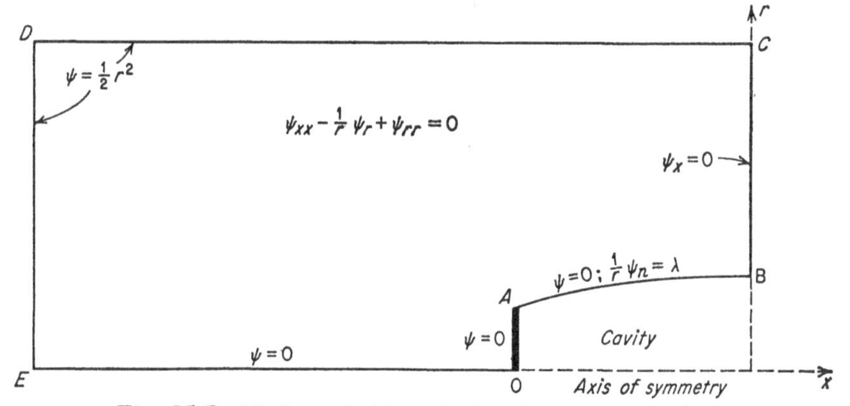


**Fig. 15.7** Mathematical formulation of cavity-flow problem.

interior mesh points. For this determination, only the condition $\psi = 0$ was imposed on $AB$. After the values of $\psi$ were determined, the next step was to compute approximate values for $r^{-1}\psi_n$. If these values were constant along $AB$, then the problem would be considered solved, with $\lambda$ being selected as the constant value. Otherwise, a new curve $AB$ would be assumed and the process repeated.

The attempt to solve the problem on the NORC computer at the U.S. Naval Proving Ground, Dahlgren, was only partially successful, largely because of difficulties in the neighborhood of the separation point $A$. Although a fine mesh was used close to this point, it was necessary to use an analytic approximation in the immediate neighborhood of $A$ based on the known two-dimensional solution. It was not possible to get the computed velocities constant to an accuracy of better than $\pm 1.5$ per cent of the average value. The fact that no systematic method was known for modifying the boundary added to the difficulties. Estimates are needed for the errors introduced by the imposition of the artificial boundaries, by the use of finite-difference methods, and by the use of approximate analytic methods at the stagnation point.

As for the numerical techniques employed, it should be mentioned that, once the curve $AB$ was specified, the mesh points and the distances to the boundary were automatically determined by the machine. There were approximately 1,150 interior mesh points. Unfortunately, the successive-overrelaxation method, while considerably faster than the Gauss-Seidel method, was much less effective than expected. In fact, since the use of the estimated best value of $\omega$ led to a divergent process, it was necessary to use a much smaller value of $\omega$ with the accompanying loss in convergence speed. It is believed that the difficulty was caused by the choice of the difference-equation representations of the differential equation for points connecting regions of a fine mesh and of a coarser mesh. When the value 1.4, which appeared to be optimal, was used for $\omega$, approximately 100 iterations were required to reduce the residuals to $10^{-5}$, starting with reasonable initial estimates. Since the NORC performed one iteration in approximately 5 sec, it was possible to treat one trial cavity boundary in less than 10 min.

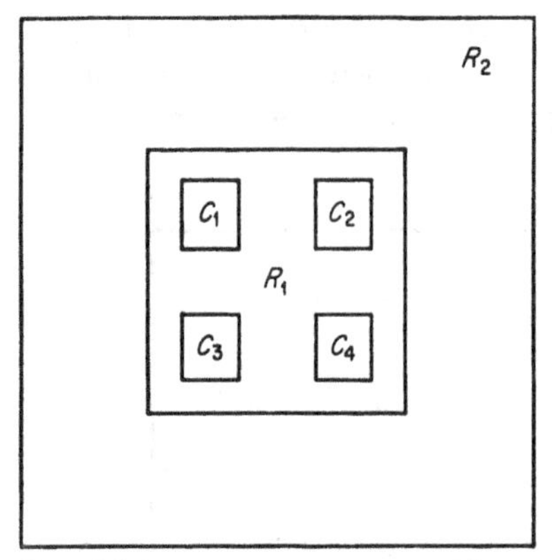

**Fig. 15.8** Nuclear-reactor problem.

*b. Nuclear-reactor Problems.* An important problem arising in the design of nuclear reactors is the determination of the neutron flux inside the reactor. Figure 15.8 shows a cross section of a very highly idealized reactor. The regions $R_1$ and $R_2$ are diffusion regions. The $C_i$, $i = 1, 2, 3, 4$, represent control-rod regions. For simplicity, we shall consider the two-group case discussed by Varga.[50] Here it is assumed that each neutron belongs either to the fast group or to the slow group. The functions $\phi(x,y)$ and $\psi(x,y)$ represent, respectively, the slow-neutron flux and the fast-neutron flux. The functions $\phi$ and $\psi$ satisfy the equations

$$(D_s\phi_x)_x + (D_s\phi_y)_y - \sigma_s\phi + S_s^f\psi = 0 \tag{15.56}$$

$$(D_f\psi_x)_x + (D_f\psi_y)_y - \sigma_f\psi + \eta S_{25}\phi = 0 \tag{15.57}$$

The coefficients $D_s$, $D_f$, $\sigma_s$, $\sigma_f$, $S_s^f$ and $S_{25}$ are assumed to be regionwise constant. The constant $\eta$ is to be determined. The last term in the first equation represents the degeneration of fast neutrons to slow neutrons; the last term in the second equation represents the creation of fast neutrons by the fission of slow neutrons.

On the outer boundary, both $\phi$ and $\psi$ vanish. Across interfaces between diffusion regions, the functions $\phi$, $\psi$, $D_s\nabla\phi$, and $D_f\nabla\psi$ are required to be continuous. On interfaces between diffusion regions and control-rod regions, both $\psi$ and $D_f\nabla\psi$ must be continuous. Although $\phi$ is not defined in a control-rod region, the condition

$$\frac{D_s}{\phi}\frac{\partial \phi}{\partial n} = c < 0$$

must be satisfied on the boundary of each such region.

In order to apply the method of finite differences, Varga[50] superimposed a network over the entire region, requiring that all interfaces lie on mesh lines. The difference equations are derived by using discrete representations of certain integrals obtained from the differential equations by means of Green's theorem. Such a procedure allows for the incorporation of the interface conditions in a natural way. The values of $\phi$ and $\psi$ at each mesh point are expressed as linear combinations of the values at the four neighboring mesh points.

Having determined two difference equations analogous to Eqs. (15.54) and (15.55), one carries out the following procedure. Starting with arbitrary initial approximations $\phi^{(0)}$, $\psi^{(0)}$, and $\eta_0$ for $\phi$, $\psi$, and $\eta$, respectively, one then solves the first equation for $\phi^{(1)}$. Here the first equation is treated as an elliptic boundary-value problem; the successive-overrelaxation method of iteration is used. Next, by means of the value of $\phi^{(1)}$ just obtained, the second equation is solved for $\psi^{(1)}$. A new value of $\eta$ is determined from

$$\eta_1 = \eta_0 \frac{(\phi_1,\phi_0)}{(\phi_1,\phi_1)}$$

where, in general, the inner product $(f,g)$ of two functions denotes the sum of the products $fg$ taken over all mesh points. The entire process is repeated starting with $\phi^{(1)}$, $\psi^{(1)}$, and $\eta_1$, and is continued until convergence is achieved.

The entire process of determining $\eta_{i+1}$ from $\eta_i$ is called an *outer iteration.* Each outer iteration consists of two sets of *inner iterations*, the inner iterations being merely iterations of the successive-overrelaxation method.

Two similar codes have been prepared at the Westinghouse Bettis Plant for the IBM 704 computer. The first, which is known as the QED code and is described in Ref. 50, is restricted to two groups. The second, which is known as the PDQ code and is described in Ref. 4, is a substantially improved version of the QED code and handles problems involving up to four groups. In both programs, the coefficients of the difference equation are computed automatically. Next, upper and lower bounds for the optimum values of $\omega$ to be used for the $n$ sets of inner iterations are computed, $n$ being the number of groups. With the value of $\omega$ chosen for each group, the iteration process begins. The PDQ program is capable of handling from 1,250 to 6,500 mesh points, depending on the size of the computer-core storage. It is reported in Ref. 4 that a typical two-group, 2,500-point problem can be solved in less than one hour.

Wachspress[51] described the CURE program for solving similar problems. The CURE program, which has also been prepared for the IBM 704 computer, can handle problems involving up to three groups. As many as 644 mesh points can be taken. The Peaceman-Rachford method of iteration is used. Because of the relatively few mesh points involved, however, it is by no means clear that the use of the Peaceman-Rachford method is preferable to the use of the successive-overrelaxation method. On the other hand, even though the basic theory for the Peaceman-Rachford method has not been shown to hold except for Laplace's equation and a rectangle, experiences in which the CURE program was used indicate that the method converges as rapidly as predicted, even for much more general cases.

*c. Thermal-ignition Problem.* Hicks, Kelso, and Davis[22] described the mathematical formulation of a thermal-ignition problem. An explosive material is placed in contact with a hot gas. Soon the temperature begins to rise very rapidly. The objective is to determine the time required for the maximum temperature to reach a certain value, known as the *ignition temperature.* A highly simplified version of the problem involves the solution of the differential equation

$$U_t = U_{xx} + e^{-1/U} \qquad 0 < x \quad t > 0$$

in the region $0 < x$, subject to the initial condition

$$U(x,0) = U_0 \qquad 0 < x$$

and to the boundary conditions

$$\begin{aligned} U_x(0,t) &= -a[b - U(0,t)] && t \geq 0 \\ \lim_{x\to\infty} U_x(x,t) &= 0 && t > 0 \end{aligned}$$

Here units have been chosen to make the coefficients of $U_{xx}$ and of $e^{-1/U}$ in the differential equation unity. Thus $U$ represents a "reduced temperature." The constants $a$, $b$, and $U_0$ represent, respectively, the reduced heat-transfer coefficient, the reduced gas temperature, and the reduced initial temperature.

In order that the problem could be attacked numerically, the second boundary condition was replaced by

$$U_x(A,t) = 0 \qquad t > 0$$

where $A$ is an appropriately chosen number. Young and Ehrlich[61] computed an approximate solution on the ORDVAC computer at the Aberdeen Proving Ground, Maryland, by using the Crank-Nicolson method as described in Sec. 15.7. In proceeding from $t$ to $t + k$, however, the function $e^{-1/U}$ was always evaluated at $t$. In this way, it was sufficient to solve a linear system with a tridiagonal matrix. Greater accuracy would undoubtedly have been obtained if the average value of $e^{-1/U}$ for $t$ and $t + k$ had been used. This improvement could probably have been accomplished with one or two iterations of the procedure that was actually used.

Other examples of elliptic and parabolic partial differential equations include a weather-prediction problem described by Charney and Phillips[7] and a thermal-melting problem described by Ehrlich.[15] In the latter problem, the material was assumed to melt upon reaching a given temperature. The molten material then remained in the system but with different thermal properties. One of the objectives of the computation was to determine the rate at which the melting occurred. In proceeding from one value of $t$ to the next, it was necessary to determine the amount of melting as well as the changes in temperature.

With the availability of machines having very large storage capacities, it is expected that the solution of problems involving three space variables will be possible. This should greatly improve the accuracy obtainable and hence the value of the computations for scientific and engineering purposes.

## 15.10 The SPADE Project for the Development of a Computer Program for Solving Elliptic and Parabolic Equations

One of the principal obstacles encountered when one attempts to solve a partial differential equation by using a high-speed computer is the time and effort required to prepare an appropriate computer program. Each of the programs described in the preceding section required several months to prepare. It would thus have been very useful to have had available a comprehensive program, or a set of programs, capable of handling, with a minimum amount of input information, any one of a large class of problems.

The preparation of such a "library" of computer programs for the IBM 704 and 709 computers was the objective of the SPADE Committee.† The programs of the library were to be capable of performing

† The SPADE Committee was made up of representatives of several computing installations, primarily on the West Coast. The organizations represented included the RAND Corporation, Space Technology Laboratories, Lockheed Aircraft Corporation, North American Aviation, AiResearch, General Electric Company, International Business Machines Corporation, California Research Corporation, and The University of Texas.

the arithmetic, logical, and bookkeeping operations involved in the solution of the class of partial differential equations by finite-difference methods. While the user still would be fully responsible for the numerical analysis, including choice of mesh size and difference equation, as well as the stability, convergence, and accuracy, the programs of SPADE would relieve him of much of the programming drudgery. The user would specify the region, mesh size, boundary conditions, differential equations, iterative method, etc. He also would prepare a small coordinating program to connect those routines of the library that are involved. Alternatively, there would be complete program "packages" available for handling frequently used classes of problems.

The problems included within the scope of the SPADE programs would involve systems of partial differential equations. There would be two independent space variables and, possibly, one independent "timelike" variable. Each equation would have the form

$$a_i \frac{\partial U_i}{\partial t} = L_i(U^{(1)},U^{(2)}, \ldots ,U^{(m)}) \qquad i = 1, 2, \ldots , m$$

where the $U^{(1)}$, $U^{(2)}$, . . . are the dependent variables, $t$ is the timelike variable, and the $L_i$ are differential expressions independent of $t$, such that the system

$$L_i(U^{(1)},U^{(2)}, \ldots ,U^{(m)}) = 0 \qquad i = 1, 2, \ldots , m$$

is elliptic. The coefficients $a_i$ may be either zero or one. It is assumed that there exist suitable difference representations for the $L_i$ of the form

$$U_0^{(k)} = \sum_{k=1}^{m} \sum_{j=1}^{p} \alpha_{k,j} U_j^{(k)} + \beta^{(k)}$$

Here $U_i^{(k)}$ designates the value of $U^{(k)}$ at the mesh point $P_i$, and $p$ is the number of neighboring points of the point $P_0$ that is under consideration. The coefficients $\alpha_{k,j}$ and $\beta^{(k)}$ may depend on $U_j^{(1)}$, $U_j^{(2)}$, . . . , $U_j^{(m)}$, $j = 1, 2, \ldots , p$, as well as on $P_0$. It may be well to emphasize here that the routines of SPADE are designed to handle the operations that are needed to solve problems of the kind now discussed, on the assumption that the usual methods apply. The responsibility for determining whether or not this is the case, however, would rest entirely with the user of the library.

The boundary conditions are of the form

$$\sum_{k=1}^{p} \left( \alpha_k^{(\nu)} U^{(k)} + \beta_k^{(\nu)} \frac{\partial U^{(k)}}{\partial n} \right) = C^{(\nu)} \qquad \nu = 1, 2, \ldots , m$$

where the coefficients $\alpha_k^{(\nu)}$, $\beta_k^{(\nu)}$, and $C^{(\nu)}$ may depend on $U^{(1)}, U^{(2)}, \ldots, U^{(m)}$, as well as on the independent variables. In addition to the boundary of the region, there may be one or more interfaces in the region, each with appropriate continuity conditions for the dependent variables and for certain of the derivatives normal to such interfaces.

The SPADE programs have been divided into three classes, or "phases." In Phase I, the coordinates and other appropriate items of information are computed for each mesh point. The input to Phase I consists of specifications of each boundary and interface as a series of *paths*. A *path* may be specified either as the arc of a conic section the equation of which is prescribed as input data, or as a series of points. In the latter case, the machine treats consecutive points as though they were joined by line segments. In addition, the mesh sizes are specified. While each mesh line is required to be parallel to a coordinate axis, the spacing need not be uniform. Furthermore, subdivisions of the mesh by factors of powers of two in certain areas are allowed. At the completion of Phase I, the information for each boundary and interior mesh point is stored on magnetic tape as a *point record*.

In Phase II, each point record is augmented to include the coefficients involved in the difference equation at the point. The routines to be prepared first require that algebraic formulas for the coefficients be provided in terms of the computed distances from the point to its neighbors and in terms of the coefficients of the differential equations. The latter would be available as subroutines; hence, the numerical coefficients could be computed automatically. More sophisticated routines might actually allow the machine to derive the difference equation automatically.

In Phase III, the difference equations would be solved by an iterative process. In the case of a time-dependent problem, the result would be merely the completion of a single time step followed by a return to Phase II for a recalculation of the coefficients. The same would be true in the nonlinear case in which the coefficients, as functions of the dependent variables, would be recomputed every few iterations. For the time-independent linear case, however, it is clear that, since the system of linear algebraic equations is completely specified by the information on magnetic tape, practically any iterative method could in principle be used, including the successive-overrelaxation method, the Peaceman-Rachford method, or other methods mentioned in Sec. 15.5.

## REFERENCES

1. Aitken, A. C., On the Iterative Solution of a System of Linear Equations, *Proc. Roy. Soc. Edinburgh, Sect. A*, vol. 63, pp. 52–60, 1950.
2. Allen, D. N. de G., "Relaxation Methods," McGraw-Hill Book Company, Inc., New York, 1954.

3. Arms, R. J., and L. D. Gates, Jr., "The Computation of an Axially Symmetric Free Boundary Problem on NORC, Part II," U.S. Naval Proving Ground Report No. 1533, Dahlgren, Va., Apr. 5, 1957.
4. Bilodeau, G. C., W. R. Cadwell, J. P. Dorsey, J. G. Fairey, R. S. Varga, "PDQ—An IBM-704 Code to Solve the Two-dimensional Few Group Neutron Diffusion Equations," Report WAPD-TM-70, Bettis Plant, Westinghouse Electric Corporation, Pittsburgh, Pa., August, 1957. Copies available from the Office of Technical Services, U.S. Department of Commerce, Washington, D.C.
5. Birkhoff, Garrett, and Richard S. Varga, Implicit Alternating Direction Methods, *Trans. Amer. Math. Soc.*, vol. 92, pp. 13–24, 1959.
6. Bruce, G. H., D. W. Peaceman, H. H. Rachford, Jr., and J. D. Rice, Calculation of Unsteady-state Gas Flow through Porous Media, *Petroleum Trans., A.I.M.E.*, vol. 198, pp. 79–92, 1953.
7. Charney, J. G., and N. A. Phillips, Numerical Integration of the Quasi-geostrophic Equations for Barotropic and Simple Baroclinic Flows, *J. Meteorol.*, vol. 10, pp. 71–99, 1953.
8. Courant, Richard, Hyperbolic Partial Differential Equations and Applications, chap. 5 in "Modern Mathematics for the Engineer," First Series, edited by E. F. Beckenbach, McGraw-Hill Book Company, Inc., New York, 1956.
9. ———, K. Friedrichs, and H. Lewy, Über die partiellen Differenzengleichungen der mathematischen Physik, *Math. Ann.*, vol. 100, pp. 32–74, 1928.
10. Crank, J., and P. Nicolson, A Practical Method for Numerical Evaluation of Solutions of Partial Differential Equations of the Heat-conduction Type, *Proc. Cambridge Philos. Soc.*, vol. 43, pp. 50–67, 1947.
11. Cuthill, Elizabeth H., and Richard S. Varga, A Method of Normalized Block Iteration, *J. Assoc. Comput. Mach.*, vol. 6, pp. 236–244, 1959.
12. Douglas, Jim, Jr., On the Numerical Integration of $\partial^2u/\partial x^2 + \partial^2u/\partial y^2 = \partial u/\partial t$ by Implicit Methods, *J. Soc. Indust. Appl. Math.*, vol. 3, pp. 42–65, 1955.
13. ———, On the Relation between Stability and Convergence in the Numerical Solution of Linear Parabolic and Hyperbolic Differential Equations, *J. Soc. Indust. Appl. Math.*, vol. 4, pp. 20–37, 1956.
14. ——— and H. Rachford, On the Numerical Solution of Heat Conduction Problems in Two and Three Space Variables, *Trans. Amer. Math. Soc.*, vol. 82, pp. 421–439, 1956.
15. Ehrlich, L. W., A Numerical Method of Solving a Heat Flow Problem with Moving Boundary, *J. Assoc. Comput. Mach.*, vol. 5, pp. 161–176, 1958.
16. Forsythe, George E., What Are Relaxation Methods? chap. 17 in "Modern Mathematics for the Engineer," First Series, edited by E. F. Beckenbach, McGraw-Hill Book Company, Inc., New York, 1956.
17. Frankel, S., Convergence Rates of Iterative Treatments of Partial Differential Equations, *Math. Tables Aids Comput.*, vol. 4, pp. 65–75, 1950.
18. Friedman, Bernard, "The Iterative Solution of Elliptic Difference Equations," A.E.C. Research and Development Report NYO-7698, Institute of Mathematical Sciences, New York University, June 1, 1957.
19. Garabedian, P. R., Estimation of the Relaxation Factor for Small Mesh Size, *Math. Tables Aids Comput.*, vol. 10, pp. 183–185, 1956.
20. Geiringer, H., On the Solution of Systems of Linear Equations by Certain Iteration Methods, pp. 365–393 in "Reissner Anniversary Volume, Contributions to Applied Mechanics," J. W. Edwards, Publisher, Inc., Ann Arbor, Mich., 1949.
21. Gerschgorin, S., Fehlerabschätzung für das Differenzenverfahren zur Lösung partieller Differentialgleichungen, *Z. Angew. Math. Mech.*, vol. 10, pp. 373–382, 1930.

22. Hicks, Bruce L., J. W. Kelso, and Julian Davis, "Mathematical Theory of the Ignition Process Considered as a Thermal Reaction," Ballistic Research Laboratories Report No. 756, Aberdeen Proving Ground, Md., 1957.
23. John, F., On Integration of Parabolic Equations by Difference Methods, *Comm. Pure Appl. Math.*, vol. 5, pp. 155–211, 1952.
24. Juncosa, M. L., and David Young, On the Order of Convergence of Solutions of a Difference Equation to a Solution of the Diffusion Equation, *J. Soc. Indust. Appl. Math.*, vol. 1, pp. 111–135, 1953.
25. ——— and ———, On the Crank-Nicolson Procedure for Solving Parabolic Partial Differential Equations, *Proc. Cambridge Philos. Soc.*, vol. 53, part 2, pp. 448–461, 1957.
26. Kahan, W., "The Rate of Convergence of the Extrapolated Gauss-Seidel Iteration," abstract of paper presented at the Conference on Matrix Computations, Wayne State University, Detroit, Mich., Sept. 4, 1957.
27. Laasonen, P., On the Degree of Convergence of Discrete Approximations for the Solutions of the Dirichlet Problem, *Ann. Acad. Sci. Fenn., Ser. A. I.*, vol. 246, pp. 1–19, 1957.
28. ———, On the Solution of Poisson's Difference Equation, *J. Assoc. Comput. Mach.*, vol. 5, pp. 370–382, 1958.
29. Lanczos, C., Solution of Systems of Linear Equations by Minimized Iterations, *J. Res. Nat. Bur. Standards*, vol. 49, pp. 33–53, 1952.
30. Lax, P. D., and R. D. Richtmyer, Survey of Stability of Linear Finite Difference Equations, *Comm. Pure Appl. Math.*, vol. 9, pp. 267–293, 1956.
31. Lehmer, Derrick H., High-speed Computing Devices and Their Applications, chap. 19 in "Modern Mathematics for the Engineer," First Series, edited by E. F. Beckenbach, McGraw-Hill Book Company, Inc., New York, 1956.
32. Leutert, W. W., On the Convergence of Unstable Approximate Solutions of the Heat Equation to the Exact Solution, *J. Math. Phys.*, vol. 30, pp. 245–251, 1952.
33. Liebmann, H., Die angenährte Ermittelung harmonischer Funktionen und konformer Abbildung, *Sitzungsberichte der mathematische-naturwissenschaftlichen Klasse der bayerischen Akademie der Wissenschaften zu München*, 1918, pp. 385–416.
34. O'Brien, G. G., M. A. Hyman, and S. Kaplan, A Study of the Numerical Solution of Partial Differential Equations, *J. Math. Phys.*, vol. 29, pp. 223–251, 1951.
35. Peaceman, D. W., and H. H. Rachford, Jr., The Numerical Solution of Parabolic and Elliptic Differential Equations, *J. Soc. Indust. Appl. Math.*, vol. 3, pp. 28–41, 1955.
36. Richardson, L. F., The Approximate Arithmetical Solution by Finite Differences of Physical Problems Involving Differential Equations, with Application to the Stresses in a Masonry Dam, *Philos. Trans. Roy. Soc. London, Ser. A.*, vol. 210, pp. 307–357, 1910.
37. Riabouchinsky, D., On Steady Fluid Motions with Free Surface, *Proc. London Math. Soc.*, vol. 19, pp. 206–215, 1920–1921.
38. Rosenbloom, P. C., "The Difference Equation Method for Solving the Dirichlet Problem," *Nat. Bur. Standards Applied Mathematics Series*, vol. 18, 1952
39. Schiffer, Menahem M., Boundary-value Problems in Elliptic Partial Differential Equations, chap. 6 in "Modern Mathematics for the Engineer," First Series, edited by E. F. Beckenbach, McGraw-Hill Book Company, Inc., New York, 1956.
40. Sheldon, J., On the Numerical Solution of Elliptic Difference Equations, *Math. Tables Aids Comput.*, vol. 9, pp. 101–112, 1955.
41. Shortley, G., Use of Tschebyscheff-polynomial Operators in the Numerical Solution of Boundary-value Problems, *J. Appl. Phys.*, vol. 24, pp. 392–396, 1953.

42. ——— and R. Weller, The Numerical Solution of Laplace's Equation, *J. Appl. Phys.*, vol. 9, pp. 334–344, 1938.
43. ———, ———, Paul Darby, and Edward H. Gamble, Numerical Solution of Axisymmetrical Problems with Applications to Electrostatics and Torsion, *J. Appl. Phys.*, vol. 18, pp. 116–129, 1947.
44. Sokolnikoff, Ivan S., The Elastostatic Boundary-value Problems, chap. 7 in "Modern Mathematics for the Engineer," First Series, edited by E. F. Beckenbach, McGraw-Hill Book Company, Inc., New York, 1956.
45. Southwell, R. V., "Relaxation Methods in Theoretical Physics," Oxford University Press, New York, 1946.
46. Stiefel, E., Recent Developments in Relaxation Techniques, in "Proceedings of the International Congress of Mathematicians," vol. 1, Amsterdam, 1954.
47. Thomas, L. H., Elliptic Problems in Linear Difference Equations over a Network, unpublished manuscript, Watson Scientific Computing Laboratory.
48. Todd, John, A Direct Approach to the Problem of Stability in the Numerical Solution of Partial Differential Equations, *Comm. Pure Appl. Math.*, vol. 9, pp. 597–612, 1956.
49. Tompkins, Charles B., Methods of Steep Descent, chap. 18 in "Modern Mathematics for the Engineer," First Series, edited by E. F. Beckenbach, McGraw-Hill Book Company, Inc., New York, 1956.
50. Varga, Richard S., "Numerical Solution of the Two-group Diffusion Equation in *x-y* Geometry," Report WAPD-159, Bettis Plant, Westinghouse Electric Corp., Pittsburgh, Pa., August, 1956.
51. Wachspress, E. L., "CURE: A Generalized Two-space-dimension Multigroup Coding for the IBM-704," Report KAPL-1724, Knolls Atomic Power Laboratory, General Electric Company, Schenectady, N.Y., Apr. 30, 1957.
52. ——— and G. J. Habetler, An Alternating Direction-implicit Iteration Technique, *J. Soc. Indust. Appl. Math.*, vol. 8, pp. 403–424, 1960.
53. Walsh, J. L., and David Young, On the Accuracy of the Numerical Solution of the Dirichlet Problem by Finite Differences, *J. Research Nat. Bur. Standards*, vol. 51, pp. 343–363, 1953.
54. ——— and ———, On the Degree of Convergence of Solutions of Difference Equations to the Solution of the Dirichlet Problem, *J. Math. Phys.*, vol. 33, pp. 80–93, 1954.
55. Wasow, W., On the Truncation Error in the Solution of Laplace's Equation by Finite Differences, *J. Res. Nat. Bur. Standards*, vol. 48, pp. 345–348, 1952.
56. ———, The Accuracy of Difference Approximations to Plane Dirichlet Problems with Piecewise Analytic Boundary Values, *Quart. Appl. Math.*, vol. 15, pp. 53–63, 1957.
57. Young, David, Iterative Methods for Solving Partial Difference Equations of Elliptic Type, *Trans. Amer. Math. Soc.*, vol. 76, pp. 92–111, 1954.
58. ———, On Richardson's Method for Solving Linear Systems with Positive Definite Matrices, *J. Math. Phys.*, vol. 32, pp. 254–255, 1954.
59. ———, Ordvac Solutions of the Dirichlet Problem, *J. Assoc. Comput. Mach.*, vol. 2, pp. 137–161, 1955.
60. ——— and Louis Ehrlich, Some Numerical Studies of Iterative Methods for Solving Elliptic Difference Equations, pp. 143–162 in "Boundary Problems in Differential Equations," University of Wisconsin Press, Madison, Wis., 1960.
61. ——— and ———, "On the Numerical Solution of Linear and Non-linear Parabolic Equations on the Ordvac," Interim Technical Report No. 18, Office of Ordnance Research Contract DA-36-034-ORD-1486, University of Maryland, February, 1956.

62. ———, L. D. Gates, Jr., R. J. Arms, and D. F. Eliezer, "The Computation of an Axially Symmetric Free Boundary Problem on NORC," U.S. Naval Proving Ground Report No. 1413, Dahlgren, Va., Dec. 16, 1955.
63. ——— and Charles H. Warlick, "On the Use of Richardson's Method for the Numerical Solution of Laplace's Equation on the ORDVAC," Ballistic Research Laboratories Memorandum Report No. 707, Aberdeen Proving Ground, Md., July, 1953.

# 16

# Circle, Sphere, Symmetrization, and Some Classical Physical Problems

GEORGE PÓLYA
PROFESSOR OF MATHEMATICS
STANFORD UNIVERSITY

## 16.1 Introduction

This chapter consists of three parts. The first part offers mainly plausible (intuitive, inductive, heuristic) considerations. Engineers and physicists use such considerations quite frequently; if they used them more explicitly, more consciously, they could use them, perhaps, more efficiently. And so, it is hoped even the tone of presentation of this first part may possibly do some good. The second part brings the main mathematical idea. The third part offers additional remarks.

### THE HEURISTIC ASPECT

## 16.2 Observations

The essential ideas with which this chapter is concerned are applicable to various problems of classical mathematical physics. Yet, for the sake of concreteness, we prefer to single out one typical problem, namely, that of the vibrating membrane. We shall discuss this particular problem most of the time, but we shall try to discuss it so that the essential ideas appear in a form readily applicable to other problems.

The tone of presentation in this first part is similar to that of Ref. 7, where also some of the points here treated are mentioned in a different context; cf. vol. 1, pages 168 to 171, and vol. 2, pages 9 to 12.

Let $\Lambda$ denote the principal frequency (which belongs to the characteristic tone of lowest pitch) of a uniform membrane stretched over the plane domain $D$. We shall consider various domains, but all the membranes we mention are supposed to have the same thickness, density, and elasticity, and also the same uniform tension when at rest. Therefore, $\Lambda$ depends only on the shape and size of the domain $D$, so that it is a functional of $D$; we postpone the formal definition of $\Lambda$ until Sec. 16.8.

The computation of the principal frequency $\Lambda$ for a given domain $D$ is a typical problem of mathematical physics. It is typical, for instance, that "exact" solutions are exceptional. There are only a few simple domains for which $\Lambda$ can be "exactly" or "explicitly" computed. These are the circle, an arbitrary section of the circle, an arbitrary rectangle, and just three triangles, namely, the triangles with angles 60°, 60°, 60°; 45°, 45°, 90°; and 30°, 60°, 90°. For other domains, such as an ellipse or an arbitrary triangle, the computation of $\Lambda$ is much more difficult: We have to deal with little-known transcendental functions or resort to approximations the error of which may be difficult to estimate.

To start or to check such approximations, it would be desirable to know more about the dependence of the frequency $\Lambda$ on the shape of the domain $D$. In order to examine this dependence, we consider, with a few additions, a table computed by Lord Rayleigh (Ref. 10, vol. 1, page 345 [211]). Table 16.1 exhibits the numerical value of the principal frequency for a dozen different shapes: the circle, four circular sectors (semicircle, quadrant, sextant, and octant—i.e., the sectors with opening 180°, 90°, 60°, and 45°, respectively), four rectangles [with sides in proportion 1:1 (square), 3:2, 2:1, 3:1], and the three triangles mentioned above. All these figures have the same area = 1; they are so arranged that the pitch increases as we read down.

**Table 16.1** Principal Frequencies of Membranes of Equal Area

| Figure | Frequency |
|---|---|
| Circle | 4.261 |
| Square | 4.443 |
| Quadrant | 4.551 |
| Sextant | 4.616 |
| Rectangle 3:2 | 4.624 |
| Octant | 4.755 |
| Equilateral triangle | 4.774 |
| Semicircle | 4.803 |
| Rectangle 2:1 | 4.967 |
| Isosceles right triangle | 4.967 |
| Triangle 30°, 60°, 90° | 5.157 |
| Rectangle 3:1 | 5.736 |

**Table 16.2** Perimeters of Figures of Equal Area

| Figure | Perimeter |
|---|---|
| Circle | 3.55 |
| Square | 4.00 |
| Quadrant | 4.03 |
| Rectangle 3:2 | 4.08 |
| Semicircle | 4.10 |
| Sextant | 4.21 |
| Rectangle 2:1 | 4.24 |
| Octant | 4.44 |
| Equilateral triangle | 4.56 |
| Rectangle 3:1 | 4.64 |
| Isosceles right triangle | 4.84 |
| Triangle 30°, 60°, 90° | 5.08 |

Table 16.1 offers a highly interesting indication. The reader should try to catch by himself the suggestion offered—this is a challenge to his faculty of observation. To strengthen that indication, we place Table 16.2 next to Table 16.1. Table 16.2 lists the same 12 figures of equal area as Table 16.1, but instead of the principal frequency $\Lambda$, it exhibits the length of the perimeter of each figure, and the figures are so ordered

that the perimeters increase as we read down. Therefore, the 12 figures are differently arranged in our two tables (but not too differently). Before passing to the next section, the reader should try to unravel by himself some of the mystery that the comparison of our two tables is about to reveal.

## 16.3 Conjectures

The main suggestion offered by Tables 16.1 and 16.2 arises from the fact that the circle tops both: Of the twelve figures considered, the circle has both the lowest principal frequency and the shortest perimeter. Now, as the reader probably knows (for full knowledge, he should also know a proof), the circle has the shortest perimeter not only among the twelve figures of equal area listed in Table 16.2 but among *all* figures of equal area. This geometrical minimum property of the circle and the resemblance between Tables 16.1 and 16.2 jointly suggest a physical minimum property of the circle: *Of all membranes of equal area, the circle has* not only the shortest perimeter but also *the lowest principal frequency.* This beautiful property of the circle was first stated by Lord Rayleigh as a conjecture (Ref. 10, vol. 1, page 339 [210]).

Yet the comparison of Tables 16.1 and 16.2 has still other suggestions to offer. Some of these indications become clearer if we collect from these tables *figures of the same kind* and rearrange them; see Table 16.3. (To the rectangles and triangles listed in Tables 16.1 and 16.2, Table 16.3

**Table 16.3** Principal Frequencies and Perimeters of Figures of Equal Area

| Figure | Principal frequency | Perimeter |
|---|---|---|
| Rectangles: | | |
| 1:1 (square) | 4.443 | 4.00 |
| 3:2 | 4.624 | 4.08 |
| 2:1 | 4.967 | 4.24 |
| 3:1 | 5.736 | 4.64 |
| $a:b \left(\frac{b}{a} \to 0\right)$ | $\pi\sqrt{\frac{a}{b}}$ | $2\sqrt{\frac{a}{b}}$ |
| Triangles: | | |
| 60°, 60°, 60° | 4.774 | 4.56 |
| 45°, 45°, 90° | 4.967 | 4.84 |
| 30°, 60°, 90° | 5.157 | 5.08 |
| 0°, 90°, 90° / 0°, 0°, 180° $\left(\frac{h}{a} \to 0\right)$ | $\pi\sqrt{\frac{a}{2h}}$ | $2\sqrt{\frac{2a}{h}}$ |
| Regular figures: | | |
| Circle | 4.261 | 3.55 |
| Square | 4.443 | 4.00 |
| Equilateral triangle | 4.774 | 4.56 |

adds their limiting cases: the "infinitely narrow" rectangle, with area 1 and sides $a$ and $b$, where $b/a$ tends to 0, and "infinitely narrow" triangles; a triangle is so called if, 1 being its area, $a$ its longest side, and $h$ its altitude perpendicular to $a$, the ratio $h/a$ tends to 0. The corresponding approximate, or "asymptotic," values of principal frequency and perimeter are given in Table 16.3.)

Observe that, within each of the three sets of figures displayed by Table 16.3, principal frequencies and perimeters *vary in the same direction.* Now, the study of the lengths of perimeters presents much easier and much more elementary problems than the study of principal frequencies. For perimeters, we can easily verify the observed regularities. For principal frequencies, the mathematical problems are much less accessible. Yet we can hope that the regularities observed remain valid beyond the narrow limits of the observational material collected in Table 16.3. And so we suspect, besides the great conjecture of Lord Rayleigh, further physical analogues of elementary geometrical facts:

*Of all triangles with a given area, the equilateral triangle has* not only the shortest perimeter, but also *the lowest principal frequency.*

*Of all quadrilaterals with a given area, the square has* not only the shortest perimeter, but also *the lowest principal frequency.*

*Consider the regular polygons with a given area. As the number of sides increases,* not only the perimeter but also *the principal frequency steadily decreases.*

## 16.4 A Line of Inquiry

The three foregoing statements are, of course, merely conjectures at this stage of our inquiry, although they are conjectures reasonably motivated by the observational material collected in our tables, by analogy, and by our whole previous consideration. This consideration also suggests a procedure of research.

Our aim is to learn something about the dependence of the principal frequency $\Lambda$ on the shape of the domain $D$. Yet there is doubtless some sort of parallelism between the frequency $\Lambda$ and the perimeter of $D$. Now the study of the length of the perimeter is more elementary and more accessible than the study of $\Lambda$. Let us, therefore, study first the dependence of the length of the perimeter on the shape of the figure in the hope that this study may bring us some useful suggestion about $\Lambda$.

This leads us to certain geometrical maximum and minimum problems. Such problems have been studied already by some ancient Greek geometers (Zenodorus, Pappus) and by innumerable mathematicians after them.

Of the various approaches known to the author, the most fruitful for our line of inquiry seems to be a highly original idea due to the Swiss geometer Jacob Steiner. He invented a geometrical transformation that we call today *symmetrization* or, more specifically, *Steiner symmetrization.* This transformation can be performed in a plane or in space. See Ref. 13, vol. 2, pages 75 to 91; see also vol. 2, pages 264 to 269, which form a part of the highly stimulating extensive memoir on geometric maxima and minima, vol. 2, pages 177 to 308.

## 16.5 Plane

In a plane, we symmetrize a figure with respect to a specified straight line, given in advance and called the *line of symmetrization.* For instance, symmetrization with respect to the line $A^*H^*$ of Fig. 16.1 transforms the

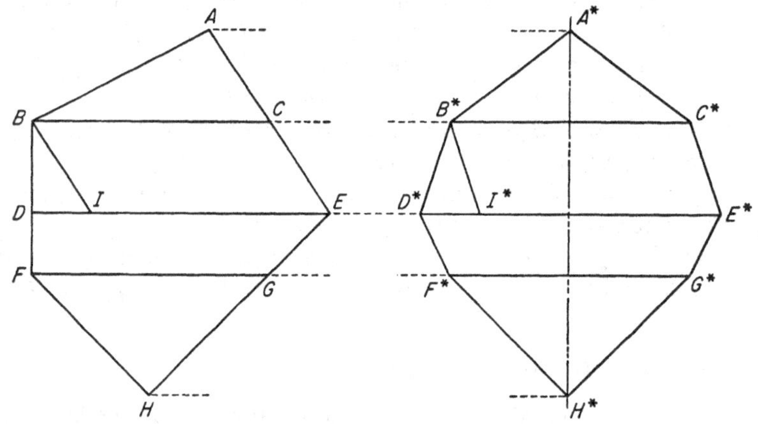


**Fig. 16.1** Steiner symmetrization.

pentagon $ABFHE$ into the octagon $A^*B^*D^*F^*H^*G^*E^*C^*$. We conceive the original figure (in our particular case, the pentagon) as consisting of "matches" (of various lengths—slender rods, line segments) parallel to each other and perpendicular to the given line of symmetrization (in our case, $A^*H^*$). Each match (segment) is pushed along its own line into a position where it is bisected by the line of symmetrization; thus, the segment $BC$ is shifted into the position $B^*C^*$. In their new positions, the segments fill the new, transformed, *symmetrized* figure: It has, by construction, a line of symmetry, the line $A^*H^*$ in the case of Fig. 16.1. We consider here only figures having not more than two boundary points in common with any line perpendicular to the line of symmetrization. This simple case is enough to bring out the decisive ideas. For the general case, see Ref. 9, page 5.

Figure 16.2 shows three successive symmetrizations. The original Fig. 16.2*a* is an irregular (general) quadrilateral with a vertical diagonal.

Symmetrized with respect to the horizontal line $\alpha$, Fig. 16.2*a* is transformed into Fig. 16.2*b*, a quadrilateral with one axis of symmetry, $\alpha$. Symmetrized with respect to the vertical line $\beta$, Fig. 16.2*b* is changed into Fig. 16.2*c*, a rhombus. Symmetrized with respect to $\gamma$, a line perpendicular to two of its sides, the rhombus (Fig. 16.2*c*) is transformed into Fig. 16.2*d*, a rectangle.

Jacob Steiner found interesting properties of symmetrization that are essential to our purpose. Here is the first:

I. *Symmetrization leaves the area unchanged.*

This property is obvious: In pushing the matches, we do not change their dimensions. It will be enough to give a more formal proof for a polygon; in fact, we could treat a curved line afterward, as a limiting

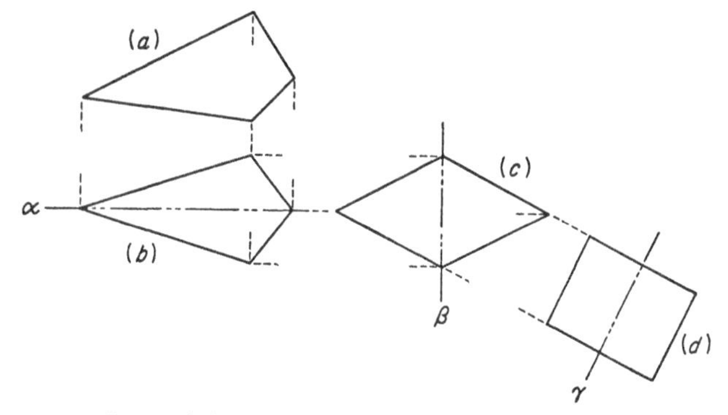


**Fig. 16.2** Repeated Steiner symmetrization.

case. We divide the polygon that we are about to symmetrize by parallels passing through its vertices and perpendicular to the line of symmetrization. Thus, the polygon is dissected into two triangles ($\triangle ABC$ and $\triangle HGF$ in Fig. 16.1) and two trapezoids ($BDEC$, $DFGE$). A triangle is symmetrized into another triangle ($\triangle ABC$ into $\triangle A^*B^*C^*$) with the same base and the same altitude and, therefore, the same area. A trapezoid is symmetrized into another trapezoid ($BDEC$ into $B^*D^*E^*C^*$) with the same lower base, the same upper base, the same altitude, and, therefore, the same area. By summing up the contributions of all parts, we prove property I.

II. *Symmetrization decreases the length of the perimeter.*

In proving this less obvious property, we can use the same parallels we have used in proving property I. We need a theorem from elementary geometry: *Of all triangles with the same base and the same altitude, the isosceles triangle has the shortest perimeter.* For instance, in Fig. 16.1, the broken line $B^*A^*C^*$, formed by two sides of the isosceles $\triangle A^*B^*C^*$, is shorter than the broken line $BAC$, formed by the two corresponding sides of $\triangle ABC$, which is transformed into $\triangle A^*B^*C^*$ by symmetrization.† The same theorem may help us when we consider the combined length of the two slanting sides of a trapezoid before and after symmetrization: It is shorter after. Thus in Fig. 16.1, we have

$$B^*I^* \| C^*E^* \qquad BI \| CE$$
$$D^*B^* + C^*E^* = D^*B^*I^* \leq DBI = DB + CE$$

since $\triangle D^*B^*I^*$ is isosceles, but $\triangle DBI$ need not be isosceles. By summing up the contributions of all triangles and trapezoids into which our two polygons, the original and the symmetrized, are divided, we see that the latter, the symmetrical one, has the shorter perimeter, and so we have proved property II.

We take the term "decreases" in the statement of the property just proved in its wider sense, in which it means "does not increase," and we keep this interpretation also in the sequel. In the present case, it is easy to point out those (trivial) exceptional cases in which the perimeter is not diminished; but it is much harder to keep track of the exceptional cases in some of the following proofs.

## 16.6 Space

In space, we symmetrize a solid with respect to a specified plane, given in advance and called the *plane of symmetrization.* Again, we conceive the original solid as consisting of matches (line segments) parallel to each other and perpendicular to the plane of symmetrization. Each match (segment) is shifted along its own line into a position where it is bisected by the plane of symmetrization. In their new positions, the segments form the new, transformed, symmetrized solid: It has, by construction, a plane of symmetry.

For instance, in Fig. 16.3, the plane of symmetrization is vertical, perpendicular to the horizontal plane of the drawing, which it intersects along the line $A^*C^*$. The original solid is a hill, represented by contour lines on the left; the symmetrized solid is a symmetric hill, halved by the vertical plane through $A^*C^*$.

† A physicist may be tempted to prove this theorem of geometry by an optical argument: Regard the line $AA^*$ as a mirror. Then $B^*A^*C^*$ is the path of light, starting from a source at $B^*$, being reflected at $A^*$, and proceeding hence to the eye of an observer at $C^*$. Yet $BAC$ cannot be a path described by light, since the lines $BA$ and $CA$ do not include the same angle with the mirror $AA^*$. And, as we know, light chooses the shortest path—hence $B^*A^*C^*$ is shorter than $BAC$. If we go back to the figure on which Heron of Alexandria, discoverer of the optical principle of the shortest path, based this principle, we obtain a beautiful proof for the geometric theorem. See Ref. 7, vol. 1, pages 142–144.

In any plane that is perpendicular to the plane of symmetrization, we have the geometric relationships described in Sec. 16.5. To the properties I and II proved in that section, there correspond the following properties, also discovered by Jacob Steiner:

III. *Symmetrization leaves the volume unchanged.*

IV. *Symmetrization decreases the surface area.*

Property III is obvious (the matches do not change their dimensions when shifted). By elementary geometry, the proof of III is easy, but

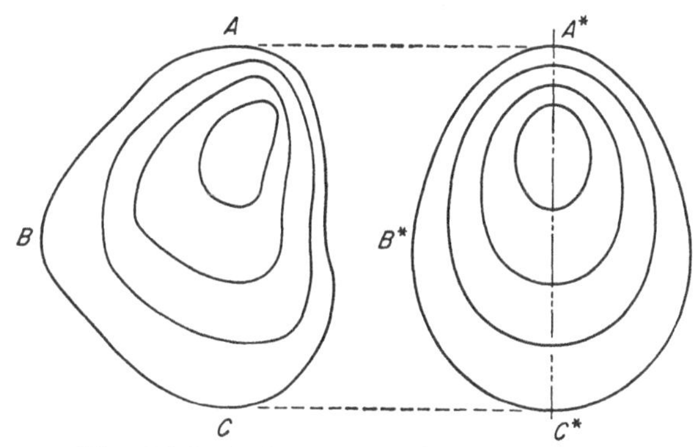


**Fig. 16.3** Steiner symmetrization in space.

that of IV is a little more troublesome. By integral calculus, the proof of III is immediate and that of IV very short, but we shall save space and omit these proofs; see Ref. 9, pages 182 to 184.

## 16.7 Applications

The following two applications of symmetrization are presented without complete rigor (it would be much more difficult to render the second rigorous than the first), yet even so they seem to offer valuable suggestions.

*a.* Figure 16.2 shows how any quadrilateral $a$ can be transformed into a rectangle $d$ by three successive symmetrizations. By symmetrizing the rectangle $d$ with respect to a perpendicular to one of its diagonals, we obtain another rhombus $e$ (draw it!). By symmetrizing this rhombus $e$ with respect to a perpendicular to two of its sides, we obtain another rectangle $f$—we pass from $e$ to $f$ as we have passed from $c$ to $d$ in Fig. 16.2. By repeating our last steps, we pass from rectangle to rhombus, from rhombus to rectangle, and so on, generating an infinite sequence of quadrilaterals

| | $c,$ | | $e,$ | | $g,$ | | . . . rhombi |
|---|---|---|---|---|---|---|---|
| $a, b$ | | | | | | | |
| | | $d,$ | | $f,$ | | $h,$ | . . . rectangles |

By the two properties proved in Sec. 16.5, each quadrilateral in this infinite sequence has the same area as, but a shorter perimeter than, the preceding quadrilateral, so that all have the same area but the perimeter decreases steadily as we advance in the sequence.

The reader is asked to concede that our infinite sequence has a limit. The limiting figure, as a limit of rhombi, must be a rhombus, and, as a limit of rectangles, it must be a rectangle; thus, being both a rhombus and a rectangle, it is a square. This limiting square still has the same area as all quadrilaterals of the sequence but a shorter perimeter than any quadrilateral of the sequence. We compare the limiting square with $a$, the initial figure of the sequence, which is an arbitrary quadrilateral, and find: *A square has a shorter perimeter than any other quadrilateral with the same area.* We have proved here a theorem of which we have supposed knowledge in formulating one of our conjectures on principal frequency at the end of Sec. 16.3.

*b*. In the plane, we are given the figure $F$, the boundary of which is an arbitrary closed curve, and we are also given two straight lines, 1 and 2, which include the angle $\theta$.

By symmetrizing $F$ with respect to 1, we obtain $F_1$.
By symmetrizing $F_1$ with respect to 2, we obtain $F_2$.
By symmetrizing $F_2$ with respect to 1, we obtain $F_3$.
By symmetrizing $F_3$ with respect to 2, we obtain $F_4$.

And so on. We thus generate the infinite sequence

$$F, \quad \begin{matrix} F_1, & F_3, & F_5, & \ldots & F_{2n-1}, & \ldots & (1) \\ F_2, & F_4, & F_6, & \ldots & F_{2n} & \ldots & (2) \end{matrix}$$

from the initial figure $F$ by successive symmetrizations with respect to the two given axes 1 and 2 alternately; $F_1, F_3, F_5, \ldots, F_{2n-1}, \ldots$ are symmetric with respect to 1, and $F_2, F_4, F_6, \ldots, F_{2n}, \ldots$ with respect to 2.

The reader is asked, as before, to concede that our infinite sequence has a limit $L$. This limiting figure $L$, as a limit of $F_{2n-1}$, must admit the line 1 as axis of symmetry; and, as a limit of $F_{2n}$, it must admit the line 2 as axis of symmetry; and so $L$ is symmetric with respect to both lines, 1 and 2.

Draw the line 3 as the mirror image of 1 with respect to 2. This line 3 —which, by the symmetry of $L$ with respect to 2, is "located in $L$ in the same way as 1 is located in $L$"—must be an axis of symmetry for $L$, just as 1 is. Draw the line 4 as the mirror image of 2 with respect to 3; by using both the result and the method of the foregoing remark, we see that also the line 4 must be an axis of symmetry for $L$; and so on. The lines so obtained, 1, 2, 3, 4, . . . , all pass through a common point, and the angle between any two successive lines is $\theta$.

This is so, independently of the choice of $\theta$. Yet we may choose $\theta$ from the start so that the ratio $\theta/\pi$ is *irrational*. Then, as is easily seen, the lines 1, 2, 3, 4, . . . are all different from each other, they are "everywhere dense," and all of them are axes of symmetry for $L$. What figure can $L$ be? It is intuitively obvious (and can be proved) that $L$ must be a circle.

Now, by the properties I and II proved in Sec. 16.5, all the figures $F$, $F_1$, $F_2$, . . . , generated by repeated symmetrization, have the same area, whereas the length of their perimeters steadily decreases as we advance in the sequence. Hence the limiting circular figure $L$ has still the same area as any one of them, for instance $F$, but $L$ has a perimeter shorter than any one of them, for instance $F$. Yet the boundary of $F$ was an arbitrary closed curve, and so we see: *A circle has a shorter perimeter than any other plane figure with the same area.* This is the theorem that has suggested, by analogy, Rayleigh's conjecture formulated in Sec. 16.3.

## THE KEY IDEA OF THE PROOF

### 16.8 Definition

Our previous work was merely heuristic; observations, conjectures, and analogies may be interesting and stimulating, they may be in some respects even more important than proofs, but they prove nothing. And between the geometric theorems that we have proved (or nearly proved) and the conjectures concerning $\Lambda$ that we wish to prove, we perceive for the moment only the loose link of analogy.

One thing is certain: If we wish to prove anything about $\Lambda$, we cannot remain on the merely intuitive level, but we must introduce, and use, a mathematical *definition* of $\Lambda$. Now, $\Lambda$ can be defined in various ways.

It is usual to define $\Lambda^2$ as the *first eigenvalue* of a boundary-value problem. There is a function $w$ vanishing along the boundary of the domain $D$, continuous on the domain plus its boundary, not vanishing identically in the interior of $D$, and in the interior of $D$ satisfying the differential equation

$$\frac{\partial^2 w}{\partial x^2} + \frac{\partial^2 w}{\partial y^2} + \Lambda^2 w = 0$$

By these conditions, $\Lambda^2$ would not yet be determined, but $\Lambda^2$ is defined as the *least* value of this kind.

For our purpose, however, it is more advantageous to define $\Lambda$ by a *minimum principle*. Let $f = f(x,y)$ be any function, defined, sufficiently "smooth," nonnegative, and not identically vanishing in the interior of

$D$, but *vanishing along the whole boundary of $D$;* then

$$\Lambda^2 \leq \frac{\iint_D (f_x^2 + f_y^2)\, dx\, dy}{\iint_D f^2\, dx\, dy} \tag{16.1}$$

Of course, $f_x$ and $f_y$ denote partial derivatives:

$$f_x = \frac{\partial f}{\partial x} \qquad f_y = \frac{\partial f}{\partial y}$$

The double integrals on the right-hand side of the inequality (16.1) are extended over the whole domain $D$. The case of equality can be attained in the relationship (16.1) (when $f = w$), so that $\Lambda^2$ is the *minimum* of the quotient (usually called the *Rayleigh ratio*) on the right-hand side of the inequality (16.1); this defines $\Lambda$. For the equivalence of this definition with the former, see Ref. 9, especially pages 89 to 91.

The definition of $\Lambda$ by a minimum principle is of great importance in practice. It yields upper approximations to $\Lambda$ and forms the basis of the widely used Rayleigh-Ritz method.[4,14]

## 16.9 From Surface Area to Dirichlet Integral

The definition of $\Lambda$ by a minimum principle is also important for our more theoretical purpose. It may direct our attention to the function $f(x,y)$, which, as we have said, vanishes along the boundary of the domain $D$ and is positive inside $D$. The surface

$$z = f(x,y)$$

and the $xy$ plane include a solid that we call the *hill;* see Fig. 16.3, on the left, where the outermost contour line, corresponding to the elevation $z = 0$, forms the boundary of the domain $D$ that we wish to call the *base* of the hill.

We symmetrize the hill (see Sec. 16.6) with respect to the vertical plane that intersects the horizontal $xy$ plane in the line $A^*C^*$. We thus obtain a new solid, the *symmetric hill* (see Fig. 16.3, on the right) contained between the $xy$ plane and a new surface the equation of which we write in the form

$$z = f^*(x,y)$$

The base of the symmetric hill, i.e., that part of its boundary that lies in the $xy$ plane, is a symmetric plane domain that we denote by $D^*$; it is surrounded by the outermost contour line on the right in Fig. 16.3. Of course, the function $f^*(x,y)$ vanishes along this outermost contour line, the boundary of $D^*$.

By property III, formulated in Sec. 16.6, the volume remains unchanged by symmetrization:

$$\iint_D f\,dx\,dy = \iint_{D^*} f^*\,dx\,dy \tag{16.2}$$

Yet, by property IV (again see Sec. 16.6) the surface area is diminished by symmetrization. The base, however, which is the part of the surface area contained in the $xy$ plane, remains unchanged by symmetrization, by virtue of property I proved in Sec. 16.5:

$$\iint_D dx\,dy = \iint_{D^*} dx\,dy \tag{16.3}$$

Thus the remaining (curved) part of the surface area is responsible for the diminishing of the whole:

$$\iint_D (1 + f_x^2 + f_y^2)^{1/2}\,dx\,dy \geq \iint_{D^*} (1 + f_x^{*2} + f_y^{*2})^{1/2}\,dx\,dy \tag{16.4}$$

We are approaching the turning point of our discussion. We shall attain it by realizing the physical significance of the numerator of the Rayleigh ratio; see the right-hand side of the inequality (16.1). This numerator is a double integral extended over the domain $D$ (the equilibrium position of the membrane), usually called the *Dirichlet integral;*[2,11] the integrand is the square of the gradient of the function $f$. The Dirichlet integral is proportional to the potential energy of the membrane, since it is proportional (as we shall see in a moment) to the change of area that the membrane undergoes when its equilibrium position (fully in the $xy$ plane) is disturbed. The change of area must be considered, as are all displacements in the classical theory of elasticity,[12] as being "very small" or "infinitesimal"; cf. Ref. 10, page 307 [194].

Starting from an arbitrary function $f$, we make it "small" by multiplying it by a "small" positive constant $\epsilon$. Let us apply the inequality (16.4) to $\epsilon f$, instead of $f$. We thus obtain

$$\iint_D [1 + \epsilon^2(f_x^2 + f_y^2)]^{1/2}\,dx\,dy \geq \iint_{D^*} [1 + \epsilon^2(f_x^{*2} + f_y^{*2})]\,dx\,dy$$

Expand both sides in powers of the small quantity $\epsilon$:

$$\iint_D \left[1 + \frac{\epsilon^2}{2}(f_x^2 + f_y^2) + \cdots\right] dx\,dy \geq \iint_{D^*} \left[1 + \frac{\epsilon^2}{2}(f_x^{*2} + f_y^{*2}) + \cdots\right] dx\,dy$$

By virtue of Eq. (16.3), the initial terms cancel. After division by $\epsilon^2$, we pass to the limit by letting $\epsilon$ tend to 0 and obtain

$$\iint_D (f_x^2 + f_y^2)\,dx\,dy \geq \iint_{D^*} (f_x^{*2} + f_y^{*2})\,dx\,dy \tag{16.5}$$

We have added a new property to the properties I, II, III, and IV that were discovered by Jacob Steiner:

V. *Symmetrization decreases the Dirichlet integral.*

The foregoing transition from the surface area to the Dirichlet integral contributes a good deal to elucidating certain analogies between geometrical and physical quantities such as we have observed at the beginning between the perimeter and the principal frequency of a membrane.†

## 16.10 A Minor Remark

Property V is the key to the physical applications of symmetrization. We need, however, one more (easy) remark for our next conclusion.

Let $\Phi(t)$ be a steadily increasing positive-valued function of the positive variable $t$; we are concerned here with such simple examples as $\Phi(t) = t^2$ or $\Phi(t) = \epsilon t$, where $\epsilon$ is a positive constant. We assert: *If the symmetrization of $f(x,y)$ yields $f^*(x,y)$, then the symmetrization of $\Phi[f(x,y)]$ yields $\Phi[f^*(x,y)]$.*

Obviously, the functions $f(x,y)$ and $\Phi[f(x,y)]$ have the same level lines: If

$$f(x,y) = \text{const}$$

then also

$$\Phi[f(x,y)] = \text{const}$$

and vice versa. Of course, the constants in these two equations are in general different: Those congruent level lines are at different elevations. Still, the two functions, $f(x,y)$ and $\Phi[f(x,y)]$, are represented by the same contour map; Fig. 16.3, made to represent the symmetrization of $f(x,y)$, can just as well represent the symmetrization of $\Phi[f(x,y)]$, and so the inspection of Fig. 16.3 renders the truth of the advanced assertion obvious.

Symmetrization, which, transforming $f(x,y)$ into $f^*(x,y)$, transforms $f^2$ into $f^{*2}$, preserves volume (Sec. 16.6, property III). Therefore, besides Eq. (16.2), we also have

$$\iint_D f^2\,dx\,dy = \iint_{D^*} f^{*2}\,dx\,dy \tag{16.6}$$

We shall need this fact in the next section.

## 16.11 Symmetrization and Principal Frequency

We have now collected the facts we need for proving the following property:

† In the seminar of Professor Timoshenko, I was about to present another proof for property V (cf. Ref. 8 or 9, pages 153 to 157) as he intervened with the remark that considerations of energy should have some bearing on the matter—and then, quite abruptly, the idea of the foregoing proof presented itself to my mind. A shorter description of this event in my paper,[5] where the foregoing argument was first presented, should not give rise to misinterpretations about the authorship of my proof; cf. Ref. 3, page 133. (A few short passages of my paper[5] have been reproduced here with the kind permission of the editor of the *Quarterly of Applied Mathematics*.)

VI. *Symmetrization decreases the principal frequency.*

We are given an arbitrary plane domain $D$. The corresponding principal frequency is defined (see Sec. 16.8) as the minimum of the Rayleigh ratio; there exists a function $f$ for which the case of equality is attained in the inequality (16.1):

$$\Lambda^2 = \frac{\iint_D (f_x{}^2 + f_y{}^2)\,dx\,dy}{\iint_D f^2\,dx\,dy} \tag{16.7}$$

We symmetrize in space with respect to a plane of symmetrization perpendicular to the $xy$ plane; see Fig. 16.3. By symmetrizing the domain $D$, we obtain the domain $D^*$ to which there corresponds the principal frequency $\Lambda^*$. This symmetrization transforms the function $f$ arising in Eq. (16.7), which vanishes along the boundary of $D$, into $f^*$, which vanishes along the boundary of $D^*$; see Sec. 16.9. By combining the relationships (16.5) and (16.6), we obtain

$$\frac{\iint_D (f_x{}^2 + f_y{}^2)\,dx\,dy}{\iint_D f^2\,dx\,dy} \geq \frac{\iint_{D^*} (f_x^{*2} + f_y^{*2})\,dx\,dy}{\iint_{D^*} f^{*2}\,dx\,dy} \tag{16.8}$$

Now $\Lambda^*$, the principal frequency corresponding to $D^*$, is defined by the minimum property stated in Sec. 16.8, and the function $f^*$, since it vanishes along the boundary of $D^*$, is appropriate for computing an upper bound for $\Lambda^*$: Just as we obtained the inequality (16.1), we similarly can establish the inequality

$$\frac{\iint_{D^*} (f_x^{*2} + f_y^{*2})\,dx\,dy}{\iint_{D^*} f^{*2}\,dx\,dy} \geq \Lambda^{*2} \tag{16.9}$$

Combination of the relationships (16.7) to (16.9) yields

$$\Lambda^2 \geq \Lambda^{*2}$$

which proves property VI.

## 16.12 Scope of the Proof

The foregoing proof hinges on the work of Sec. 16.9, the symmetrization of the Dirichlet integral. Therefore, it applies essentially also to some other physical and geometrical quantities defined, as $\Lambda$, by a minimum principle involving the Dirichlet integral (in the plane or in space). We mention here only two such quantities:

The torsional rigidity $P$ of a uniform elastic cylinder the cross section of which is the domain $D$

The electrostatic capacity $C$ of a metallic plate (with vanishing thickness) coextensive with the domain $D$

For more details and more analogous quantities, see Ref. 9, especially pages 1 to 3.

It is convenient to add here a few more notations:

The area $A$ of $D$

The length $L$ of the perimeter of $D$

The polar moment of inertia $I$ of $D$ with respect to an axis perpendicular to $D$ and passing through its center of gravity

Properties I and II and (what is the main point) the result proved in Sec. 16.9 are contained in the following comprehensive proposition:

THEOREM 16.1. *Symmetrization leaves $A$ unchanged, decreases $L$, $I$, $\Lambda$, and $C$, and increases $P$.*

For $A$ and $L$, see Sec. 16.5; for $\Lambda$, see Sec. 16.11. The proof for $P$ and $C$ is closely related to the proof for $\Lambda$; the proof for $I$ is much simpler. For details concerning Theorem 16.1 and the rest of this section, see Ref. 9, pages 151 to 161 and *passim.*

In possession of Theorem 16.1, we perceive that the argument $b$ in Sec. 16.7 extends much beyond its original scope. In fact, it comes pretty close to proving the following comprehensive proposition:

THEOREM 16.2. *Of all plane domains with a given area $A$, the circle yields the lowest value for $L$, $I$, $\Lambda$, and $C$ and the highest value for $P$.*

In so far as this proposition refers to $P$, the domain in question is supposed to be simply connected. The argument $b$ of Sec. 16.7 can be modified, developed, and rendered rigorous in various ways; see especially Ref. 9, pages 189 to 193. There is an extremely short and elementary argument for $I$.[6]

Also the argument $a$ in Sec. 16.7 extends beyond its original scope and proves, in fact, the following result:

THEOREM 16.3. *Of all quadrilaterals with a given area $A$, the square yields the lowest value for $L$, $I$, $\Lambda$, and $C$ and the highest value for $P$.*

For some of these quantities, the argument $a$ of Sec. 16.7 can be greatly simplified; see Exercises 5 and 6 below. There is an analogous theorem for triangles of which we have stated a particular case (concerning $\Lambda$) as a conjecture toward the end of Sec. 16.3; we leave the proof as another exercise to the reader (Exercise 7).

Yet we have no means at our disposal for proving an analogous proposition for pentagons. The reason is that, by symmetrizing a polygon, we increase, in most cases, the number of vertices. For instance, in Fig. 16.1, the original polygon has five vertices, but the new polygon, obtained by symmetrization, has eight vertices. In dealing with triangles or quadrilaterals we can, by a suitable choice of the line of symmetrization, avoid increasing the number of vertices; yet, in dealing with pentagons or hexagons or higher polygons, we cannot. And so decades may pass before the last conjecture stated in Sec. 16.3 (about regular polygons) will be proved.

Yet we can extend the foregoing considerations to three-dimensional problems without essential difficulties. We thus obtain, among others, the following theorems (see Refs. 8 and 9, *passim*):

*Symmetrization decreases the electrostatic capacity of a condenser.*

*Of all solids with a given volume, the sphere has the smallest electrostatic capacity.*

*Of all tetrahedra with a given volume, the regular tetrahedron has the smallest electrostatic capacity.*

## ADDITIONAL REMARKS

### 16.13 Alternative Symmetrization

We are given a plane domain $D$. By an appropriate straight line $l$, which is *not* a line of symmetry for $D$, we cut $D$ into two "unequal halves." We adjoin to each "half" its mirror image with respect to $l$ and thus obtain two symmetric domains $D'$ and $D''$. (In the top row of Fig. 16.4, the two "halves" of the trapezoid $D$, along with their images in the dashed line, form the rhombus $D'$ and the hexagon $D''$.) Now, we face an alternative: We have to choose between $D'$ and $D''$; if an appropriate choice leads us to one of these two figures, we say that it has been derived from $D$ by *alternative symmetrization.*†

Let $A$, $A'$, and $A''$ denote the area, and $L$, $L'$, and $L''$ the length of the perimeter, of $D$, $D'$, and $D''$, respectively. Obviously,

$$A' + A'' = 2A \qquad L' + L'' = 2L$$

If the "unequal halves" into which $D$ has been cut by $l$ are of equal area, then

$$A' = A'' = A$$

† The geometric operation for which we have introduced the term "alternative symmetrization" has been employed by Steiner; see Ref. 13, vol. 2, pages 193 and 194, 299, etc. Its first application to a physical problem is due to R. Courant, Ref. 1. Further physical applications have been sketched by L. E. Payne and H. F. Weinberger; see two abstracts in *Bulletin of the American Mathematical Society,* vol. 59, pages 244 and 363, 1953. The application of alternative symmetrization to the proof of the uniqueness in Rayleigh's problem has been briefly indicated by Payne and Weinberger in a conversation with the author and is published here with their kind permission; of course, the author assumes full responsibility for the exposition in Sec. 16.14, for which he had to supply the details by himself. The remarks in Sec. 16.15 have been known to the author since about 1950. (Cf. the abstract of an article by J. Hersch, "Une symétrisation differénte de celle de Steiner," *Enseignement Math.,* vol. 5, pp. 219–220, 1959.)

If the notation is so chosen that $L' \leq L''$, then also $L' \leq L$; we select $D'$ and in so doing we derive from $D$, by alternative symmetrization, a symmetric figure with the same area and a shorter, or possibly equal, perimeter. The perimeter will definitely be shortened if $L' \neq L''$.

We can put this procedure to good use if we wish to determine the figure with a given area $A$ the perimeter of which is a minimum. Any straight line that bisects the area of the desired figure must also bisect its

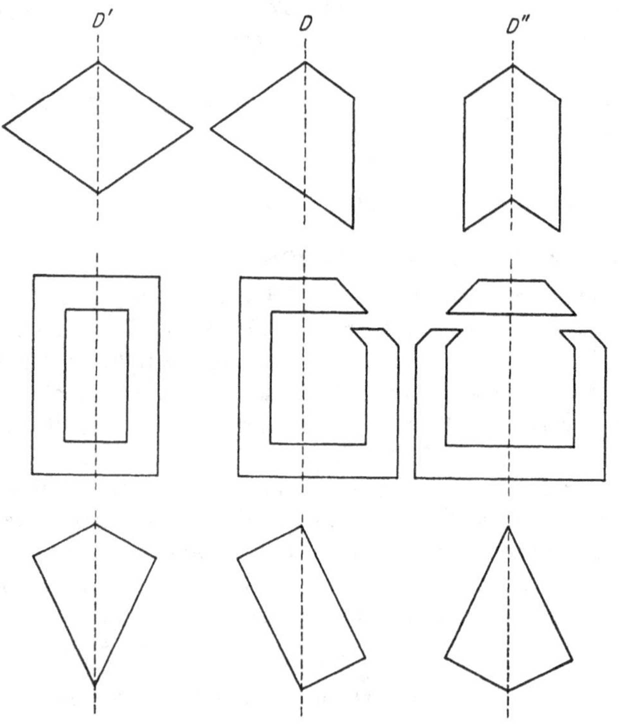


**Fig. 16.4** Alternative symmetrization.

perimeter; if this were not so, alternative symmetrization would yield a figure with the same area $A$ and a shorter perimeter, which would contradict the definition of the desired figure. This hints very strongly that the desired figure (if it exists) must be the circle.

The argument just sketched can be developed more fully (especially, we can take care of more intricate possibilities; cf. the middle row of Fig. 16.4) and, what is more interesting for us, it can be adapted to some physical problems.

## 16.14 Uniqueness

"Of all plane domains with a given area, the circle has the lowest principal frequency." (Cf. Theorem 16.2 in Sec. 16.12.) Yet is the circle the *only* figure with that given area and the minimum principal frequency? Or is there some noncircular figure that has the same area *and* the same principal frequency as the circle? Steiner symmetrization, the transition from the surface area to the Dirichlet integral, and the connected ideas sketched in the foregoing argument give no obvious answer to this question and for this reason the formulation of Theorem 16.2 leaves a gap: Intentionally, it leaves open the question of the uniqueness of the extremal figure. Fortunately, alternative symmetrization can fill this gap.

If a domain $D$ is not a circle, we have to show that there is another domain $\bar{D}$ with the same area as $D$ but a lower principal frequency than $D$; being given $D$, we have to construct such a domain $\bar{D}$.

If a connected domain (consisting of one piece) is neither a circle nor an annulus contained between two concentric circles, it cannot have a line of symmetry in every direction. Let us assume that our domain $D$ (for example, the trapezoid in the top row of Fig. 16.4) has no vertical line of symmetry. Yet there is a vertical line $l$ that bisects the area of $D$; this can be shown by a continuity consideration. Alternative symmetrization performed on $D$ with respect to this line $l$ yields two symmetric domains $D'$ and $D''$ (the rhombus and the nonconvex hexagon in our example), each of which is equal in area to $D$.

We consider now the function $f$ for which equality is attained in the inequality (16.1) of Sec. 16.8. We define two new functions $f'$ and $f''$ (the primes do not denote derivatives!). The function $f'$ is defined in the domain $D'$. In the "half" of $D'$ that coincides with a certain "half" of $D$, the function $f'$ coincides with the function $f$, and in any two points of $D'$ that are mirror images of each other with respect to the line $l$, the function $f'$ takes the same value. Visibly, the function $f'$ vanishes along the boundary of $D'$ (as $f$ vanishes along the boundary of $D$) and $f'$ is continuous throughout $D'$, in particular along the line $l$, although its normal derivative on $l$ may not exist. The function $f''$ is analogously defined in $D''$; as $f'$ extends $f$ from one "half" of $D$ into the full domain $D'$ by symmetry, so $f''$ extends $f$ from the other "half" of $D$ into $D''$.

Since equality is attained by our $f$ in the inequality (16.1), we have Eq. (16.7) of Sec. 16.11. By the definition of $f'$ and $f''$, we can write Eq. (16.7) in the form

$$\Lambda^2 = \frac{\iint_{D'} (f_x'^2 + f_y'^2)\,dx\,dy + \iint_{D''} (f_x''^2 + f_y''^2)\,dx\,dy}{\iint_{D'} f'^2\,dx\,dy + \iint_{D''} f''^2\,dx\,dy} \tag{16.10}$$

by first multiplying the numerator and the denominator on the right-hand side of Eq. (16.7) by 2.

Consider the two Rayleigh ratios:

$$\frac{\iint_{D'} (f_x'^2 + f_y'^2)\,dx\,dy}{\iint_{D'} f'^2\,dx\,dy} \qquad \frac{\iint_{D''} (f_x''^2 + f_y''^2)\,dx\,dy}{\iint_{D''} f''^2\,dx\,dy} \tag{16.11}$$

By an elementary property of fractions, the ratios (16.11) cannot both be larger than the right-hand side of Eq. (16.10). We assume the notation so chosen that

$$\Lambda^2 \geq \frac{\iint_{D'} (f_x'^2 + f_y'^2)\,dx\,dy}{\iint_{D'} f'^2\,dx\,dy} \tag{16.12}$$

Now, as we have observed above, $f'$ is continuous in $D'$ and vanishes along the boundary of $D'$. Therefore, the right-hand side of the inequality (16.12) yields an upper bound for $\Lambda'^2$ [of course, $\Lambda'$ is the principal frequency of $D'$; we have to apply the inequality (16.1) to $D'$ instead of $D$]. Yet more is true; we actually have

$$\frac{\iint_{D'} (f_x'^2 + f_y'^2)\,dx\,dy}{\iint_{D'} f'^2\,dx\,dy} > \Lambda'^2 \tag{16.13}$$

Exclusion of the equality in the relationship (16.13) ($>$ instead of $\geq$) is the cardinal point of our argument.

Equality in the relationship (16.13) could be attained only if $f'$ represented the true shape of the vibrating membrane stretched over $D'$ (if it satisfied the corresponding differential equation). As such, $f'$ would have continuous partial derivatives (even of higher order) and, since it coincides with $f$ in one half of $D'$, it would, by virtue of the differential equation, fully coincide with $f$. Therefore, $D'$ would also coincide with $D$, and so $D$ would have $l$ as a line of symmetry—yet this is precisely the circumstance that we have excluded from the start. Thus, we have fully proved the inequality (16.13), with $>$, not only with $\geq$.

Obviously, the inequalities (16.12) and (16.13) now yield

$$\Lambda^2 > \Lambda'^2$$

and so $D'$ is a domain such as we have desired: It has the same area as $D$ but a lower principal frequency than $D$.

We can take care of more complicated cases such as the one represented by the middle row of Fig. 16.4, extend the argument (with appropriate changes) to $P$ and $C$, and so we can prove that the circle, *and only the circle,* yields the extremal value in all five cases considered in Theorem 16.2. The extension to three dimensions presents no additional difficulty.†

## 16.15 Where the Alternative Symmetrization Leaves No Alternative

We have still to discuss the case illustrated by the bottom row of Fig. 16.4. The rectangle and the vertical should represent to us any domain $D$ with a *center of symmetry* and a line $l$ passing through this center that is *not* a line of symmetry for $D$. Alternative symmetrization applied to $D$ with respect to $l$ yields two *congruent* figures. The argument of the foregoing Sec. 16.14 and elementary considerations lead us to the following counterpart of Theorem 16.1 (in which "symmetrization" means "Steiner symmetrization" and the terms "decrease" and "increase" are used in the wide sense):

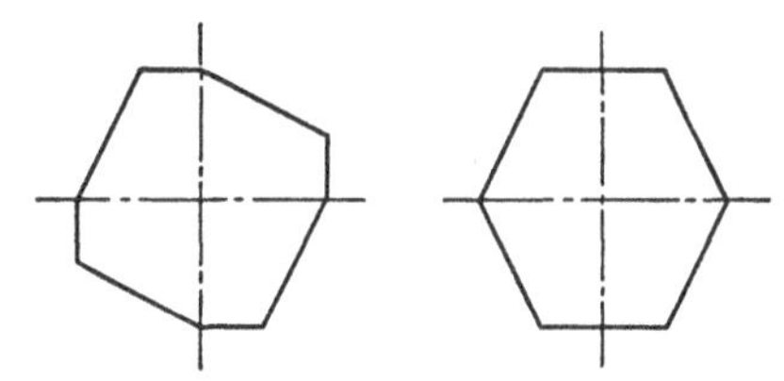

**Fig. 16.5** Higher rotational symmetrization.

*Alternative symmetrization of a domain that has a center of symmetry, with respect to a line that passes through this center but is not an axis of symmetry of the domain, leaves $A$ and $L$ unchanged, decreases $I$ or leaves it unchanged, definitely decreases $\Lambda$ and $C$, and definitely increases $P$.*

We shall not stop to explain in detail the extension of the foregoing remark to centers of symmetry of higher order, which is hinted by Fig. 16.5. See, however, Exercise 9, below.

† The process hinted by Fig. 16.4 may yield $n$ *connected* symmetric domains $D'$, $D''$, . . . , $D^{(n)}$ ($n = 2$ in the top row, $n = 3$ in the middle row). None of these $n$ domains has, however, a larger area than $D$. We replace Eq. (16.10) by another in which there are $n$ terms both in the numerator and in the denominator of the right-hand side, and we consider correspondingly $n$ Rayleigh ratios, not only two as in formulas (16.11). Among these $n$ ratios, there will be one (belonging to a domain that, with suitable notation, may be called $D'$) for which the inequality (16.12) holds. If the area of $D'$ happens to be less than that of $D$, we "inflate" $D'$ to a similar figure that has the same area as $D$; this operation diminishes the principal frequency. Eventually, by the argument that leads to the inequality (16.13), we obtain a domain with the same area as, but a lower principal frequency than, $D$, *quod erat faciendum.* We can take care of the case of the annulus in various ways—for instance, by Steiner symmetrization followed by alternative symmetrization.

## 16.16 One More Inequality Suggested by Observation

Consider, in Table 16.3, the quotients of which the denominator is a number in the last column and the numerator the corresponding number in the preceding column. All these quotients are pretty close to 1 and none is greater than the value $\pi/2$ yielded by the infinitely narrow rectangle. For this last statement to be true not only for the cases collected in Table 16.3 but generally, we would have

$$\Lambda < \frac{\pi L}{2A}$$

This inequality will be proved by the author in a paper that makes essential use of a foregoing research of E. Makai.

### EXERCISES

**1.** By Steiner symmetrization, show that an ellipse is transformed into an ellipse and an ellipsoid into an ellipsoid.

**2.** *Steiner symmetrization, in plane or space.* The figure $F$, which has a center of symmetry, is transformed by symmetrization into $F^*$. Show that $F^*$ also has a center of symmetry.

**3.** *Steiner symmetrization in the plane.* The figure $F$, which has two axes of symmetry perpendicular to each other, is transformed by symmetrization into $F^*$. Show that $F^*$ also has two axes of symmetry perpendicular to each other.

**4.** *Steiner symmetrization in the plane.* The figure $F$, which has three different axes of symmetry, is transformed by symmetrization into $F^*$. Show by an example that $F^*$ need not have more than one axis of symmetry.

**5.** Prove that of all quadrilaterals of equal area, the square minimizes the perimeter.

In showing this you need not the whole (infinite) process $a$ of Sec. 16.7 (only its first steps, indicated by Fig. 16.2), provided that you can prove *directly* that the square has a shorter perimeter than any other *rectangle* with the same area (which can be done by very simple algebra).

**6.** Prove that of all quadrilaterals with a given area, the square has the lowest principal frequency.

In showing this, you need not the whole (infinite) process $a$ of Sec. 16.7, provided that you use the expression

$$\Lambda^2 = \pi^2\left(\frac{1}{a^2} + \frac{1}{b^2}\right)$$

for the principal frequency of a rectangle with sides $a$ and $b$. (It is not so easy to eliminate the infinite process from the proof for the full Theorem 16.3 of Sec. 16.12.)

**7.** Show that of all triangles with a given area, the equilateral triangle has the lowest principal frequency.

(Steiner symmetrization of $\triangle ABC$ with respect to the perpendicular bisector of $AB$ yields the isosceles $\triangle A'B'C'$ with $A' = A$, $B' = B$, $A'C' = B'C'$. Now use the perpendicular bisector of $A'C'$ as line of symmetrization, then that of $A''B''$, then that of $A'''C'''$, and so on, alternately.)

**8.** Let $a$ denote the horizontal side and $h$ the vertical side of a rectangle. The sum of the two parallel sides of an isosceles trapezoid is $2a$; the altitude (perpendicular to these parallel sides) is $h$. A parallelogram has the same base $a$ and the same altitude $h$ as the rectangle, and its four angles are equal to the four angles of the isosceles trapezoid (the four angles are, of course, differently disposed around the two figures). Show that, of these three figures, the last one (the parallelogram) has the highest principal frequency.

**9.** The octagon in Fig. 16.5 has a center of symmetry of order 4 (coincides with itself when rotated about this center through the angle $2\pi/4$). We obtain the hexagon in Fig. 16.5 from the octagon by replacing two quarters of the octagon by their respective mirror images. Show that, in the transition from the octagon to the hexagon, $A$, $L$, and $I$ remain unchanged, $\Lambda$ and $C$ are decreased, and $P$ is increased.

## REFERENCES


1. Courant, R., Beweis des Satzes, dass von allen homogenen Membranen gegebenen Umfanges und gegebener Spannung die kreisförmige den tiefsten Grundton Gibt, *Math. Z.*, vol. 1, pp. 321–328, 1918.
2. Hestenes, Magnus R., Elements of the Calculus of Variations, chap. 4 in "Modern Mathematics for the Engineer," First Series, edited by E. F. Beckenbach, McGraw-Hill Book Company, Inc., New York, 1956.
3. Jenkins, J. J., "Univalent Functions and Conformal Mapping," *Ergeb. der Mathematik*, new series, vol. 18, Springer-Verlag, Berlin, 1958.
4. Morrey, Charles B., Jr., Nonlinear Methods, chap. 16 in "Modern Mathematics for the Engineer," First Series, edited by E. F. Beckenbach, McGraw-Hill Book Company, Inc., New York, 1956.
5. Pólya, G., Torsional Rigidity, Principal Frequency, Electrostatic Capacity and Symmetrization, *Quart. Appl. Math.*, vol. 6, pp. 267–277, 1948.
6. ———, Remarks on the Foregoing Paper, *J. Math. Phys.*, vol. 31, pp. 55–57, 1952.
7. ———, "Mathematics and Plausible Reasoning," Princeton University Press, Princeton, N.J., 1954.
8. ——— and G. Szegö, Inequalities for the Capacity of a Condenser, *Amer. J. Math.*, vol. 67, pp. 1–32, 1945.
9. ——— and ———, "Isoperimetric Inequalities in Mathematical Physics," Princeton University Press, Princeton, N.J., 1951.
10. Rayleigh, Lord J. W. S., "The Theory of Sound," 2d ed., London, 1894. Reprinted by Dover Publications, New York, 1955. In references, the section number in square brackets follows the numbers of the pages quoted.
11. Schiffer, Menahem M., Boundary-value Problems in Elliptic Partial Differential Equations, chap. 6 in "Modern Mathematics for the Engineer," First Series, edited by E. F. Beckenbach, McGraw-Hill Book Company, Inc., New York, 1956.
12. Sokolnikoff, Ivan S., The Elastostatic Boundary-value Problems, chap. 7 in "Modern Mathematics for the Engineer," First Series, edited by E. F. Beckenbach, McGraw-Hill Book Company, Inc., New York, 1956.
13. Steiner, Jacob, "Gesammelte Werke," G. Reimer, Berlin, 1881–1882.
14. Tompkins, Charles B., Methods of Steep Descent, chap. 18 in "Modern Mathematics for the Engineer," First Series, edited by E. F. Beckenbach, McGraw-Hill Book Company, Inc., New York, 1956.

# Name Index

# Subject Index